Introduction to Particle Technology – Second Edition

Introduction to Particle Technology

SECOND EDITION

Martin Rhodes
Monash University, Australia

John Wiley & Sons, Ltd

Telephone (+44) 1243 779777

Email (for orders and customer service enquiries): cs-books@wiley.co.uk
Visit our Home Page on www.wileyeurope.com or www.wiley.com

Other Wiley Editorial Offices

John Wiley & Sons Inc., 111 River Street, Hoboken, NJ 07030, USA
Jossey-Bass, 989 Market Street, San Francisco, CA 94103-1741, USA
Wiley-VCH Verlag GmbH, Boschstr. 12, D-69469 Weinheim, Germany
John Wiley & Sons Australia Ltd, 42 McDougall Street, Milton, Queensland 4064, Australia
John Wiley & Sons (Asia) Pte Ltd, 2 Clementi Loop #02-01, Jin Xing Distripark, Singapore 129809
John Wiley & Sons Ltd, 6045 Freemont Blvd, Mississauga, Ontario L5R 4J3, Canada

Wiley also publishes its books in a variety of electronic formats. Some content that appears in print may not be available in electronic books.

Library of Congress Cataloging-in-Publication Data

Rhodes, Martin.
Introduction to particle technology / Martin Rhodes. – 2nd ed.
p. cm.
Includes bibliographical references and index.
ISBN 978-0-470-01427-1 (cloth) – ISBN 978-0-470-01428-8 (pbk.)
1. Particles. I. Title.
TP156.P3R48 2008
620′.43–dc22 2007041699

British Library Cataloguing in Publication Data

A catalogue record for this book is available from the British Library
ISBN 978-0-470-01427-1 (Cloth) ISBN 978-0-470-01428-8 (Paper)

Typeset in 10/12 pt Palatino by Thomson Digital, India.

Contents

About the Contributors

Dr Karen P. Hapgood holds a BE and PhD in Chemical Engineering from the University of Queensland, Australia. Her PhD was on granulation processes and she continues to research in this area and related areas of powder technology. Karen worked for Merck & Co., USA for 5 years, where she worked on designing, troubleshooting and scaling up tablet and capsule manufacturing processes. Karen is currently a Senior Lecturer at the Department of Chemical Engineering at Monash University, Australia.

George Vincent Franks holds a Bachelor's degree in Materials Science and Engineering from MIT (1985) and a PhD in Materials Engineering from the University of California at Santa Barbara (1997). George worked for 7 years in the ceramic processing industry as a process development engineer for Norton Company and Ceramic Process Systems Incorporated. His industrial work focused mainly on near net shape forming of ceramic green bodies and non-oxide ceramic firing. He has research and teaching experience at the Universities of Melbourne and Newcastle in Australia. George is currently Associate Professor in the Department of Chemical and Biomolecular Engineering at the University of Melbourne and the recently formed Australian Mineral Science Research Institute. His research interests include, mineral processing, particularly flocculation, advanced ceramics powder processing, colloid and surface chemistry, ion specific effects, alumina surfaces and suspension rheology.

Jennifer Sinclair Curtis received her BS degree in chemical engineering from Purdue University and PhD in chemical engineering from Princeton University. Jennifer has an internationally recognized research program in the development and validation of numerical models for the prediction of particle flow phenomena. Jennifer was a recipient of the NSF Presidential Young Investigator Award, the Eminent Overseas Lectureship Award by the Institution of Engineers in Australia and ASEE's Sharon Keillor Award for Women in Engineering. She currently serves on the Editorial Advisory Board of the AIChE Journal, Powder Technology, and the Journal of Pharmaceutical Development and Technology. Jennifer has also served as a Trustee of the non-profit Computer Aids for Chemical Engineering Corporation and on the National Academy of Engineering's Committee on Engineering Education. Jennifer is currently Professor and Chair of the Chemical Engineering Department at the University of Florida.

Martin Rhodes holds a Bachelor's degree in chemical engineering and a PhD in particle technology from Bradford University in the UK, industrial experience in chemical and combustion engineering and many years experience as an academic at Bradford and Monash Universities. He has research interests in various aspects of gas fluidization and particle technology, areas in which he has many refereed publications in journals and international conference proceedings. Martin is on the editorial boards of Powder Technology and KONA and on the advisory board of Advanced Powder Technology. Martin has a keen interest in particle technology education and has published books and CDROM on Laboratory Demonstrations and directed continuing education courses for industry in the UK and Australia. He was co-founder of the Australasian Particle Technology Society. Martin has a Personal Chair in the Department of Chemical Engineering at Monash University, Australia, where he is presently Head of Department.

Preface to the Second Edition

It is 10 years since the publication of the first edition of Introduction to Particle Technology. During that time many colleagues from around the world have provided me with comments for improving the text. I have taken these comments into consideration in preparing the second edition. In addition, I have broadened the coverage of particle technology topics – in this endeavour I am grateful to my co-authors Jennifer Sinclair Curtis and George Franks, who have enabled the inclusion of chapters on Slurry Transport and Colloids and Fine Particles, and Karen Hapgood, who permitted an improved chapter on size enlargement and granulation. I have also included a chapter on the Health Effects of Fine Powders – covering both beneficial and harmful effects. I am also indebted to colleagues Peter Wypych, Lyn Bates, Derek Geldart, Peter Arnold, John Sanderson and Seng Lim for contributing case studies for Chapter 16.

Martin Rhodes
Balnarring, December 2007

Preface to the Second Edition

Preface to the First Edition

Particle Technology

Particle technology is a term used to refer to the science and technology related to the handling and processing of particles and powders. Particle technology is also often described as powder technology, particle science and powder science. Powders and particles are commonly referred to as bulk solids, particulate solids and granular solids. Today particle technology includes the study of liquid drops, emulsions and bubbles as well as solid particles. In this book only solid particles are covered and the terms particles, powder and particulate solids will be used interchangeably.

The discipline of particle technology now includes topics as diverse as the formation of aerosols and the design of bucket elevators, crystallization and pneumatics transport, slurry filtration and silo design. A knowledge of particle technology may be used in the oil industry to design the catalytic cracking reactor which produces gasoline from oil or it may be used in forensic science to link the accused with the scene of crime. Ignorance of particle technology may result in lost production, poor product quality, risk to health, dust explosion or storage silo collapse.

Objective

The objective of this textbook is to introduce the subject of particle technology to students studying degree courses in disciplines requiring knowledge of the processing and handling of particles and powders. Although the primary target readership is amongst students of chemical engineering, the material included should form the basis of courses on particle technology for students studying other disciplines including mechanical engineering, civil engineering, applied chemistry, pharmaceutics, metallurgy and minerals engineering.

A number of key topics in particle technology are studied giving the fundamental science involved and linking this, wherever possible, to industrial practice. The coverage of each topic is intended to be exemplary rather than exhaustive. This is not intended to be a text on unit operations in powder technology for chemical engineers. Readers wishing to know more about the industrial practice and equipment for handling and processing are referred to the various handbooks of powder technology which are available.

The topics included have been selected to give coverage of broad areas within easy particle technology: characterization (size analysis), processing (fluidized beds granulation), particle formation (granulation, size reduction), fluid-particle-separation (filtration, settling, gas cyclones), safety (dust explosions), transport (pneumatic transport and standpipes). The health hazards of fine particles or dusts are not covered. This is not to suggest in any way that this topic is less important than others. It is omitted because of a lack of space and because the health hazards associated with dusts are dealt with competently in the many texts on Industrial or Occupational Hygiene which are now available. Students need to be aware however, that even chemically inert dusts or 'nuisance dust' can be a major health hazard. Particularly where products contain a significant proportion of particles under 10 μm and where there is a possibility of the material becoming airborne during handling and processing. The engineering approach to the health hazard of fine powders should be strategic wherever possible; aiming to reduce dustiness by agglomeration, to design equipment for containment of material and to minimize exposure of workers.

The topics included demonstrate how the behaviour of powders is often quite different from the behaviour of liquids and gases. Behaviour of particulate solids may be surprising and often counter-intuitive when intuition is based on our experience with fluids. The following are examples of this kind of behaviour:
When a steel ball is placed at the bottom of a container of sand and the container is vibrated in a vertical plane, the steel ball will rise to the surface.

A steel ball resting on the surface of a bed of sand will sink swiftly if air is passed upward through the sand causing it to become fluidized.

Stirring a mixture of two free-flowing powders of different sizes may result in segregation rather than improved mixture quality.

Engineers and scientists are use to dealing with liquids and gases whose properties can be readily measured, tabulated and even calculated. The boiling point of pure benzene at one atmosphere pressure can be safely relied upon to remain at 80.1 °C. The viscosity of water at 20 °C can be confidently predicted to be 0.001 Pa s. The thermal conductivity of copper at 100 °C is 377 W/m·K. With particulate solids, the picture is quite different. The flow properties of sodium bicarbonate powder, for example, depends not only on the particle size distribution, the particle shape and surface properties, but also on the humidity of atmosphere and the state of the compaction of the powder. These variables are not easy to characterize and so their influence on the flow properties is difficult to predict with any confidence.

In the case of particulate solids it is almost always necessary to rely on performing appropriate measurements on the actual powder in question rather than relying on tabulated data. The measurements made are generally measurements of bulk properties, such as shear stress, bulk density, rather than measurements of fundamental properties such as particle size, shape and density. Although this is the present situation, in the not too distant future, we will be able to rely on sophisticated computer models for simulation of particulate systems. Mathematical modelling of particulate solids behaviour is a rapidly developing area of research around the world, and with increased computing power and better visualization software, we will soon be able to link fundamental particle properties directly to bulk powder behaviour. It will even be possible to predict, from first principles, the influence of the presence of gases and liquids within the powder or to incorporate chemical reaction.

Particle technology is a fertile area for research. Many phenomena are still unexplained and design procedures rely heavily on past experience rather than on fundamental understanding. This situation presents exciting challenges to researchers from a wide range of scientific and engineering disciplines around the world. Many research groups have websites which are interesting and informative at levels ranging from primary schools to serious researchers. Students are encouraged to visit these sites to find out more about particle technology. Our own website at Monash University can be accessed via the Chemical Engineering Department web page at http://www.eng.monash.edu.au/chemeng/

Martin Rhodes
Mount Eliza, May 1998

Introduction

Particulate materials, powders or bulk solids are used widely in all areas of the process industries, for example in the food processing, pharmaceutical, biotechnology, oil, chemical, mineral processing, metallurgical, detergent, power generation, paint, plastics and cosmetics industries. These industries involve many different types of professional scientists and engineers, such as chemical engineers, chemists, biologists, physicists, pharmacists, mineral engineers, food technologists, metallurgists, material scientists/engineers, environmental scientists/engineers, mechanical engineers, combustion engineers and civil engineers. Some figures give an indication of the significance of particle technology in the world economy: for the DuPont company, whose business covers chemicals, agricultural, pharmaceuticals, paints, dyes, ceramics, around two-thirds of its products involve particulate solids (powders, crystalline solids, granules, flakes, dispersions or pastes); around 1% of all electricity generated worldwide is used in reducing particle size; the impact of particulate products to the US economy was estimated to be US$ 1 trillion.

Some examples of the processing steps involving particles and powder include particle formation processes (such as crystallization, precipitation, granulation, spray drying, tabletting, extrusion and grinding), transportation processes (such as pneumatic and hydraulic transport, mechanical conveying and screw feeding) and mixing, drying and coating processes. In addition, processes involving particulates require reliable storage facilities and give rise to health and safety issues, which must be satisfactorily handled. Design and operation of these many processes across this wide range of industries require a knowledge of the behaviour of powders and particles. This behaviour is often counterintuitive, when intuition is based on our knowledge of liquids and gases. For example, actions such as stirring, shaking or vibrating, which would result in mixing of two liquids, are more likely to produce size segregation in a mixture of free-flowing powders of different sizes. A storage hopper holding 500 t of powder may not deliver even 1 kg when the outlet valve is opened unless the hopper has been correctly designed. When a steel ball is placed at the bottom of a container of sand and the container is vibrated in the vertical plane, the steel ball will rise to the surface. This steel ball will then sink swiftly to the bottom again if air is passed upwards through the sand causing it to be fluidized.

Engineers and scientists are used to dealing with gases and liquids, whose properties can be readily measured, tabulated or even calculated. The boiling point of pure benzene at atmospheric pressure can be safely assumed to remain at 80.1 °C. The thermal conductivity of copper can always be relied upon to be 377 W/m·K at 100 °C. The viscosity of water at 20 °C can be confidently expected to be 0.001 Pa s. With particulate solids, however, the situation is quite different. The flow properties of sodium bicarbonate powder, for example, depend not only of the particle size distribution, but also on particle shape and surface properties, the humidity of the surrounding atmosphere and the state of compaction of the powder. These variables are not easy to characterize and so their influence on the flow properties of the powder is difficult to predict or control with any confidence. Interestingly, powders appear to have some of the behavioural characteristics of the three phases, solids, liquids and gases. For example, like gases, powders can be compressed; like liquids, they can be made to flow, and like solids, they can withstand some deformation.

The importance of knowledge of the science of particulate materials (often called particle or powder technology) to the process industries cannot be overemphasized. Very often, difficulties in the handling or processing powders are ignored or overlooked at the design stage, with the result that powder-related problems are the cause of an inordinate number of production stoppages. However, it has been demonstrated that the application of even a basic understanding of the ways in which powders behave can minimize these processing problems, resulting in less downtime, improvements in quality control and environmental emissions.

This text is intended as an introduction to particle technology. The topics included have been selected to give coverage of the broad areas of particle technology: characterization (size analysis), processing (granulation, fluidization), particle formation (granulation, size reduction), storage and transport (hopper design, pneumatic conveying, standpipes, slurry flow), separation (filtration, settling, cyclones), safety (fire and explosion hazards, health hazards), engineering the properties of particulate systems (colloids, respirable drugs, slurry rheology). For each of the topics studied, the fundamental science involved is introduced and this is linked, where possible, to industrial practice. In each chapter there are worked examples and exercises to enable the reader to practice the relevant calculations and and a 'Test Yourself' section, intended to highlight the main concepts covered. The final chapter includes some case studies–real examples from the process industries of problems that arose and how they were solved.

A website with laboratory demonstrations in particle technology, designed to accompany this text, is available. This easily navigated resource incorporates many video clips of particle and powder phenomena with accompanying explanatory text. The videos bring to life many of the phenomena that I have tried to describe here in words and diagrams. For example, you will see: fluidized beds (bubbling, non-bubbling, spouted) in action; core flow and mass flow in hoppers, size segregation during pouring, vibration and rolling; pan granulation of fine powders, a coal dust explosion; a cyclone separator in action; dilute and dense phase pneumatic conveying. The website will aid the reader in understanding particle technology and is recommended as a useful adjunct to this text.

1

Particle Size Analysis

1.1 INTRODUCTION

In many powder handling and processing operations particle size and size distribution play a key role in determining the bulk properties of the powder. Describing the size distribution of the particles making up a powder is therefore central in characterizing the powder. In many industrial applications a single number will be required to characterize the particle size of the powder. This can only be done accurately and easily with a mono-sized distribution of spheres or cubes. Real particles with shapes that require more than one dimension to fully describe them and real powders with particles in a range of sizes, mean that in practice the identification of single number to adequately describe the size of the particles is far from straightforward. This chapter deals with how this is done.

1.2 DESCRIBING THE SIZE OF A SINGLE PARTICLE

Regular-shaped particles can be accurately described by giving the shape and a number of dimensions. Examples are given in Table 1.1.

The description of the shapes of irregular-shaped particles is a branch of science in itself and will not be covered in detail here. Readers wishing to know more on this topic are referred to Hawkins (1993). However, it will be clear to the reader that no single physical dimension can adequately describe the size of an irregularly shaped particle, just as a single dimension cannot describe the shape of a cylinder, a cuboid or a cone. Which dimension we do use will in practice depend on (a) what property or dimension of the particle we are able to measure and (b) the use to which the dimension is to be put.

If we are using a microscope, perhaps coupled with an image analyser, to view the particles and measure their size, we are looking at a projection of the shape of the particles. Some common diameters used in microscope analysis are statistical diameters such as Martin's diameter (length of the line which bisects the particle

Introduction to Particle Technology - 2nd Edition Martin Rhodes

Table 1.1 Regular-shaped particles

Shape	Sphere	Cube	Cylinder	Cuboid	Cone
Dimensions	Radius	Side length	Radius and height	Three side lengths	Radius and height

image), Feret's diameter (distance between two tangents on opposite sides of the particle) and shear diameter (particle width obtained using an image shearing device) and equivalent circle diameters such as the projected area diameter (area of circle with same area as the projected area of the particle resting in a stable position). Some of these diameters are described in Figure 1.1. We must

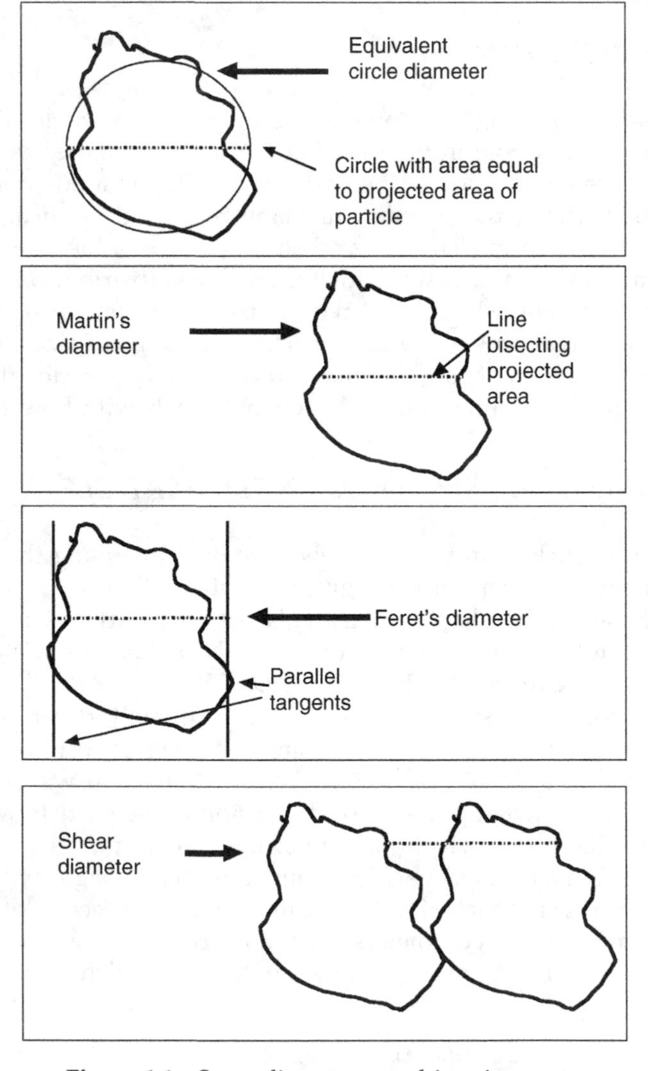

Figure 1.1 Some diameters used in microscopy

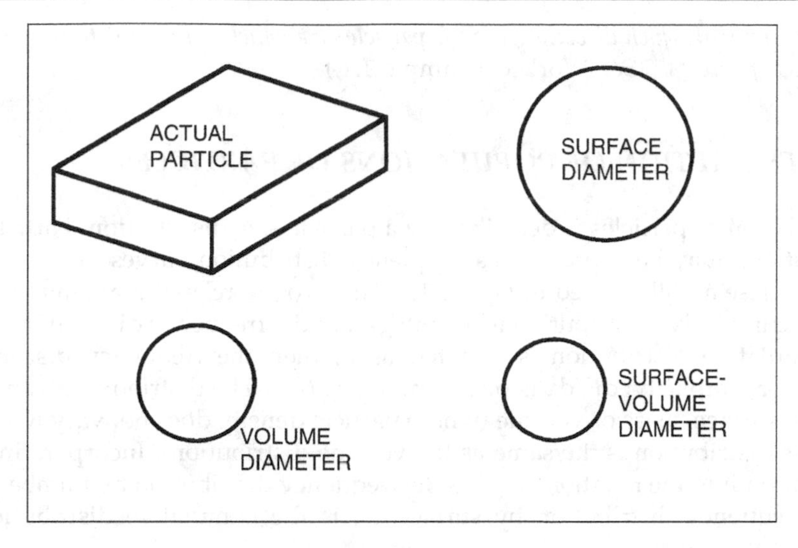

Figure 1.2 Comparison of equivalent sphere diameters

remember that the orientation of the particle on the microscope slide will affect the projected image and consequently the measured equivalent sphere diameter.

If we use a sieve to measure the particle size we come up with an equivalent sphere diameter, which is the diameter of a sphere passing through the same sieve aperture. If we use a sedimentation technique to measure particle size then it is expressed as the diameter of a sphere having the same sedimentation velocity under the same conditions. Other examples of the properties of particles measured and the resulting equivalent sphere diameters are given in Figure 1.2.

Table 1.2 compares values of these different equivalent sphere diameters used to describe a cuboid of side lengths 1, 3, 5 and a cylinder of diameter 3 and length 1.

The volume equivalent sphere diameter or equivalent volume sphere diameter is a commonly used equivalent sphere diameter. We will see later in the chapter that it is used in the Coulter counter size measurements technique. By definition, the equivalent volume sphere diameter is the diameter of a sphere having the same volume as the particle. The surface-volume diameter is the one measured when we use permeametry (see Section 1.8.4) to measure size. The surface-volume (equivalent sphere) diameter is the diameter of a sphere having the same surface to volume ratio as the particle. *In practice it is important to use the method of*

Table 1.2 Comparison of equivalent sphere diameters

Shape	Sphere passing the same sieve aperture, x_p	Sphere having the same volume, x_v	Sphere having same surface the area, x_s	Sphere having the same surface to volume ratio, x_{sv}
Cuboid	3	3.06	3.83	1.95
Cylinder	3	2.38	2.74	1.80

size measurement which directly gives the particle size which is relevant to the situation or process of interest. (See Worked Example 1.1.)

1.3 DESCRIPTION OF POPULATIONS OF PARTICLES

A population of particles is described by a particle size distribution. Particle size distributions may be expressed as frequency distribution curves or cumulative curves. These are illustrated in Figure 1.3. The two are related mathematically in that the cumulative distribution is the integral of the frequency distribution; i.e. if the cumulative distribution is denoted as F, then the frequency distribution $\mathrm{d}F/\mathrm{d}x$. For simplicity, $\mathrm{d}F/\mathrm{d}x$ is often written as $f(x)$. The distributions can be by number, surface, mass or volume (where particle density does not vary with size, the mass distribution is the same as the volume distribution). Incorporating this information into the notation, $f_N(x)$ is the frequency distribution by number, $f_S(x)$ is the frequency distribution by surface, F_S is the cumulative distribution by

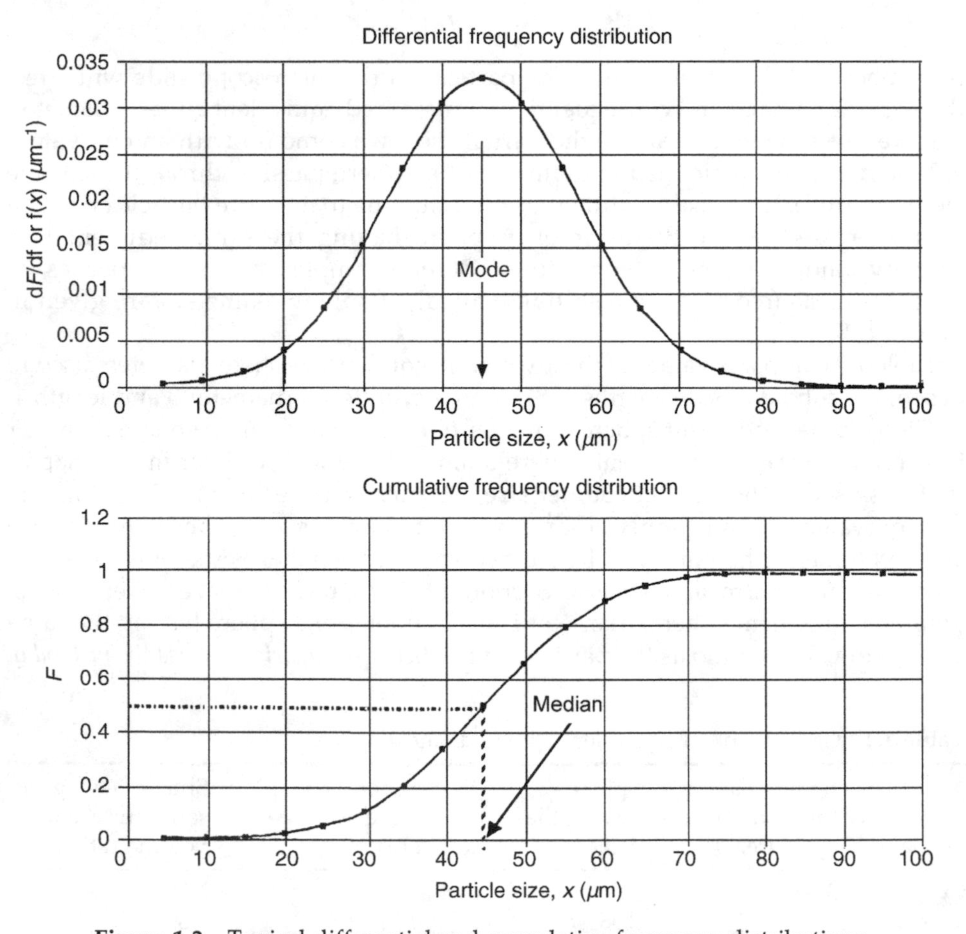

Figure 1.3 Typical differential and cumulative frequency distributions

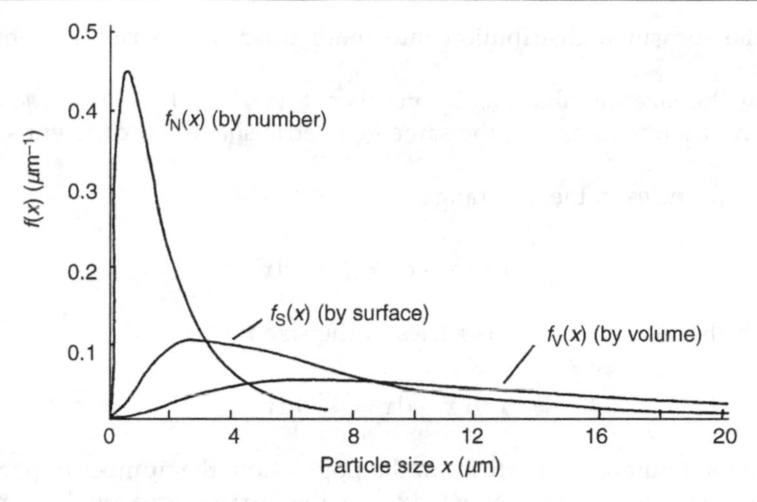

Figure 1.4 Comparison between distributions

surface and F_M is the cumulative distribution by mass. In reality these distributions are smooth continuous curves. However, size measurement methods often divide the size spectrum into size ranges or classes and the size distribution becomes a histogram.

For a given population of particles, the distributions by mass, number and surface can differ dramatically, as can be seen in Figure 1.4.

A further example of difference between distributions for the same population is given in Table 1.3 showing size distributions of man-made objects orbiting the earth (*New Scientist*, 13 October 1991).

The number distribution tells us that only 0.2% of the objects are greater than 10 cm. However, these larger objects make up 99.96% of the mass of the population, and the 99.3% of the objects which are less than 1.0 cm in size make up only 0.01% of the mass distribution. Which distribution we would use is dependent on the end use of the information.

1.4 CONVERSION BETWEEN DISTRIBUTIONS

Many modern size analysis instruments actually measure a number distribution, which is rarely needed in practice. These instruments include software to

Table 1.3 Mass and number distributions for man-made objects orbiting the earth

Size (cm)	Number of objects	% by number	% by mass
10–1000	7000	0.2	99.96
1–10	17 500	0.5	0.03
0.1–1.0	3 500 000	99.3	0.01
Total	3 524 500	100.00	100.00

convert the measured distribution into more practical distributions by mass, surface, etc.

Relating the size distributions by number, $f_N(x)$, and by surface, $f_S(x)$ for a population of particles having the same geometric shape but different size:

Fraction of particles in the size range

$$x \text{ to } x + dx = f_N(x)dx$$

Fraction of the total surface of particles in the size range

$$x \text{ to } x + dx = f_S(x)dx$$

If N is the total number of particles in the population, the number of particles in the size range x to $x + dx = Nf_N(x)dx$ and the surface area of these particles $= (x^2\alpha_S)Nf_N(x)dx$, where α_S is the factor relating the linear dimension of the particle to its surface area.

Therefore, the fraction of the total surface area contained on these particles $[f_S(x)dx]$ is:

$$\frac{(x^2\alpha_S)Nf_N(x)dx}{S}$$

where S is the total surface area of the population of particles.

For a given population of particles, the total number of particles, N, and the total surface area, S are constant. Also, assuming particle shape is independent of size, α_S is constant, and so

$$f_S(x) \propto x^2 f_N(x) \quad \text{or} \quad f_S(x) = k_S x^2 f_N(x) \tag{1.1}$$

where

$$k_S = \frac{\alpha_S N}{S}$$

Similarly, for the distribution by volume

$$f_V(x) = k_V x^3 f_N(x) \tag{1.2}$$

where

$$k_V = \frac{\alpha_V N}{V}$$

where V is the total volume of the population of particles and α_V is the factor relating the linear dimension of the particle to its volume.

And for the distribution by mass

$$f_m(x) = k_m \rho_p x^3 f_N(x) \tag{1.3}$$

where

$$k_m = \frac{\alpha_V \rho_p N}{V}$$

assuming particle density ρ_p is independent of size.

The constants k_S, k_V and k_m may be found by using the fact that:

$$\int_0^{\infty} f(x)\mathrm{d}x = 1 \tag{1.4}$$

Thus, when we convert between distributions it is necessary to make assumptions about the constancy of shape and density with size. Since these assumptions may not be valid, the conversions are likely to be in error. Also, calculation errors are introduced into the conversions. For example, imagine that we used an electron microscope to produce a number distribution of size with a measurement error of ±2%. Converting the number distribution to a mass distribution we triple the error involved (i.e. the error becomes ±6%). For these reasons, conversions between distributions are to be avoided wherever possible. This can be done by choosing the measurement method which gives the required distribution directly.

1.5 DESCRIBING THE POPULATION BY A SINGLE NUMBER

In most practical applications, we require to describe the particle size of a population of particles (millions of them) by a single number. There are many options available; the mode, the median, and several different means including arithmetic, geometric, quadratic, harmonic, etc. Whichever expression of central tendency of the particle size of the population we use must reflect the property or properties of the population of importance to us. We are, in fact, modelling the real population with an artificial population of mono-sized particles. This section deals with calculation of the different expressions of central tendency and selection of the appropriate expression for a particular application.

The *mode* is the most frequently occurring size in the sample. We note, however, that for the same sample, different modes would be obtained for distributions by number, surface and volume. The mode has no practical significance as a measure of central tendency and so is rarely used in practice.

The *median* is easily read from the cumulative distribution as the 50% size; the size which splits the distribution into two equal parts. In a mass distribution, for example, half of the particles by mass are smaller than the median size. Since the

Table 1.4 Definitions of means

$g(x)$	Mean and notation
x	arithmetic mean, $\bar{x}_{\mathrm{a}}$
x^2	quadratic mean, $\bar{x}_{\mathrm{q}}$
x^3	cubic mean, $\bar{x}_{\mathrm{c}}$
$\log x$	geometric mean, $\bar{x}_{\mathrm{g}}$
$1/x$	harmonic mean, $\bar{x}_{\mathrm{h}}$

median is easily determined, it is often used. However, it has no special significance as a measure of central tendency of particle size.

Many different *means* can be defined for a given size distribution; as pointed out by Svarovsky (1990). However, they can all be described by:

$$g(\bar{x}) = \frac{\int_0^1 g(x)\mathrm{d}F}{\int_0^1 \mathrm{d}F} \quad \text{but} \quad \int_0^1 \mathrm{d}F = 1 \quad \text{and so} \quad g(\bar{x}) = \int_0^1 g(x)\mathrm{d}F \qquad (1.5)$$

where $\bar{x}$ is the mean and g is the weighting function, which is different for each mean definition. Examples are given in Table 1.4.

Equation (1.5) tells us that the mean is the area between the curve and the $F(x)$ axis in a plot of $F(x)$ versus the weighting function $g(x)$ (Figure 1.5). In fact, graphical determination of the mean is always recommended because the distribution is more accurately represented as a continuous curve.

Each mean can be shown to conserve two properties of the original population of particles. For example, the arithmetic mean of the surface distribution conserves the surface and volume of the original population. This is demonstrated in Worked Example 1.3. This mean is commonly referred to as the surface-volume mean or the Sauter mean. The arithmetic mean of the number

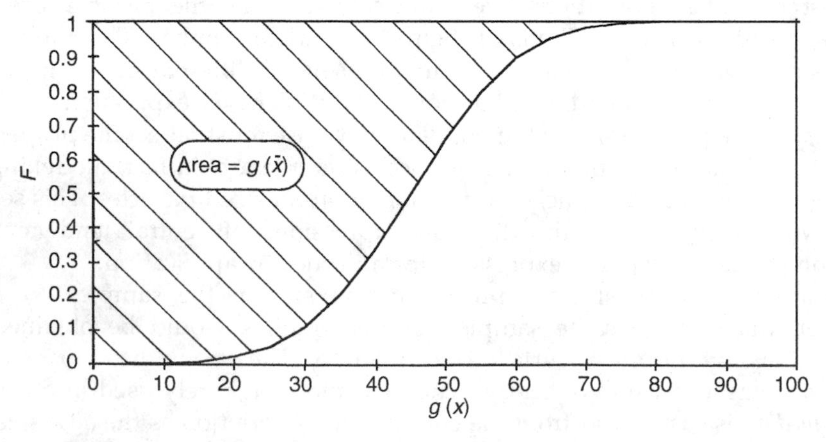

Figure 1.5 Plot of cumulative frequency against weighting function $g(x)$. Shaded area is $g(\bar{x}) = \int_0^1 g(x)\,\mathrm{d}F$

distribution $\bar{x}_{aN}$ conserves the number and length of the original population and is known as the number-length mean $\bar{x}_{NL}$:

$$\text{number-length mean, } \bar{x}_{NL} = \bar{x}_{aN} = \frac{\int_0^1 x\,dF_N}{\int_0^1 dF_N} \tag{1.6}$$

As another example, the quadratic mean of the number distribution $\bar{x}_{qN}$ conserves the number and surface of the original population and is known as the number-surface mean $\bar{x}_{NS}$:

$$\text{number-surface mean, } \bar{x}_{NS}^2 = \bar{x}_{qN}^2 = \frac{\int_0^1 x^2\,dF_N}{\int_0^1 dF_N} \tag{1.7}$$

A comparison of the values of the different means and the mode and median for a given particle size distribution is given in Figure 1.6. This figure highlights two points: (a) that the values of the different expressions of central tendency can vary significantly; and (b) that two quite different distributions could have the same

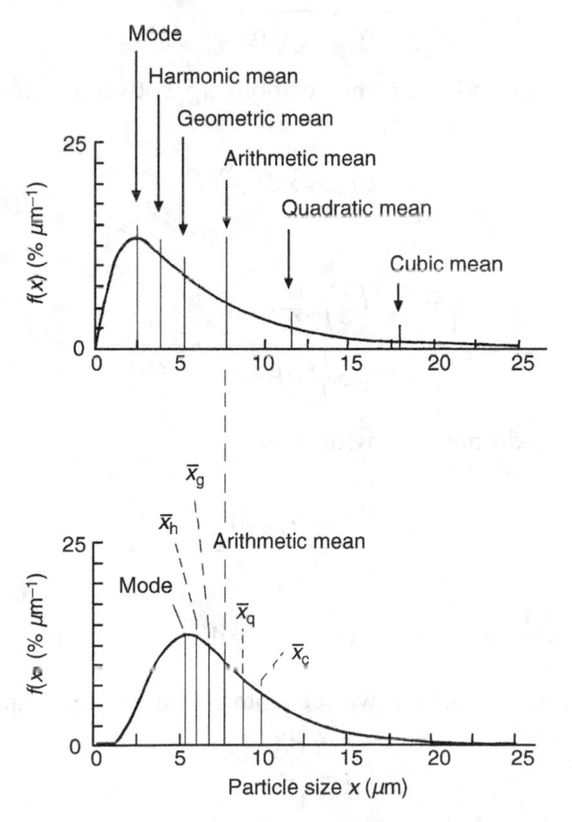

Figure 1.6 Comparison between measures of central tendency. Adapted from Rhodes (1990). Reproduced by permission

arithmetic mean or median, etc. If we select the wrong one for our design correlation or quality control we may be in serious error.

So how do we decide which mean particle size is the most appropriate one for a given application? Worked Examples 1.3 and 1.4 indicate how this is done.

For Equation (1.8), which defines the surface-volume mean, please see Worked Example 1.3.

1.6 *EQUIVALENCE OF MEANS*

Means of different distributions can be equivalent. For example, as is shown below, the arithmetic mean of a surface distribution is equivalent (numerically equal to) the harmonic mean of a volume (or mass) distribution:

$$\text{arithmetic mean of a surface distribution, } \bar{x}_{aS} = \frac{\int_0^1 x\,dF_S}{\int_0^1 dF_S} \tag{1.9}$$

The harmonic mean $\bar{x}_{hV}$ of a volume distribution is defined as:

$$\frac{1}{\bar{x}_{hV}} = \frac{\int_0^1 \left(\frac{1}{x}\right) dF_V}{\int_0^1 dF_V} \tag{1.10}$$

From Equations (1.1) and (1.2), the relationship between surface and volume distributions is:

$$dF_v = x\,dF_s \frac{k_v}{k_s} \tag{1.11}$$

hence

$$\frac{1}{\bar{x}_{hV}} = \frac{\int_0^1 \left(\frac{1}{x}\right) x \frac{k_v}{k_s} dF_s}{\int_0^1 x \frac{k_v}{k_s} dF_s} = \frac{\int_0^1 dF_s}{\int_0^1 x\,dF_s} \tag{1.12}$$

(assuming k_s and k_v do not vary with size)
and so

$$\bar{x}_{hV} = \frac{\int_0^1 x\,dF_s}{\int_0^1 dF_s}$$

which, by inspection, can be seen to be equivalent to the arithmetic mean of the surface distribution $\bar{x}_{aS}$ [Equation (1.9)].

Recalling that $dF_s = x^2 k_S dF_N$, we see from Equation (1.9) that

$$\bar{x}_{aS} = \frac{\int_0^1 x^3 dF_N}{\int_0^1 x^2 dF_N}$$

which is the surface-volume mean, $\bar{x}_{SV}$ [Equation (1.8) - see Worked Example 1.3].

Summarizing, then, the surface-volume mean may be calculated as the arithmetic mean of the surface distribution or the harmonic mean of the volume distribution. The practical significance of the equivalence of means is that it permits useful means to be calculated easily from a single size analysis.

The reader is invited to investigate the equivalence of other means.

1.7 COMMON METHODS OF DISPLAYING SIZE DISTRIBUTIONS

1.7.1 Arithmetic-normal Distribution

In this distribution, shown in Figure 1.7, particle sizes with equal differences from the arithmetic mean occur with equal frequency. Mode, median and arithmetic mean coincide. The distribution can be expressed mathematically by:

$$\frac{\mathrm{d}F}{\mathrm{d}x} = \frac{1}{\sigma\sqrt{2\pi}}\exp\left[-\frac{(x-\bar{x})^2}{2\sigma^2}\right] \tag{1.13}$$

where σ is the standard deviation.

To check for a arithmetic-normal distribution, size analysis data is plotted on normal probability graph paper. On such graph paper a straight line will result if the data fits an arithmetic-normal distribution.

1.7.2 Log-normal Distribution

This distribution is more common for naturally occurring particle populations. An example is shown in Figure 1.8. If plotted as $\mathrm{d}F/\mathrm{d}(\log x)$ versus x, rather than $\mathrm{d}F/\mathrm{d}x$ versus x, an arithmetic-normal distribution in $\log x$ results (Figure 1.9). The mathematical expression describing this distribution is:

$$\frac{\mathrm{d}F}{\mathrm{d}z} = \frac{1}{\sigma_z\sqrt{2\pi}}\exp\left[-\frac{(z-\bar{z})^2}{2\sigma_z^2}\right] \tag{1.14}$$

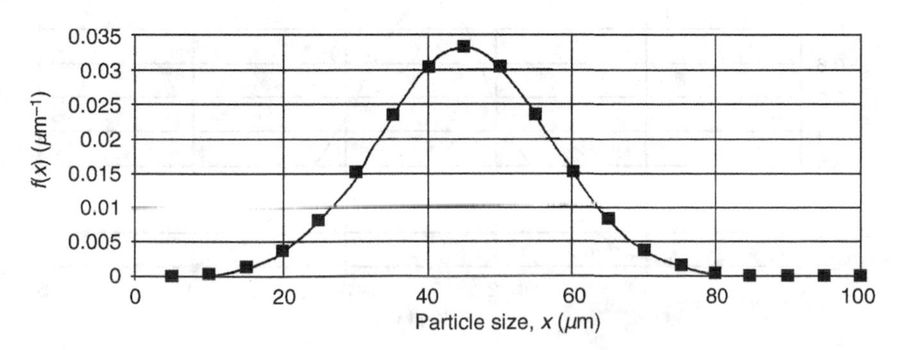

Figure 1.7 Arithmetic-normal distribution with an arithmetic mean of 45 and standard deviation of 12

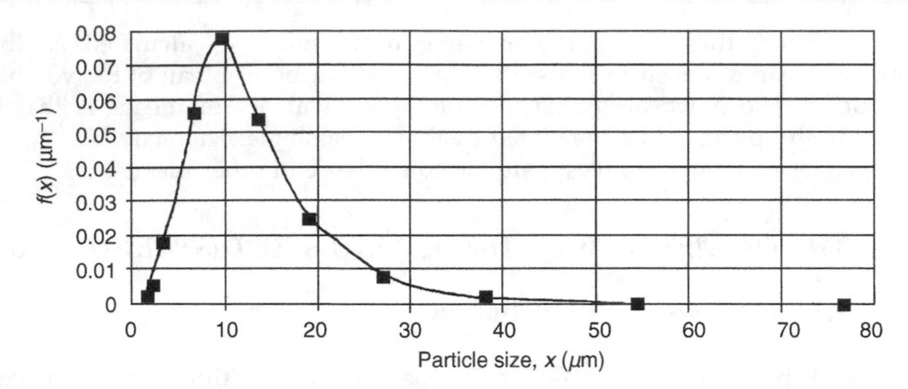

Figure 1.8 Log-normal distribution plotted on linear coordinates

where $z = \log x$, $\bar{z}$ is the arithmetic mean of $\log x$ and σ_z is the standard deviation of $\log x$.

To check for a log-normal distribution, size analysis data are plotted on log-normal probability graph paper. Using such graph paper, a straight line will result if the data fit a log-normal distribution.

1.8 METHODS OF PARTICLE SIZE MEASUREMENT

1.8.1 Sieving

Dry sieving using woven wire sieves is a simple, cheap method of size analysis suitable for particle sizes greater than 45 μm. Sieving gives a mass distribution and a size known as the sieve diameter. Since the length of the particle does not hinder its passage through the sieve apertures (unless the particle is extremely elongated), the sieve diameter is dependent on the maximum width and maximum thickness of the

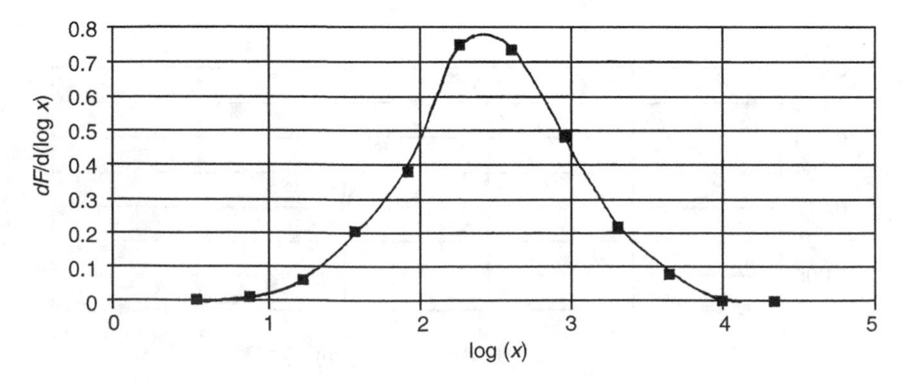

Figure 1.9 Log-normal distribution plotted on logarithmic coordinates

particle. The most common modern sieves are in sizes such that the ratio of adjacent sieve sizes is the fourth root of two (eg. 45, 53, 63, 75, 90, 107 μm). If standard procedures are followed and care is taken, sieving gives reliable and reproducible size analysis. Air jet sieving, in which the powder on the sieve is fluidized by a jet or air, can achieve analysis down to 20 μm. Analysis down to 5 μm can be achieved by wet sieving, in which the powder sample is suspended in a liquid.

1.8.2 Microscopy

The optical microscope may be used to measure particle sizes down to 5 μm. For particles smaller than this diffraction causes the edges of the particle to be blurred and this gives rise to an apparent size. The electron microscope may be used for size analysis below 5 μm. Coupled with an image analysis system the optical microscope or electron microscope can readily give number distributions of size and shape. Such systems calculate various diameters from the projected image of the particles (e.g. Martin's, Feret's, shear, projected area diameters, etc.). Note that for irregular-shaped particles, the projected area offered to the viewer can vary significantly depending on the orientation of the particle. Techniques such as applying adhesive to the microscope slide may be used to ensure that the particles are randomly orientated.

1.8.3 Sedimentation

In this method, the rate of sedimentation of a sample of particles in a liquid is followed. The suspension is dilute and so the particles are assumed to fall at their single particle terminal velocity in the liquid (usually water). Stokes' law is assumed to apply ($Re_p < 0.3$) and so the method using water is suitable only for particles typically less than 50 μm in diameter. The rate of sedimentation of the particles is followed by plotting the suspension density at a certain vertical position against time. The suspension density is directly related to the cumulative undersize and the time is related to the particle diameter via the terminal velocity. This is demonstrated in the following:

Referring to Figure 1.10, the suspension density is sampled at a vertical distance, h below the surface of the suspension. The following assumptions are made:

- The suspension is sufficiently dilute for the particles to settle as individuals (i.e. not hindered settling – see Chapter 3).
- Motion of the particles in the liquid obeys Stokes' law (true for particles typically smaller than 50 μm).
- Particles are assumed to accelerate rapidly to their terminal free fall velocity U_T so that the time for acceleration is negligible.

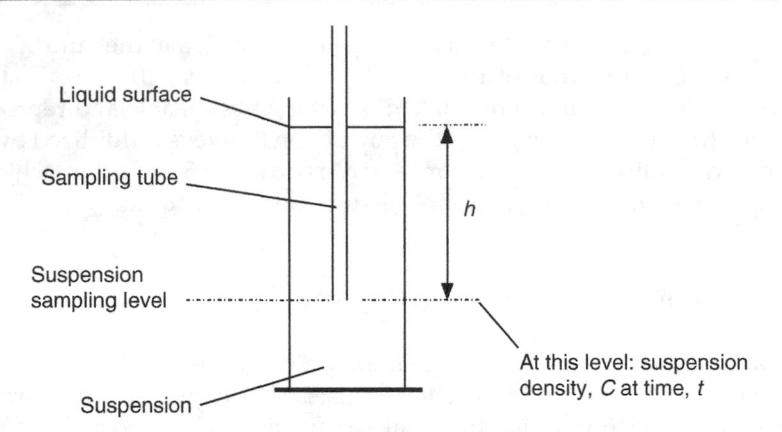

Figure 1.10 Size analysis by sedimentation

Let the original uniform suspension density be C_0. Let the suspension density at the sampling point be C at time t after the start of settling. At time t all those particles travelling faster than h/t will have fallen below the sampling point. The sample at time t will therefore consist only of particles travelling a velocity $\leq h/t$. Thus, if C_0 is representative of the suspension density for the whole population, then C represents the suspension density for all particles which travel at a velocity $\leq h/t$, and so C/C_0 is the mass fraction of the original particles which travel at a velocity $\leq h/t$. That is,

$$\text{cumulative mass fraction} = \frac{C}{C_0}$$

All particles travel at their terminal velocity given by Stokes' law [Chapter 2, Equation (2.13)]:

$$U_T = \frac{x^2(\rho_p - \rho_f)g}{18\,\mu}$$

Thus, equating U_T with h/t, we determine the diameter of the particle travelling at our cut-off velocity h/t. That is,

$$x = \left[\frac{18\,\mu h}{t(\rho_p - \rho_f)g}\right]^{1/2} \tag{1.15}$$

Particles smaller than x will travel slower than h/t and will still be in suspension at the sampling point. Corresponding values of C/C_0 and x therefore give us the cumulative mass distribution. The particle size measured is the Stokes' diameter, i.e. the diameter of a sphere having the same terminal settling velocity in the Stokes region as the actual particle.

A common form of this method is the Andreason pipette which is capable of measuring in the range 2–100 μm. At size below 2 μm, Brownian motion causes significant errors. Increasing the body force acting on the particles by centrifuging the suspension permits the effects of Brownian motion to be reduced so that particle sizes down to 0.01 μm can be measured. Such a device is known as a pipette centrifuge.

The labour involved in this method may be reduced by using either light absorption or X-ray absorption to measure the suspension density. The light absorption method gives rise to a distribution by surface, whereas the X-ray absorption method gives a mass distribution.

1.8.4 Permeametry

This is a method of size analysis based on fluid flow through a packed bed (see Chapter 6). The Carman–Kozeny equation for laminar flow through a randomly packed bed of uniformly sized spheres of diameter x is [Equation 6.9]:

$$\frac{(-\Delta p)}{H} = 180\frac{(1-\varepsilon)^2}{\varepsilon^3}\frac{\mu U}{x^2}$$

where $(-\Delta p)$ is the pressure drop across the bed, ε is the packed bed void fraction, H is the depth of the bed, μ is the fluid viscosity and U is the superficial fluid velocity. In Worked Example 1.3, we will see that, when we are dealing with non-spherical particles with a distribution of sizes, the appropriate mean diameter for this equation is the surface-volume diameter $\bar{x}_{SV}$, which may be calculated as the arithmetic mean of the surface distribution, $\bar{x}_{aS}$.

In this method, the pressure gradient across a packed bed of known voidage is measured as a function of flow rate. The diameter we calculate from the Carman–Kozeny equation is the arithmetic mean of the surface distribution (see Worked Example 6.1 in Chapter 6).

1.8.5 Electrozone Sensing

Particles are held in supension in a dilute electrolyte which is drawn through a tiny orifice with a voltage applied across it (Figure 1.11). As particles flow through the orifice a voltage pulse is recorded.

The amplitude of the pulse can be related to the volume of the particle passing the orifice. Thus, by electronically counting and classifying the pulses according to amplitude this technique can give a number distribution of the equivalent volume sphere diameter. The lower size limit is dictated by the smallest practical orifice and the upper limit is governed by the need to maintain particles in suspension. Although liquids more viscous than water may be used to reduce sedimentation, the practical range of size for this method is 0.3–1000 μm. Errors are introduced if more that one particle passes through the orifice at a time and so dilute suspensions are used to reduce the likelihood of this error.

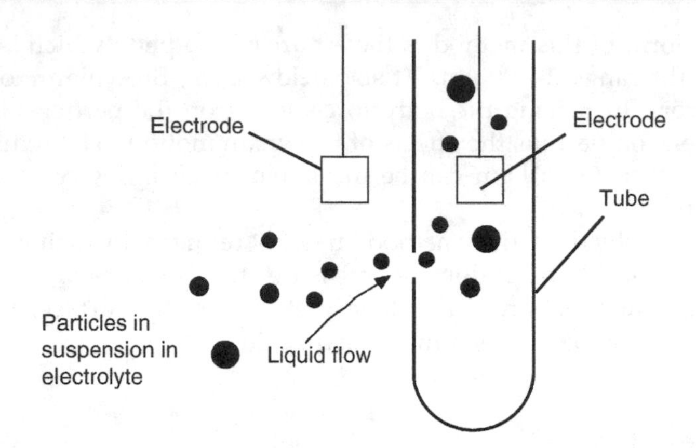

Figure 1.11 Schematic of electrozone sensing apparatus

1.8.6 Laser Diffraction

This method relies on the fact that for light passing through a suspension, the diffraction angle is inversely proportional to the particle size. An instrument would consist of a laser as a source of coherent light of known fixed wavelength (typically 0.63 μm), a suitable detector (usually a slice of photosensitive silicon with a number of discrete detectors, and some means of passing the sample of particles through the laser light beam (techniques are available for suspending particles in both liquids and gases are drawing them through the beam).

To relate diffraction angle with particle size, early instruments used the Fraunhofer theory, which can give rise to large errors under some circumstances (e.g. when the refractive indices of the particle material and suspending medium approach each other). Modern instruments use the Mie theory for interaction of light with matter. This allows particle sizing in the range 0.1–2000 μm, provided that the refractive indices of the particle material and suspending medium are known.

This method gives a volume distribution and measures a diameter known as the laser diameter. Particle size analysis by laser diffraction is very common in industry today. The associated software permits display of a variety of size distributions and means derived from the original measured distribution.

1.9 SAMPLING

In practice, the size distribution of many tonnes of powder are often assumed from an analysis performed on just a few grams or milligrams of sample. The importance of that sample being representative of the bulk powder cannot be overstated. However, as pointed out in Chapter 11 on mixing and segregation, most powder handling and processing operations (pouring, belt conveying, handling in bags or drums, motion of the sample bottle, etc.) cause particles to segregate according to size and to a lesser extent density and shape. This natural tendency to segregation means that extreme care must be taken in sampling.

There are two golden rules of sampling:

1. The powder should be in motion when sampled.

2. The whole of the moving stream should be taken for many short time increments.

Since the eventual sample size used in the analysis may be very small, it is often necessary to split the original sample in order to achieve the desired amount for analysis. These sampling rules must be applied at every step of sampling and sample splitting.

Detailed description of the many devices and techniques used for sampling in different process situations and sample dividing are outside the scope of this chapter. However, Allen (1990) gives an excellent account, to which the reader is referred.

1.10 WORKED EXAMPLES

WORKED EXAMPLE 1.1

Calculate the equivalent volume sphere diameter x_v and the surface-volume equivalent sphere diameter x_{sv} of a cuboid particle of side length 1, 2, 4 mm.

Solution

The volume of cuboid $= 1 \times 2 \times 4 = 8\,\text{mm}^3$

The surface area of the particle $= (1 \times 2) + (1 \times 2) + (1 + 2 + 1 + 2) \times 4 = 28\,\text{mm}^2$

The volume of sphere of diameter x_v is $\pi x_v^3/6$

Hence, diameter of a sphere having a volume of $8\,\text{mm}^3$, $x_v = 2.481\,\text{mm}$

The *equivalent volume sphere diameter* x_v of the cuboid particle is therefore $x_v = 2.481\,\text{mm}$

The surface to volume ratio of the cuboid particle $= \dfrac{28}{8} = 3.5\,\text{mm}^2/\text{mm}^3$

The surface to volume ratio for a sphere of diameter x_{sv} is therefore $6/x_{sv}$

Hence, the diameter of a sphere having the same surface to volume ratio as the particle $= 6/3.5 = 1.714\,\text{mm}$

The *surface-volume equivalent sphere diameter* of the cuboid, $x_{sv} = 1.714\,\text{mm}$

WORKED EXAMPLE 1.2

Convert the surface distribution described by the following equation to a cumulative volume distribution:

$$F_S = (x/45)^2 \quad \text{for} \quad x \leq 45\,\mu\text{m}$$
$$F_S = 1 \quad \text{for} \quad x > 45\,\mu\text{m}$$

Solution

From Equations (1.1)–(1.3),

$$f_v(x) = \frac{k_v}{k_s} x f_s(x)$$

Integrating between sizes 0 and x:

$$F_v(x) = \int_0^x \left(\frac{k_v}{k_s}\right) x f_s(x) \mathrm{d}x$$

Noting that $f_s(x) = \mathrm{d}F_s/\mathrm{d}x$, we see that

$$f_s(x) = \frac{\mathrm{d}}{\mathrm{d}x}\left(\frac{x}{45}\right)^2 = \frac{2x}{(45)^2}$$

and our integral becomes

$$F_v(x) = \int_0^x \left(\frac{k_v}{k_s}\right) \frac{2x^2}{(45)^2} \mathrm{d}x$$

Assuming that k_v and k_s are independent of size,

$$F_v(x) = \left(\frac{k_v}{k_s}\right) \int_0^x \frac{2x^2}{(45)^2} \mathrm{d}x$$
$$= \frac{2}{3}\left[\frac{x^3}{(45)^2}\right] \frac{k_v}{k_s}$$

k_v/k_s may be found by noting that $F_v(45) = 1$; hence

$$\frac{90}{3}\frac{k_v}{k_s} = 1 \quad \text{and so} \quad \frac{k_v}{k_s} = 0.0333$$

Thus, the formula for the volume distribution is

$$F_v = 1.096 \times 10^{-5} x^3 \quad \text{for} \quad x \leq 45\,\mu\text{m}$$
$$F_v = 1 \quad \text{for} \quad x > 45\,\mu\text{m}$$

WORKED EXAMPLE 1.3

What mean particle size do we use in calculating the pressure gradient for flow of a fluid through a packed bed of particles using the Carman–Kozeny equation (see Chapter 6)?

Solution

The Carman–Kozeny equation for laminar flow through a randomly packed bed of particles is:

$$\frac{(-\Delta p)}{L} = K\frac{(1-\varepsilon)^2}{\varepsilon^3}S_v^2\,\mu U$$

where S_v is the specific surface area of the bed of particles (particle surface area per unit particle volume) and the other terms are defined in Chapter 6. If we assume that the bed voidage is independent of particle size, then to write the equation in terms of a mean particle size, we must express the specific surface, S_v, in terms of that mean. The particle size we use must give the same value of S_v as the original population or particles. Thus the mean diameter $\bar{x}$ must conserve the surface and volume of the population; that is, the mean must enable us to calculate the total volume from the total surface of the particles. This mean is the surface-volume mean $\bar{x}_{sv}$

$$\bar{x}_{sv} \times (\text{total surface}) \times \frac{\alpha_v}{\alpha_s} = (\text{total volume})\left(\text{eg. for spheres, } \frac{\alpha_v}{\alpha_s} = \frac{1}{6}\right)$$

$$\text{and therefore } \bar{x}_{sv}\int_0^\infty f_s(x)\mathrm{d}x \cdot \frac{k_v}{k_s} = \int_0^\infty f_v(x)\mathrm{d}x$$

$$\text{Total volume of particles, } \mathrm{V} = \int_0^\infty x^3\alpha_V \mathrm{N} f_N(x)\mathrm{d}x$$

$$\text{Total surface area of particles, } \mathrm{S} = \int_0^\infty x^2\alpha_S \mathrm{N} f_N(x)\mathrm{d}x$$

$$\text{Hence, } \bar{x}_{sv} = \frac{\alpha_s}{\alpha_v}\frac{\int_0^\infty x^3\alpha_v \mathrm{N} f_N(x)\mathrm{d}x}{\int_0^\infty x^2\alpha_s \mathrm{N} f_N(x)\mathrm{d}x}$$

Then, since α_V, α_S and N are independent of size, x,

$$\bar{x}_{sv} = \frac{\int_0^\infty x^3 f_N(x)\mathrm{d}x}{\int_0^\infty x^2 f_N(x)\mathrm{d}x} = \frac{\int_0^1 x^3\mathrm{d}F_N}{\int_0^1 x^2\mathrm{d}F_N}$$

This is the definition of the mean which conserves surface and volume, known as the surface-volume mean, $\bar{x}_{SV}$.

So

$$\bar{x}_{SV} = \frac{\int_0^1 x^3\mathrm{d}F_N}{\int_0^1 x^2\mathrm{d}F_N} \tag{1.8}$$

The correct mean particle diameter is therefore the surface-volume mean as defined above. (We saw in Section 1.6 that this may be calculated as the arithmetic mean of the surface distribution $\bar{x}_{aS}$, or the harmonic mean of the volume distribution.) Then in the Carman–Kozeny equation we make the following substitution for S_v:

$$S_v = \frac{1}{\bar{x}_{SV}} \frac{k_s}{k_v}$$

e.g. for spheres, $S_v = 6/\bar{x}_{SV}$.

WORKED EXAMPLE 1.4 (AFTER SVAROVSKY, 1990)

A gravity settling device processing a feed with size distribution $F(x)$ and operates with a grade efficiency $G(x)$. Its total efficiency is defined as:

$$E_T = \int_0^1 G(x) \mathrm{d} F_M$$

How is the mean particle size to be determined?

Solution

Assuming plug flow (see Chapter 3), $G(x) = U_T A/Q$ where, A is the settling area, Q is the volume flow rate of suspension and U_T is the single particle terminal velocity for particle size x, given by (in the Stokes region):

$$U_T = \frac{x^2(\rho_p - \rho_f)g}{18\,\mu} \qquad \text{(Chapter 2)}$$

hence

$$E_T = \frac{Ag(\rho_p - \rho_f)}{18\,\mu Q} \int_0^1 x^2 \mathrm{d} F_M$$

where $\int_0^1 x^2 \mathrm{d} F_M$ is seen to be the definition of the quadratic mean of the distribution by mass $\bar{x}_{qM}$ (see Table 1.4).

This approach may be used to determine the correct mean to use in many applications.

WORKED EXAMPLE 1.5

A Coulter counter analysis of a cracking catalyst sample gives the following cumulative volume distribution:

Channel	1	2	3	4	5	6	7	8
% volume differential	0	0.5	1.0	1.6	2.6	3.8	5.7	8.7

Channel	9	10	11	12	13	14	15	16
% volume differential	14.3	22.2	33.8	51.3	72.0	90.9	99.3	100

(a) Plot the cumulative volume distribution versus size and determine the median size.

(b) Determine the surface distribution, giving assumptions. Compare with the volume distribution.

(c) Determine the harmonic mean diameter of the volume distribution.

(d) Determine the arithmetic mean diameter of the surface distribution.

Solution

With the Coulter counter the channel size range differs depending on the tube in use. We therefore need the additional information that in this case channel 1 covers the size range 3.17 μm to 4.0 μm, channel 2 covers the range 4.0 μm to 5.04 μm and so on up to channel 16, which covers the range 101.4 μm to 128 μm. The ratio of adjacent size range boundaries is always the cube root of 2. For example,

$$\sqrt[3]{2} = \frac{4.0}{3.17} = \frac{5.04}{4.0} = \frac{128}{101.4}, \text{ etc.}$$

The resulting lower and upper sizes for the channels are shown in columns 2 and 3 of Table 1W5.1.

Table 1W5.1 Size distribution data associated with Worked Example 1.5

1 Channel number	2 Lower size of range μm	3 Upper size of range μm	4 Cumulative per cent undersize	5 F_v	6 $1/x$	7 Cumulative area under F_v versus $1/x$	8 F_s	9 Cumulative area under F_s versus x
1	3.17	4.00	0	0	0.2500	0.0000	0.0000	0.0000
2	4.00	5.04	0.5	0.005	0.1984	0.0011	0.0403	0.1823
3	5.04	6.35	1	0.01	0.1575	0.0020	0.0723	0.3646
4	6.35	8.00	1.6	0.016	0.1250	0.0029	0.1028	0.5834
5	8.00	10.08	2.6	0.026	0.0992	0.0040	0.1432	0.9480
6	10.08	12.70	3.8	0.038	0.0787	0.0050	0.1816	1.3855
7	12.70	16.00	5.7	0.057	0.0625	0.0064	0.2299	2.0782
8	16.00	20.16	8.7	0.087	0.0496	0.0081	0.2904	3.1720
9	20.16	25.40	14.3	0.143	0.0394	0.0106	0.3800	5.2138
10	25.40	32.00	22.2	0.222	0.0313	0.0134	0.4804	8.0942
11	32.00	40.32	33.8	0.338	0.0248	0.0166	0.5973	12.3236
12	40.32	50.80	51.3	0.513	0.0197	0.0205	0.7374	18.7041
13	50.80	64.00	72	0.72	0.0156	0.0242	0.8689	26.2514
14	64.00	80.63	90.9	0.909	0.0124	0.0268	0.9642	33.1424
15	80.63	101.59	99.3	0.993	0.0098	0.0277	0.9978	36.2051
16	101.59	128.00	100	1	0.0078	**0.0278**	1.0000	**36.4603**

Note: Based on arithmetic means of size ranges.

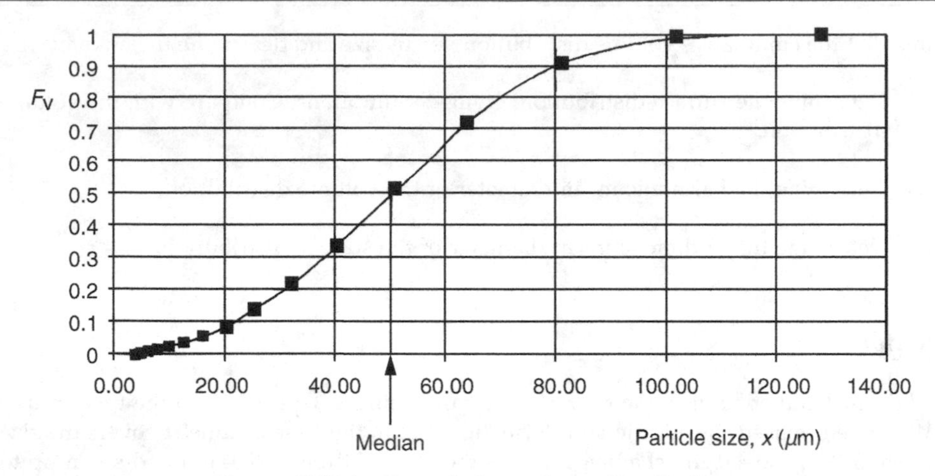

Figure 1W5.1 Cumulative volume distribution

(a) The cumulative undersize distribution is shown numerically in column 5 of Table 1W5.1 and graphically in Figure 1W5.1. By inspection, we see that the median size is 50 μm (b), i.e. 50% by volume of the particles is less than 50 μm.

(b) The surface distribution is related to the volume distribution by the expression:

$$f_s(x) = \frac{f_v(x)}{x} \times \frac{k_s}{k_v} \quad \text{(from [Equations(1.1) and (1.2)]}$$

Recalling that $f(x) = dF/dx$ and integrating between 0 and x:

$$\frac{k_s}{k_v}\int_0^x \frac{1}{x}\frac{dF_v}{dx}dx = \int_0^x \frac{dF_s}{dx}dx$$

or

$$\frac{k_s}{k_v}\int_0^x \frac{1}{x}dF_v = \int_0^x dF_s = F_s(x)$$

(assuming particle shape is invariant with size so that k_s/k_v is constant).

So the surface distribution can be found from the area under a plot of $1/x$ versus F_v multiplied by the factor k_s/k_v (which is found by noting that $\int_{x=0}^{x=\infty} dF_s = 1$).

Column 7 of Table 1W5.1 shows the area under $1/x$ versus F_v. The factor k_s/k_v is therefore equal to 0.0278. Dividing the values of column 7 by 0.0278 gives the surface distribution F_s shown in column 8. The surface distribution is shown graphically in Figure 1W5.2. The shape of the surface distribution is quite different from that of the volume distribution; the smaller particles make up a high proportion of the total surface. The median of the surface distribution is around 35 μm, i.e. particles under 35 μm contribute 50% of the total surface area.

(c) The harmonic mean of the volume distribution is given by:

$$\frac{1}{\bar{x}_{hV}} = \int_0^1 \left(\frac{1}{x}\right)dF_v$$

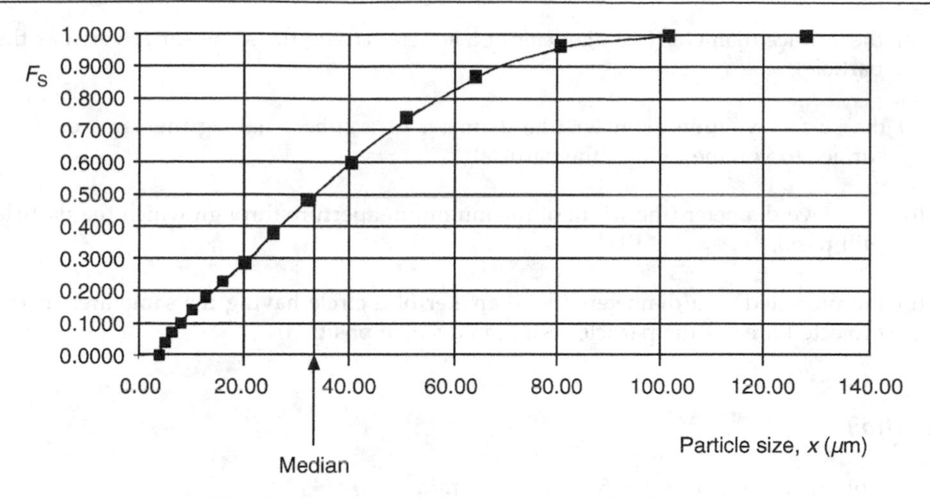

Figure 1W5.2 Cumulative surface distribution

This can be calculated graphically from a plot of F_v versus $1/x$ or numerically from the tabulated data in column 7 of Table 1W5.1. Hence,

$$\frac{1}{\bar{x}_{hV}} = \int_0^1 \left(\frac{1}{x}\right) dF_v = 0.0278$$

and so, $\bar{x}_{hV} = 36\,\mu m$.

We recall that the harmonic mean of the volume distribution is equivalent to the surface-volume mean of the population.

(d) The arithmetic mean of the surface distribution is given by:

$$\bar{x}_{aS} = \int_0^1 x\, dF_s$$

This may be calculated graphically from our plot of F_s versus x (Figure 1W5.2) or numerically using the data in Table 1W5.1. This area calculation as shown in Column 9 of the table shows the cumulative area under a plot of F_s versus x and so the last figure in this column is equivalent to the above integral.

Thus:

$$\bar{x}_{as} = 36.4\,\mu m$$

We may recall that the arithmetic mean of the surface distribution is also equivalent to the surface-volume mean of the population. This value compares well with the value obtained in (c) above.

WORKED EXAMPLE 1.6

Consider a cuboid particle $5.00 \times 3.00 \times 1.00$ mm. Calculate for this particle the following diameters:

(a) the volume diameter (the diameter of a sphere having the same volume as the particle);

(b) the surface diameter (the diameter of a sphere having the same surface area as the particle);

(c) the surface-volume diameter (the diameter of a sphere having the same external surface to volume ratio as the particle);

(d) the sieve diameter (the width of the minimum aperture through which the particle will pass);

(e) the projected area diameters (the diameter of a circle having the same area as the projected area of the particle resting in a stable position).

Solution

(a) Volume of the particle $= 5 \times 3 \times 1 = 15\,\text{mm}^3$

Volume of a sphere $= \dfrac{\pi x_v^3}{6}$

Thus volume diameter, $x_v = \sqrt[3]{\dfrac{15 \times 6}{\pi}} = 3.06\,\text{mm}$

(b) Surface area of the particle $= 2 \times (5 \times 3) + 2 \times (1 \times 3) + 2 \times (1 \times 5) = 46\,\text{mm}^2$

Surface area of sphere $= \pi x_s^2$

Therefore, surface diameter, $x_s = \sqrt{\dfrac{46}{\pi}} = 3.83\,\text{mm}$

(c) Ratio of surface to volume of the particle $= 46/15 = 3.0667$

For a sphere, surface to volume ratio $= \dfrac{6}{x_{sv}}$

Therefore, $x_{sv} = \dfrac{6}{3.0667} = 1.96\,\text{mm}$

(d) The smallest square aperture through which this particle will pass is 3 mm. Hence, the sieve diameter, $x_p = 3\,\text{mm}$

(e) This particle has three projected areas in stable positions:

$$\text{area } 1 = 3\,\text{mm}^2;\ \text{area } 2 = 5\,\text{mm}^2;\ \text{area } 3 = 15\,\text{mm}^2$$

$$\text{area of circle} = \frac{\pi x^2}{4}$$

hence, projected area diameters:

$$\text{projected area diameter } 1 = 1.95\,\text{mm};$$

$$\text{projected area diameter } 2 = 2.52\,\text{mm};$$

$$\text{projected area diameter } 3 = 4.37\,\text{mm}.$$

TEST YOURSELF

1.1 Define the following equivalent sphere diameters: ***equivalent volume diameter, equivalent surface diameter, equivalent surface-volume diameter***. Determine the values of each one for a cuboid of dimensions 2 mm × 3 mm × 6 mm.

1.2 List three types of distribution that might be used in expressing the range of particle sizes contained in a given sample.

1.3 If we measure a number distribution and wish to convert it to a surface distribution, what assumptions have to be made?

1.4 Write down the mathematical expression defining (a) the quadratic mean and (b) the harmonic mean.

1.5 For a give particle size distribution, the mode, the arithmetic mean, the harmonic mean and the quadratic mean all have quite different numberical values. How do we decide which mean is appropriate for describing the powder's behaviour in a given process?

1.6 What are the golden rules of sampling?

1.7 When using the sedimentation method for determination of particle size distribution, what assumptions are made?

1.8 In the electrozone sensing method of size analysis, (a) what equivalent sphere particle diameter is measured and (b) what type of distribution is reported?

EXERCISES

1.1 For a regular cuboid particle of dimensions 1.00 × 2.00 × 6.00 mm, calculate the following diameters:

(a) the equivalent volume sphere diameter;

(b) the equivalent surface sphere diameter;

(c) the surface-volume diameter (the diameter of a sphere having the same external surface to volume ratio as the particle);

(d) the sieve diameter (the width of the minimum aperture through which the particle will pass),

(e) the projected area diameters (the diameter of a circle having the same area as the projected area of the particle resting in a stable position).

[Answer: (a) 2.84 mm; (b) 3.57 mm; (c) 1.80 mm; (d) 2.00 mm; (e) 2.76 mm, 1.60 mm and 3.91 mm.]

1.2 Repeat Exercise 1.1 for a regular cylinder of diameter 0.100 mm and length 1.00 mm.

[Answer: (a) 0.247 mm; (b) 0.324 mm; (c) 0.142 mm; (d) 0.10 mm; (e) 0.10 mm (unlikely to be stable in this position) and 0.357 mm.]

1.3 Repeat Exercise 1.1 for a disc-shaped particle of diameter 2.00 mm and length 0.500 mm.

[Answer: (a) 1.44 mm; (b) 1.73 mm; (c) 1.00 mm; (d) 2.00 mm; (e) 2.00 mm and 1.13 mm (unlikely to be stable in this position).]

1.4 1.28 g of a powder of particle density $2500\,\text{kg/m}^3$ are charged into the cell of an apparatus for measurement of particle size and specific surface area by permeametry. The cylindrical cell has a diameter of 1.14 cm and the powder forms a bed of depth 1 cm. Dry air of density $1.2\,\text{kg/m}^3$ and viscosity 18.4×10^{-6} Pa s flows at a rate of $36\,\text{cm}^3/\text{min}$ through the powder (in a direction parallel to the axis of the cylindrical cell) and producing a pressure difference of 100 mm of water across the bed. Determine the surface-volume mean diameter and the specific surface of the powder sample.

(Answer: 20 μm; $120\,\text{m}^2/\text{kg}$.)

1.5 1.1 g of a powder of particle density $1800\,\text{kg/m}^3$ are charged into the cell of an apparatus for measurement of particle size and specific surface area by permeametry. The cylindrical cell has a diameter of 1.14 cm and the powder forms a bed of depth 1 cm. Dry air of density $1.2\,\text{kg/m}^3$ and viscosity 18.4×10^{-6} Pa s flows through the powder (in a direction parallel to the axis of the cylindrical cell). The measured variation in pressure difference across the bed with changing air flow rate is given below:

Air flow (cm^3/min)	20	30	40	50	60
Pressure difference across the bed (mm of water)	56	82	112	136	167

Determine the surface-volume mean diameter and the specific surface of the powder sample.

(Answer: 33 μm; $100\,\text{m}^2/\text{kg}$.)

1.6 Estimate the (a) arithmetic mean, (b) quadratic mean, (c) cubic mean, (d) geometric mean and (e) harmonic mean of the following distribution.

Size	2	2.8	4	5.6	8	11.2	16	22.4	32	44.8	64	89.6
cumulative % undersize	0.1	0.5	2.7	9.6	23	47.9	73.8	89.8	97.1	99.2	99.8	100

[Answer: (a) 13.6; (b) 16.1; (c) 19.3; (d) 11.5; (e) 9.8.]

1.7 The following volume distribution was derived from a sieve analysis

Size (μm)	37–45	45–53	53–63	63–75	75–90	90–106	106–126	126–150	150–180	180–212
Volume % in range	0.4	3.1	11	21.8	27.3	22	10.1	3.9	0.4	0

(a) Estimate the arithmetic mean of the volume distribution.

From the volume distribution derive the number distribution and the surface distribution, giving assumptions made.
Estimate:

(b) the mode of the surface distribution;

(c) the harmonic mean of the surface distribution.

Show that the arithmetic mean of the surface distribution conserves the surface to volume ratio of the population of particles.

[Answer: (a) 86 μm; (b) 70 μm; (c) 76 μm.]

2

Single Particles in a Fluid

This chapter deals with the motion of single solid particles in fluids. The objective here is to develop an understanding of the forces resisting the motion of any such particle and provide methods for the estimation of the steady velocity of the particle relative to the fluid. The subject matter of the chapter will be used in subsequent chapters on the behaviour of suspensions of particles in a fluid, fluidization, gas cyclones and pneumatic transport.

2.1 *MOTION OF SOLID PARTICLES IN A FLUID*

The drag force resisting very slow steady relative motion (creeping motion) between a rigid sphere of diameter x and a fluid of infinite extent, of viscosity μ is composed of two components (Stokes, 1851):

$$\text{a pressure drag force, } F_\text{p} = \pi x \mu \text{U} \quad (2.1)$$

$$\text{a shear stress drag force, } F_\text{s} = 2\pi x \mu \text{U} \quad (2.2)$$

$$\text{Total drag force resisting motion, } F_\text{D} = 3\pi x \mu U \quad (2.3)$$

where U is the relative velocity.

This is known as Stokes' law. Experimentally, Stokes' law is found to hold almost exactly for single particle Reynolds number, $Re_\text{p} \leq 0.1$, within 9% for $Re_\text{p} \leq 0.3$, within 3% for $Re_\text{p} \leq 0.5$ and within 9 % for $Re_\text{p} \leq 1.0$, where the single particle Reynolds number is defined in Equation (2.4).

$$\text{Single particle Reynolds number, } Re_\text{p} = xU\rho_\text{f}/\mu \quad (2.4)$$

$$\text{A drag coefficient, } C_\text{D} \text{ is defined as } C_\text{D} = R'/\left(\frac{1}{2}\rho_\text{f}U^2\right) \quad (2.5)$$

where R' is the force per unit projected area of the particle.

Introduction to Particle Technology - 2nd Edition Martin Rhodes

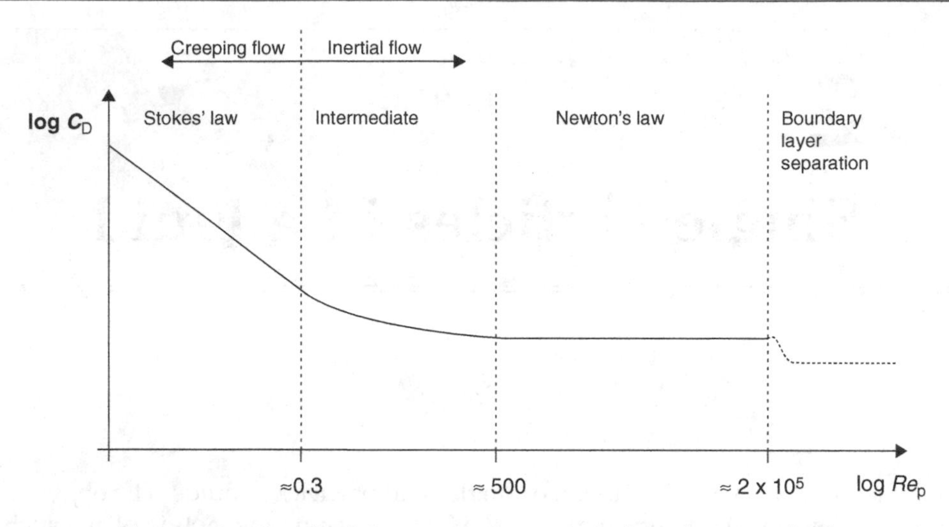

Figure 2.1 Standard drag curve for motion of a sphere in a fluid

Thus, for a sphere:

$$R' = F_D \Big/ \left(\frac{\pi x^2}{4}\right) \tag{2.6}$$

and, Stokes' law, in terms of this drag coefficient, becomes:

$$C_D = 24/Re_p \tag{2.7}$$

At higher relative velocities, the inertia of the fluid begins to dominate (the fluid must accelerate out of the way of the particle). Analytical solution of the Navier–Stokes equations is not possible under these conditions. However, experiments give the relationship between the drag coefficient and the particle Reynolds number in the form of the so-called standard drag curve (Figure 2.1). Four regions are identified: the Stokes' law region; the Newton's law region in which drag coefficient is independent of Reynolds number; an intermediate region between the Stokes and Newton regions; and the boundary layer separation region. The Reynolds number ranges and drag coefficient correlations for these regions are given in Table 2.1.

The expression given for C_D in the intermediate region in Table 2.1 is that of Schiller and Naumann (1933), which fits the data with an accuracy of around ±7% in the intermediate range. Several correlations have been proposed for C_D

Table 2.1 Reynolds number ranges for single particle drag coefficient correlations

Region	Stokes	Intermediate	Newton's law
Re_p range	<0.3	$0.3 < Re_p < 500$	$500 < Re_p < 2 \times 10^5$
C_D	$24/Re_p$	$\frac{24}{Re_p}(1 + 0.15\,Re_p^{0.687})$	~ 0.44

over the entire range; the one presented in Equation (2.8) is that of Haider and Levenspiel (1989), which is claimed to fit the data with a root mean square deviation of 0.024.

$$C_D = \frac{24}{Re_p}\left(1 + 0.1806\, Re_p^{0.6459}\right) + \left(\frac{0.4251}{1 + \dfrac{6880.95}{Re_p}}\right) \tag{2.8}$$

2.2 PARTICLES FALLING UNDER GRAVITY THROUGH A FLUID

The relative motion under gravity of particles in a fluid is of particular interest. In general, the forces of buoyancy, drag and gravity act on the particle:

$$\text{gravity} - \text{buoyancy} - \text{drag} = \text{acceleration force} \tag{2.9}$$

A particle falling from rest in a fluid will initially experience a high acceleration as the shear stress drag, which increases with relative velocity, will be small. As the particle accelerates the drag force increases, causing the acceleration to reduce. Eventually a force balance is achieved when the acceleration is zero and a maximum or terminal relative velocity is reached. This is known as the single particle terminal velocity.

For a spherical particle, Equation (2.9) becomes

$$\frac{\pi x^3}{6}\rho_p g - \frac{\pi x^3}{6}\rho_f g - R'\frac{\pi x^2}{4} = 0 \tag{2.10}$$

Combining Equation (2.10) with Equation (2.5):

$$\frac{\pi x^3}{6}(\rho_p - \rho_f)g - C_D\frac{1}{2}\rho_f U_T^2\frac{\pi x^2}{4} = 0 \tag{2.11}$$

where U_T is the single particle terminal velocity. Equation (2.11) gives the following expression for the drag coefficient under terminal velocity conditions:

$$C_D = \frac{4}{3}\frac{gx}{U_T^2}\left[\frac{(\rho_p - \rho_f)}{\rho_f}\right] \tag{2.12}$$

Thus in the Stokes' law region, with $C_D = 24/Re_p$, the single particle terminal velocity is given by:

$$U_T = \frac{x^2(\rho_p - \rho_f)g}{18\mu} \tag{2.13}$$

Note that in the Stokes' law region the terminal velocity is proportional to the square of the particle diameter.

In the Newton's law region, with $C_D = 0.44$, the terminal velocity is given by:

$$U_T = 1.74\left[\frac{x(\rho_p - \rho_f)g}{\rho_f}\right]^{1/2} \tag{2.14}$$

Note that in this region the terminal velocity is independent of the fluid viscosity and proportional to the square root of the particle diameter.

In the intermediate region no explicit expression for U_T can be found. However, in this region, the variation of terminal velocity with particle and fluid properties is approximately described by:

$$U_T \propto x^{1.1},\ (\rho_p - \rho_f)^{0.7},\ \rho_f^{-0.29},\ \mu^{-0.43}$$

Generally, when calculating the terminal velocity for a given particle or the particle diameter for a given velocity, it is not known which region of operation is relevant. One way around this is to formulate the dimensionless groups, $C_D Re_p^2$ and C_D/Re_p :

- *To calculate U_T, for a given size x.* Calculate the group

$$C_D Re_p^2 = \frac{4}{3}\frac{x^3\rho_f(\rho_p - \rho_f)g}{\mu^2} \tag{2.15}$$

which is independent of U_T
(Note that $C_D Re_p^2 = \frac{4}{3}Ar$, where Ar is the Archimedes number.)

For given particle and fluid properties, $C_D Re_p^2$ is a constant and will therefore produce a straight line of slope −2 if plotted on the logarithmic coordinates (log C_D versus log Re_p) of the standard drag curve. The intersection of this straight line with the drag curve gives the value of Re_p and hence U_T (Figure 2.2).

- *To calculate size x, for a given U_T.* Calculate the group

$$\frac{C_D}{Re_p} = \frac{4}{3}\frac{g\mu(\rho_p - \rho_f)}{U_T^3\rho_f^2} \tag{2.16}$$

which is independent of particle size x.

For a given terminal velocity, particle density and fluid properties, C_D/Re_p is constant and will produce a straight line of slope +1 if plotted on the logarithmic coordinates (log C_D versus log Re_p) of the standard drag curve. The intersection of this straight line with the drag curve gives the value of Re_p and hence, x (Figure 2.2). An alternative to this graphical method, but based on the same approach, is to use tables of corresponding values of Re_p, C_D, $C_D Re_p^2$, and C_D/Re_p. For example, see Perry and Green (1984).

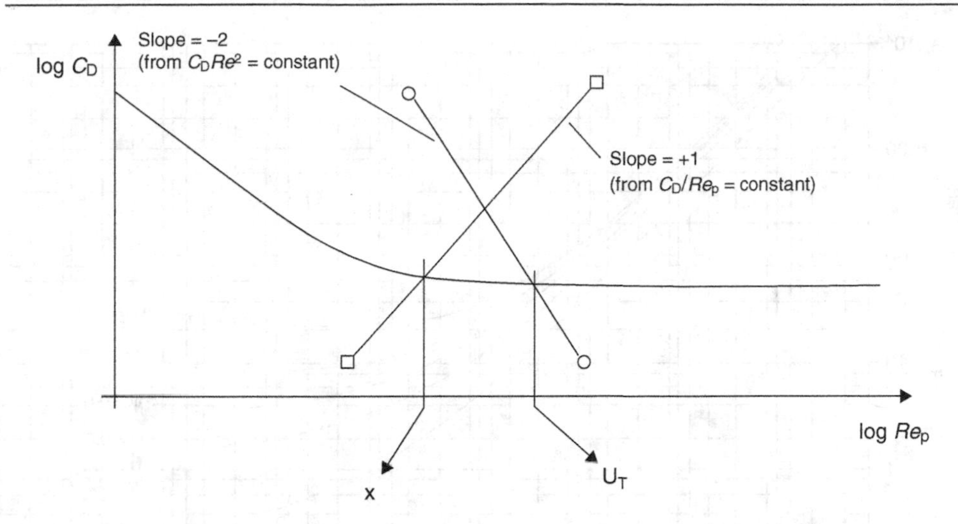

Figure 2.2 Method for estimating terminal velocity for a given size of particle and vice versa (Note: Re_p is based on the equivalent volume sphere diameter, x_v)

2.3 *NON-SPHERICAL PARTICLES*

The effect of shape of non-spherical particles on their drag coefficient has proved difficult to define. This is probably due to the difficulty in describing particle shape for irregular particles. Engineers and scientist often require a single number to describe the shape of a particle. One simple approach is to describe the shape of a particle in terms of its sphericity, the ratio of the surface area of a sphere of volume equal to that of the particle to the surface area of the particle. For example, a cube of side one unit has a volume of 1 (cubic units) and a surface area of 6 (square units). A sphere of the same volume has a diameter, x_v of 1.24 units. The surface area of a sphere of diameter 1.24 units is 4.836 units. The sphericity of a cube is therefore 0.806 (=4.836/6).

Shape affects drag coefficient far more in the intermediate and Newton's law regions than in the Stokes' law region. It is interesting to note that in the Stokes' law region particles fall with their longest surface nearly parallel to the direction of motion, whereas, in the Newton's law region particles present their maximum area to the oncoming fluid.

For non-spherical particles the particle Reynolds number is based on the equal-volume sphere diameter, i.e. the diameter of the sphere having the same volume as that of the particle. Figure 2.3 (after Brown *et al.*, 1950) shows drag curves for particles of different sphericities. This covers regular and irregular particles. The plot should be used with caution, since sphericity on its own may not be sufficient in some cases to describe the shape for all types of particles.

Small particles in gases and all common particles in liquids quickly accelerate to their terminal velocity. As an example, a 100 μm particle falling from rest in water requires 1.5 ms to reach its terminal velocity of 2 mm/s. Table 2.2 gives

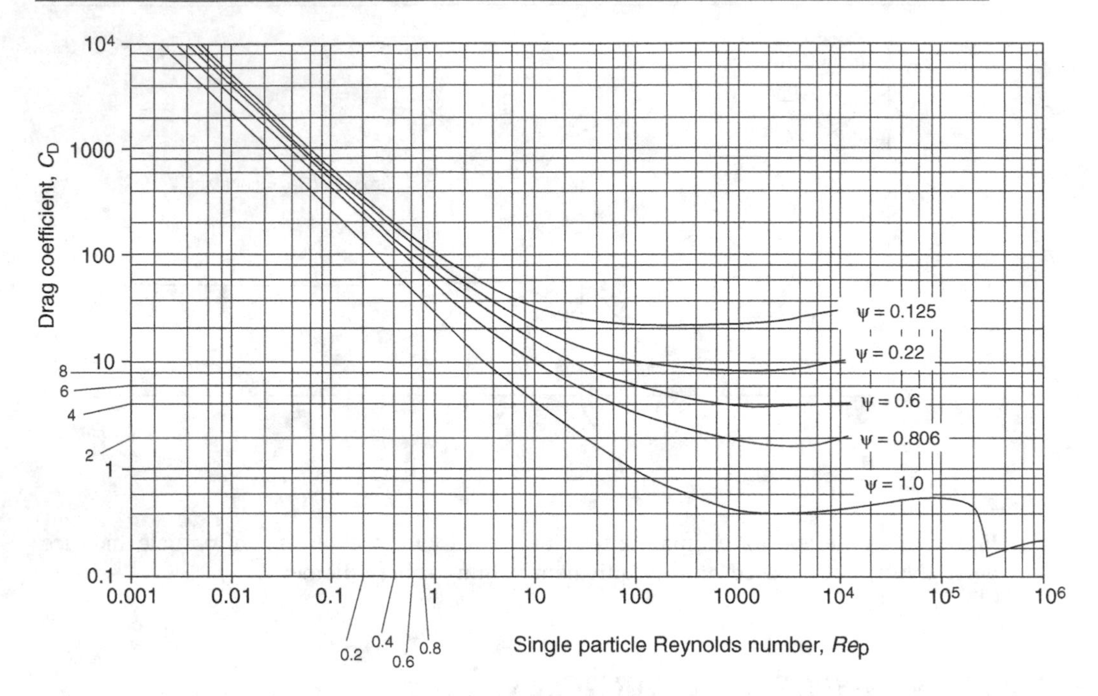

Figure 2.3 Drag coefficient C_D versus Reynolds number Re_p for particles of sphericity φ ranging from 0.125 to 1.0. (Note Re_p and C_D are based on the equivalent volume diameter)

some interesting comparisons of terminal velocities, acceleration times and distances for sand particles falling from rest in air.

2.4 EFFECT OF BOUNDARIES ON TERMINAL VELOCITY

When a particle is falling through a fluid in the presence of a solid boundary the terminal velocity reached by the particle is less than that for an infinite fluid. In practice, this is really only relevant to the falling sphere method of measuring liquid viscosity, which is restricted to the Stokes' region. In the case of a particle falling along the axis of a vertical pipe this is described by a wall factor, f_w, the ratio of the velocity in the pipe, U_D to the velocity in an infinite fluid, U_∞. The correlation of Francis (1933) for f_w is given in Equation (2.17).

$$f_w = \left(1 - \frac{x}{D}\right)^{2.25} \quad Re_p \leq 0.3; \quad x/D \leq 0.97 \tag{2.17}$$

Table 2.2 Sand particles falling from rest in air (particle density, 2600 kg/m^3)

Size	Time to each 99% of U_T(s)	U_T (m/s)	Distance travelled in this time (m)
30 μm	0.033	0.07	0.00185
3 mm	3.5	14	35
3 cm	11.9	44	453

2.5 FURTHER READING

For further details on the motion of single particles in fluids (accelerating motion, added mass, bubbles and drops, non-Newtonian fluids) the reader is referred to Coulson and Richardson (1991), Clift *et al.* (1978) and Chhabra (1993).

2.6 WORKED EXAMPLES

WORKED EXAMPLE 2.1

Calculate the upper limit of particle diameter x_{max} as a function of particle density ρ_p for gravity sedimentation in the Stokes' law regime. Plot the results as x_{max} versus ρ_p over the range $0 \leq \rho_p \leq 8000\,\text{kg}/m^3$ for settling in water and in air at ambient conditions. Assume that the particles are spherical and that Stokes' law holds for $Re_p \leq 0.3$.

Solution

The upper limit of particle diameter in the Stokes' regime is governed by the upper limit of single particle Reynolds number:

$$Re_p = \frac{\rho_f x_{max} U}{\mu} = 0.3$$

In gravity sedimentation in the Stokes' regime particles accelerate rapidly to their terminal velocity. In the Stokes' regime the terminal velocity is given by Equation (2.13):

$$U_T = \frac{x^2(\rho_p - \rho_f)g}{18\mu}$$

Solving these two equations for x_{max} we have

$$x_{max} = \left[0.3 \times \frac{18\mu^2}{g(\rho_p - \rho_f)\rho_f}\right]^{1/3}$$

$$= 0.82 \times \left[\frac{\mu^2}{(\rho_p - \rho_f)\rho_f}\right]^{1/3}$$

Thus, for air (density 1.2 kg/m^3 and viscosity 1.84×10^{-5} Pa s):

$$x_{max} = 5.37 \times 10^{-4}\left[\frac{1}{(\rho_p - 1.2)}\right]^{1/3}$$

And for water (density 1000 kg/m^3 and viscosity 0.001 Pa s):

$$x_{max} = 8.19 \times 10^{-4}\left[\frac{1}{(\rho_p - 1000)}\right]^{1/3}$$

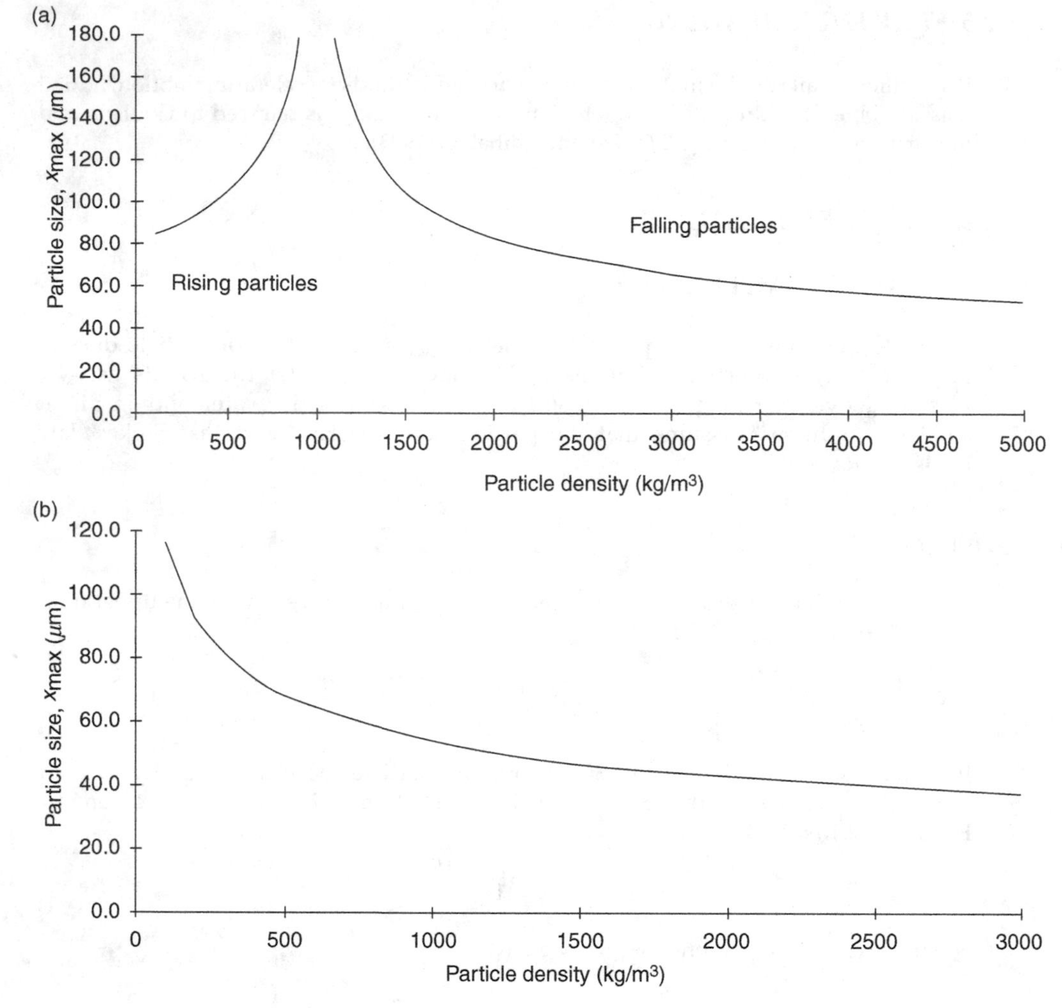

Figure 2W1.1 (a) Limiting particle size for Stokes' law in water. (b) Limiting particle size for Stokes' law in air

These equations for x_{max} as a function are plotted in Figure 2W1.1 for particle densities greater than and less than the fluid densities.

WORKED EXAMPLE 2.2

A gravity separator for the removal from water of oil droplets (assumed to behave as rigid spheres) consists of a rectangular chamber containing inclined baffles as shown schematically in Figure 2W2.1.

(a) Derive an expression for the ideal collection efficiency of this separator as a function of droplet size and properties, separator dimensions, fluid properties and fluid velocity (assumed uniform).

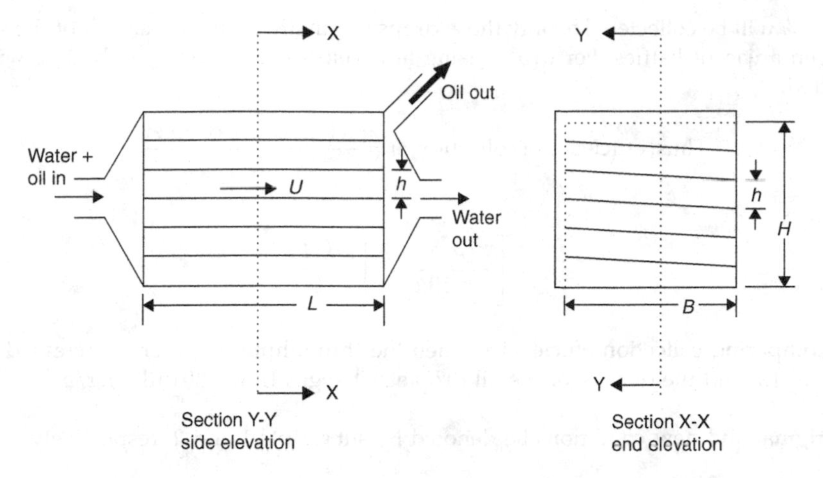

Figure 2W2.1 Schematic diagram of oil–water separator

(b) Hence, calculate the per cent change in collection efficiency when the throughput of water is increased by a factor of 1.2 and the density of the oil droplets changes from 750 to 800 kg/m^3.

Solution

(a) Referring to Figure 2W2.1, we will assume that all particles rising to the under surface of a baffle will be collected. Therefore, any particle which can rise a distance h or greater in the time required for it to travel the length of the separator will be collected. Let the corresponding minimum vertical droplet velocity be $U_{T_{min}}$.

Assuming uniform fluid velocity and negligible relative velocity between drops and fluid in the horizontal direction,

$$\text{drop residence time, } t = L/U$$

$$\text{Then } U_{T_{min}} = hU/L$$

Assuming that the droplets are small enough for Stokes' law to apply and that the time and distance for acceleration to terminal velocity is negligible, then droplet velocity will be given by Equation (2.13):

$$U_T = \frac{x^2(\rho_p - \rho_f)g}{18\mu}$$

This is the minimum velocity for drops to be collected whatever their original position between the baffles. Thus

$$U_{T_{min}} = \frac{x^2(\rho_p - \rho_f)g}{18\mu} = \frac{hU}{L}$$

Assuming that drops of all sizes are uniformly distributed over the vertical height of the separator, then drops rising a distance less than h in the time required for them to travel the length of the separator will be fractionally collected depending on their original vertical position between two baffles. Thus for droplets rising at a velocity of $0.5U_{T_{min}}$

only 50% will be collected, i.e. only those drops originally in the upper half of the space between adjacent baffles. For drops rising at a velocity of $0.25U_{T_{min}}$ only 25% will be collected.

$$\text{Thus, efficiency of collection, } \eta = \frac{\text{actual } U_T \text{ for droplet}}{U_{T_{min}}}$$

and so,

$$\eta = \left[\frac{x^2(\rho_p - \rho_f)g}{18\mu}\right] \Big/ \frac{hU}{L}$$

(b) Comparing collection efficiencies when the throughput of water is increased by a factor of 1.2 and the density of the oil droplets changes from 750 to 850 kg/m^3.

Let original and new conditions be denoted by subscripts 1 and 2, respectively.

Increasing throughput of water by a factor of 1.2 means that $U_2/U_1 = 1.2$.

Therefore from the expression for collection efficiency derived above:

$$\frac{\eta_2}{\eta_1} = \left(\frac{\rho_{p2} - \rho_f}{U_2}\right) \Big/ \left(\frac{\rho_{p1} - \rho_f}{U_1}\right)$$

$$\frac{\eta_2}{\eta_1} = \left(\frac{850 - 1000}{750 - 1000}\right) \times \frac{1}{1.2} = 0.5$$

The decrease in collection efficiency is therefore 50%.

WORKED EXAMPLE 2.3

A sphere of diameter 10 mm and density 7700 kg/m^3 falls under gravity at terminal conditions through a liquid of density 900 kg/m^3 in a tube of diameter 12 mm. The measured terminal velocity of the particle is 1.6 mm/s. Calculate the viscosity of the fluid. Verify that Stokes' law applies.

Solution

To solve this problem, we first convert the measured terminal velocity to the equivalent velocity which would be achieved by the sphere in a fluid of infinite extent. Assuming Stokes' law we can determine the fluid viscosity. Finally we check the validity of Stokes' law.

Using the Francis wall factor expression [Equation (2.17)]:

$$\frac{U_{T_\infty}}{U_{T_D}} = \frac{1}{(1 - x/D)^{2.25}} = 56.34$$

Thus, terminal velocity for the particle in a fluid of infinite extent,

$$U_{T_\infty} = U_{T_D} \times 56.34 = 0.0901 \text{ m/s}$$

Equating this value to the expression for U_{T_∞} in the Stokes' regime [Equation (2.13)]:

$$U_{T_\infty} = \frac{(10 \times 10^{-3})^2 \times (7700 - 900) \times 9.81}{18\mu}$$

Hence, fluid viscosity, $\mu = 4.11$ Pa s.

Checking the validity of Stokes' law:

$$\text{Single particle Reynolds number, } Re_p = \frac{x\rho_f U}{\mu} = 0.197$$

Re_p is less than 0.3 and so the assumption that Stokes' law holds is valid.

WORKED EXAMPLE 2.4

A mixture of spherical particles of two materials A and B is to be separated using a rising stream of liquid. The size range of both materials is 15–40 μm. (a) Show that a complete separation is not possible using water as the liquid. The particle densities for materials A and B are 7700 and 2400 kg/m^3, respectively. (b) Which fluid property must be changed to achieve complete separation? Assume Stokes' law applies.

Solution

(a) First, consider what happens to a single particle introduced into the centre of a pipe in which a fluid is flowing upwards at a velocity U which is uniform across the pipe cross-section. We will assume that the particle is small enough to consider the time and distance for its acceleration to terminal velocity to be negligible. Referring to Figure 2W4.1(a), if the fluid velocity is greater than the terminal velocity of the particle U_T, then the particle will move upwards; if the fluid velocity is less than U_T the particle will fall and if the fluid velocity is equal to U_T the particle will remain at the same vertical position. In each case the velocity of the particle relative to the pipe wall is $(U - U_T)$. Now consider introducing two particles of different size and density having terminal velocities U_{T_1} and U_{T_2}. Referring to Figure 2W4.1(b), at low fluid velocities $(U < U_{T_2} < U_{T_1})$, both particles will fall. At high fluid velocities $(U > U_{T_1} > U_{T_2})$, both particles will be carried upwards. At intermediate fluid velocities $(U_{T_1} > U > U_{T_2})$, particle 1 will fall and particle 2 will rise. Thus we have the basis of a separator according to particle size and density. From the analysis above we see that to be able to completely separate particles A and B, there must be no overlap between the ranges of terminal velocity for the particles; i.e. all sizes of the denser material A must have terminal velocities which are greater than all sizes of the less dense material B.

Assuming Stokes' law applies, Equation (2.13), with fluid density and viscosity 1000 kg/m^3 and 0.001 Pa s, respectively, gives

$$U_T = 545x^2(\rho_p - 1000)$$

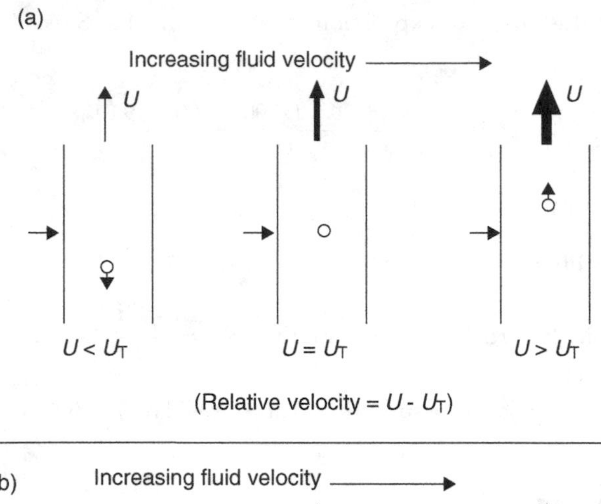

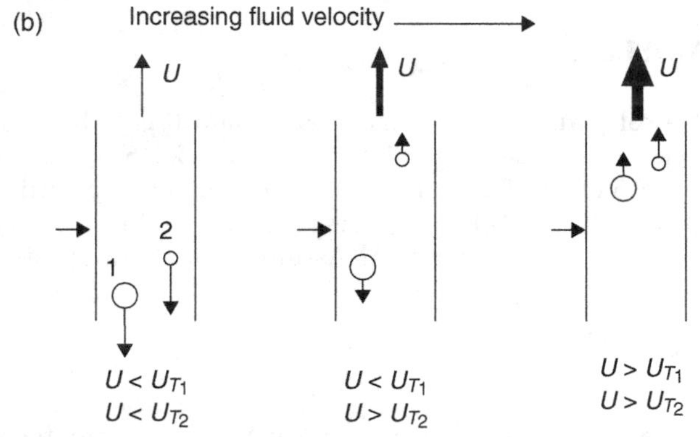

Figure 2W4.1 Relative motion of particles in a moving fluid

Based on this equation, the terminal velocities of the extreme sizes of particles A and B are:

Size (μm) →	15	40
U_{T_A} (mm/s)	0.82	5.84
U_{T_B} (mm/s)	0.17	1.22

We see that there is overlap of the ranges of terminal velocities. We can therefore select no fluid velocity which would completely separate particles A and B.

(b) Inspecting the expression for terminal velocity in the Stokes' regime [Equation (2.13)] we see that changing the fluid viscosity will have no effect on our ability to separate the particles, since change in viscosity will change the terminal velocities of all particles in the same proportion. However, changing the fluid density will have a different effect on particles of different density and this is the effect we are looking for. The critical condition for separation of particles A and B is when the terminal velocity of the smallest A particle is equal to the terminal velocity of the largest B particle.

$$U_{T_{B40}} = U_{T_{A15}}$$

Hence,

$$545 \times x_{40}^2 \times (2400 - \rho_f) = 545 \times x_{15}^2 \times (7700 - \rho_f)$$

From which, critical minimum fluid density $\rho_f = 1533\,\text{kg/m}^3$.

WORKED EXAMPLE 2.5

A sphere of density $2500\,\text{kg/m}^3$ falls freely under gravity in a fluid of density $700\,\text{kg/m}^3$ and viscosity $0.5 \times 10^{-3}\,\text{Pa s}$. Given that the terminal velocity of the sphere is 0.15 m/s, calculate its diameter. What would be the edge length of a cube of the same material falling in the same fluid at the same terminal velocity?

Solution

In this case we know the terminal velocity, U_T, and need to find the particle size x. Since we do not know which regime is appropriate, we must first calculate the dimensionless group C_D/Re_p [Equation (2.16)]:

$$\frac{C_D}{Re_p} = \frac{4}{3}\frac{g\mu(\rho_p - \rho_f)}{U_T^3 \rho_f^2}$$

hence,

$$\frac{C_D}{Re_p} = \frac{4}{3}\left[\frac{9.81 \times (0.5 \times 10^{-3}) \times (2500 - 700)}{0.15^3 \times 700^2}\right]$$

$$\frac{C_D}{Re_p} = 7.12 \times 10^{-3}$$

This is the relationship between drag coefficent C_D and single particle Reynolds number Re_p for particles of density $2500\,\text{kg/m}^3$ having a terminal velocity of 0.15 m/s in a fluid of density $700\,\text{kg/m}^3$ and viscosity $0.5 \times 10^{-3}\,\text{Pa s}$. Since C_D/Re_p is a constant, this relationship will give a straight line of slope +1 when plotted on the log–log coordinates of the standard drag curve.

For plotting the relationship:

Re_p	C_D
100	0.712
1000	7.12
10 000	71.2

These values are plotted on the standard drag curves for particles of different sphericity (Figure 2.3). The result is shown in Figure 2W5.1.

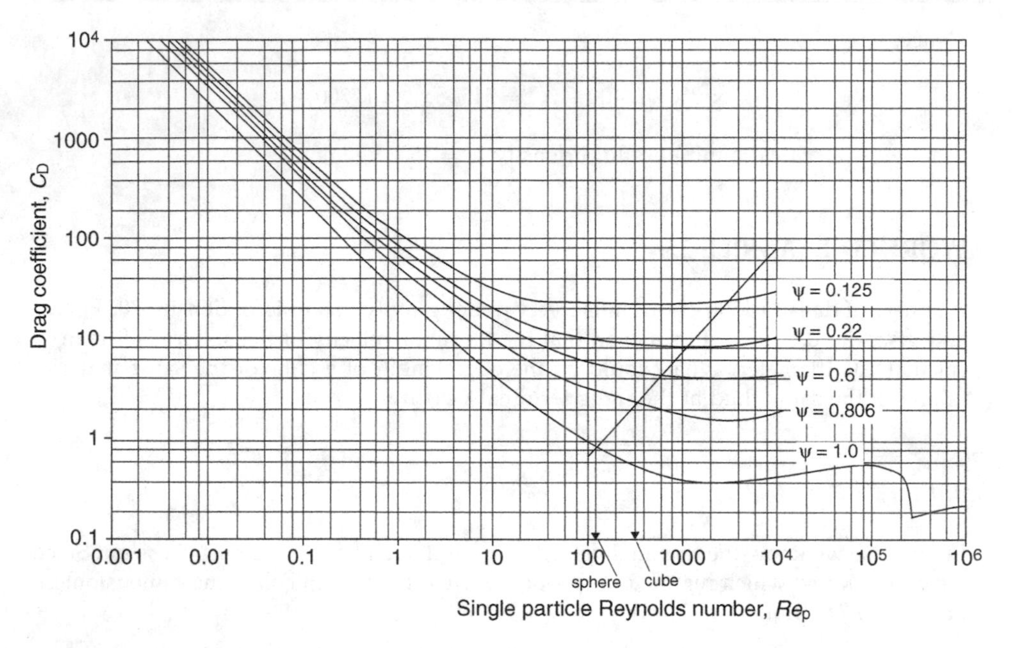

Figure 2W5.1 Drag coefficient C_D versus Reynolds number Re_p.

Where the plotted line intersects the standard drag curve for a sphere ($\psi = 1$), $Re_p = 130$.

The diameter of the sphere may be calculated from:

$$Re_p = 130 = \frac{\rho_f x_v U_T}{\mu}$$

Hence, sphere diameter $x_V = 619\,\mu m$.

For a cube having the same terminal velocity under the same conditions, the same C_D versus Re_p relationship applies, only the standard drag curve is that for a cube ($\psi = 0.806$).

Cube sphericity

For a cube of side 1 unit, the volume is 1 cubic unit and the surface area is 6 square units. If x_v is the diameter of a sphere having the same volume as the cube, then

$$\frac{\pi x_v^3}{6} = 1.0 \text{ which gives } x_v = 1.24 \text{ units}$$

$$\text{Therefore, sphericity } \psi = \frac{\text{surface area of a sphere of volume equal to the particle}}{\text{surface area of the particle}}$$

$$\psi = \frac{4.836}{6} = 0.806$$

At the intersection of this standard drag curve with the plotted line, $Re_p = 310$.

Recalling that the Reynolds number in this plot uses the equivalent volume sphere diameter,

$$x_v = \frac{310 \times (0.5 \times 10^{-3})}{0.15 \times 700} = 1.48 \times 10^{-3}\,\text{m}$$

And so the volume of the particle is $\frac{\pi x_v^3}{6} = 1.66 \times 10^{-9}\,\text{m}^3$

Giving a cube side length of $(1.66 \times 10^{-9})^{1/3} = 1.18 \times 10^{-3}\,\text{m}(1.18\,\text{mm})$

WORKED EXAMPLE 2.6

A particle of equivalent volume diameter 0.5 mm, density 2000 kg/m^3 and sphericity 0.6 falls freely under gravity in a fluid of density 1.6 kg/m^3 and viscosity 2×10^{-5} Pa s. Estimate the terminal velocity reached by the particle.

Solution

In this case we know the particle size and are required to determine its terminal velocity without knowing which regime is appropriate. The first step is, therefore, to calculate the dimensionless group $C_D Re_p^2$:

$$C_D Re_p^2 = \frac{4}{3}\frac{x^3 \rho_f (\rho_p - \rho_f) g}{\mu^2}$$

$$= \frac{4}{3}\left[\frac{(0.5 \times 10^{-3})^3 \times 1.6 \times (2000 - 1.6) \times 9.81}{(2 \times 10^{-5})^2}\right]$$

$$= 13069$$

This is the relationship between drag coefficient C_D and single particle Reynolds number Re_p for particles of size 0.5 mm and density 2000 kg/m^3 falling in a fluid of density 1.6 kg/m^3 and viscosity 2×10^{-5} Pa s. Since $C_D Re_p^2$ is a constant, this relationship will give a straight line of slope −2 when plotted on the log–log coordinates of the standard drag curve.

For plotting the relationship:

Re_p	C_D
10	130.7
100	1.307
1000	0.013

These values are plotted on the standard drag curves for particles of different sphericity (Figure 2.3). The result is shown in Figure 2W6.1.

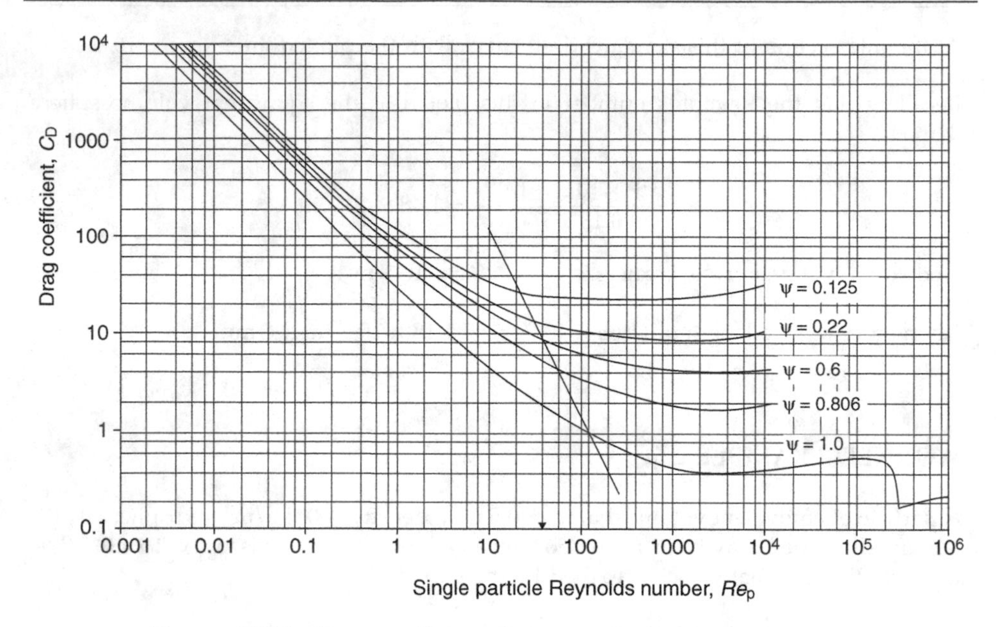

Figure 2W6.1 Drag coefficient C_D versus Reynolds number Re_p.

Where the plotted line intersects the standard drag curve for a sphericity of $0.6 (\psi = 0.6)$, $Re_p = 40$.

The terminal velocity U_T may be calculated from

$$Re_p = 40 = \frac{\rho_f x_v U_T}{\mu}$$

Hence, terminal velocity, $U_T = 1.0\,\text{m/s}$.

TEST YOURSELF

2.1 The total drag force acting on a particle is a function of:

(a) the particle drag coefficient;

(b) the projected area of the particle;

(c) both (a) and (b);

(d) none of the above.

2.2 The particle drag coefficient is a function of:

(a) the particle Reynolds number;

(b) the particle sphericity;

(c) both (a) and (b);

(d) none of the above.

2.3 Stokes drag assumes:

(a) the drag coefficient is constant;

(b) the particle Reynolds number is constant;

(c) the drag force is constant;

(d) none of the above.

2.4 The particle Reynolds number is not a function of:

(a) the particle density;

(b) the pipe diameter;

(c) both (a) and (b);

(d) none of the above.

2.5 The force on a particle due to buoyancy depends on:

(a) the particle density;

(b) the particle size;

(c) both (a) and (b);

(d) none of the above.

2.6 The force on a particle due to gravity depends on:

(a) the particle density;

(b) the particle size;

(c) both (a) and (b);

(d) none of the above.

2.7 The particle Reynolds number is dependent on:

(a) the particle density;

(b) the fluid density;

(c) both (a) and (b);

(d) none of the above.

2.8 For particle in suspension in a fluid flowing in a pipe, as the pipe size increases (all other things held constant), the particle Reynolds number:

(a) increases;

(b) decreases;

(c) remains the same.

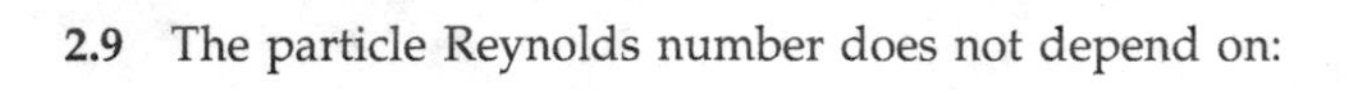
2.9 The particle Reynolds number does not depend on:

(a) particle density;

(b) velocity;

(c) viscosity;

(d) none of the above.

2.10 The units of viscosity are:

(a) length;

(b) mass/length;

(c) mass × length/time;

(d) none of the above.

EXERCISES

2.1 The settling chamber, shown schematically in Figure 2E1.1, is used as a primary separation device in the removal of dust particles of density 1500 kg/m^3 from a gas of density 0.7 kg/m^3 and viscosity 1.90×10^{-5} Pa s.

(a) Assuming Stokes' law applies, show that the efficiency of collection of particles of size x is given by the expression

$$\text{collection efficiency, } \eta_x = \frac{x^2 g(\rho_p - \rho_f)L}{18\mu HU}$$

where U is the uniform gas velocity through the parallel-sided section of the chamber. State any other assumptions made.

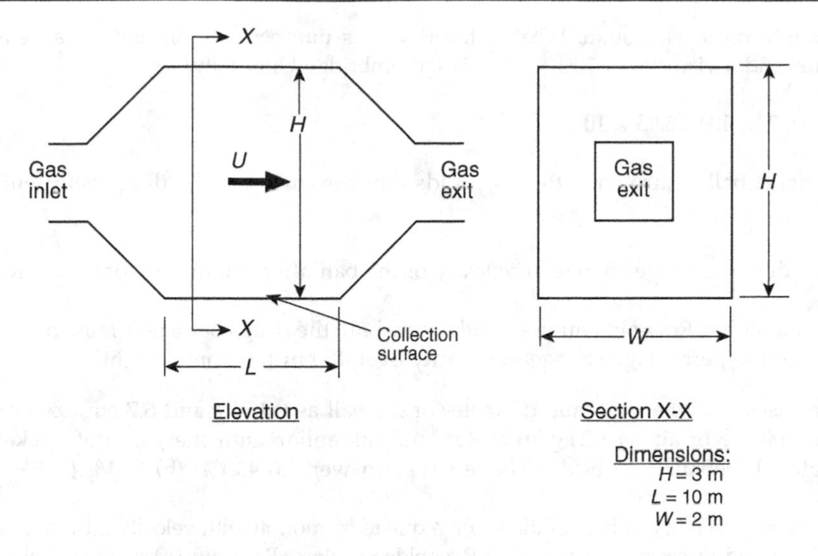

Figure 2E1.1 Schematic diagram of settling chamber

(b) What is the upper limit of particle size for which Stokes' law applies?

(c) When the volumetric flow rate of gas is $0.9\,m^3/s$, and the dimensions of the chamber are those shown in Figure 2E1.1, determine the collection efficiency for spherical particles of diameter 30 μm [Answer: (b) 57 μm; (c) (86%).]

2.2 A particle of equivalent sphere volume diameter 0.2 mm, density $2500\,kg/m^3$ and sphericity 0.6 falls freely under gravity in a fluid of density $1.0\,kg/m^3$ and viscosity 2×10^{-5} Pa s. Estimate the terminal velocity reached by the particle. (Answer: 0.6 m/s.)

2.3 Spherical particles of density $2500\,kg/m^3$ and in the size range 20–100 μm are fed continuously into a stream of water (density, $1000\,kg/m^3$ and viscosity, 0.001 Pa s) flowing upwards in a vertical, large diameter pipe. What maximum water velocity is required to ensure that no particles of diameter greater than 60 μm are carried upwards with water?

(Answer: 2.9 mm/s.)

2.4 Spherical particles of density $2000\,kg/m^3$ and in the size range 20–100 μm are fed continuously into a stream of water (density, $1000\,kg/m^3$ and viscosity, 0.001 Pa s) flowing upwards in a vertical, large diameter pipe. What maximum water velocity is required to ensure that no particles of diameter greater than 50 μm are carried upwards with the water?

(Answer: 1.4 mm/s.)

2.5 A particle of equivalent volume diameter 0.3 mm, density $2000\,kg/m^3$ and sphericity 0.6 falls freely under gravity in a fluid of density $1.2\,kg/m^3$ and viscosity 2×10^{-5} Pa s. Estimate the terminal velocity reached by the particle.

(Answer: 0.67 m/s.)

2.6 Assuming that a car is equivalent to a flat plate 1.5 m square, moving normal to the airstream, and with a drag coefficient, $C_D = 1.1$, calculate the power required for steady motion at 100 km/h on level ground. What is the Reynolds number? For air assume a density of 1.2 kg/m^3 and a viscosity of 1.71×10^{-5} Pa s. (Cambridge University)

(Answer: 32.9 kW; 2.95×10^6.)

2.7 A cricket ball is thrown with a Reynolds number such that the drag coefficient is 0.4 ($Re \approx 10^5$).

(a) Find the percentage change in velocity of the ball after 100 m horizontal flight in air.

(b) With a higher Reynolds number and a new ball, the drag coefficient falls to 0.1. What is now the percentage change in velocity over 100 m horizontal flight?

(In both cases take the mass and diameter of the ball as 0.15 kg and 6.7 cm, respectively, and the density of air as 1.2 kg/m^3.) Readers unfamiliar with the game of cricket may substitute a baseball. (Cambridge University) [Answer: (a) 43.1%; (b) 13.1%.]

2.8 The resistance F of a sphere of diameter x, due to its motion with velocity u through a fluid of density ρ and viscosity μ, varies with Reynolds number ($Re = \rho u x/\mu$) as given below:

$\log_{10}Re$	2.0	2.5	3.0	3.5	4.0
$C_D = \dfrac{F}{\frac{1}{2}\rho u^2(\pi x^2/4)}$	1.05	0.63	0.441	0.385	0.39

Find the mass of a sphere of 0.013 m diameter which falls with a steady velocity of 0.6 m/s in a large deep tank of water of density 1000 kg/m^3 and viscosity 0.0015 Pa s. (Cambridge University)

(Answer: 0.0021 kg.)

2.9 A particle of 2 mm in diameter and density of 2500 kg/m^3 is settling in a stagnant fluid in the Stokes' flow regime.

(a) Calculate the viscosity of the fluid if the fluid density is 1000 kg/m^3 and the particle falls at a terminal velocity of 4 mm/s.

(b) What is the drag force on the particle at these conditions?

(c) What is the particle drag coefficient at these conditions?

(d) What is the particle acceleration at these conditions?

(e) What is the apparent weight of the particle?

2.10 Starting with the force balance on a single particle at terminal velocity, show that:

$$C_D = \frac{4}{3}\frac{gx}{U_T^2}\left[\frac{\rho_p - \rho_f}{\rho_f}\right]$$

where the symbols have their usual meaning.

2.11 A spherical particle of density 1500 kg/m^3 has a terminal velocity of 1 cm/s in a fluid of density 800 kg/m^3 and viscosity 0.001 Pa s. Estimate the diameter of the particle.

2.12 Estimate the largest diameter of spherical particle of density 2000 kg/m^3 which would be expected to obey Stokes' law in air of density 1.2 kg/m^3 and viscosity 18×10^{-6} Pa s.

3

Multiple Particle Systems

3.1 SETTLING OF A SUSPENSION OF PARTICLES

When many particles flow in a fluid in close proximity to each other the motion of each particle is influenced by the presence of the others. The simple analysis for the fluid particle interaction for a single particle is no longer valid but can be adapted to model the multiple particle system.

For a suspension of particles in a fluid, Stokes' law is assumed to apply but an effective suspension viscosity and effective average suspension density are used:

$$\text{effective viscosity, } \mu_e = \mu / f(\varepsilon) \tag{3.1}$$

$$\text{average suspension density, } \rho_{ave} = \varepsilon \rho_f + (1 - \varepsilon)\rho_p \tag{3.2}$$

where ε is the voidage or volume fraction occupied by the fluid. The effective viscosity of the suspension is seen to be equal to the fluid viscosity, μ modified by a function $f(\varepsilon)$ of the fluid volume fraction.

The drag coefficient for a single particle in the Stokes' law region was shown in Chapter 2. to be given by $C_D = 24/Re_p$. Substituting the effective viscosity and average density for the suspension, Stokes' law becomes

$$C_D = \frac{24}{Re_p} = \frac{24\mu_e}{U_{rel}\rho_{ave}x} \tag{3.3}$$

where $C_D = R' \Big/ \left(\frac{1}{2}\rho_{ave}U_{rel}^2\right)$ and U_{rel} is the relative velocity of the particle to the fluid.

Introduction to Particle Technology - 2nd Edition Martin Rhodes

Under terminal velocity conditions for a particle falling under gravity in a suspension, the force balance,

$$\text{drag force} = \text{weight} - \text{upthrust}$$

becomes

$$\left(\frac{\pi x^2}{4}\right)\frac{1}{2}\rho_{\text{ave}}U_{\text{rel}}^2 C_{\text{D}} = (\rho_{\text{p}} - \rho_{\text{ave}})\left(\frac{\pi x^3}{6}\right)g \tag{3.4}$$

giving

$$U_{\text{rel}} = (\rho_{\text{p}} - \rho_{\text{ave}})\frac{x^2 g}{18\mu_{\text{e}}} \tag{3.5}$$

Substituting for average density ρ_{ave} and effective viscosity μ_{e} of the suspension, we obtain the following expression for the terminal falling velocity for a particle in a suspension:

$$U_{\text{rel}_{\text{T}}} = (\rho_{\text{p}} - \rho_{\text{f}})\frac{x^2 g}{18\mu}\varepsilon f(\varepsilon) \tag{3.6}$$

Comparing this with the expression for the terminal free fall velocity of a single particle in a fluid [Equation (2.13)], we find that

$$U_{\text{rel}_{\text{T}}} = U_{\text{T}}\varepsilon f(\varepsilon) \tag{3.7}$$

$U_{\text{rel}_{\text{T}}}$ is known as the particle settling velocity in the presence of other particles or the hindered settling velocity.

In the following analysis, it is assumed that the fluid and the particles are incompressible and that the volume flowrates, Q_{f} and Q_{p}, of the fluid and the particles are constant.

We define U_{fs} and U_{ps} as the superficial velocities of the fluid and particles, respectively:

$$\text{superficial fluid velocity, } U_{\text{fs}} = \frac{Q_{\text{f}}}{A} \tag{3.8}$$

$$\text{superficial particle velocity, } U_{\text{ps}} = \frac{Q_{\text{p}}}{A} \tag{3.9}$$

where A is the vessel cross-sectional area.

Under isotropic conditions the flow areas occupied by the fluid and the particles are:

$$\text{flow area occupied by the fluid, } A_{\text{f}} = \varepsilon A \tag{3.10}$$

$$\text{flow area occupied by the particles, } A_{\text{p}} = (1 - \varepsilon)A \tag{3.11}$$

And so continuity gives:

$$\text{for the fluid}: Q_f = U_{ps}A = U_f A\varepsilon \quad (3.12)$$

$$\text{for the particles}: Q_p = U_{ps}A = U_p A(1-\varepsilon) \quad (3.13)$$

hence the actual velocities of the fluid and the particles, U_f and U_p are given by:

$$\text{actual velocity of the fluid}, U_f = U_{fs}/\varepsilon \quad (3.14)$$

$$\text{actual velocity of the particles}, U_p = U_{ps}/(1-\varepsilon) \quad (3.15)$$

3.2 *BATCH SETTLING*

3.2.1 Settling Flux as a Function of Suspension Concentration

When a batch of solids in suspension are allowed to settle, say in a measuring cylinder in the laboratory, there is no net flow through the vessel and so

$$Q_p + Q_f = 0 \quad (3.16)$$

hence

$$U_p(1-\varepsilon) + U_f\varepsilon = 0 \quad (3.17)$$

and

$$U_f = -U_p\frac{(1-\varepsilon)}{\varepsilon} \quad (3.18)$$

In hindered settling under gravity the relative velocity between the particles and the fluid ($U_p - U_f$) is U_{rel_T}. Thus using the expression for U_{rel_T} found in Equation (3.7), we have

$$U_p - U_f = U_{rel_T} = U_T\varepsilon f(\varepsilon) \quad (3.19)$$

Combining Equation (3.19) with Equation (3.18) gives the following expression for U_p, the hindered settling velocity of particles relative to the vessel wall in batch settling:

$$U_p = U_T\varepsilon^2 f(\varepsilon) \quad (3.20)$$

The effective viscosity function, $f(\varepsilon)$, was shown theoretically to be

$$f(\varepsilon) = \varepsilon^{2.5} \quad (3.21)$$

for uniform spheres forming a suspension of solid volume fraction less than $0.1[(1-\varepsilon) \le 0.1]$.

Richardson and Zaki (1954) showed by experiment that for $Re_p < 0.3$ (under Stokes' law conditions where drag is independent of fluid density),

$$U_p = U_T \varepsilon^{4.65} [\text{giving } f(\varepsilon) = \varepsilon^{2.65}] \tag{3.22}$$

and for $Re_p > 500$ (under Newton's law conditions where drag is independent of fluid viscosity)

$$U_p = U_T \varepsilon^{2.4} [\text{giving } f(\varepsilon) = \varepsilon^{0.4}] \tag{3.23}$$

In general, the Richardson and Zaki relationship is given as:

$$U_p = U_T \varepsilon^n \tag{3.24}$$

Khan and Richardson (1989) recommend the use of the following correlation for the value of exponent n over the entire range of Reynolds numbers:

$$\frac{4.8 - n}{n - 2.4} = 0.043 Ar^{0.57} \left[1 - 2.4 \left(\frac{x}{D} \right)^{0.27} \right] \tag{3.25}$$

where Ar is the Archimedes number $[x^3 \rho_f (\rho_p - \rho_f) g / \mu^2]$ and x is the particle diameter and D is the vessel diameter. The most appropriate particle diameter to use here is the surface-volume mean.

Expressed as a volumetric solids settling flux, U_{ps}, Equation (3.24) becomes

$$U_{ps} = U_p (1 - \varepsilon) = U_T (1 - \varepsilon) \varepsilon^n \tag{3.26}$$

or, dimensionless particle settling flux,

$$\frac{U_{ps}}{U_T} = (1 - \varepsilon) \varepsilon^n \tag{3.27}$$

Taking first and second derivates of Equation (3.27) demonstrates that a plot of dimensionless particle settling flux versus suspension volumetric concentration, $1 - \varepsilon$ has a maximum at $\varepsilon = n/(n+1)$ and an inflection point at $\varepsilon = (n-1)/(n+1)$. The theoretical form of such a plot is therefore that shown in Figure 3.1.

3.2.2 Sharp Interfaces in Sedimentation

Interfaces or discontinuities in concentration occur in the sedimentation or settling of particle suspensions.

In the remainder of this chapter, for convenience, the symbol C will be used to represent the particle volume fraction $1 - \varepsilon$. Also for convenience the particle volume fraction will be called the concentration of the suspension.

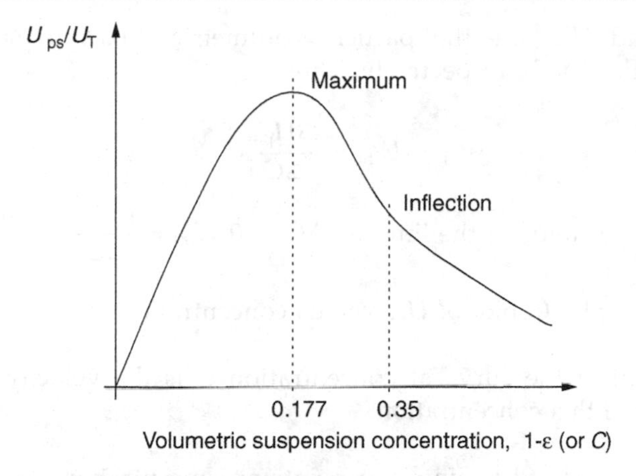

Figure 3.1 Variation of dimensionless settling flux with suspension concentration, based on Equation (3.27) (for $Re_p < 0.3$, i.e. $n = 4.65$)

Consider Figure 3.2, which shows the interface between a suspension of concentration C_1 containing particles settling at a velocity U_{p_1} and a suspension of concentration C_2 containing particles settling at a velocity U_{p_2}.

The interface is falling at a velocity U_{int}. All velocities are measured relative to the vessel walls. Assuming incompressible fluid and particles, the mass balance over the interface gives

$$(U_{p_1} - U_{int})C_1 = (U_{p_2} - U_{int})C_2$$

hence

$$U_{int} = \frac{U_{p_1}C_1 - U_{p_2}C_2}{C_1 - C_2} \tag{3.28}$$

or, since U_pC is particle volumetric flux, U_{ps}, then:

$$U_{int} = \frac{U_{ps_1} - U_{ps_2}}{C_1 - C_2} \tag{3.29}$$

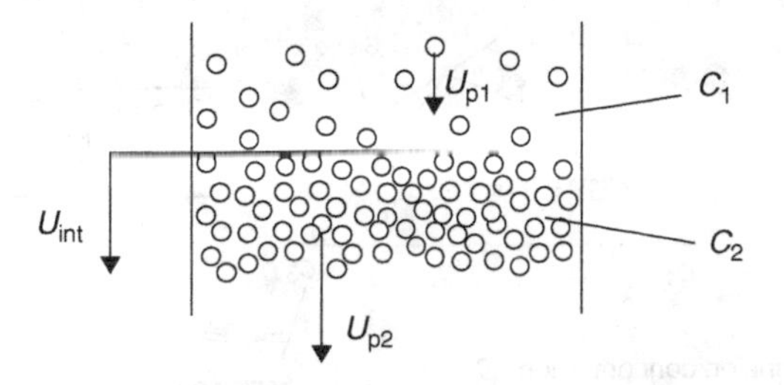

Figure 3.2 Concentration interface in sedimentation

where U_{ps_1} and U_{ps_2} are the particle volumetric fluxes in suspensions of concentration C_1 and C_2, respectively. Thus,

$$U_{int} = \frac{\Delta U_{ps}}{\Delta C} \tag{3.30}$$

$$\text{and, in the limit as } \Delta C \to 0, U_{int} = \frac{dU_{ps}}{dC} \tag{3.31}$$

Hence, on a flux plot (a plot of U_{ps} versus concentration):

(a) The gradient of the curve at concentration C is the velocity of a layer of suspension of this concentration.

(b) The slope of a chord joining two points at concentrations C_1 and C_2 is the velocity of a discontinuity or interface between suspensions of these concentrations.

This is illustrated in Figure 3.3.

3.2.3 The Batch Settling Test

The simple batch settling test can supply all the information for the design of a thickener for separation of particles from a fluid. In this test a suspension of particles of known concentration is prepared in a measuring cylinder. The cylinder is shaken to thoroughly mix the suspension and then placed upright to allow the suspension to settle. The positions of the interfaces which form are monitored in time. Two types of settling occur depending on the initial concentration of the suspension. The first type of settling is depicted in Figure 3.4 (Type 1 settling). Three zones of constant concentration are formed. These are: zone A,

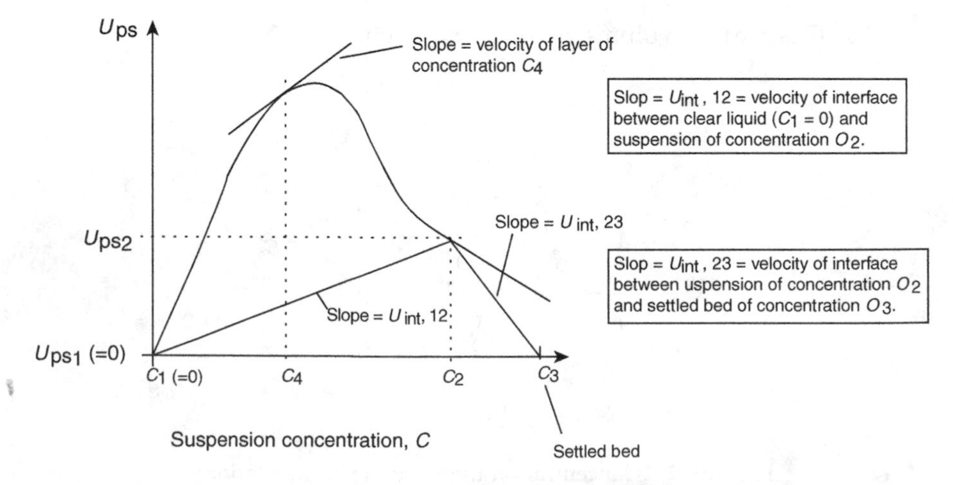

Figure 3.3 Determination of interface and layer velocities from a batch flux plot

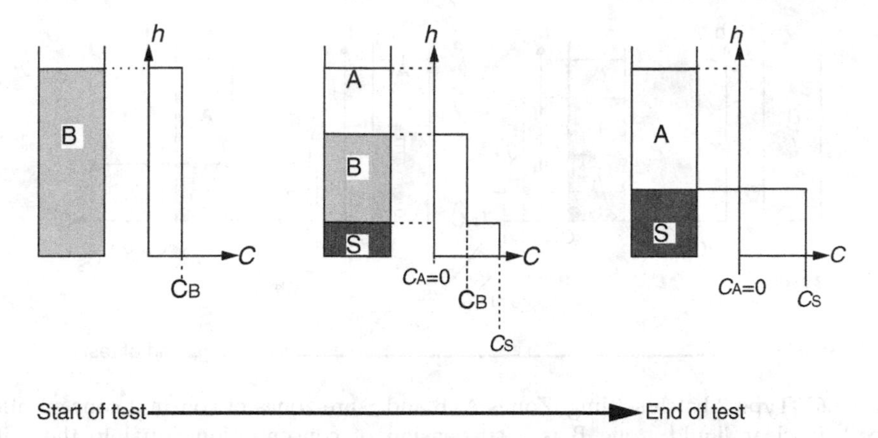

Figure 3.4 Type 1 batch settling. Zones A, B and S are zones of constant concentration. Zone A is a clear liquid; zone B is a suspension of concentration equal to the initial suspension concentration; zone S is a suspension of settled bed or sediment concentration

clear liquid ($C = 0$); zone B, of concentration equal to the initial suspension concentration (C_B); and zone S, the sediment concentration (C_S). Figure 3.5 is a typical plot of the height of the interfaces AB, BS and AS with time for this type of settling. On this plot the slopes of the lines give the velocities of the interfaces. For example, interface AB descends at constant velocity, interface BS rises at constant velocity. The test ends when the descending AB meets the rising BS forming an interface between clear liquid and sediment (AS) which is stationary.

In the second type of settling (Type 2 settling), shown in Figure 3.6, a zone of variable concentration, zone E, is formed in addition to the zones of constant concentration (A, B and S). The suspension concentration within zone E varies with position. However, the minimum and maximum concentrations within this zone, $C_{E_{min}}$ and $C_{E_{max}}$, are constant. Figure 3.7 is a typical plot of the height of the interfaces AB, BE_{min}, $E_{max}S$ and AS with time for this type of settling.

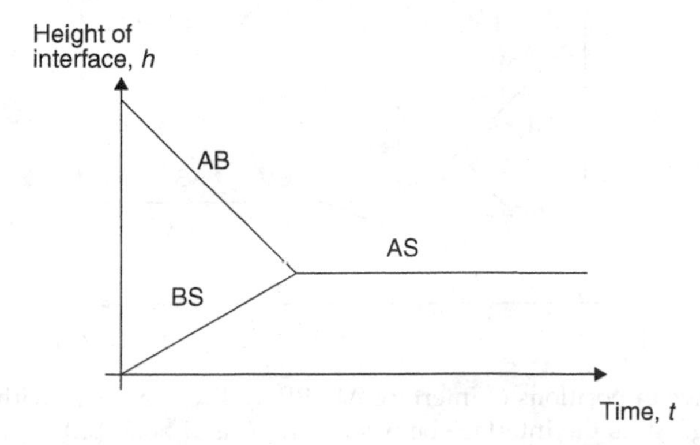

Figure 3.5 Change in positions of interface AB, BS and AS with time in Type 1 batch settling (e.g. AB is the interface between zone A and zone B; see Figure 3.4)

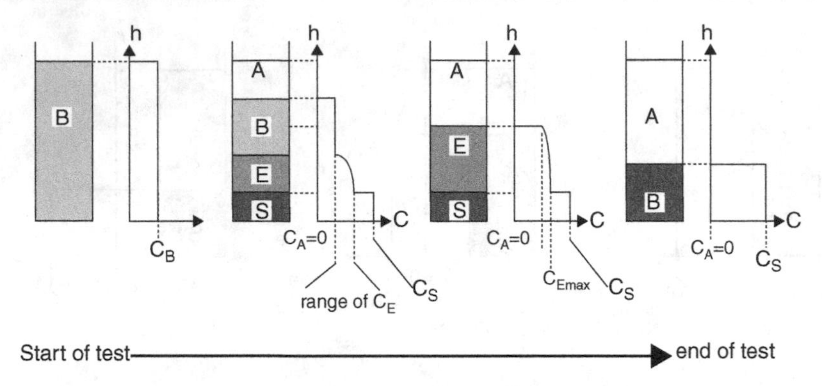

Figure 3.6 Type 2 batch settling. Zones A, B and S are zones of constant concentration. Zone A is clear liquid; zone B is a suspension of concentration equal to the initial suspension concentration; zone S is a suspension of settled bed concentration. Zone E is a zone of variable concentration

The occurrence of Type 1 or Type 2 settling depends on the initial concentration of the suspension, C_B. In simple terms, if an interface between zone B and a suspension of concentration greater than C_B but less that C_S rises faster than the interface between zones B and S then a zone of variable concentration will form. Examination of the particle flux plot enables us to determine which type of settling is occurring. Referring to Figure 3.8, a tangent to the curve is drawn through the point ($C = C_S$, $U_{ps} = 0$). The concentration at the point of tangent is C_{B_2}. The concentration at the point of intersection of the projected tangent with the curve is C_{B_1}. Type 1 settling occurs when the initial suspension concentration is less than C_{B_1} and greater than C_{B_2}. Type 2 settling occurs when the initial suspension concentration lies between C_{B_1} and C_{B_2}. Strictly, beyond C_{B_2}, Type 1 settling will again occur, but this is of little practical significance.

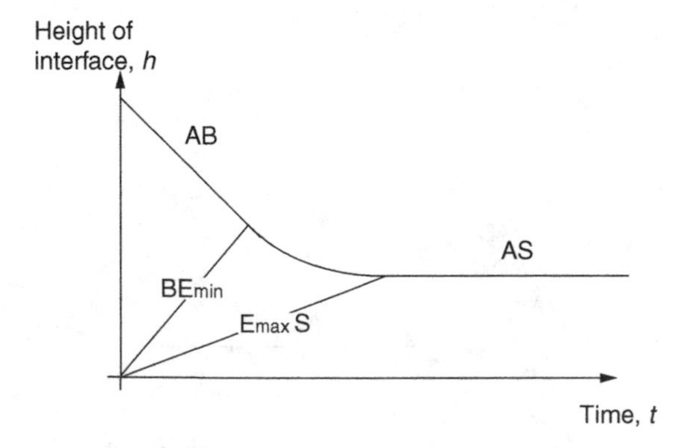

Figure 3.7 Change in positions of interface AB, BE_{min}, $E_{max}S$ and AS with time in Type 2 batch settling (e.g. AB is the interface between zone A and zone B. BE_{min} is the interface between zone B and the lowest suspension concentration in the variable zone E; see Figure 3.6)

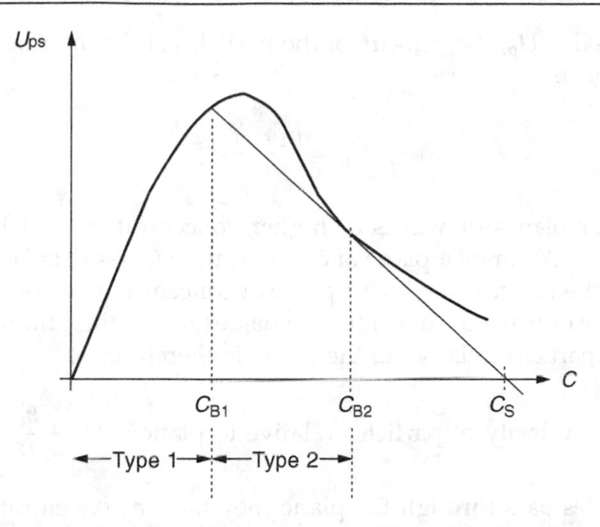

Figure 3.8 Determining if settling will be Type 1 or Type 3. A line through C_S tangent to the flux curve gives C_{B_1} and C_{B_2}. Type 2 settling occurs when initial suspension concentration is between C_{B_1} and C_{B_2}

3.2.4 Relationship Between the Height–Time Curve and the Flux Plot

Following the AB interface in the simple batch settling test gives rise to the height–time curve shown in Figure 3.9 (Type 1 settling). In fact, there will be a family of such curves for different initial concentrations. The following analysis permits the derivation of the particle flux plot from the height–time curve.

Referring to Figure 3.9, at time t the interface between clear liquid and suspension of concentration C is at a height h from the base of the vessel and velocity of the interface is the slope of the curve at this time:

$$\text{velocity of interface} = \frac{dh}{dt} = \frac{h_1 - h}{t} \tag{3.32}$$

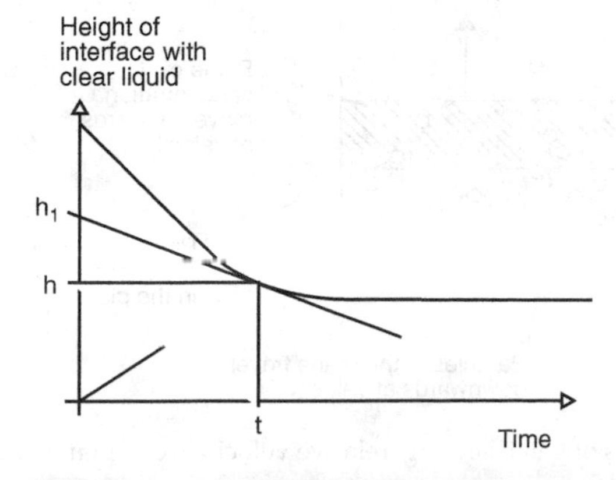

Figure 3.9 Analysis of batch settling test

This is also equal to U_p, the velocity of the particles at the interface relative to the vessel wall. Hence,

$$U_p = \frac{h_1 - h}{t} \tag{3.33}$$

Now consider planes or waves of higher concentration which rise from the base of the vessel. At time t a plane of concentration C has risen a distance h from the base. Thus the velocity at which a plane of concentration C rises from the base is h/t. This plane or wave of concentration passes up through the suspension. The velocity of the particles relative to the plane is therefore:

$$\text{velocity of particles relative to plane} = U_p + \frac{h}{t} \tag{3.34}$$

As the particles pass through the plane they have a concentration, C (refer to Figure 3.10). Therefore, the volume of particles which have passed through this plane in time t is

$$= \text{area} \times \text{velocity of particles} \times \text{concentration} \times \text{time}$$

$$= A\left(U_p + \frac{h}{t}\right)Ct \tag{3.35}$$

But, at time t this plane is interfacing with the clear liquid, and so at this time all the particles in the test have passed through the plane.

$$\text{The total volume of all the particles in the test} = C_B h_0 A \tag{3.36}$$

where h_0 is the initial suspension height.

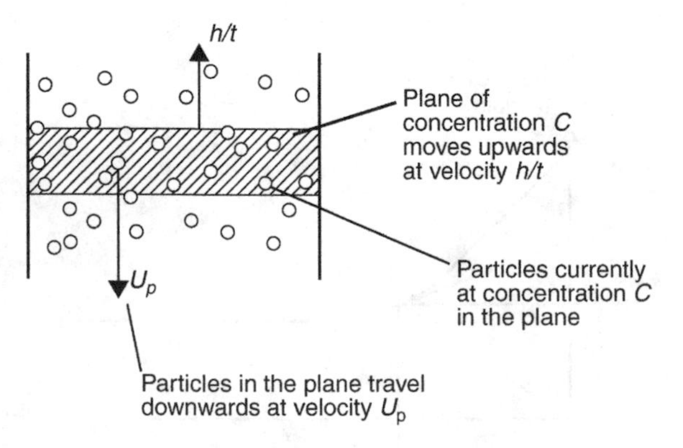

Figure 3.10 Analysis of batch settling; relative velocities of a plane of concentration C and the particles in the plane

Therefore,

$$C_B h_0 A = A\left(U_p + \frac{h}{t}\right)Ct \tag{3.37}$$

hence, substituting for U_p from Equation (3.33), we have

$$C = \frac{C_B h_0}{h_1} \tag{3.38}$$

3.3 CONTINUOUS SETTLING

3.3.1 Settling of a Suspension in a Flowing Fluid

We will now look at the effects of imposing a net fluid flow on to the particle settling process with a view to eventually producing a design procedure for a thickener. This analysis follows the method suggested by Fryer and Uhlherr (1980).

Firstly, we will consider a settling suspension flowing downwards in a vessel. A suspension of solids concentration $(1 - \varepsilon_F)$ or C_F is fed continuously into the top of a vessel of cross-sectional area A at a volume flow rate Q (Figure 3.11): The suspension is drawn off from the base of the vessel at the same rate. At a given axial position, X, in the vessel let the local solids concentration be $(1 - \varepsilon)$ or C and the volumetric fluxes of the solids and the fluid be U_{ps} and U_{fs}, respectively. Then assuming incompressible fluid and solids, continuity gives

$$Q = (U_{ps} + U_{fs})A \tag{3.39}$$

At position X, the relative velocity between fluid and particles, U_{rel} is given by

$$U_{rel} = \frac{U_{ps}}{1-\varepsilon} - \frac{U_{fs}}{\varepsilon} \tag{3.40}$$

Our analysis of batch settling gave us the following expression for this relative velocity:

$$U_{rel} = U_T \varepsilon f(\varepsilon) \tag{3.7}$$

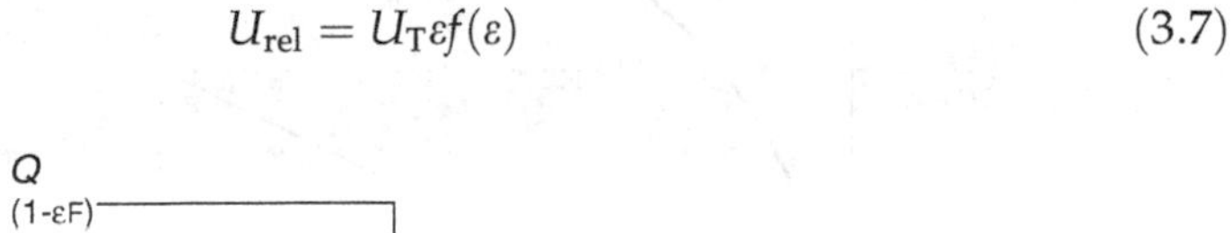
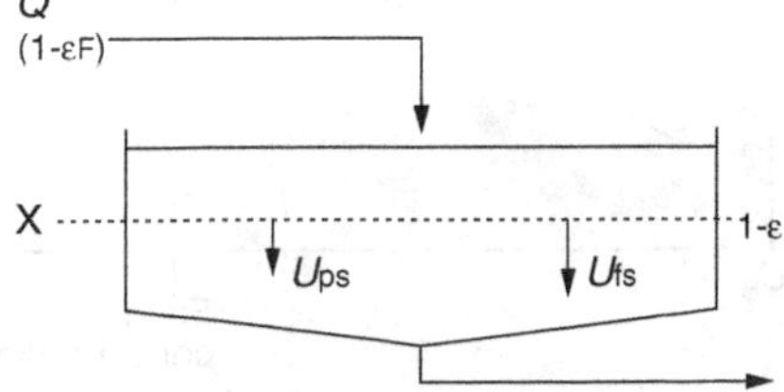

Figure 3.11 Continuous settling; downflow only

and so, combining Equations (3.39), (3.40) and (3.7) we have

$$U_{ps} = \frac{Q(1-\varepsilon)}{A} + U_T\varepsilon^2(1-\varepsilon)f(\varepsilon) \tag{3.41}$$

or

total solids flux = flux due to bulk flow + flux due to settling

We can use this expression to convert our batch flux plot into a continuous total downward flux plot. Referring to Figure 3.12, we plot a line of slope Q/A through the origin to represent the bulk flow flux and then add this to the batch flux plot to give the continuous total downward flux plot. Now, in order to graphically determine the solids concentration at level X in the vessel we apply the mass balance between feed and the point X. Reading up from the feed concentration C_F to the bulk flow line gives the value of the volumetric particle flux fed to the vessel, QC_F/A. By continuity this must also be the total flux at level X or any level in the vessel. Hence, reading across from the flux of QC_F/A to the continuous total flux curve, we may read off the particle concentration in the vessel during downward flow, which we will call C_B. (The subscript B will eventually refer to the 'bottom' section of the continuous thickener.) In downward flow the value of C_B will always be lower than the feed concentration C_F, since the solids velocity is greater in downward flow than in the feed (concentration × velocity = flux).

A similar analysis applied to upward flow of a particle suspension in a vessel gives total downward particle flux,

$$U_{ps} = U_T\varepsilon^2 f(\varepsilon) - \frac{Q(1-\varepsilon)}{A} \tag{3.42}$$

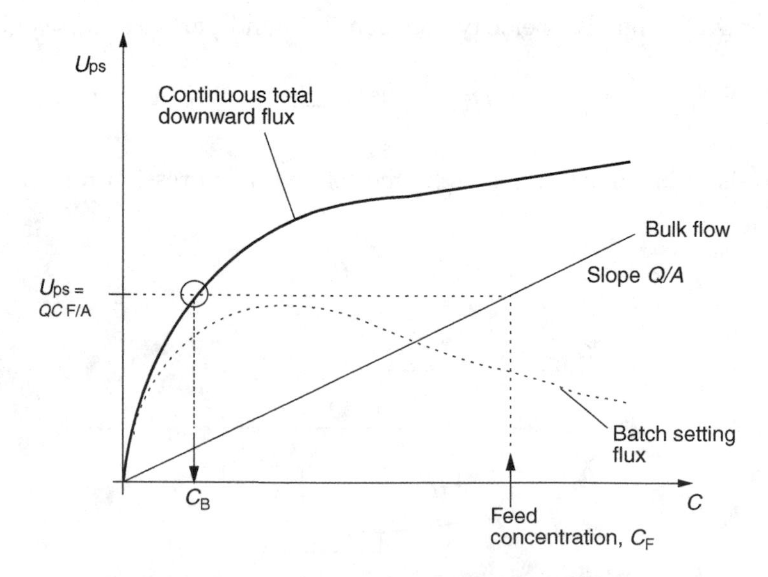

Figure 3.12 Total flux plot for settling in downward flow

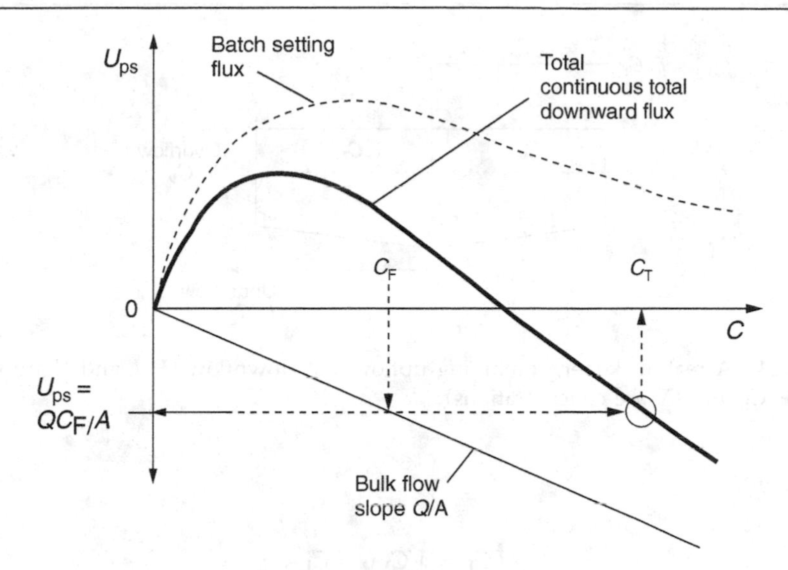

Figure 3.13 Total flux plot for settling in upward flow

or

total solids flux = flux due to settling – flux due to bulk flow

Hence, for upward flow, we obtain the continuous total flux plot by subtracting the straight line representing the flux due to bulk flow from the batch flux curve (Figure 3.13). Applying the material balance as we did for downward flow, we are able to graphically determine the particle concentration in the vessel during upward flow of fluid, C_T. (The subscript T refers to the 'top' section of the continuous thickener.) It will be seen from Figure 3.13 that the value of particle concentration for the upward-flowing suspension, C_T, is always greater than the feed concentration, C_F. This is because the particle velocity during upward flow is always less than that in the feed.

3.3.2 A Real Thickener (with Upflow and Downflow Sections)

Consider now a real thickener shown schematically in Figure 3.14. The feed suspension of concentration C_F is fed into the vessel at some point intermediate between the top and bottom of the vessel at a volume flow rate, F. An 'underflow' is drawn off at the base of the vessel at a volume flow rate, L, and concentration C_L. A suspension of concentration C_V overflows at a volume flow rate V at the top of the vessel (this flow is called the 'overflow'). Let the mean particle concentrations in the bottom (downflow) and top (upflow) sections be C_B and C_T, respectively. The total and particle material balances over the thickener are:

Total:

$$F = V + L \tag{3.43}$$

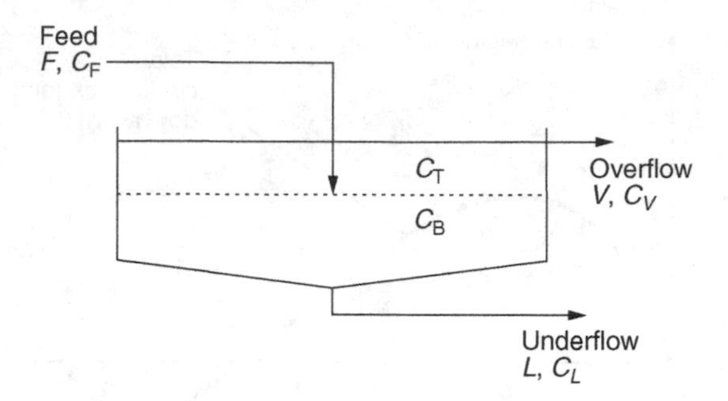

Figure 3.14 A real thickener, combining upflow and downflow (*F*, *L* and *V* are volume flows; C_F, C_L and C_V are concentrations)

Particle:

$$FC_F = VC_V + LC_L \quad (3.44)$$

These material balances link the total continuous flux plots for the upflow and downflow sections in the thickener.

3.3.3 Critically Loaded Thickener

Figure 3.15 shows flux plots for a 'critically loaded' thickener. The line of slope F/A represents the relationship between feed concentration and feed flux for a volumetric feed rate, *F*. The material balance equations [Equations (3.43) and (3.44)] determine that this line intersects the curve for the total flux in the downflow section when the total flux in the upflow section is zero. Under critical loading conditions the feed concentration is just equal to the critical value giving rise to a feed flux equal to the total continuous flux that the downflow section can deliver at that concentration. Thus the combined effect of bulk flow and settling in the downflow section provides a flux equal to that of the feed. Under these conditions, since all particles fed to the thickener can be dealt with by the downflow section, the upflow flux is zero. The material balance then dictates that the concentration in the downflow section, C_B, is equal to C_F and the underflow concentration, C_L is FC_F/L. The material balance may be performed graphically and is shown in Figure 3.15. From the feed flux line, the feed flux at a feed concentration, C_F is $U_{ps} = FC_F/A$. At this flux the concentration in the downflow section is $C_B = C_F$. The downflow flux is exactly equal to the feed flux and so the flux in the upflow section is zero. In the underflow, where there is no sedimentation, the underflow flux, LC_L/A, is equal to the downflow flux. At this flux the underflow concentration, C_L is determined from the underflow line.

Figure 3.15 indicates that under critical conditions there are two possible solutions for the concentration in the upflow section, C_T. One solution, the obvious one, is $C_T = 0$; the other is $C_T = C_B$. In this second situation a fluidized bed of particles at concentration C_B with a distinct surface is observed in the upflow section.

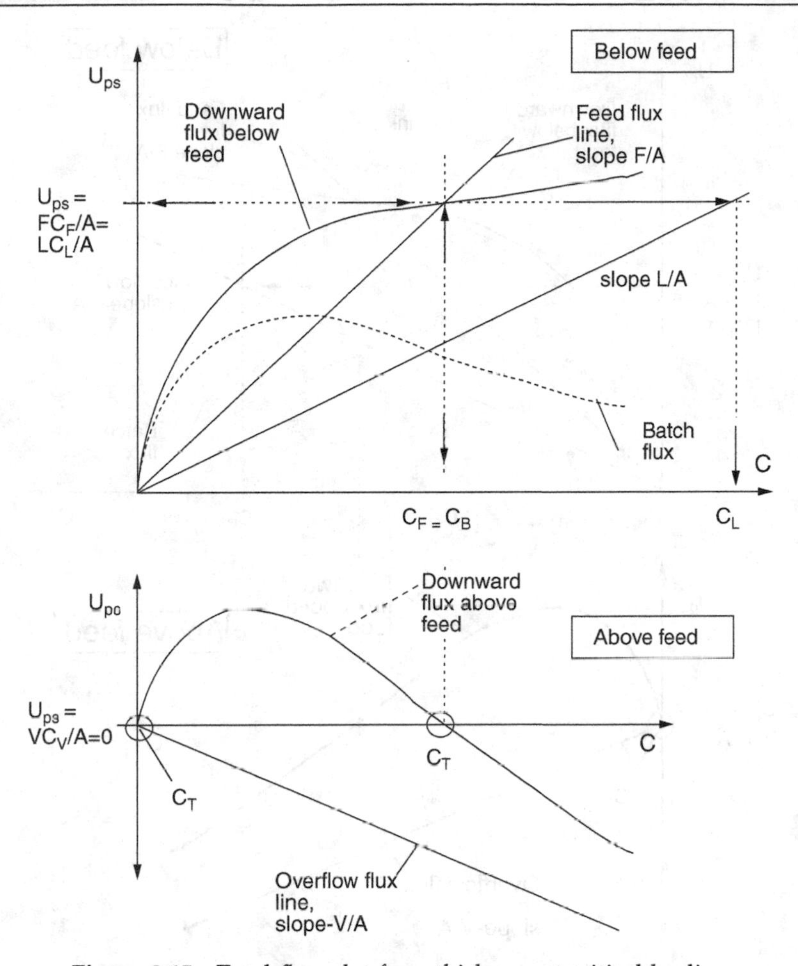

Figure 3.15 Total flux plot for a thickener at critical loading

3.3.4 Underloaded Thickener

When the feed concentration C_F is less than the critical concentration the thickener is said to be underloaded. This situation is depicted in Figure 3.16. Here the feed flux, FC_F/A, is less than the maximum flux due to bulk flow and settling which can be provided by the downflow section. The flux in the upflow section is again zero ($C_T = C_V = 0; VC_V/A = 0$). The graphical mass balance shown in Figure 3.16 enables C_D and C_L to be determined (feed flux = downflow section flux = underflow flux).

3.3.5 Overloaded Thickener

When the feed concentration C_F is greater than the critical concentration, the thickener is said to be overloaded. This situation is depicted in Figure 3.17. Here

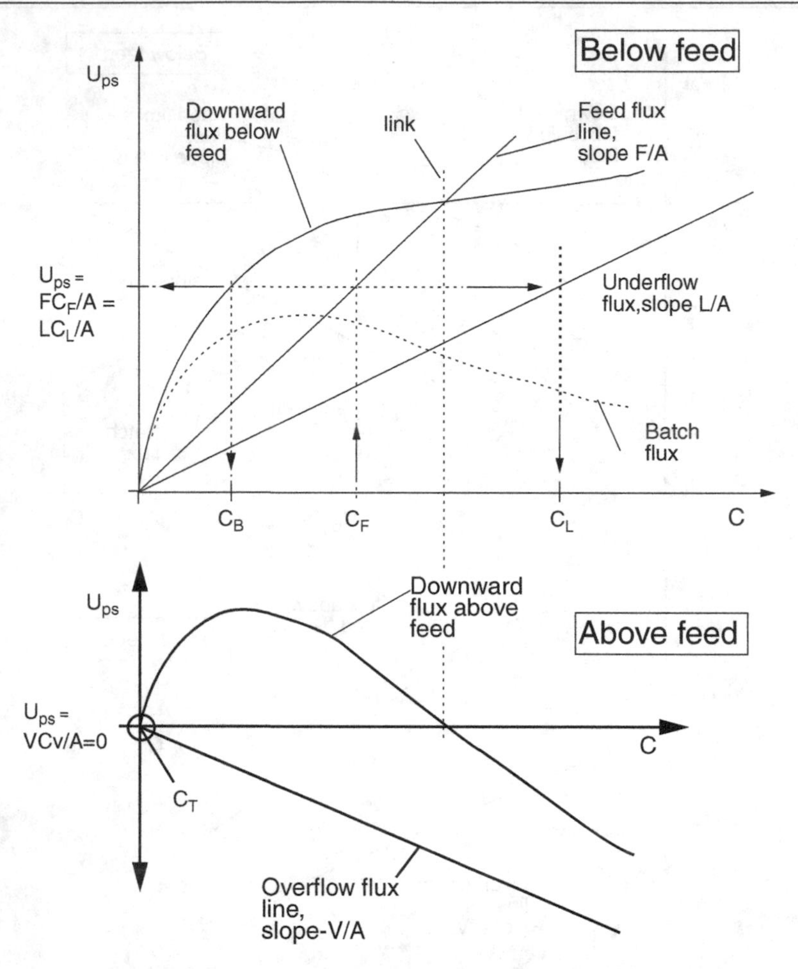

Figure 3.16 Total flux plot for an underloaded thickener

the feed flux, FC_F/A, is greater than the maximum flux due to bulk flow and settling provided by the downflow section. The excess flux must pass through the upflow section and out through the overflow. The graphical material balance is depicted in Figure 3.17. At the feed concentration C_F, the difference between the feed flux and the total flux in the downflow section gives the excess flux which must pass through the upflow section. This flux applied to the upflow section graph gives the value of the concentration in the upflow section, C_T, and the overflow concentration, C_V (upflow section flux = overflow flux).

3.3.6 Alternative Form of Total Flux Plot

A common form of continuous flux plot is that exhibiting a minimum total flux shown under critical conditions in Figure 3.18. With this alternative flux plot the critical loading condition occurs when the feed concentration gives rise to a flux

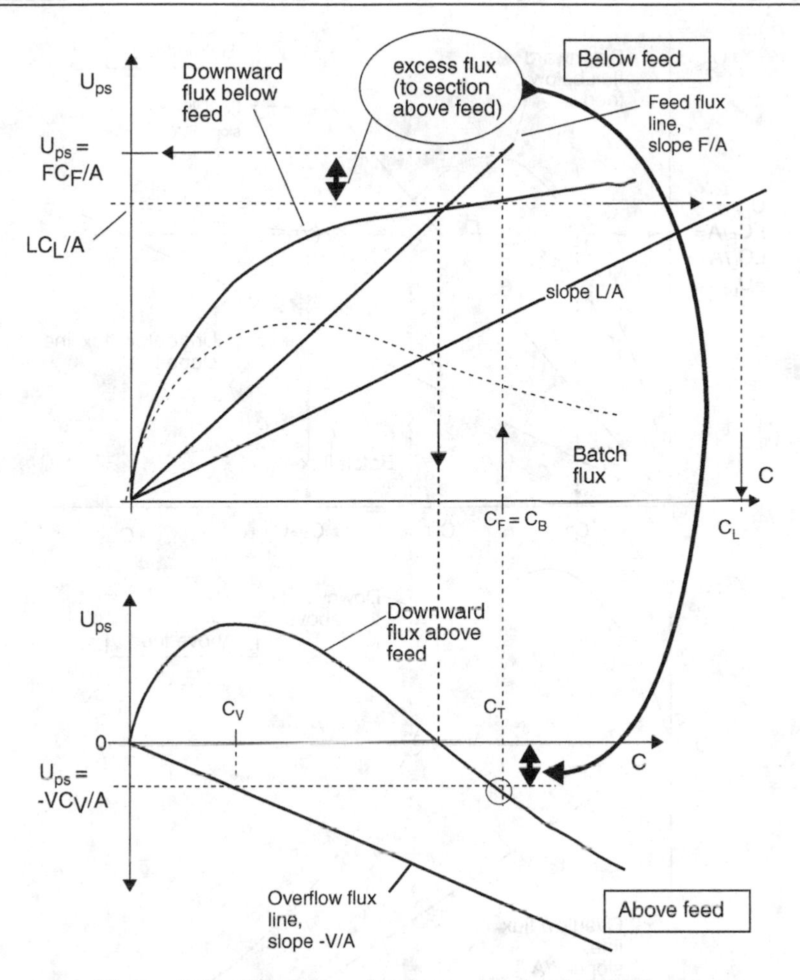

Figure 3.17 Total flux plot for an overloaded thickener

equal to this minimum in the total flux curve. The downflow section cannot operate in a stable manner above this flux. Under critical conditions the upflow flux is again zero and the graphical material balance depicted in Figure 3.18 gives the values of C_B and C_L. It will be noted that under these conditions there are two possible values of C_B; these may coexist in the downflow section with a discontinuity between them at any position between the feed level and the underflow.

Figure 3.19 shows this alternative flux plot in an overloaded situation. For the graphical solution in this case, the excess flux must be read from the flux axis of the downflow section plot and applied to the upflow section plot in order to determine the value of C_T and C_V. Note that in this case although there are theoretically two possible values of C_B, in practice only the higher value can stably coexist with the higher concentration region, C_T, above it.

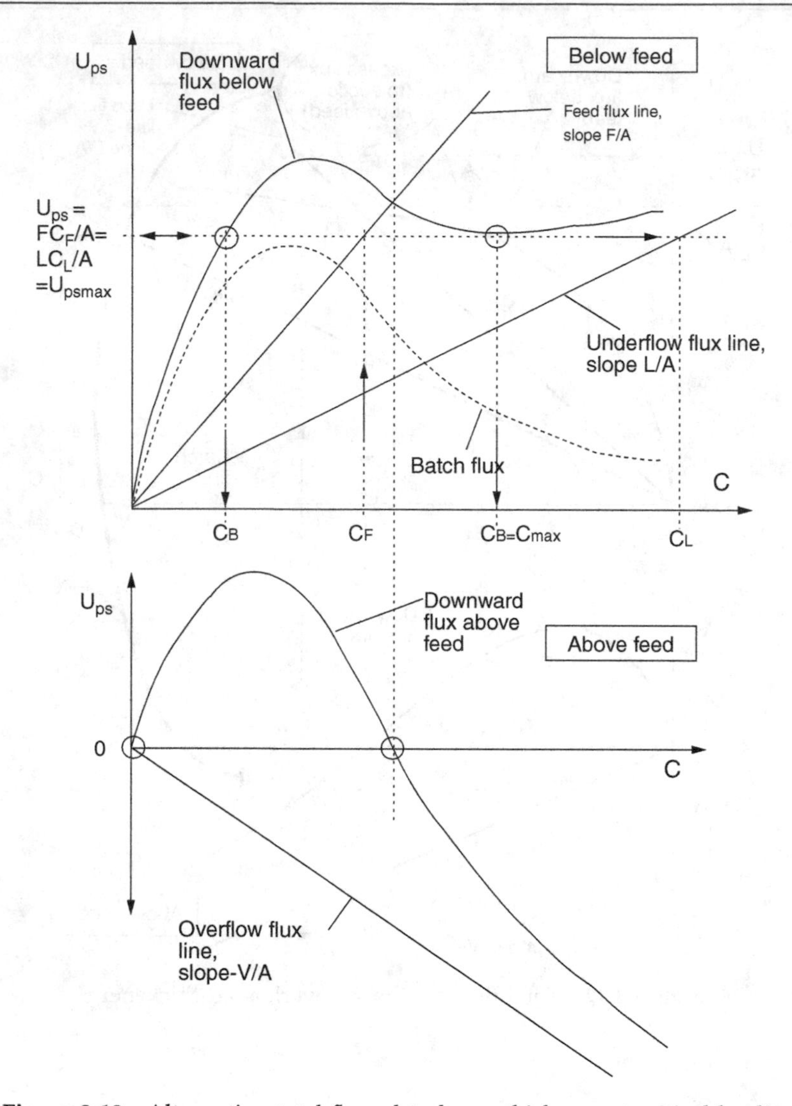

Figure 3.18 Alternative total flux plot shape; thickener at critical loading

3.4 *WORKED EXAMPLES*

WORKED EXAMPLE 3.1

A height–time curve for the sedimentation of a suspension, of initial suspension concentration 0.1, in vertical cylindrical vessel is shown in Figure 3W1.1. Determine:

(a) the velocity of the interface between clear liquid and suspension of concentration 0.1;

(b) the velocity of the interface between clear liquid and a suspension of concentration 0.175;

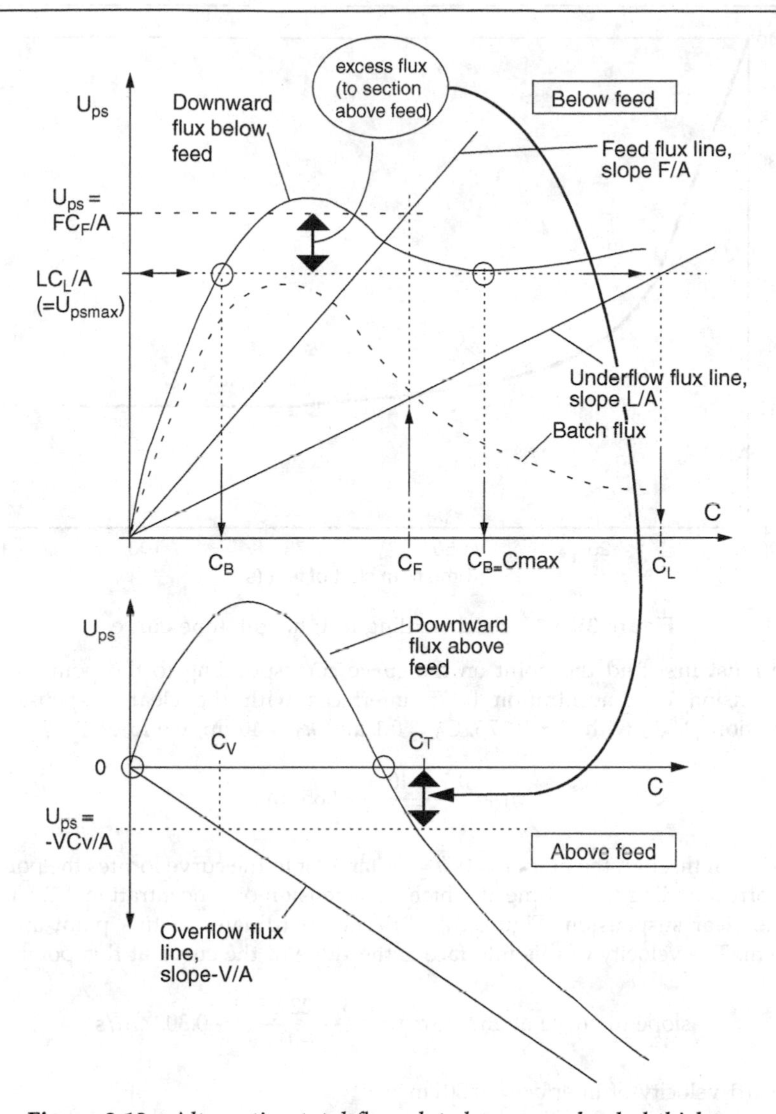

Figure 3.19 Alternative total flux plot shape; overloaded thickener

(c) the velocity at which a layer of concentration 0.175 propagates upwards from the base of the vessel;

(d) the final sediment concentration.

Solution

(a) Since the initial suspension concentration is 0.1, the velocity required in this question is the velocity of the AB interface. This is given by the slope of the straight portion of the height–time curve.

$$\text{Slope} = \frac{20 - 40}{15 - 0} = 1.333\,\text{cm/s}$$

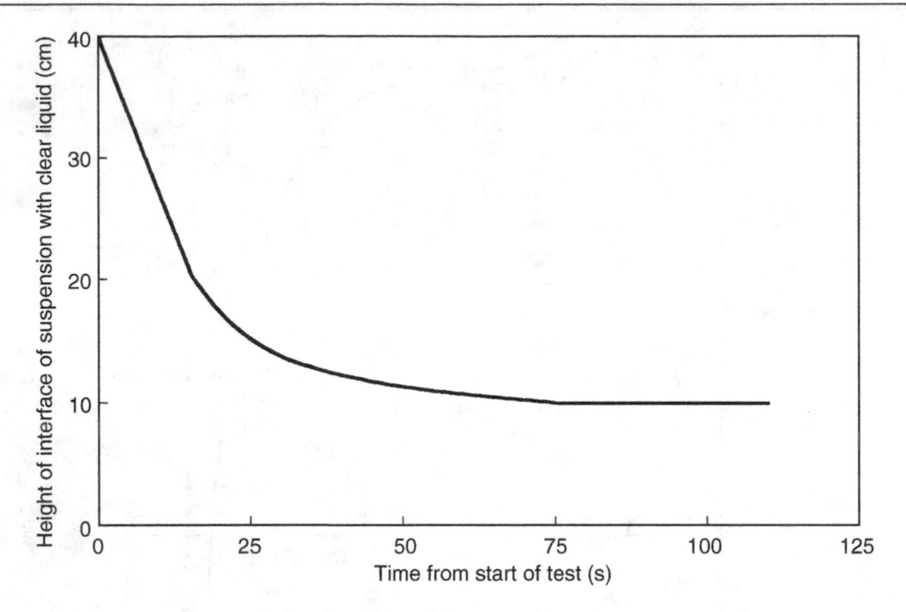

Figure 3W1.1 Batch settling test; height–time curve

(b) We must first find the point on the curve corresponding to the point at which a suspension of concentration 0.175 interfaces with the clear suspension. From Equation (3.38), with $C = 0.175$, $C_B = 0.1$ and $h_0 = 40\,\text{cm}$, we find:

$$h_1 = \frac{0.1 \times 40}{0.175} = 22.85\,\text{cm}$$

A line drawn through the point $t = 0, h = h_1$ tangent to the curve locates the point on the curve corresponding to the time at which a suspension of concentration 0.175 interfaces with the clear suspension (Figure 3W1.2). The coordinates of this point are $t = 26\,\text{s}$, $h = 15\,\text{cm}$. The velocity of this interface is the slope of the curve at this point:

$$\text{slope of curve at } 26\,\text{s}, 15\,\text{cm} = \frac{15 - 22.85}{26 - 0} = -0.302\,\text{cm/s}$$

downward velocity of interface= 0.30 cm/s

(c) From the consideration above, after 26 s the layer of concentration 0.175 has just reached the clear liquid interface and has travelled a distance of 15 cm from the base of the vessel in this time.

$$\text{Therefore, upward propagation velocity of this layer} = \frac{h}{t} = \frac{15}{16} = 0.577\,\text{cm/s}$$

(d) To find the concentration of the final sediment we again use Equation (3.38). The value of h_1 corresponding to the final sediment (h_{1s}) is found by drawing a tangent to the part of the curve corresponding to the final sediment and projecting it to the h axis.

In this case $h_{1S} = 10\,\text{cm}$ and so from Equation (3.38),

$$\text{final sediment concentration}, C = \frac{C_0 h_0}{h_{1S}} = \frac{0.1 \times 4.0}{10} = 0.4$$

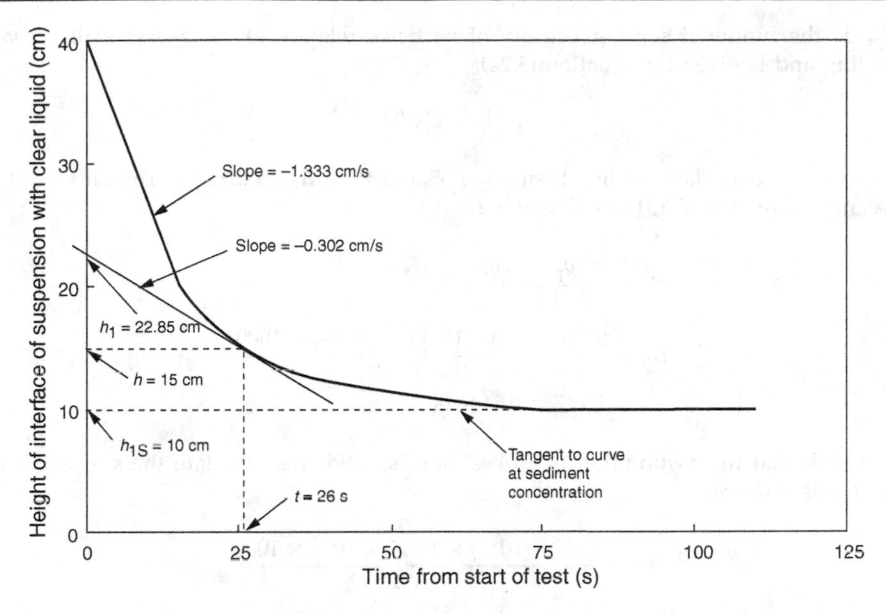

Figure 3W1.2 Batch settling test

WORKED EXAMPLE 3.2

A suspension in water of uniformly sized sphere (diameter 150 μm, density 1140 kg/m^3 has a solids concentration of 25% by volume. The suspension settles to a bed of solids concentration of 55% by volume. Calculate:

(a) the rate at which the water/suspension interface settles;

(b) the rate at which the sediment/suspension interface rises (assume water properties: density, 1000 kg/m^3; viscosity, 0.001 Pa s).

Solution

(a) Solids concentration of initial suspension, $C_B = 0.25$

Equation (3.28) allows us to calculate the velocity of interfaces between suspensions of different concentrations.

The velocity of the interface between initial suspension (B) and clear liquid (A) is therefore:

$$U_{\text{int,AB}} = \frac{U_{\text{pA}}C_A - U_{\text{pB}}C_B}{C_A - C_B}$$

Since $C_A = 0$, the equation reduces to

$$U_{\text{int,AB}} = U_{\text{pB}}$$

U_{pB} is the hindered settling velocity of particles relative to the vessel wall in batch settling and is given by Equation (3.24):

$$U_p = U_T \varepsilon^n$$

Assuming Stokes' law applies, then $n = 4.65$ and the single particle terminal velocity is given by Equation (2.13) (see Chapter 2):

$$U_T = \frac{x^2(\rho_p - \rho_f)g}{18\mu}$$

$$U_T = \frac{9.81 \times (150 \times 10^{-6})^2 \times (1140 - 1000)}{18 \times 0.001}$$

$$= 1.717 \times 10^{-3}\ \text{m/s}$$

To check that the assumption of Stokes' law is valid, we calculate the single particle Reynolds number:

$$Re_p = \frac{(150 \times 10^{-6}) \times 1.717 \times 10^{-3} \times 1000}{0.001}$$

$= 0.258$, which is less than the limiting value for Stokes' law (0.3) and so the assumption is valid.

The voidage of the initial suspension, $\varepsilon_B = 1 - C_B = 0.75$

$$\text{hence, } U_{pB} = 1.717 \times 10^{-3} \times 0.75^{4.65}$$

$$= 0.45 \times 10^{-3}\ \text{m/s}$$

Hence, the velocity of the interface between the initial suspension and the clear liquid is 0.45 mm/s. The fact that the velocity is positive indicates that the interface is moving downwards.

(b) Here again we apply Equation (3.28) to calculate the velocity of interfaces between suspensions of different concentrations.

The velocity of the interface between initial suspension (B) and sediment (S) is therefore

$$U_{\text{int, BS}} = \frac{U_{pB}C_B - U_{pS}C_S}{C_B - C_S}$$

With $C_B = 0.25$ and $C_S = 0.55$ and since the velocity of the sediment, U_{pS} is zero, we have:

$$U_{\text{int,BS}} = \frac{U_{pB}0.25 - 0}{0.25 - 0.55} = -0.833U_{pB}$$

And from part (a), we know that $U_{pB} = 0.45$mm/s, and so $U_{\text{int,BS}} = -0.375$ mm/s.

The negative sign signifies that the interface is moving upwards. So, the interface between initial suspension and sediment is moving upwards at a velocity of 0.375 mm/s.

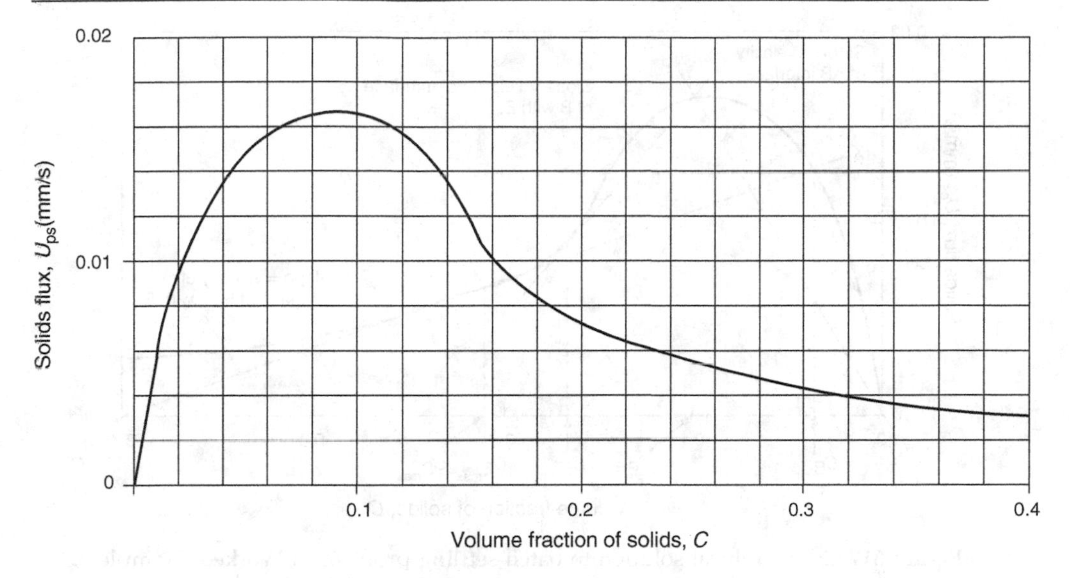

Figure 3W3.1 Batch flux plot

WORKED EXAMPLE 3.3

For the batch flux plot shown in Figure 3W3.1, the sediment has a solids concentration of 0.4 volume fraction of solids.

(a) Determine the range of initial suspension concentrations over which a zone of variable concentration is formed under batch settling conditions.

(b) Calculate and plot the concentration profile after 50 min in a batch settling test of a suspension with an initial concentration 0.1 volume fraction of solids, and initial suspension height of 100 cm.

(c) At what time will the settling test be complete?

Solution

(a) Determine the range of initial suspension concentrations by drawing a line through the point $C = C_S = 0.4$, $U_{ps} = 0$ tangent to the batch flux curve. This is shown as line XC_S in Figure 3W3.2. The range of initial suspension concentrations for which a zone of variable concentration is formed in batch settling (Type 2 settling) is defined by $C_{B_{min}}$ and $C_{B_{max}}$. $C_{B_{min}}$ is the value of C at which the line XC_S intersects the settling curve and $C_{B_{max}}$ is the value of C at the tangent. From Figure 3W3.2, we see that $C_{B_{min}} = 0.036$ and $C_{B_{max}} = 0.21$.

(b) To calculate the concentration profile we must first determine the velocities of the interfaces between the zones A, B, E and S and hence find their positions after 50 min.

The line AB in Figure 3W3.2 joins the point representing A the clear liquid (0, 0) and the point B representing the initial suspension (0.1, U_{ps}). The slope of line AB is equal to the

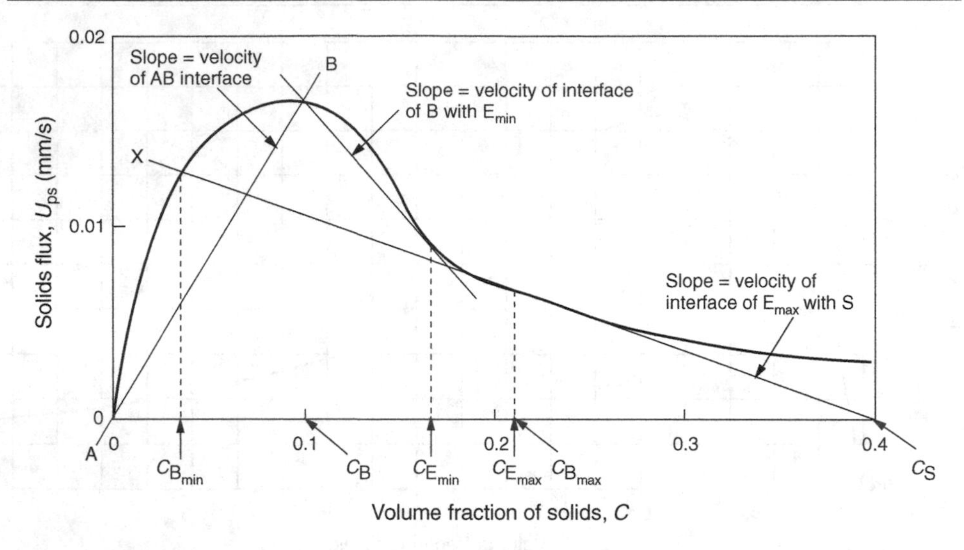

Figure 3W3.2 Graphical solution to batch settling problem in Worked Example 3.3

velocity of the interface between zones A and B. From Figure 3W3.2, $U_{\text{int, AB}} = +0.166\,\text{mm/s}$ or $+1.00\,\text{cm/min}$.

The slope of the line from point B tangent to the curve is equal to the velocity of the interface between the initial suspension B and the minimum value of the variable concentration zone C_{Emin}.

From Figure 3W3.2,

$$U_{\text{int, BE}_{\text{min}}} = -0.111\,\text{mm/s or} - 0.66\,\text{cm/min}$$

The slope of the line tangent to the curve and passing through the point representing the sediment (point $C = C_S = 0.4$, $U_{ps} = 0$) is equal to the velocity of the interface between the maximum value of the variable concentration zone $C_{E_{\text{max}}}$ and the sediment.

From Figure 3W3.2,

$$U_{\text{int, E}_{\text{max}}\text{S}} = -0.0355\,\text{mm/s or} - 0.213\,\text{cm/min}$$

Therefore, after 50 min the distances travelled by the interfaces will be:

AB interface	50.0 cm(1.00 × 50)downwards
BE_{min}interface	33.2 cm upwards
E_{max}S interface	10.6 cm upwards

Therefore, the positions of the interfaces (distance from the base of the test vessel) after 50 min will be

AB interface	50.0 cm
BE_{min}interface	33.2 cm
E_{max}S interface	10.6 cm

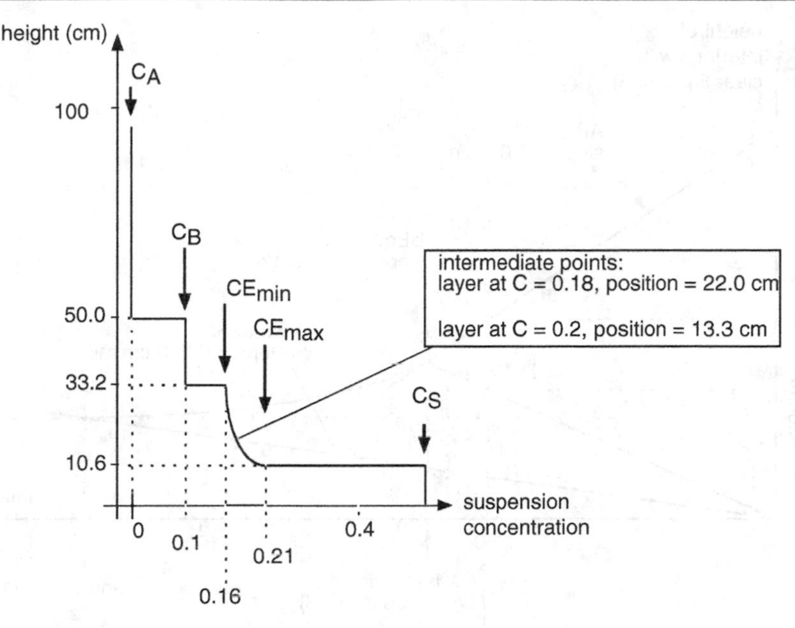

Figure 3W3.3 Sketch of concentration profile in batch settling test vessel after 50 min

From Figure 3W3.2 we determine the minimum and maximum values of suspension concentration in the variable zone

$$C_{E_{min}} = 0.16$$
$$C_{E_{max}} = 0.21$$

Using this information we can plot the concentration profile in the test vessel 50 min after the start of the test. A sketch of the profile is shown in Figure 3W3.3. The shape of the concentration profile within the variable concentration zone may be determined by the following method. Recalling that the slope of the batch flux plot (Figure 3W3.1) at a value of suspension concentration C is the velocity of a layer of suspension of that concentration, we find the slope at two or more values of concentration and then determine the positions of these layers after 50 min:

- Slope of batch flux plot at $C = 0.18$ is 0.44 cm/ min upwards.

  Hence, position of a layer of concentration 0.18 after 50 min is 22.0 cm from the base.

- Slope of batch flux plot at $C = 0.20$, is 0.27 cm/ min upwards.

  Hence, position of a layer of concentration 0.20 after 50 min is 13.3 cm from the base.

These two points are plotted on the concentration profile in order to determine the shape of the profile within the zone of variable concentration.

Figure 3W3.4 is a sketched plot of the height–time curve for this test constructed from the information above. The shape of the curved portion of the curve can again be

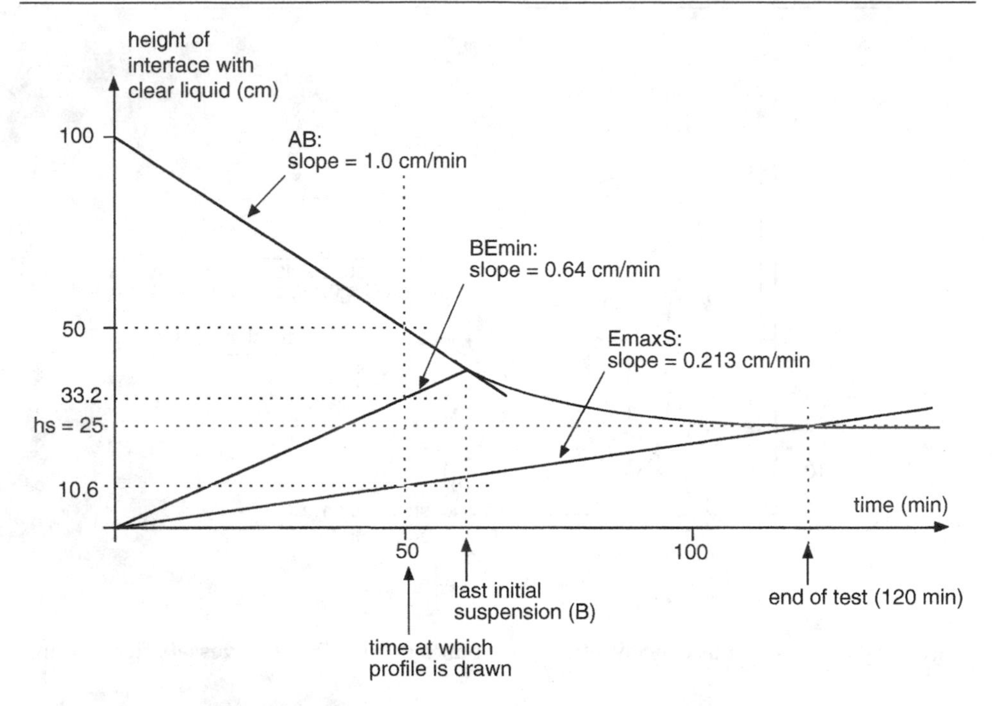

Figure 3W3.4 Sketch of height–time curve for the batch settling test in Worked Example 3.3

determined by plotting the positions of two or more layers of suspension of different concentration. The initial suspension concentration zone (B) ends when the AB line intersects the BE_{min} line, both of which are plotted from a knowledge of their slopes.

The time for the end of the test is found in the following way. The end of the test is when the position of the $E_{max}S$ interface coincides with the height of the final sediment. The height of the final sediment may be found using Equation (3.38) [see part (d) of Worked Example 3.1]:

$$C_S h_S = C_B h_0$$

where h_S is the height of the final sediment and h_0 is the initial height of the suspension (at the start of the test). With $C_S = 0.4$, $C_B = 0.1$ and $h_0 = 100\,\text{cm}$, we find that $h_S = 25\,\text{cm}$. Plotting h_S on Figure 3W3.4, we find that the $E_{max}S$ line intersects the final sediment line at about 120 min and so the test ends at this time.

WORKED EXAMPLE 3.4

Using the flux plot shown in Figure 3W4.1:

(a) Graphically determine the limiting feed concentration for a thickener of area $100\,\text{m}^2$ handling a feed rate of $0.019\,\text{m}^3/\text{s}$ and an underflow rate of $0.01\,\text{m}^3/\text{s}$. Under these conditions what will be the underflow concentration and the overflow concentration?

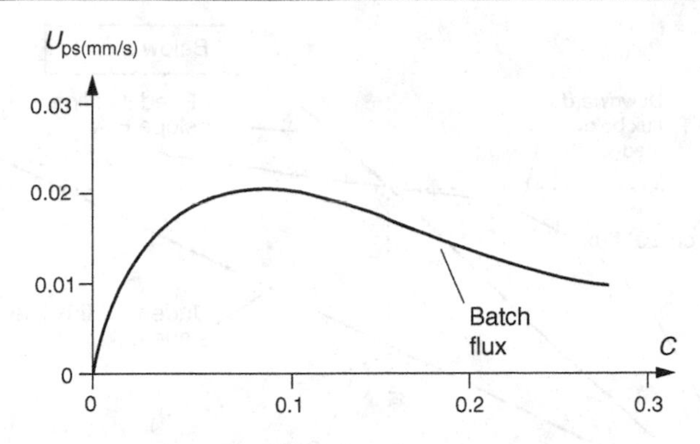

Figure 3W4.1 Batch flux plot

(b) Under the same flow conditions as above, the feed concentration is increased to 0.2. Estimate the solids concentration in the overflow, in the underflow, in the upflow section and in the downflow section of the thickener.

Solution

(a) Feed rate, $F = 0.019\,\text{m}^3/\text{s}$

Underflow rate, $L = 0.01\,\text{m}^3/\text{s}$

Material balance gives, overflow rate, $V = F - L = 0.009\,\text{m}^3/\text{s}$

Expressing these flows as fluxes based on the thickener area ($A = 100\,\text{m}^2$):

$$\frac{F}{A} = 0.19\,\text{mm/s}$$
$$\frac{L}{A} = 0.10\,\text{mm/s}$$
$$\frac{V}{A} = 0.09\,\text{mm/s}$$

The relationships between bulk flux and suspension concentration are then:

$$\text{Feed flux} = C_F\left(\frac{F}{A}\right)$$

$$\text{Flux in underflow} = C_L\left(\frac{L}{A}\right)$$

$$\text{Flux in overflow} = C_V\left(\frac{V}{A}\right)$$

Lines of slope F/A, L/A and $-V/A$ drawn on the flux plot represent the fluxes in the feed, underflow and overflow, respectively (Figure 3W4.2). The total flux plot for the section below the feed point is found by adding the batch flux plot to the underflow flux line. The total flux plot for the section above the feed point is found by adding the batch

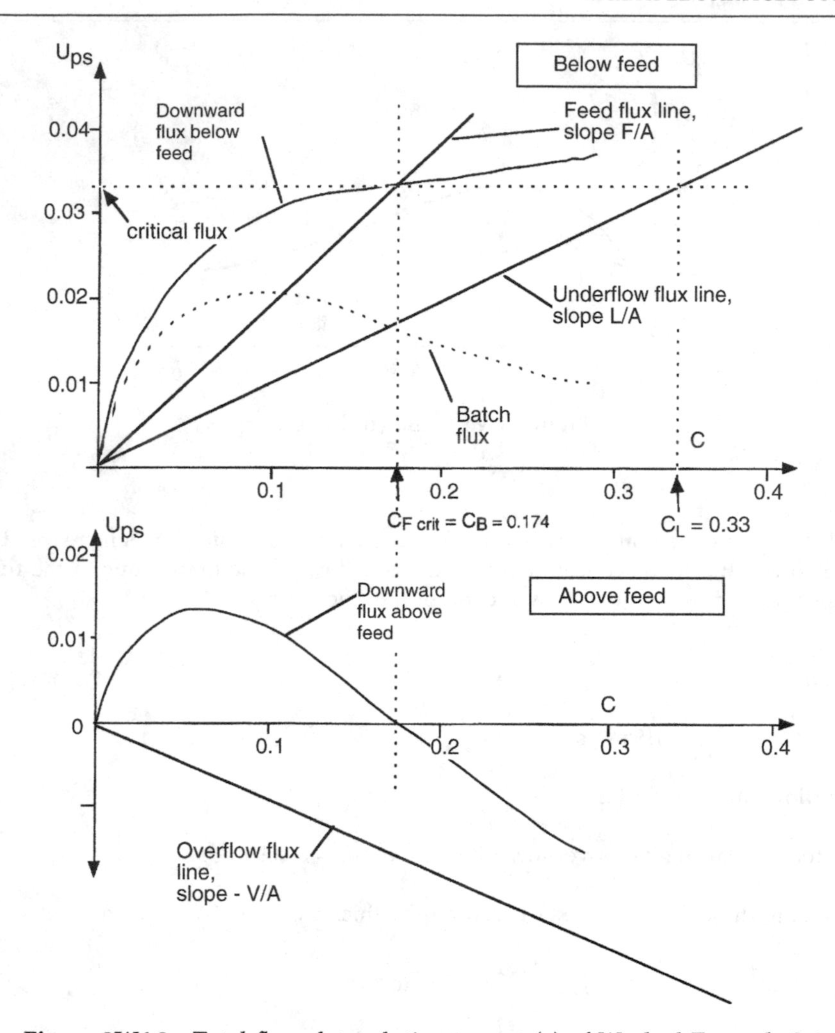

Figure 3W4.2 Total flux plot: solution to part (a) of Worked Example 3.4

flux plot to the overflow flux line (which is negative since it is an upward flux). These plots are shown in Figure 3W4.2.

The critical feed concentration is found where the feed flux line intersects the plot of total flux in the section below the feed (Figure 3.W4.2). This gives a critical feed flux of 0.0335 mm/s. The downflow section below the feed point is unable to take a flux greater than this. The corresponding feed concentration is $C_{F_{crit}} = 0.174$.

The concentration in the downflow section, C_B is also 0.174.

The corresponding concentration in the underflow is found where the critical flux line intersects the underflow flux line. This gives $C_L = 0.33$.

(b) Referring now to Figure 3W4.3, if the feed flux is increased to 0.2, we see that the corresponding feed flux is 0.038 mm/s. At this feed concentration the downflow section

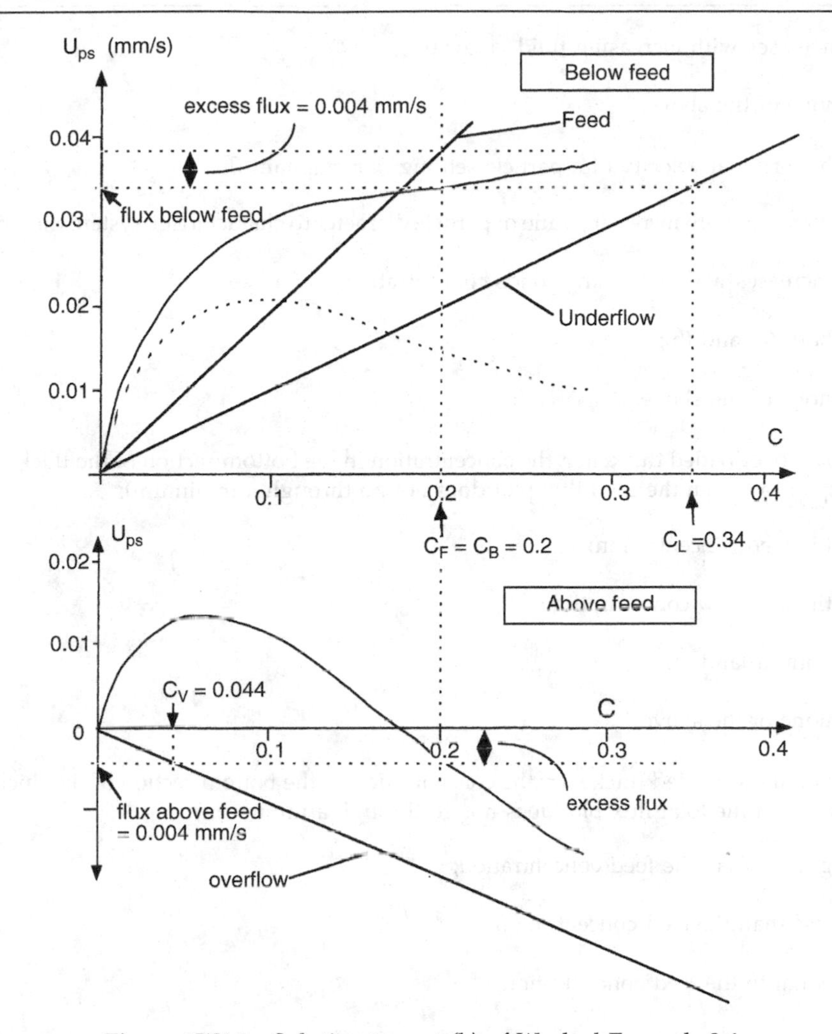

Figure 3W4.3 Solution to part (b) of Worked Example 3.4

is only able to take a flux of 0.034 mm/s and gives an underflow concentration, $C_L = 0.34$. The excess flux of 0.004 mm/s passes into the upflow section. This flux in the upflow section gives a concentration, $C_T = 0.2$ and a corresponding concentration, $C_V = 0.044$ in the overflow.

TEST YOURSELF

3.1 The particle settling velocity in a fluid–particle suspension:

(a) increases with increasing ratio of particle diameter to characteristic system dimension;

(b) increases with increasing particle concentration;

(c) increases with increasing fluid viscosity;

(d) none of the above.

3.2 The terminal velocity of a particle settling in a stagnant fluid:

(a) increases with increasing ratio of particle diameter to characteristic system dimension;

(b) increases with increasing solids concentration;

(c) both (a) and (b);

(d) none of the above.

3.3 In an overloaded thickener, the concentration in the bottom section of the thickener is equal to (when the total flux plot does not go through a minimum):

(a) the feed concentration;

(b) the overflow concentration;

(c) both (a) and (b);

(d) none of the above.

3.4 In an underloaded thickener, the concentration in the bottom section of the thickener is (when the total flux plot does not go through a minimum):

(a) greater than the feed concentration;

(b) less than the feed concentration;

(c) equal to the feed concentration.

3.5 In an underloaded thickener, the concentration in the overflow is (when the total flux plot does not go through a minimum):

(a) greater than the feed concentration;

(b) less than the feed concentration;

(c) equal to the feed concentration.

3.6 In an underloaded thickener, the concentration in the underflow is (when the total flux plot does not go through a minimum):

(a) greater than the concentration in the bottom section;

(b) less than the concentration in the bottom section;

(c) equal to the concentration in the bottom section.

3.7 When a particle reaches terminal velocity:

(a) the particle acceleration is constant;

(b) the particle acceleration is zero;

(c) the particle acceleration equals the apparent weight of the particle;

(d) none of the above.

3.8 Which of the following do not influence hindered settling velocity?

(a) particle density;

(b) particle size;

(c) particle suspension concentration;

(d) none of the above.

EXERCISES

3.1 A suspension in water of uniformly sized spheres of diameter 100 μm and density 1200 kg/m^3 has a solids volume fraction of 0.2. The suspension settles to a bed of solids volume fraction 0.5. (For water, density is 1000 kg/m^3 and viscosity is 0.001 Pa s.)

The single particle terminal velocity of the spheres in water may be taken as 1.1 mm/s.

Calculate:

(a) the velocity at which the clear water/suspension interface settles;

(b) the velocity at which the sediment/suspension interface rises.

[Answer: (a) 0.39 mm/s; (b) 0.26 mm/s.]

3.2 A height–time curve for the sedimentation of a suspension in a vertical cylindrical vessel is shown in Figure 3E2.1. The initial solids concentration of the suspension is 150 kg/m^3.

Determine:

(a) the velocity of the interface between clear liquid and suspension of concentration 150 kg/m^3;

(b) the time from the start of the test at which the suspension of concentration 240 kg/m^3 is in contact with the clear liquid;

(c) the velocity of the interface between the clear liquid and suspension of concentration 240 kg/m^3;

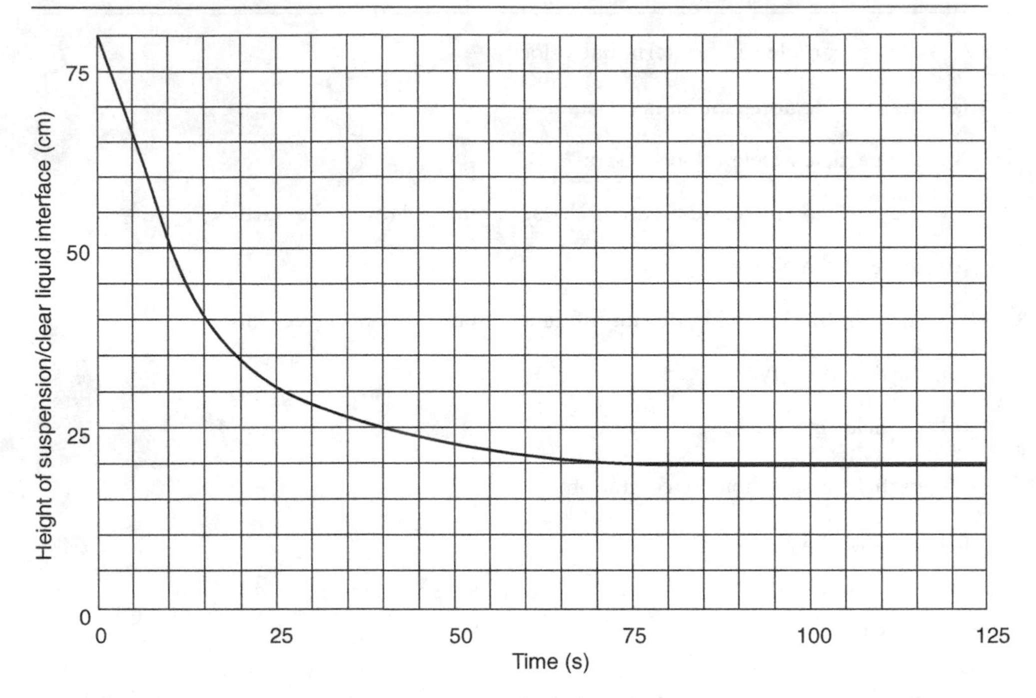

Figure 3E2.1 Batch settling test results. Height–time curve for use in Exercises 3.2 and 3.4

(d) the velocity at which a layer of concentration 240 kg/m^3 propagates upwards from the base of the vessel;

(e) the concentration of the final sediment.

[Answer: (a) 2.91 cm/s; (b) 22 s; (c) 0.77 cm/s downwards; (d) 1.50 cm/s upwards; (e) 600 kg/m^3.]

3.3 A suspension in water of uniformly sized spheres of diameter 90 μm and density 1100 kg/m^3 has a solids volume fraction of 0.2. The suspension settles to a bed of solids volume fraction 0.5. (For water, density is 1000 kg/m^3 and viscosity is 0.001 Pa s.)

The single particle terminal velocity of the spheres in water may be taken as 0.44 mm/s.

Calculate:

(a) the velocity at which the clear water/suspension interface settles;

(b) the velocity at which the sediment/suspension interface rises.

[Answer: (a) 0.156 mm/s; (b) 0.104 mm/s.]

3.4 A height–time curve for the sedimentation of a suspension in a vertical cylindrical vessel is shown in Figure 3E2.1. The initial solids concentration of the suspension is 200 kg/m^3.

Determine:

(a) the velocity of the interface between clear liquid and suspension of concentration $200\,kg/m^3$;

(b) the time from the start of the test at which the suspension of concentration $400\,kg/m^3$ is in contact with the clear liquid;

(c) the velocity of the interface between the clear liquid and suspension of concentration $400\,kg/m^3$;

(d) the velocity at which a layer of concentration $400\,kg/m^3$ propagates upwards from the base of the vessel;

(e) the concentration of the final sediment.

[Answers: (a) 2.9 cm/s downwards; (b) 32.5 s; (c) 0.40 cm/s downwards; (d) 0.846 cm/s upwards; (e) $800\,kg/m^3$.]

3.5

(a) Spherical particles of uniform diameter 40 μm and particle density $2000\,kg/m^3$ form a suspension of solids volume fraction 0.32 in a liquid of density $880\,kg/m^3$ and viscosity 0.0008 Pa s. Assuming Stokes' law applies, calculate (i) the sedimentation velocity and (ii) the sedimentation volumetric flux for this suspension.

(b) A height–time curve for the sedimentation of a suspension in a cylindrical vessel is shown in Figure 3E5.1. The initial concentration of the suspension for this test is $0.12\,m^3/m^3$.

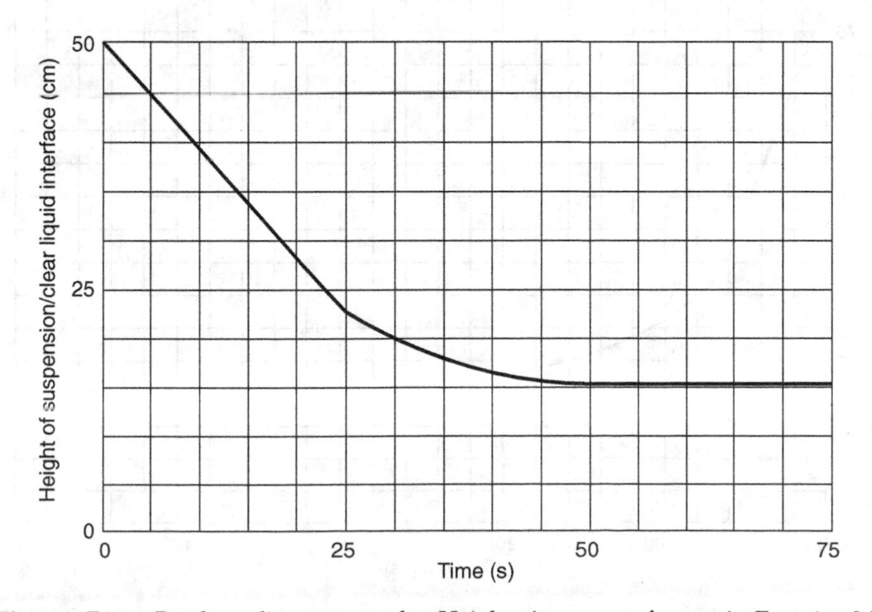

Figure 3E5.1 Batch settling test results. Height–time curve for use in Exercise 3.5

Calculate:

(i) the velocity of the interface between clear liquid and a suspension of concentration, $0.12\,m^3/m^3$;

(ii) the velocity of the interface between clear liquid and a suspension of concentration $0.2\,m^3/m^3$;

(iii) the velocity at which a layer of concentration, $0.2\,m^3/m^3$ propagates upwards from the base of the vessel;

(iv) the concentration of the final sediment;

(v) the velocity at which the sediment propagates upwards from the base.

[Answer: (a)(i) 0.203 mm/s, (ii) 0.065 mm/s, (b)(i) 1.11 cm/s downwards, (ii) 0.345 cm/s downwards, (iii) 0.514 cm/s upwards, (iv) 0.4, (v) 0.30 cm/s upwards.]

3.6 A height–time curve for the sedimentation of a suspension in a vertical cylindrical vessel is shown in Figure 2E6.1. The initial solids concentration of the suspension is $100\,kg/m^3$.

Determine:

(a) the velocity of the interface between clear liquid and suspension of concentration $100\,kg/m^3$;

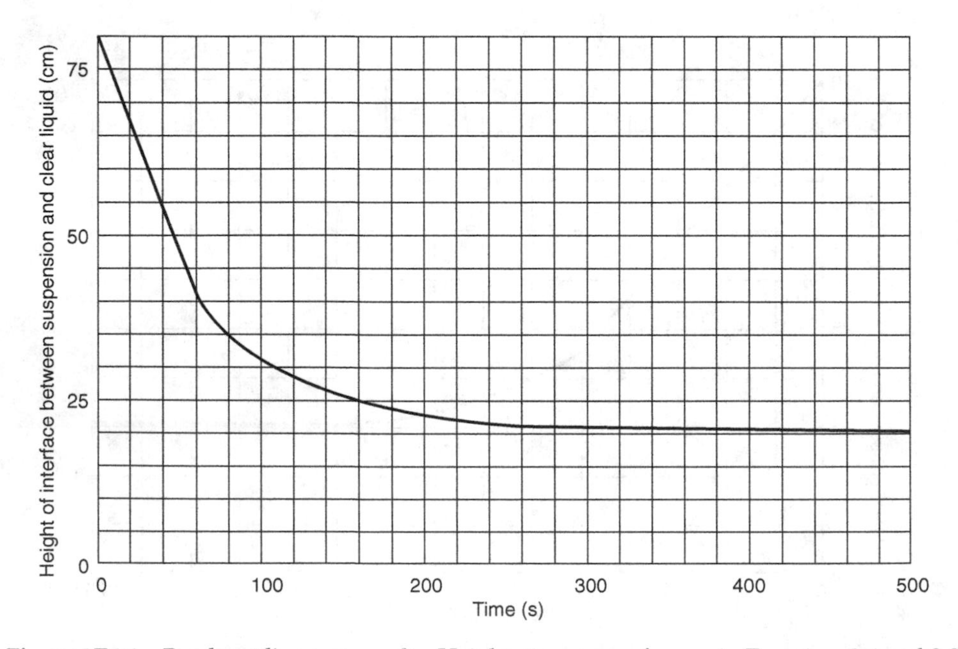

Figure 3E6.1 Batch settling test results. Height–time curve for use in Exercises 3.6 and 3.8

(b) the time from the start of the test at which the suspension of concentration 200 kg/m^3 is in contact with the clear liquid;

(c) the velocity of the interface between the clear liquid and suspension of concentration 200 kg/m^3;

(d) the velocity at which a layer of concentration 200 kg/m^3 propagates upwards from the base of the vessel;

(e) the concentration of the final sediment.

[Answer: (a) 0.667 cm/s downwards; (b) 140 s; (c) 0.0976 cm/s downwards; (d) 0.189 cm/s upwards; (e) 400 kg/m^3.]

3.7 A suspension in water of uniformly sized spheres of diameter 80 μm and density 1300 kg/m^3 has a solids volume fraction of 0.10. The suspension settles to a bed of solids volume fraction 0.4. (For water, density is 1000 kg/m^3 and viscosity is 0.001 Pa s.)

The single particle terminal velocity of the spheres under these conditions is 1.0 mm/s.

Calculate:

(a) the velocity at which the clear water/suspension interface settles;

(b) the velocity at which the sediment/suspension interface rises.

[Answer: (a) 0.613 mm/s; (b) 0.204 mm/s.]

3.8 A height–time curve for the sedimentation of a suspension in a vertical cylindrical vessel is shown in Figure 3E6.1. The initial solids concentration of the suspension is 125 kg/m^3.

Determine:

(a) the velocity of the interface between clear liquid and suspension of concentration 125 kg/m^3;

(b) the time from the start of the test at which the suspension of concentration 200 kg/m^3 is in contact with the clear liquid;

(c) the velocity of the interface between the clear liquid and suspension of concentration 200 kg/m^3;

(d) the velocity at which a layer of concentration 200 kg/m^3 propagates upwards from the base of the vessel;

(e) the concentration of the final sediment.

[Answer: (a) 0.667 cm/s downwards; (b) 80 s; (c) 0.192 cm/s downwards; (d) 0.438 cm/s upwards; (e) 500 kg/m^3.]

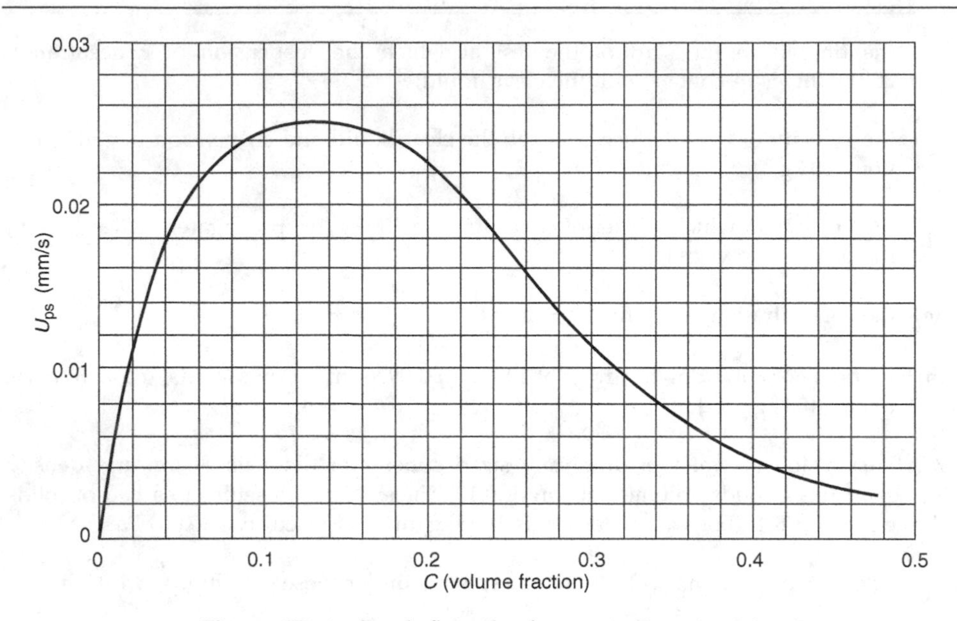

Figure 3E9.1 Batch flux plot for use in Exercise 3.9

3.9 Use the batch flux plot in Figure 3E9.1 to answer the following questions. (Note that the sediment concentration is 0.44 volume fraction.)

(a) Determine the range of initial suspension concentration over which a variable concentration zone is formed under batch settling conditions.

(b) For a batch settling test using a suspension with an initial concentration 0.18 volume fraction and initial height 50 cm, determine the settling velocity of the interface between clear liquid and suspension of concentration 0.18 volume fraction.

(c) Determine the position of this interface 20 min after the start of this test.

(d) Produce a sketch showing the concentration zones in the settling test 20 min after the start of this test.

[Answer: (a) 0.135 to 0.318; (b) 0.80 cm/ min; (c) 34 cm from base; (d) BE interface is 12.5 cm from base.]

3.10 Consider the batch flux plot shown in (Figure 3W3.1). Given that the final sediment concentration is 0.36 volume fraction:

(a) determine the range of initial suspension concentration over which a variable concentration zone is formed under batch settling conditions;

(b) calculate and sketch the concentration profile after 40 min of the batch settling test with an initial suspension concentration of 0.08 and an initial height of 100 cm;

(c) estimate the height of the final sediment and the time at which the test is complete.

[Answers: (a) 0.045 to 0.20; (c) 22.2 cm; 83 min.]

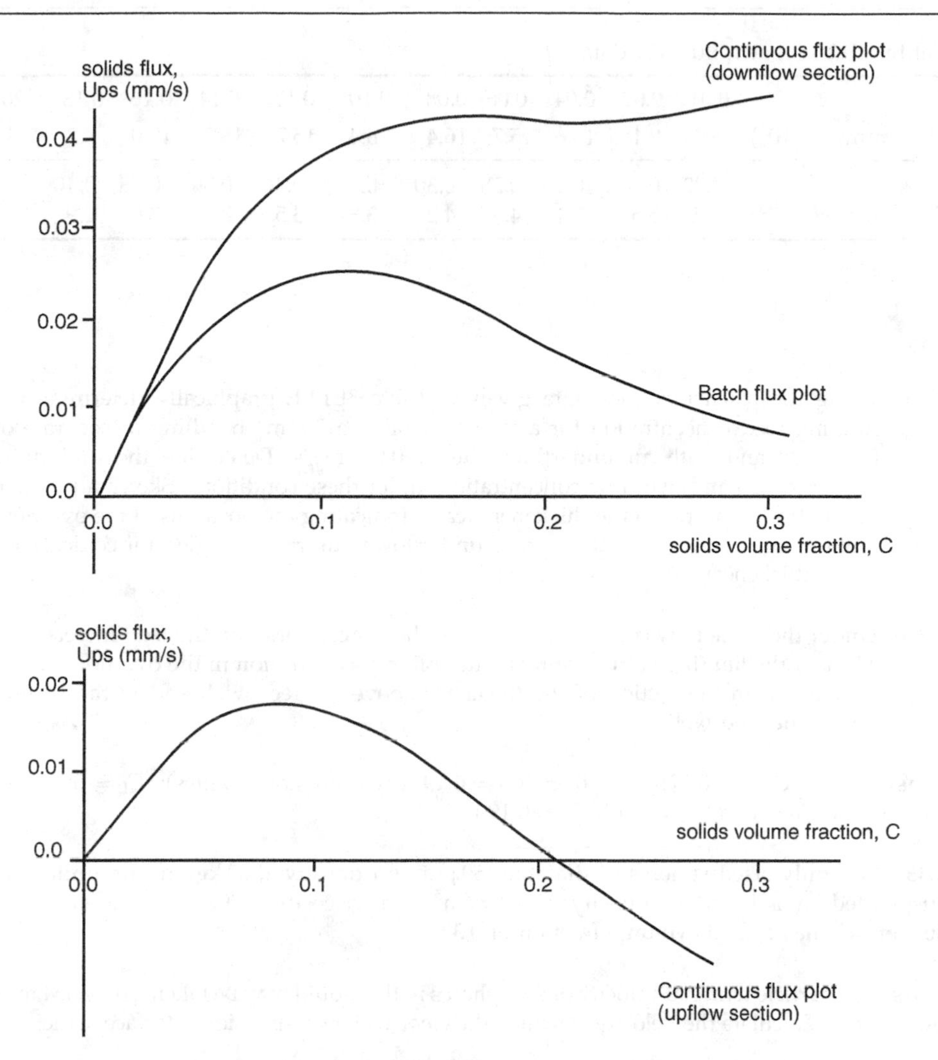

Figure 3E11.1 Total flux plot for use in Exercise 3.11

3.11 The batch and continuous flux plots supplied in Figure 3E11.1 are for a thickener of area 200 m^2 handling a feed rate of 0.04 m^3/s and an underflow rate of 0.025 m^3/s.

(a) Using these plots, graphically determine the critical or limiting feed concentration for this thickener.

(b) Given that if the feed concentration is 0.18 m^3/m^3, determine the solids concentrations in the overflow, underflow, in the regions above and below the feed well.

(c) Under the same flow rate conditions in the same thickener, the feed concentration increases to 0.24. Estimate the new solids concentration in the overflow and the underflow once steady state has been reached.

[Answer: (a) 0.21; (b) $C_V = 0$, $C_T = 0$, $C_L = 0.29$, $C_B = 0.087$; (c) $C_V = 0.08$, $C_L = 0.34$.]

Table 3E12.1 Batch flux test data

C	0.01	0.02	0.04	0.06	0.08	0.10	0.12	0.14	0.16	0.18	0.20
Flux, mm/s($\times 10^3$)	5.0	9.1	13.6	15.7	16.4	16.4	15.7	13.3	10.0	8.3	7.3
C	0.22	0.24	0.26	0.28	0.30	0.32	0.34	0.36	0.38	0.40	
Flux, mm/s($\times 10^3$)	7.7	5.6	5.1	4.5	4.2	3.8	3.5	3.3	3.0	2.9	

3.12

(a) Using the batch flux plot data given in Table 3E12.1, graphically determine the limiting feed concentration for a thickener of area 300 m^3 handling a feed rate of 0.03 m^3/s and with an underflow rate of 0.015 m^3/s. Determine the underflow concentration and overflow concentration under these conditions. Sketch a possible concentration profile in the thickener clearly indicating the positions of the overflow launder, the feed well and the point of underflow withdrawal (neglect the conical base of the thickener).

(b) Under the same flow conditions as above, the concentration in the feed increases to 110% of the limiting value. Estimate the solids concentration in the overflow, in the underflow, in the section of the thickener above the feed well and in the section below the feed well.

[Answer: (a) $C_{Fcrit} = 0.17$; $C_B = 0.05$, $C_B = 0.19$ (two possible values); $C_L = 0.34$; (b) $C_V = 0.034$; $C_L = 0.34$; $C_T = 0.19$; $C_B = 0.19$.]

3.13 Uniformly sized spheres of diameter 50 μm and density 1500 kg/m^3 are uniformly suspended in a liquid of density 1000 kg/m^3 and viscosity 0.002 Pa s. The resulting suspension has a solids volume fraction of 0.30.

The single particle terminal velocity of the spheres in this liquid may be taken as 0.00034 m/s ($Re_p < 0.3$). Calculate the velocity at which the clear water/suspension interface settles.

3.14 Calculate the settling velocity of glass spheres having a diameter of 155 μm in water at 293K. The slurry contains 60 wt % solids. The density of the glass spheres is 2467 kg/m^3.

How does the settling velocity change if the particles have a sphericity of 0.3 and an equivalent diameter of 155 μm?

3.15 Develop an expression to determine the time it takes for a particle settling in a liquid to reach 99% of its terminal velocity.

3.16 A suspension in water of uniformly sized spheres (diameter 150 μm and density 1140 kg/m^3) has a solids concentration of 25% by volume. The suspension settles to a bed of solids concentration 62% by volume. Calculate the rate at which the spheres settle in the suspension. Calculate the rate at which the settled bed height rises.

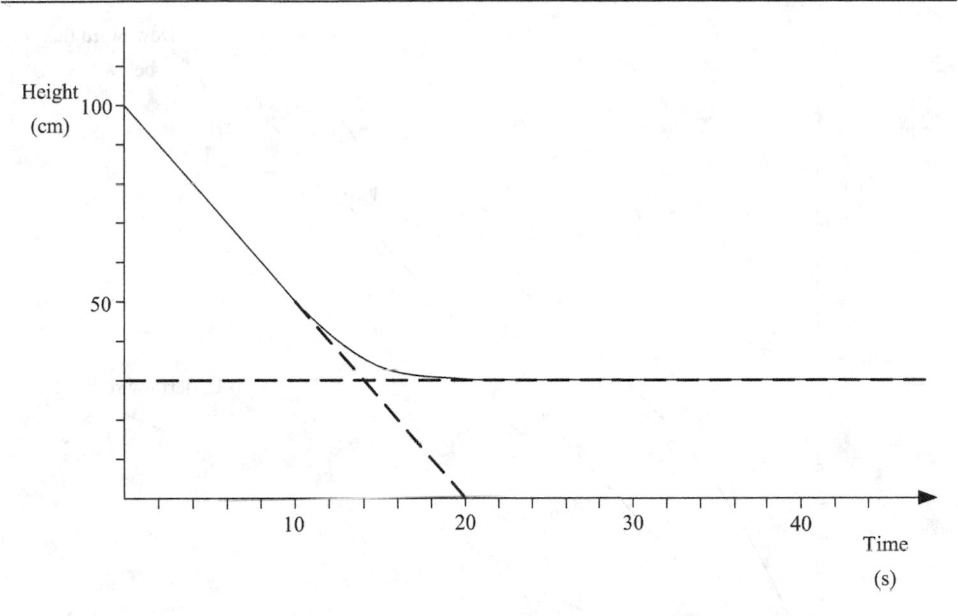

Figure 3E18.1 Plot of height of clear liquid interface versus time during settling test for use in Exercise 3.18

3.17 If 20 μm particles with a density of 2000 kg/m^3 are suspended in a liquid with a density of 900 kg/m^3 at a concentration of 50 kg/m^3, what is the solids volume fraction of the suspension? What is the bulk density of the suspension?

3.18 Given Figure 3E18.1 for the height–time curve for the sedimentation of a suspension in a vertical cylindrical vessel with an initial uniform solids concentration of 100 kg/m^3:

(a) What is the velocity at which the sediment/suspension interface rises?

(b) What is the velocity of the interface between the clear liquid and suspension of concentration 133 kg/m^3?

(c) What is the velocity at which a layer of concentration 133 kg/m^3 propagates upwards from the base of the vessel?

(d) At what time does the sediment/suspension interface start rising?

(e) At what time is the concentration of the suspension in contact with the clear liquid no longer 100 kg/m^3?

3.19 Given Figure 3E19.1 for the fluxes below the feed in a thickener of area 300 m^2 and a feed solids volume concentration of 0.1:

(a) What is the concentration of solids in the top section of the thickener?

(b) What is the concentration of solids in the bottom section of the thickener?

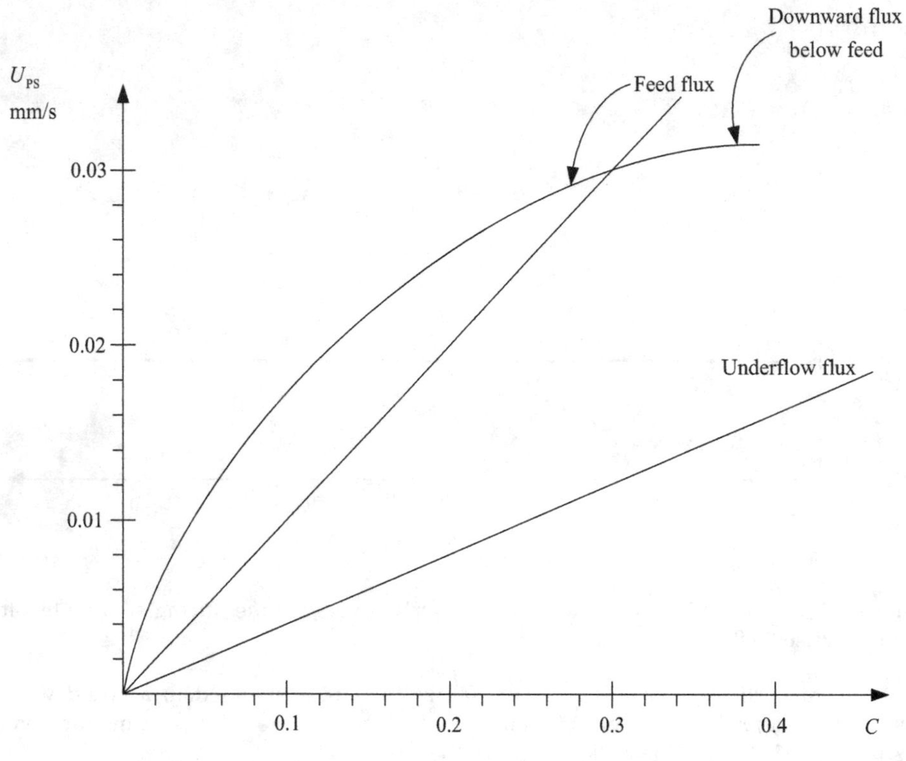

Figure 3E19.1 Total flux plot for use in Exercise 3.19

(c) What is the concentration of solids exiting the thickener?

(d) What is the flux due to bulk flow below the feed?

(e) What is the flux due to settling below the feed?

4

Slurry Transport

Jennifer Sinclair Curtis

4.1 INTRODUCTION

A slurry is a mixture of a liquid and solid particles. The term 'sludge' typically refers to a highly concentrated slurry containing fine particulate material. Each year, vast tonnages of slurries are pumped. Slurries are often used to transport coal, phosphates and minerals. Dredging of sand and silt in the maintenance of waterways is another example of solids handled in slurry form. In most slurries, the liquid phase is water. However, coal–oil and coal–methanol fuels are examples of slurries made up with liquids other than water.

The mixture density of a slurry, ρ_m, in terms of the volume fraction of solids C_v is given by:

$$\rho_m = C_v \rho_s + (1 - C_v)\rho_f \tag{4.1}$$

where ρ_f is the density of the liquid and ρ_s is the density of the solid particles. In terms of the weight or mass fraction of solids C_w, the mixture density is equivalently expressed as:

$$\frac{1}{\rho_m} = \frac{C_w}{\rho_s} + \frac{1 - C_w}{\rho_f} \tag{4.2}$$

4.2 FLOW CONDITION

In general, the flow behaviour of slurries or slurry transport can be classified into the following flow conditions:

Introduction to Particle Technology - 2nd Edition Martin Rhodes

(a) Homogeneous flow. In homogeneous flow, the particles are uniformly distributed over the pipeline cross-section. The particles settle very slowly and remain in suspension. The solids markedly influence the flow properties of the carrying liquid. Homogeneous flow is encountered in slurries of high concentration and finer particle sizes (typically less than 75 μm). For a given fine particle size, a liquid-particle mixture with higher solids concentration is more likely to result in a homogeneous suspension. This is due to the hindered settling effect (Chapter 3). Homogeneous slurries are also referred to as 'non-settling slurries'. Typical slurries that show minimal tendency to settle are sewage sludge, drilling muds, detergent slurries, fine coal slurries, paper pulp suspensions, and concentrated suspensions of fine limestone which are fed to cement kilns.

(b) Heterogeneous flow. In heterogeneous flow, there is a pronounced concentration gradient across the pipeline cross-section. Heterogeneous slurries range from fine particles fully suspended, but with significant concentration gradients, to rapidly settling large particles. Slurries in heterogeneous flow show a marked tendency to settle. In addition, the presence of the solid particles has a minimal effect on the flow properties of the carrying liquid. Examples include slurries consisting of coarse coal, potash or rocks.

(c) Saltation regime. In this regime of flow, particles begin to slide, roll and/or jump along the pipe bottom. This saltating layer of particles may accumulate and a moving bed of particles then develops along the pipe bottom. A liquid flow layer, separate from the solid particles, exists above the moving bed of particles.

The pressure drop required to pump slurries in pipelines is of prime concern to engineers. The pressure gradient required to pump a slurry of a given concentration for varying operating conditions is usually expressed in the form of graph. Such a graph is often referred to as the hydraulic characteristic of the slurry and displays a log–log plot of pressure drop versus superficial velocity, defined as volumetric flowrate divided by the cross-sectional area of the pipe. The hydraulic characteristic is similar to the Zenz diagram used to display the flow behaviour of gas–solid flows (Chapter 8). The hydraulic characteristic, shown in Figure 4.1, illustrates the modes of flow occurring in settling and non-settling slurries.

For settling slurries, at relatively high velocities the particles are carried in suspension and the flow behaviour approximates that of a homogeneous suspension. As long as the velocity is maintained above the 'standard velocity', the particle concentration gradient is minimized. As the superficial velocity is reduced below this transition or standard velocity, particle concentration gradients develop and the flow becomes heterogeneous.

The minimum point on the hydraulic characteristic curve for a settling slurry corresponds to the 'critical deposition velocity'. This is the flow velocity when particles begin to settle out. Good slurry transport design dictates that the pipe diameter and/or pump are selected so that the velocity in the pipeline over the

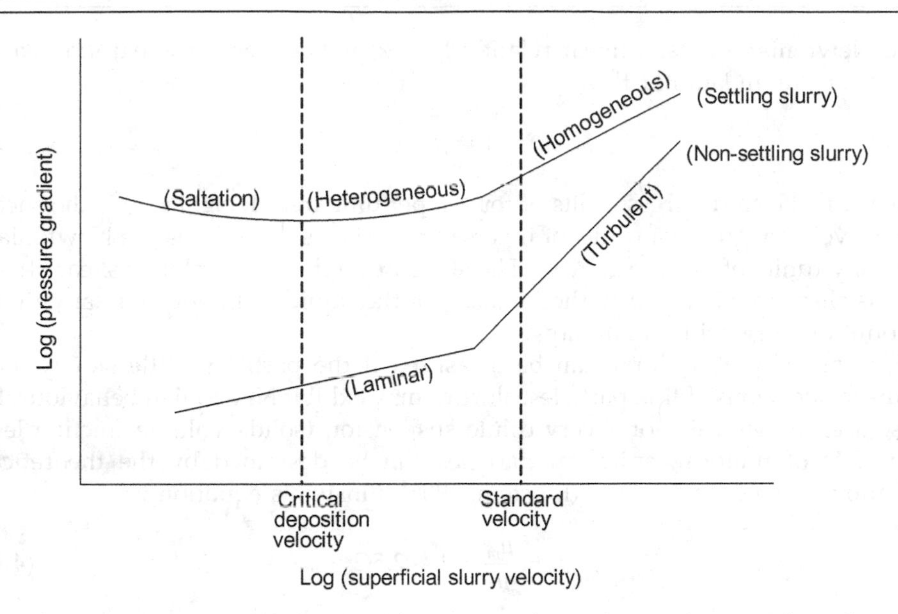

Figure 4.1 Examples of hydraulic characteristics for settling and non-settling slurries

range of operating conditions prevents the solid particles from settling out. Hence, the velocity is to be maintained just above the critical deposition velocity. As the velocity is decreased below the critical deposition velocity, a sliding, rolling or moving bed of deposited particles accumulates and this causes a steady increase in the pressure drop.

Non-settling, homogeneous slurries can be pumped through a pipeline either in laminar or turbulent flow. The hydraulic characteristic of a non-settling slurry typically exhibits a change of slope of the characteristic, reflecting the transition from laminar to turbulent flow as the superficial velocity is increased. Settling slurries typically do not show such an abrupt change in pressure drop with increasing velocity.

4.3 *RHEOLOGICAL MODELS FOR HOMOGENEOUS SLURRIES*

Although no slurry is ever perfectly homogeneous or non-settling, the homogeneous condition is a limiting form of flow behaviour that actual slurries can approach. The solid particles settle so slowly that continuum fluid models can be used to describe the flow behaviour of slurries. Like liquids, non-settling slurries may exhibit either Newtonian or non-Newtonian flow behaviour. At higher solids concentration, homogeneous slurry mixtures behave essentially as single-phase liquids with flow properties markedly different from that of the original liquid (before the solids are added). Viscometric techniques can generally be employed to analyse the flow properties of non-settling slurries in much the same way as these techniques are applied to single-phase liquids.

In Newtonian fluids, a linear relationship exists between the shear stress and the shear rate in laminar flow.

$$\tau = \mu\dot{\gamma} \tag{4.3}$$

where τ is the shear stress (units of force F per unit area A, e.g. Pa), $\dot{\gamma}$ is the shear rate or velocity gradient (units of inverse time t, e.g. s^{-1}) and μ is the Newtonian viscosity (units of Ft/A, e.g. Pa s). The slope of the line on a plot of shear stress versus shear rate is equal to the viscosity of the liquid μ. In addition, any finite amount of stress will initiate flow.

The viscosity of a slurry can be measured if the particles settle slowly. For dilute suspensions of fine particles, slurries may exhibit Newtonian behaviour. In this case, the viscosity of a very dilute suspension (solids volume fraction less than 2%) of uniform, spherical particles can be described by the theoretical equation derived by Einstein (Einstein, 1906). Einstein's equation is:

$$\mu_r = \frac{\mu_m}{\mu_f} = 1 + 2.5C_v \tag{4.4}$$

where μ_r is the relative viscosity or the ratio of the slurry viscosity μ_m to that of the single-phase fluid μ_f. When the particles are non-spherical or the particle concentration increases, the factor 2.5 typically increases. In fact, deviations from Equation (4.4) can be quite pronounced in these cases.

At high particle concentrations, slurries are often non-Newtonian. For non-Newtonian fluids, the relationship between the shear stress and shear rate, which describes the rheology of the slurry, is not linear and/or a certain minimum stress is required before flow begins. The power-law, Bingham plastic and Herschel–Bulkley models are various models used to describe the flow behaviour of slurries in which these other types of relationships between the shear stress and shear rate exist. Although less common, some slurries also display time-dependent flow behaviour. In these cases, the shear stress can decrease with time when the shear rate is maintained constant (thixotropic fluid) or can increase with time when the shear rate is maintained constant (rheopectic fluid). Milk is an example of a non-settling slurry which behaves as a thixotropic liquid.

4.3.1 Non-Newtonian Power-law Models

In power-law fluids, the relationship between the shear stress and shear rate is nonlinear and a finite amount of stress will initiate flow. The mathematical model (sometimes called the Ostwald-de-Waele equation)

$$\tau = k\dot{\gamma}^n \tag{4.5}$$

describes the relationship between the shear stress and shear rate where k and n are constants. Note that the units of the consistency index k ($N\,s^n/m^2$ in SI units) depend on the value of the dimensionless flow behaviour index n, which indicates the amount of deviation from Newtonian behaviour. For Newtonian fluids, $n = 1$, and k is equal to the Newtonian viscosity.

In power-law fluids, an 'apparent viscosity', μ_{app}, is defined in a similar manner to a Newtonian fluid,

$$\mu_{app} = \frac{\tau}{\dot{\gamma}} = k\dot{\gamma}^{n-1} \tag{4.6}$$

The apparent viscosity, μ_{app}, is equal to the slope of a line from the origin to a point on the shear stress–shear rate curve; it decreases or increases as the shear rate increases. Hence, the term 'viscosity' for a non-Newtonian fluid has no meaning unless the shear rate is specified. In shear-thinning (or pseudoplastic) slurries, the apparent viscosity decreases as the shear rate increases and the value for n is less than one. In shear-thickening (or dilatant) slurries the apparent viscosity increases as the shear rate increases and the value of n is greater than 1 (Figure 4.2).

In shear-thinning slurries, since the apparent viscosity decreases as the shear rate increases, the velocity profile in a pipe becomes increasingly blunt, tending towards a plug flow profile as n decreases. In shear-thickening slurries, the velocity profile becomes more pointed, with larger velocity gradients throughout the flow domain. For extreme dilatant fluids ($n \to \infty$), the velocity varies almost linearly with the radial position in the pipe. This is illustrated in Figure 4.3.

Most slurries are shear-thinning. It is hypothesized that this shear-thinning behaviour is due to the formation of particulate aggregates which provide a lower resistance to flow than fully dispersed particles.

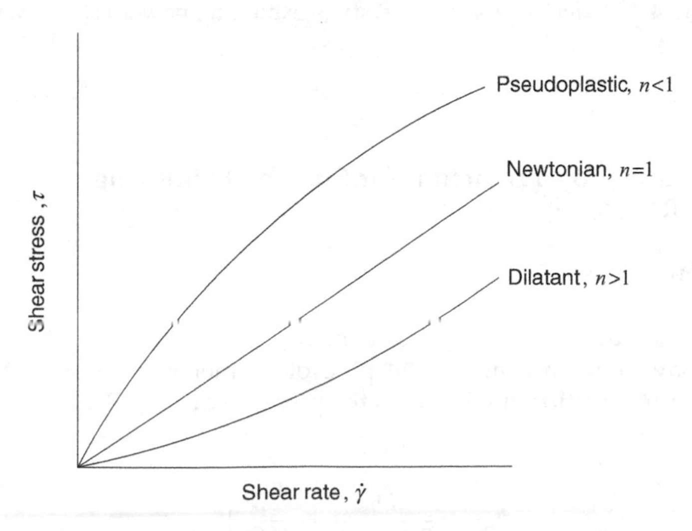

Figure 4.2 Power-law models of non-Newtonian fluids

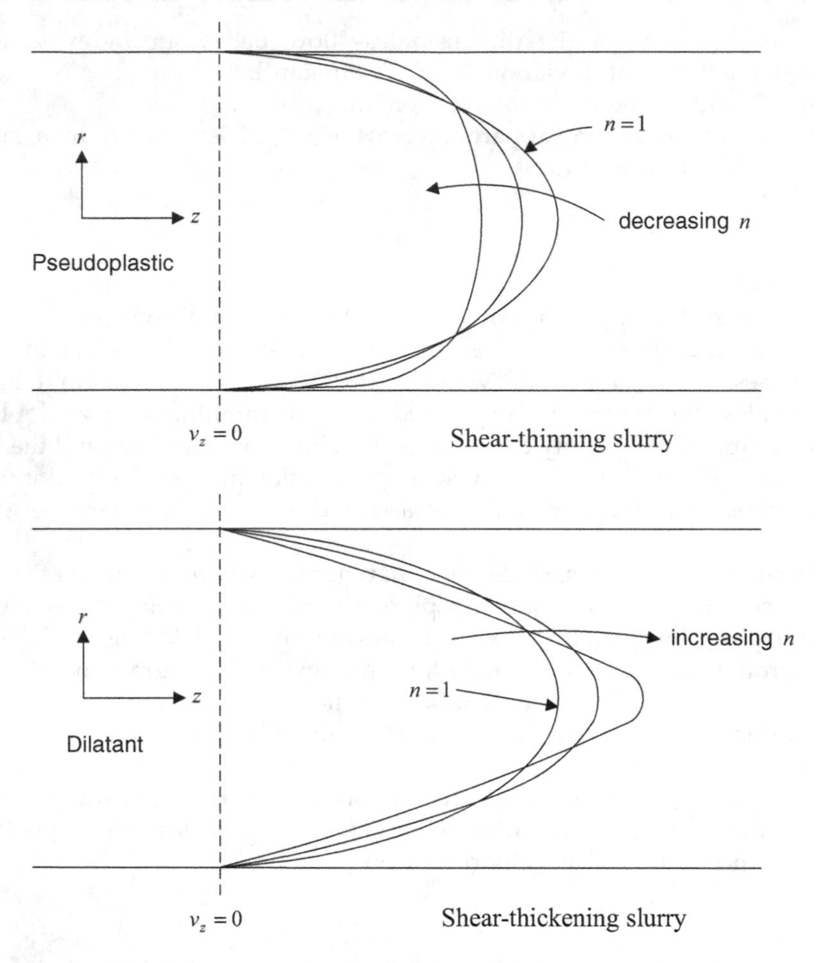

Figure 4.3 Velocity profiles for slurries exhibiting power-law rheology

4.3.2 Pressure Drop Prediction for Slurries Exhibiting Power-law Rheology

Laminar flow

The fluid momentum balance applied to the case of laminar, fully developed flow of a power-law fluid in a horizontal pipe of diameter D yields the following expression for the relationship between the pressure drop, $\Delta P/L$, and the average flow velocity, v_{AV}:

$$v_{AV} = \frac{Dn}{2(3n+1)}\left(\frac{D\Delta P}{4kL}\right)^{1/n} \quad (4.7)$$

or rearranged equivalently as

$$\frac{\Delta P}{L} = \frac{4k}{D}\left[\frac{2v_{\mathrm{AV}}(3n+1)}{Dn}\right]^n \tag{4.8}$$

Note that for $n = 1$ and $k = \mu$, Equations (4.7) and (4.8) reduce to the familiar Hagen–Poiseuille equation which describes the pressure drop–velocity relationship for the laminar flow of a Newtonian fluid.

$$\frac{\Delta P}{L} = \frac{32\mu v_{\mathrm{AV}}}{D^2} \tag{4.9}$$

Friction factor method – Laminar and Turbulent flow

The friction factor method can also be used to determine the pressure drop for homogeneous slurries. It is applied in the same way as for Newtonian fluids but the friction factor depends on a new definition for the Reynolds number.

Since the friction factor in a horizontal pipeline is related to the friction head loss, h_f (units of length), by

$$h_f = 2f_f\left(\frac{L}{D}\right)\frac{v_{\mathrm{AV}}^2}{g} \tag{4.10}$$

and the pressure drop in a *horizontal* pipeline is related to the friction head loss by the modified Bernoulli equation,

$$h_f = \frac{\Delta P}{\rho_m g} \tag{4.11}$$

then the relationship between pressure drop and friction factor is as follows:

$$\Delta P = 2f_f\rho_m\left(\frac{L}{D}\right)v_{\mathrm{AV}}^2 \tag{4.12}$$

In these equations, f_f is the Fanning friction factor

For a more general case of vertical flow in a pipeline network with varying pipe cross-sectional area and head losses and gains due to pumps (positive head), turbines (negative head), valves and fittings (negative head), the modified Bernoulli equation gives:

$$\sum H - \sum h_f = -\frac{\Delta P}{\rho_m g} + \Delta z + \Delta\left(\frac{v_{\mathrm{AV}}^2}{2g}\right) \tag{4.13}$$

where $\Delta z = z_2 - z_1$ is the vertical height difference between the exit (2) and the entrance (1) and $\sum H$ is the sum of all head (units of length) owing to pumps, turbines, valves or fittings in the pipeline network.

The generalized Reynolds number is defined in terms of an effective viscosity, μ_e

$$Re^* = \frac{\rho_m D v_{AV}}{\mu_e} \tag{4.14}$$

In order to define the effective viscosity for a power-law fluid, the expression for the viscosity of a Newtonian fluid is extended to a non-Newtonian fluid. For a Newtonian fluid, rearranging the Hagen–Poiseuille equation gives

$$\mu = \frac{D\Delta P}{4L} \bigg/ \frac{8v_{AV}}{D} = \tau_0 \bigg/ \frac{8v_{AV}}{D} \tag{4.15}$$

where τ_o is the shear stress at the pipe wall. The power-law effective viscosity, μ_e, is defined in a similar manner,

$$\mu_e = \tau_o \bigg/ \frac{8v_{AV}}{D} \tag{4.16}$$

and, the wall shear stress for a power-law fluid is given by rearranging Equation (4.8):

$$\tau_0 = \frac{\Delta PD}{4L} = k\left[\frac{2v_{AV}(3n+1)}{Dn}\right]^n = \frac{k(3n+1)^n}{(4n)^n}\left(\frac{8v_{AV}}{D}\right)^n \tag{4.17}$$

Thus, combining Equations (4.16) and (4.17), the power-law effective viscosity is given by:

$$\mu_e = \frac{k(3n+1)^n}{(4n)^n}\left(\frac{8v_{AV}}{D}\right)^{n-1} \tag{4.18}$$

and the generalized Reynolds number for a power-law fluid is:

$$Re^* = \frac{8\rho_m D^n v_{AV}^{2-n}}{k}\left[\frac{n}{(6n+2)}\right]^n \tag{4.19}$$

For slurries exhibiting power-law fluid rheology, the transition velocity from laminar to turbulent flow is governed by the flow behaviour index n of the slurry. The equation proposed by Hanks and Ricks (1974) gives an estimate of this transition velocity in terms of the generalized Reynolds number Re^*.

$$Re^*_{\text{transition}} = \frac{6464n}{(1+3n)^n}(2+n)^{\left(\frac{2+n}{1+n}\right)}\left(\frac{1}{1+3n}\right)^{2-n} \tag{4.20}$$

For Newtonian fluids and $n = 1$, the generalized Reynolds Re^* is 2100 at the transition between laminar and turbulent flow.

For laminar flow,

$$f_f = \frac{16}{Re^*}$$

For turbulent flows, Dodge and Metzner (1959) developed the following equation for the power-law fluids in smooth pipes based on a semitheoretical analysis:

$$\frac{1}{\sqrt{f_f}} = \frac{4}{n^{0.75}} \log\left(Re^* \sqrt{f_f^{2-n}}\right) - \frac{0.4}{\sqrt{n}} \quad (4.21)$$

In turbulent flow, the Fanning friction factor f_f for the slurry described by a power-law fluid model depends on both the generalized Reynolds number Re^* and the flow behaviour index n.

4.3.3 Non-Newtonian Yield Stress Models

Some slurries require a minimum stress, τ_y, before flow initiates. This minimum stress is known as the yield stress for the slurry. For example, freshly poured concrete does not flow along an inclined surface until a specific angle relative to the horizontal is reached. Other examples of slurries which exhibit a yield stress include paints and printing inks. In these cases, there is a critical film thickness below which these slurries will not flow under the action of gravity.

The behaviour of slurries which exhibit a yield stress can be represented by a model in which the relationship between the effective stress $\tau - \tau_y$ and the shear rate is either linear, as in Newtonian fluids (Bingham plastic model), or follows a power-law, as in pseudoplastic or dilatant fluids (Herschel–Bulkley model or yield power-law model). The shear stress–shear rate relationship for these models is shown in Figure 4.4.

In the Bingham plastic model, the yield stress τ_y and the plastic viscosity μ_p (the slope of the line on the shear stress–shear rate plot in Figure 4.4) characterize the slurry.

$$\tau = \tau_y + \mu_p \dot{\gamma} \quad (4.22)$$

Since

$$\mu_p = \frac{\tau - \tau_y}{\dot{\gamma}} \quad (4.23)$$

the apparent viscosity of a Bingham fluid is

$$\mu_{app} = \frac{\tau}{\dot{\gamma}} = \mu_p + \frac{\tau_y}{\dot{\gamma}} \quad (4.24)$$

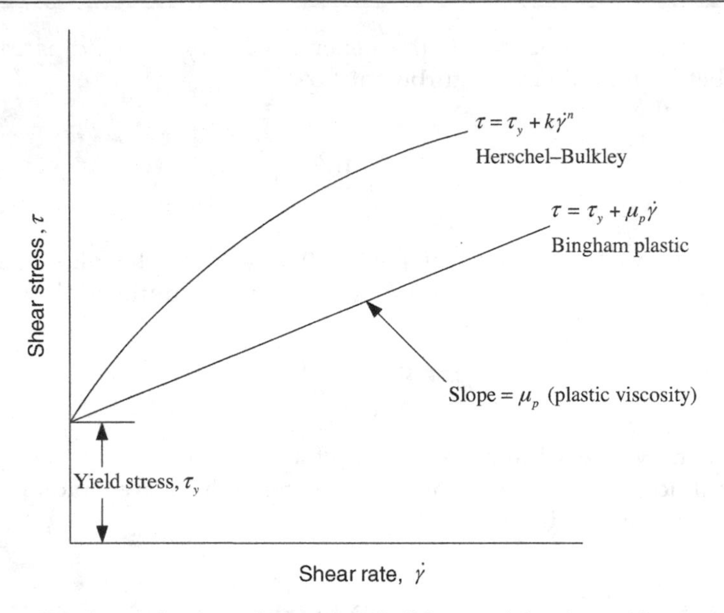

Figure 4.4 Bingham plastic and Herschel–Bulkley models of non-Newtonian fluids

In the Herschel–Bulkley model, the yield stress, consistency index k, and the flow behaviour index n characterize the slurry.

$$\tau = \tau_y + k\dot{\gamma}^n \tag{4.25}$$

The apparent viscosity is

$$\mu_{\text{app}} = \frac{\tau_y}{\dot{\gamma}} + k\dot{\gamma}^{n-1} \tag{4.26}$$

Hence, the Bingham plastic model is a special case of the Hershel–Bulkley model when $n = 1$. Slurries following the Bingham plastic model will be discussed in more detail in the next section.

In pipe flow, the shape of the velocity profile for slurries exhibiting yield stress is more complex than that for slurries exhibiting power-law rheology (Figure 4.5). A plug flow region, where $dv_z/dr = 0$, exists between $r = 0$ and $r = R^*$, where $0 < R^* < R$. In this central region of the pipe, the shear stress τ_{rz} is lower than the yield stress value τ_y. Outside of this central region, where $R^* < r < R$, the shear stress τ_{rz} is greater than the yield stress τ_y, $\tau_{rz} > \tau_y$. At the wall, the wall shear stress $\tau_{rz}|_{r=R}$ is given by the same relationship as Newtonian fluids.

$$\tau_{rz}|_{r=R} = \tau_0 = \frac{\Delta PR}{2L} \tag{4.27}$$

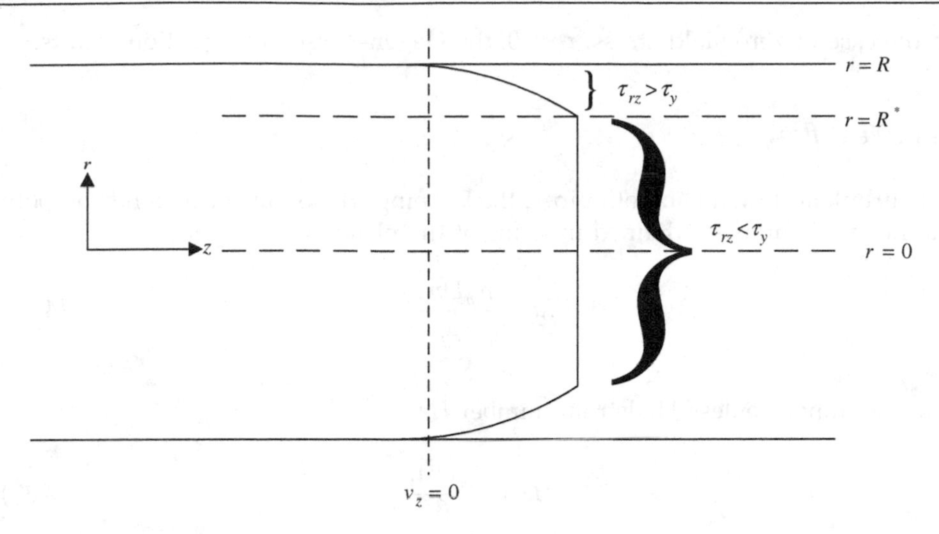

Figure 4.5 Velocity profile for slurries with a yield stress

4.3.4 Pressure Drop Prediction for Slurries Exhibiting Bingham Plastic Rheology

Laminar flow

As with slurries following a power-law flow model, it is necessary to reliably predict the pressure drop in a horizontal pipe of diameter D under laminar, fully developed flow conditions. A fundamental analysis of the Bingham plastic model yields the following expression for the mean velocity in terms of the yield stress τ_y and the wall shear stress τ_0.

$$v_{AV} = \frac{R\tau_0}{4\mu_p}\left[1 - \frac{4}{3}\frac{\tau_y}{\tau_0} + \frac{1}{3}\left(\frac{\tau_y}{\tau_0}\right)^4\right] \tag{4.28}$$

If the ratio τ_y/τ_0 is small, as it often is, the mean velocity for the Bingham slurry can be approximated by the linear expression

$$v_{AV} \approx \frac{D}{8\mu_p}\left(\tau_0 - \frac{4}{3}\tau_y\right) \tag{4.29}$$

By expressing τ_0 in terms of ΔP and rearranging, the relationship between the pressure drop and the average velocity can be obtained

$$\frac{\Delta P}{L} = \frac{32\mu_p v_{AV}}{D^2} + \frac{16\tau_y}{3D} \tag{4.30}$$

In the case of zero yield stress, $\tau_y = 0$, the Hagen–Poiseuille equation results.

Turbulent flow

For turbulent flows in smooth pipes, the Fanning friction factor depends on both the Reynolds number, defined in terms of the plastic viscosity μ_p,

$$Re = \frac{\rho_m D v_{AV}}{\mu_p} \tag{4.31}$$

and the dimensionless Hedstrom number He

$$He = \frac{\rho_m D^2 \tau_y}{\mu_p^2} \tag{4.32}$$

The Hedstrom number is the product of the Reynolds number and the ratio of the internal strain property of the fluid (τ_y/μ_p) to the shear strain conditions prevailing in the pipe v_{AV}/D.

Hence, estimation of the pressure drop in turbulent, horizontal slurry flow in a pipe, based on the method developed by Hedstrom, involves using the friction factor chart given in Figure 4.6 and Equation (4.12).

For a Bingham plastic slurry, the transition from laminar to turbulent flow depends on the Hedstrom number. The critical Reynolds number, which allows

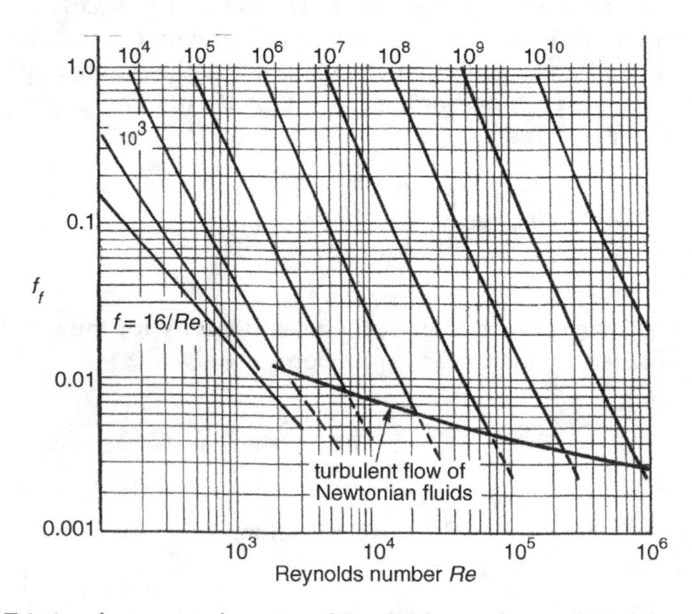

Figure 4.6 Friction factor as a function of Reynolds number and Hedstrom number for slurries exhibiting Bingham plastic rheology

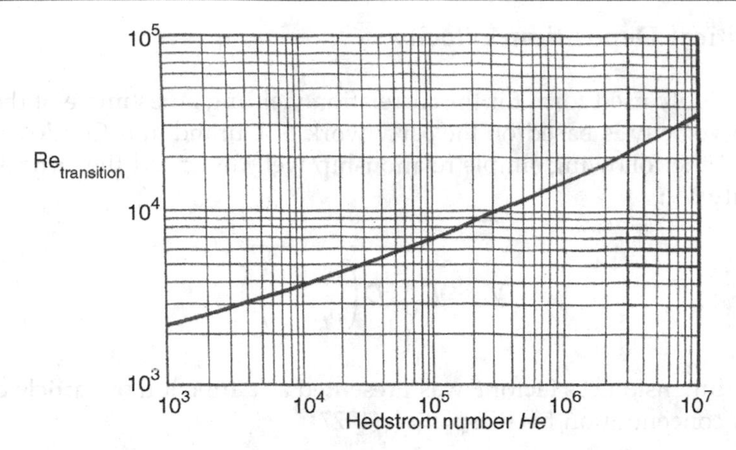

Figure 4.7 Transition Reynolds number as a function of Hedstrom number for slurries exhibiting Bingham plastic rheology

the calculation of the average velocity of the slurry at transition, is estimated by an empirical relationship given in Figure 4.7.

4.4 *HETEROGENEOUS SLURRIES*

When slurries exhibit non-homogeneous flow behaviour, prediction of the critical deposition velocity is an essential component of pipeline design. If a pipeline is operated at or below the critical deposition velocity, pipeline blockage can result as particles settle and form a stationary bed on the bottom of a horizontal pipe. In addition to maintaining the operating velocity greater than the deposition velocity, pipeline design requires prediction of the pressure gradient. Since both of these – the critical deposition velocity and the pressure drop – depend on a host of variables including particle size, pipe diameter, particle concentration, particle density, etc., the deposition velocity and pressure drop are estimated via correlations that summarize the results of experimental investigations. Because many of these correlations are based on experiments that do not report all relevant variables, the values obtained by these correlations should be regarded as estimates and one should not expect their predictive ability to be much better than ±20%. In addition, caution should be employed in the use of any correlation due to limitations in the database on which the correlation was derived. In the case of heterogeneous slurry flow, the majority of the published experimental data are for pipe sizes less than 150 mm (significantly smaller than modern industrial practice), fairly monodispersed particle mixtures, and for water-based slurries at ambient temperature. Hence, for other flow situations, pilot plant test results or prior experience with similar slurries are used to estimate the deposition velocity and pressure drop.

4.4.1 Critical Deposition Velocity

The most widely used form for the correlation yielding an estimate of the critical deposition velocity is based on the early work of Durand and Condolios (1954). They found the following simple relationship well described the critical deposition velocity V_C:

$$V_C = F\sqrt{gD\left(\frac{\rho_s}{\rho_f} - 1\right)}$$

where the dimensionless factor F was presented as a function of particle diameter and solids concentration by Wasp *et al.* (1977)

$$F = 1.87C_v^{0.186}\left(\frac{x}{D}\right)^{1/6}$$

More recently, Gillies *et al.* (2000) have shown that F is best represented in terms of the particle Archimedes number Ar $[Ar = \frac{4}{3}x^3\rho_f(\rho_s - \rho_f)g/\mu_f^2]$

$$80 < Ar < 160 \qquad F = 0.197\,Ar^{0.4}$$

$$160 < Ar < 540 \qquad F = 1.19\,Ar^{0.045}$$

$$Ar > 540 \qquad F = 1.78\,Ar^{-0.019}$$

It should be emphasized that the above relationships between F and the particle Archimedes number do not include the influence of variables such as particle shape or particle concentration so the correlation yields only an approximate value for the deposition velocity V_C.

For Archimedes number less than 80, the correlation of Wilson and Judge (1976) should be used.

$$Ar < 80 \quad F = \sqrt{2}\left[2 + 0.3\log_{10}\left(\frac{x}{DC_D}\right)\right]$$

$$1 \times 10^{-5} < \frac{x}{DC_D} < 1 \times 10^{-3}$$

4.5 *COMPONENTS OF A SLURRY FLOW SYSTEM*

The main elements of a slurry transport system are shown in Figure 4.8.

4.5.1 Slurry Preparation

Initially the slurry must be prepared by physical and/or chemical processing in order to achieve the proper slurry characteristics for effective transport. Slurry

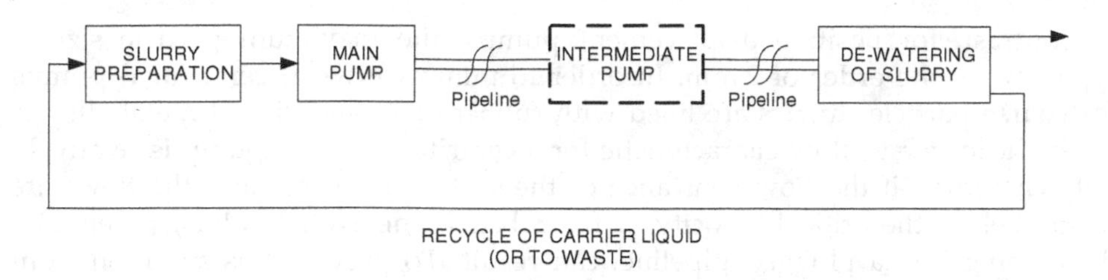

Figure 4.8 Components of a slurry conveying system

preparation involves slurrification, or addition of the liquid phase to the dry solids. In addition, chemical treatment for corrosion inhibition, or modification of slurry rheology by particle size reduction, are often aspects of slurry preparation.

In the grinding process, the desired particle size is small enough such that the generated slurry is homogeneous and easily transported, but not too small that the slurry is difficult to dewater. If the particle size is too coarse, a heterogeneous slurry will result necessitating higher pumping velocities and, consequently, higher energy costs. In addition, higher pipeline wear rates are associated with larger particles. Size reduction of larger bulk solids is typically accomplished by crushing or grinding, such as with jaw crushers (Chapter 12), until the particle size is about 2 mm. Further particle size reduction is then performed in rod mills or ball mills (Chapter 12). If significant size reduction of the solid is required, the cost of slurry preparation can be one of the two largest components (rivaling the pumping cost) of the total cost of slurry transport.

The typical practice is to add the liquid phase to the solids in agitated tanks, forming a slurry with a slightly higher concentration than the ultimate pipeline concentration. Final adjustments to the slurry concentration are then made by the addition of more liquid as the slurry enters the pipeline.

4.5.2 Pumps

Several types of pumps are useful for handling slurries. The selection of a pump for a specific slurry transport line is based on the discharge pressure requirement and the particle characteristics (particle size and abrasivity). The pumps that are used are either positive displacement pumps or centrifugal (rotodynamic) pumps.

For discharge pressures under approximately 45 bar, centrifugal pumps offer an economic advantage over positive displacement pumps. Due to the lower working pressure, the application of centrifugal pumps is generally restricted to shorter distances; they are typically used for in-plant transportation of slurries. The efficiency of a centrifugal pump is low due to the robust nature of the impeller design; the impellers and casings have wide flow passages. Efficiencies of 65% are common for centrifugal pumps compared with efficiencies of 85–90% for positive displacement pumps. The wide flow passages, however, enable the transport of very large particles, even up to 150 mm in size, in centrifugal pumps.

In contrast, for positive displacement pumps, the maximum particle size is typically on the order of 2 mm. In order to minimize wear, centrifugal pumps for coarse-particle slurries are lined with rubber or wear-resistant metal alloys.

The head versus flow characteristic for a centrifugal slurry pump is relatively flat. Therefore, if the flow resistance of the system increases and the flow rate drops below the critical velocity, a fixed bed of deposited solids, potentially developing into a plugged pipeline, can result. To prevent this situation from occurring, most centrifugal pumps have variable speed drives to maintain the flow rate.

For slurry transport systems requiring discharge pressures greater than 45 bar, only positive displacement or reciprocating pumps are technically feasible. These pumps fall into two main categories: plunger type and piston type. The choice of which type to employ depends upon the abrasivity of the slurry. The plunger pump and the piston pump are similar in construction in that both have a plunger or a piston that is being caused to pass back and forth in a chamber. In plunger pumps, the plunger reciprocates through packing, displacing liquid through cylinders in which there is considerable radial clearance (Figure 4.9). The two valves alternate open and closed as the plunger moves up and down. Plunger-type pumps are always 'single-acting' in that only one end of the plunger is used to drive the fluid. For slurry applications, the plunger is continuously flushed with clear liquid during the suction stroke to greatly reduce internal wear.

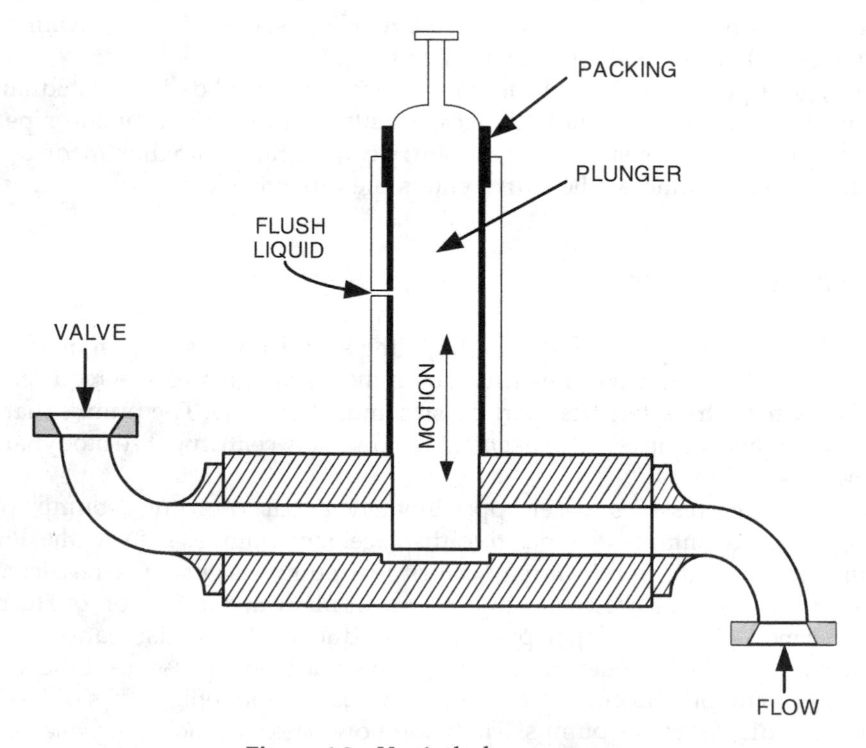

Figure 4.9 Vertical plunger pump

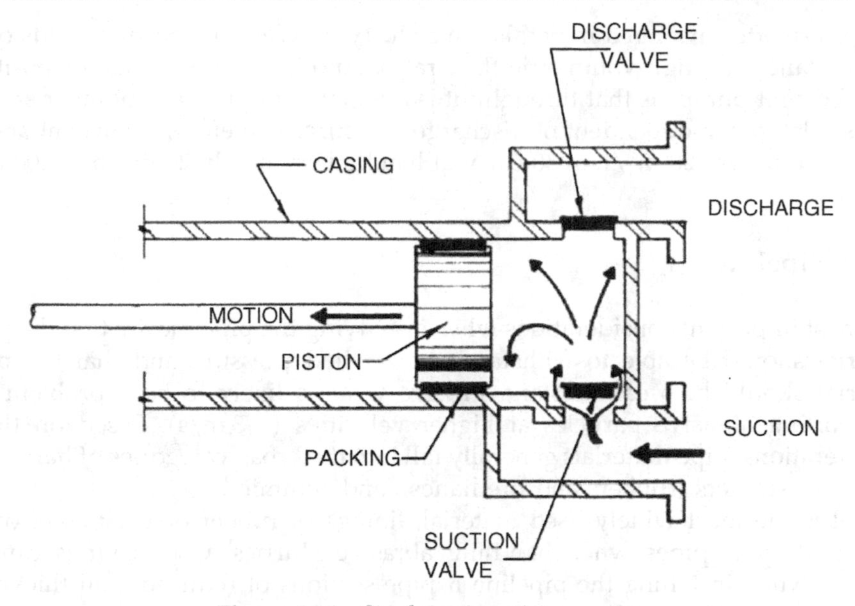

Figure 4.10 Single-acting piston pump

Piston pumps may be either single- or double-acting (Figures 4.10 and 4.11). In double-acting piston pumps, both sides of the piston are used to move the fluid. In piston pumps, as in plunger pumps, the valves (two for single-acting and four for double-acting) alternate open and closed as the piston moves back and forth. In double-acting piston pumps, both suction and discharge are accomplished with the movement of the piston in a single direction. Since the throughput of positive displacement pumps is much lower than centrifugal pumps, these

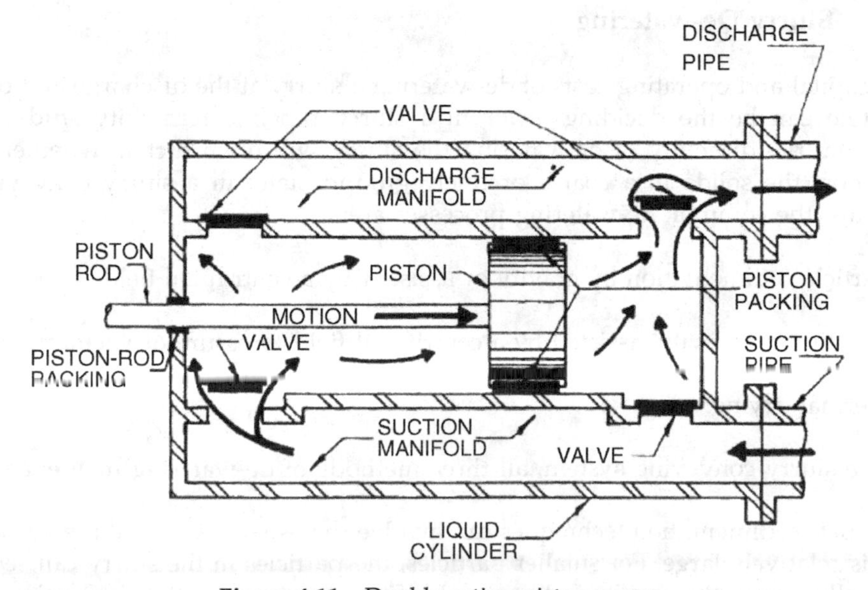

Figure 4.11 Double-acting piston pump

pumps are often arranged in parallel in a slurry line for transporting solids over long distances at high volumetric flow rates. Another characteristic of positive displacement pumps is that throughput is a function of piston or plunger speed, and is relatively independent of discharge pressure. Therefore, a constant-speed pump that moves 20 m^3/h at 30 bar will handle very nearly 20 m^3/h at 200 bar.

4.5.3 Pipeline

The most important considerations when specifying the pipeline are that the pipe material should be able to withstand the applied pressure and that the pipe material should be wear-resistant. Erosive wear is likely to be a problem for transporting abrasive particles at higher velocities ($> 3\,m/s$). Based on these considerations, pipe materials generally fall into the broad categories of hardened metals, elastomers (rubbers and urethanes), and ceramics.

Steel is the most widely used material; linings of rubber or plastic are often used with steel pipes when handling abrasive slurries. Cost savings can be realized when installing the pipeline if pipe sections of reduced wall thickness are used where the pipeline pressure is lower.

Rubber and urethane tend to wear better than metals. However, high temperatures or the presence of oils or chemicals may render this option not feasible. Ceramics are the most wear resistant materials, but they are low in toughness and impact strength. In addition, ceramic pipelines are typically the most costly. For this reason, ceramics are often used as liners, particularly in localized areas of high wear, such as pipe bends or in centrifugal pumps.

4.5.4 Slurry De-watering

The capital and operating costs of de-watering a slurry at the discharge end of a pipeline can be the deciding factor in a slurry pipeline feasibility study. In addition, the difficulty of de-watering the slurry will often dictate whether to transport the solids in a coarse or finely ground state. In a slurry conveying systems, the common de-watering processes are:

1. particle sedimentation by gravity or assisted by a centrifugal field;

2. filtration by gravity, assisted by a centrifugal field, pressure or vacuum;

3. thermal drying.

In one slurry conveying system, all three methods of de-watering may even be used.

Particle sedimentation techniques can involve the use of a screen if the particle size is relatively large. For smaller particles, the particles in the slurry can settle naturally due to the gravity in large tanks. For continuous settling operations, a thickener (Chapter 3) is employed. Solids settle into the conical bottom and are directed to a central outlet using a series of revolving rakes. Clear liquid is discharged from the top of the thickener.

Hydrocyclones are also used for liquid–solid separation. Hydrocyclones are similar in design and operation to gas cyclones; the slurry is fed tangentially to the hydrocyclone under pressure. The resulting swirling action subjects particles to high centrifugal force. The overflow of the hydrocyclone will carry predominantly clear liquid and the underflow will contain the remaining liquid and the solids. Some small fraction of the particles will be discharged in the overflow; this fraction will depend on the particle size range in the slurry and the cut size of the hydrocyclone.

4.6 *FURTHER READING*

For further reading on slurry flow, the reader is referred to the following:

Brown, N.P. and N.I. Heywood (1991), *Slurry Handling Design of Solid-Liquid Systems,* Elsevier Applied Science, London.

Shook, C.A. and M.C. Roco (1991), *Slurry Flow: Principles and Practice,* Butterworth-Heinemann, Boston.

Wilson, K.C., Addie, G.R. and R. Clift (1992), *Slurry Transport Using Centrifugal Pumps,* Elsevier Applied Science, London.

4.7 *WORKED EXAMPLES*

WORKED EXAMPLE 4.1

A slurry exhibiting power-law flow behaviour is flowing through 15 m of pipe having an inside diameter of 5 cm at an average velocity of 0.07 m/s. The density of the slurry is $1050\ \mathrm{kg/m^3}$ and its flow index and consistency index are $n = 0.4$ and $k = 13.4\ \mathrm{N\,s^{0.4}/m^2}$, respectively. Calculate the pressure drop in the pipeline.

Solution

First we check to see if the flow is laminar or turbulent by calculating the generalized Reynolds number Re^* for the specified flow conditions. We then compare this Reynolds number to the generalized Reynolds number at which the transition from laminar to turbulent flow occurs.

From Equation (4.20)

$$Re^*_{\mathrm{transition}} = \frac{6464 \times 0.4}{[1+3(0.4)]^{0.4}}(2+0.4)^{\left(\frac{2+0.4}{1+0.4}\right)}\left[\frac{1}{1+3(0.4)}\right]^{2-0.4}$$

$$Re^*_{\mathrm{transition}} = 2400$$

The generalized Re^* for the flow conditions in the pipe is found from Equation (4.19),

$$Re^* = \frac{8 \times 1050\,\text{kg/m}^3 \times (0.05\,\text{m})^{0.4} \times (0.07\,\text{m/s})^{1.6}}{13.4\text{N}\,\text{s}^{0.4}/\text{m}^2} \left(\frac{0.4}{4.4}\right)^{0.4}$$

$$Re^* = 1.03 < 2400 \Rightarrow \text{flow is laminar}$$

The pressure drop can then be calculated using Equation (4.8).

$$\frac{\Delta P}{15\,\text{m}} = \frac{4 \times 13.4\,\text{N}\,\text{s}^{0.4}/\text{m}^2}{0.05\,\text{m}} \left[\frac{2 \times 0.07\,\text{m/s} \times 2.2}{(0.05\,\text{m} \times 0.4)}\right]^{0.4}$$

$$\Delta P = 48\,000\,\text{N/m}^2$$

Using the friction factor method,

$$f_f = \frac{16}{Re^*} = \frac{16}{1.03} = 15.53$$

and from Equation (4.12)

$$\Delta P = 2 \times 15.53 \times 1050\,\text{kg/m}^3 \times \frac{15\,\text{m}}{0.05\,\text{m}} \times (0.07\,\text{m/s})^2$$

$$\Delta P = 48\,000\,\text{N/m}^2$$

Therefore, the pipeline pressure drop is 48 kPa.

WORKED EXAMPLE 4.2

A slurry with a density of $2000\,\text{kg/m}^3$, a yield stress of $0.5\,\text{N/m}^2$, and a plastic viscosity of 0.3 Pa s is flowing in a 1.0 cm diameter pipe which is 5 m long. A pressure driving force of 4 kPa is being used. Calculate the flow rate of the slurry. Is the flow laminar or turbulent?

Solution

We will first assume that the flow is laminar and then go back and check this assumption. For laminar flow of a Bingham plastic fluid,

$$v_{AV} = \frac{R\tau_0}{4\mu_p} \left[1 - \frac{4}{3}\frac{\tau_y}{\tau_0} + \frac{1}{3}\left(\frac{\tau_y}{\tau_0}\right)^4\right]$$

where the wall shear stress τ_0

$$\tau_0 = \frac{\Delta PR}{2L}$$

$$\tau_0 = \frac{(4000\,\text{N/m}^2)(0.005\,\text{m})}{2(5.0\,\text{m})} = 2.0\,\text{N/m}^2$$

Then,

$$v_{AV} = \frac{0.005\,\text{m} \times 2.0\,\text{N/m}^2}{4 \times 0.3\,\text{N s/m}^2}\left[1 - \frac{4}{3}\left(\frac{0.5}{2.0}\right) + \frac{1}{3}\left(\frac{0.5}{2.0}\right)^4\right]$$

$$v_{AV} = 0.0056\,\text{m/s}$$

And,

$$Q = v_{AV}\frac{\pi D^2}{4} = (0.0056\,\text{m/s})\left[\frac{\pi}{4}(0.01\,\text{m})^2\right]$$

$$Q = 4.37 \times 10^{-7}\,\text{m}^3/\text{s}$$

Now check to see if the flow is laminar by computing *Re* and *He*.

$$Re = \frac{\rho_m D v_{AV}}{\mu_p} = \frac{(2000\,\text{kg/m}^3)(0.01\,\text{m})(0.0056\,\text{m/s})}{0.3\,\text{kg/m/s}}$$

$$Re = 0.37$$

$$He = \frac{\rho_m D^2 \tau_y}{\mu_p^2} = \frac{(2000\,\text{kg/m}^3)(0.01\,\text{m})^2(0.5\,\text{kg/m/s}^2)}{(0.3\,\text{kg/m/s})^2}$$

$$He = 1.1$$

Given this *He* and *Re*, the flow is laminar.

WORKED EXAMPLE 4.3

The following rheology test results were obtained for a mineral slurry containing 60% solids by weight. Which rheological model describes this slurry, and what are the appropriate rheological properties for this slurry?

Shear rate (s^{-1})	Shear stress (Pa)
0	5.80
0	5.91
0	6.00
1	6.06
10	6.52
15	6.76
25	7.29
40	8.00
45	8.24

Solution

The shear stress at zero shear rate is 6.00 Pa. Hence there is a yield stress equal to 6.00 Pa. In order to determine whether the slurry behaves as a Bingham fluid or if it follows the Herschel–Bulkley model, we need to plot $\tau - \tau_y$ versus shear rate.

$\tau - \tau_y$ (Pa)	Shear rate (s^{-1})
0.06	1
0.52	10
0.76	15
1.29	25
2.00	40
2.24	45

This plot yields a straight line with slope equal to 0.05 Pa s which is the value of the plastic viscosity μ_p.

WORKED EXAMPLE 4.4

A coal–water slurry with 65% volume fraction coal (coal specific gravity = 2.5) is pumped at a rate of 3.41 m^3/h from a storage tank through a 50 m long, 1.58 cm inside diameter horizontal pipe to a boiler. The storage tank is at 1 atm pressure and the slurry must be fed to the boiler at a gauge pressure of 1.38 bar. If this slurry behaves as a Bingham plastic fluid with a yield stress of 80 Pa and a plastic viscosity of 0.2 Pa s, what is the required pumping power?

Solution

First check to see if flow is laminar or turbulent by computing *Re* and *He*. Compute mixture density ρ_m

$$\rho_m = C_w \rho_s + (1 - C_w)\rho_f$$
$$\rho_m = (0.65)(2500\,\mathrm{kg/m^3}) + (0.35)(1000\,\mathrm{kg/m^3})$$
$$\rho_m = 1975\,\mathrm{kg/m^3}$$

Compute the velocity v_{AV} given the volumetric flow rate Q.

$$Q = v_{av}\pi\frac{D^2}{4}, \text{ so } v_{AV} = \frac{3.41}{3600} \times \frac{1}{\frac{\pi}{4}(0.0158)^2} = 4.83\,\mathrm{m/s}$$

Calculate the Reynolds number

$$Re = \frac{\rho_m D v_{AV}}{\mu_p} = \frac{(1975\,\mathrm{kg/m^3})(0.0158\,\mathrm{m})(4.83\,\mathrm{m/s})}{(0.2\,\mathrm{kg/m/s})} = 754$$

Calculate the Hedstrom number He

$$He = \frac{\rho_m D^2 \tau_y}{\mu_p^2} = \frac{1975 \times 0.0158^2 \times 80}{0.2^2} = 986$$

Given $Re = 754$ and $He = 986$, the flow is laminar.

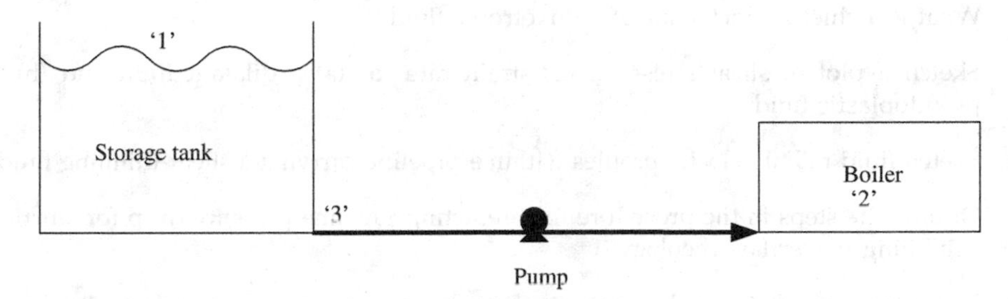

Figure 4W4.1

Applying Equation (4.13) to the process between points '1' and '2' (Figure 4W4.1)

$$H_{\text{gained by pump}} - h_f = \frac{p_2 - p_1}{\rho_m g} + \frac{v_{AV}^2}{2g}$$

The head loss due to friction in the pipeline alone is related to the pressure drop between points '3' and '2'

$$h_f = \frac{p_3 - p_2}{\rho_m g} = \frac{1}{\rho_m g}\left(\frac{32\mu_p v_{AV} L}{D^2} + \frac{16\tau_y L}{3D}\right)$$

So:

$$h_f = \frac{1}{1975 \times 9.81}\left(\frac{32 \times 0.2 \times 4.83 \times 50}{0.0158^2} + \frac{16 \times 80 \times 50}{3 \times 0.0158}\right) = 390\,\text{m}$$

$$H_{\text{gained by pump}} = 390 + \frac{1.38 \times (1.013 \times 10^5)}{1975 \times 9.81} + \frac{4.83^2}{2 \times 9.81} = 398\,\text{m}$$

Head gained by pump is related to the power required $\dot{W}$

$$H_{\text{gained by pump}} = \frac{\dot{W}}{\rho_m g Q} = 398\,\text{m}$$

Therefore, $\dot{W} = 1975 \times 9.81 \times (3.41 \div 3600) \times 398 = 7300\,\text{J/s}$

Therefore, pumping power required is 7.3 kW.

TEST YOURSELF

4.1 What are the chief distinguishing characteristics of homogeneous flow and heterogeneous flow in slurries?

4.2 What is meant by the term 'critical deposition velocity' in reference to a setting slurry?

4.3 Name three models that might describe the rheological behaviour of non-setting supensions at high concentrations.

4.4 What is a chief characteristic of a thixotropic fluid?

4.5 Sketch a plot of shear stress versus strain rate for (a) a dilatant fluid and (b) a pseudoplastic fluid.

4.6 Sketch fluid radial velocity profiles within a pipeline carrying a shear-thinning fluid.

4.7 Outline the steps in the procedure for predicting pipeline pressure drop for slurries exhibiting power-law rheology.

4.8 How might one distinguish between a slurry behaving as a Bingham plastic fluid and a Herschel–Bulkley fluid?

4.9 Define the Hedstrom number. How is this number used in prediction of pipeline pressure drop for slurries exhibiting Bingham plastic rheology?

4.10 What steps might be typically involved in preparation of a slurry for transport by pipeline?

4.11 What type of pump would be used in pipeline transport for an abrasive, non-settling slurry requiring pressures up to 60 bar?

4.12 How is erosive wear in slurry pipelines combated?

EXERCISES

4.1 Samples of a phosphate slurry mixture are analysed in a lab. The following data describe the relationship between the shear stress and the shear rate:

Shear rate, $\dot{\gamma}(s^{-1})$	Shear stress, τ(Pa)
25	38
75	45
125	48
175	51
225	53
325	55.5
425	58
525	60
625	62
725	63.2
825	64.3

The slurry mixture is non-Newtonian. If it is considered a power-law slurry, what is the relationship of the viscosity to the shear rate?

(Answer: $\tau = 23.4\dot{\gamma}^{0.15}$.)

4.2 Verify Equation (4.7).

4.3 Verify Equation (4.28).

4.4 A slurry behaving as a pseudoplastic fluid is flowing through a smooth round tube having an inside diameter of 5 cm at an average velocity of 8.5 m/s. The density of the slurry is 900 kg/m^3 and its flow index and consistency index are $n = 0.3$ and $k = 3.0\,\mathrm{N\,s^{0.3}/m^2}$. Calculate the pressure drop for (a) 50 m length of horizontal pipe and (b) 50 m length of vertical pipe with the flow moving against gravity.

(Answer: (a) 338 kPa; (b) 779 kPa.)

4.5 The concentration of a water-based slurry sample is to be found by drying the slurry in an oven. Determine the slurry weight concentration given the following data:

Weight of container plus dry solids	0.31 kg
Weight of container plus slurry	0.48 kg
Weight of container	0.12 kg

Determine the density of the slurry if the solid specific gravity is 3.0.

(Answer: 1546 kg/m^3.)

4.6 A coal-water slurry has a specific gravity of 1.3. If the specific gravity of coal is 1.65, what is the weight percent of coal in the slurry? What is the volume percent coal?

(Answer: 58.6%, 46%.)

4.7 The following rheology test results were obtained for a mineral slurry containing 60% solids by weight. Which rheological model describes the slurry and what are the appropriate rheological properties for this slurry?

Shear rate (s^{-1})	Shear stress (Pa)
0	4.0
0.1	4.03
1	4.2
10	5.3
15	5.8
25	6.7
40	7.8
45	8.2

(Answer: Herschel–Bulkley model: $k = 0.20$, $n = 0.81$.)

4.8 A mud slurry is drained from a tank through a 15.24 m long horizontal plastic hose. The hose has an elliptical cross-section, with a major axis of 101.6 mm and a minor axis of 50.8 mm. The open end of the hose is 3.05 m below the level in the tank. The mud is a Bingham plastic with a yield stress of 10 Pa, a plastic viscosity of 50 cp, and a density of 1400 kg/m^3.

(a) At what velocity will water drain from the hose?

(b) At what velocity will the mud drain from the hose?

(Answer: (a) 3.65 m/s; (b) 3.20 m/s.)

4.9 A coal slurry is found to behave as a power-law fluid with a flow index 0.3, a specific gravity 1.5, and an apparent viscosity of 0.07 Pa s at a shear rate 100 s^{-1}.

(a) What volumetric flow rate of this fluid would be required to reach turbulent flow in a 12.7 mm inside diameter smooth pipe which is 4.57 m long?

(b) What is the pressure drop (in Pa) in the pipe under these conditions?

(Answer: (a) 0.72 m^3/h; (b) 19.6 kPa.]

4.10 A mud slurry is draining from the bottom of a large tank through a 1 m long vertical pipe with a 1 cm inside diameter. The open end of the pipe is 4 m below the level in the tank. The mud behaves as a Bingham plastic with a yield stress of 10 N/m^2, an apparent viscosity of 0.04 kg/m/s, and a density of 1500 kg/m^3. At what velocity will the mud slurry drain from the hose?

(Answer: 3.5 m/s.)

4.11 A mud slurry is draining in laminar flow from the bottom of a large tank through a 5 m long horizontal pipe with a 1 cm inside diameter. The open end of the pipe is 5 m below the level in the tank. The mud is a Bingham plastic with a yield stress of 15 N/m^2, an apparent viscosity of 0.06 kg/m/s, and a density of 2000 kg/m^3. At what velocity will the mud slurry drain from the hose?

(Answer: 0.6 m/s.)

5 Colloids and Fine Particles

George V. Franks

5.1 INTRODUCTION

The importance of colloids and fine particles has been the focus of increased attention with the emergence of nanotechnology and microfluidics although it has historically been very important in many fields including paints, ceramics, foods, minerals, paper, biotechnology and other industries. The primary factor that distinguishes colloids and fine particles from larger particles is that the ratio of surface area to mass is very large for the small particles. The behaviour of fine particles is dominated by surface forces rather than body forces. Colloids are very fine particles with one or more linear dimension between about 1 nm and 10 μm suspended in a fluid. The dominance of surface forces is exhibited in the cohesive nature of fine particles, high viscosity of concentrated suspensions and slow sedimentation of dispersed colloidal suspensions.

The ratio of the surface area to the volume of a spherical particle can be calculated from the diameter of the sphere as follows (where x is the particle diameter):

$$\frac{\text{Surface Area}}{\text{Volume}} = \frac{\pi x^2}{\frac{\pi}{6}x^3} = \frac{6}{x} \tag{5.1}$$

The mass of a particle is directly related to its volume by its density. As particle size decreases the influence of surface forces (described in Section 5.3) dominate the behaviour of the powders and suspensions relative to body forces which depend on the particles' mass. Body forces are easy to understand because they are simply a result of Newton's laws of motion, where the force depends on the mass and acceleration via $F = ma$. The quintessential example of a body force is that of a particle settling under the influence of gravity ($F = mg$) as described in Chapter 2.

Introduction to Particle Technology - 2nd Edition Martin Rhodes

The mass of a 10 nm diameter silica particle is 1.4×10^{-21} kg. Because the mass of fine particles and colloids is so small, the magnitude of their body forces is less than the magnitude of the forces acting between their surfaces. These surface forces are the result of a number of physico-chemical interactions such as van der Waals, electrical double layer, bridging and steric forces, which will be described in detail in the following sections. It is these forces that control the behaviour of fine powders and colloidal suspensions which are discussed in detail in the sections that follow. The understanding of these forces is not as simple as body forces since they depend upon specific chemical interactions at the surface of the particles. A basic understanding of this specialist information is described in Section 5.3 and detailed knowledge is available in the colloid and surface chemistry textbooks (Hiemenz and Rajagopolan, 1997; Hunter, 2001; Israelechvili, 1992).

Surface forces may result in either attraction or repulsion between two particles depending on the material of which the particles are composed, the fluid type and the distance between the particles. Generally, if nothing is done to control the interaction between particles, they will be attracted to each other due to van der Waals forces which are always present. (The few rare cases where the van der Waals forces are repulsive are described in Section 5.3.1.) The dominance of attraction is the reason why fine powders in air are usually cohesive.

Another result of the small mass of colloidal particles is that they behave somewhat like molecules when dispersed in liquids. For example, they diffuse through the liquid and move randomly due to the phenomenon known as Brownian motion.

5.2 *BROWNIAN MOTION*

When colloids are dispersed in a liquid they are influenced by hydrodynamic body forces as described in Chapter 2. They also experience a phenomenon known as Brownian motion. Thermal energy from the environment causes the molecules of the liquid to vibrate. These vibrating molecules collide with each other and with the surface of the particles. The random nature of the collisions causes the particles to move in a random walk as shown in Figure 5.1. The phenomenon is named after Robert Brown who first observed the behaviour in the motion of pollen grains in water in 1827 (Perrin, 1913).

A simple application of a kinetic model allows us to determine the influence of key parameters on the average velocity of particles in suspension. Consider that the thermal energy of the environment is transferred to the particles as kinetic energy. The average thermal energy is $\frac{3}{2}kT$ (where k is Boltzmann's constant $= 1.381 \times 10^{-23}$ J/K and T is the temperature in Kelvin). If one ignores drag, collisions and other factors, the average velocity $\bar{v}$ of the particle can be estimated by equating the kinetic energy $\frac{1}{2}mv^2$ (where m is the mass of the particle) with the thermal energy as follows:

$$\bar{v} = \sqrt{\frac{3kT}{m}} \tag{5.2}$$

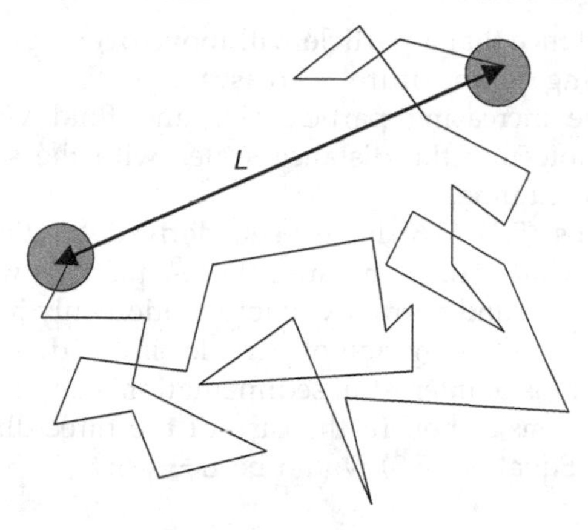

Figure 5.1 Illustration of the random walk of a Brownian particle. The distance the particle has moved over a period of time is L

This simple analysis cannot be used to determine the actual distance of the particle from its original position because it does not move in a straight line (see Figure 5.1 and below), but, it does show that either increasing temperature or decreasing the particles mass increases Brownian motion.

Thermodynamic principles dictate that the lowest free energy state (greatest entropy) of a suspension is a uniform distribution of particles throughout the volume of the fluid. Thus, the random walk of a particle due to Brownian motion provides a mechanism for the particles to arrange themselves uniformly throughout the volume of the fluid. The result is diffusion of particles from regions of high concentration to regions of lower concentration. Einstein and Smoluchowski used statistical analysis of the one-dimensional random walk to determine the average (root mean square) distance that a Brownian particle moves as a function of time (Einstein, 1956).

$$L = \sqrt{2\alpha t} \tag{5.3}$$

where α is the diffusion coefficient. Einstein further developed a relationship for the diffusion coefficient that accounted for the hydrodynamic frictional drag on a spherical particle:

$$\alpha f = kT \tag{5.4}$$

where the frictional coefficient f is defined as F_D/U, where U is the relative velocity between the particle and fluid. Then for creeping laminar flow we find from Stokes' law [Equation (2.3)] that $f = 3\pi x\mu$ so that:

$$\alpha = kT/3\pi x\mu \tag{5.5}$$

and

$$L = \sqrt{\frac{2kT}{3\pi x\mu}t} \tag{5.6}$$

Thus the average distance that a particle will move over a period of time can be determined. Increasing temperature increases the distance travelled over a period of time while increasing particle size and fluid viscosity reduce the distance travelled. Note that the distance scales with the square root of time rather than linearly with time.

Note that Equations (5.3)–(5.6) have been derived for the case of the one-dimensional random walk. This is because these equations will be used later in analysis of sedimentation under gravity where motion only in one direction (one dimension) is of interest. Only motion of particles in the direction of the applied gravitational force field is of interest in sedimentation; lateral motion in the other two orthogonal directions is not. In the case of the three-dimensional random walk, the analogy to Equation (5.3) would be $L = \sqrt{6\alpha t}$.

5.3 SURFACE FORCES

Surface forces ultimately arise from the summation of intermolecular interaction forces between all the molecules (or atoms) in the particles acting across the intervening medium. The intermolecular (and interatomic) forces are manifestations of electromagnetic interactions between the atoms of the material (Israelechvili, 1991). Comprehensive understanding of such forces is described in great detail in colloid and surface chemistry texts (Hiemenz and Rajagopolan, 1997; Hunter, 2001; Israelechvili, 1992). Discussion here is limited to a few of the forces with most technological significance.

In general, the force between two particles (F) may be either attractive or repulsive. The force depends upon the surface to surface separation distance (D) between the particles and the potential energy (V) at that separation distance. The relationship between the force and potential energy is that the force is the negative of the gradient of the potential energy with respect to distance.

$$F = -\frac{\mathrm{d}V}{\mathrm{d}D} \tag{5.7}$$

Typical potential energy and force versus separation distance relationships for fine particles are shown schematically in Figure 5.2. Thermodynamics dictates that the pair of particles move to the separation distance that results in the lowest energy configuration. A force between the particles will result if the particles are at any other separation distance. There is always a strong repulsive force at zero separation distance that prevents the particles from occupying the same space. When there is attraction (at all other separation distances), the particles reside in a potential energy well (minimum in energy) at an equilibrium separation distance (in contact) [Figure 5.2 (a) and (b)]. In some cases, a repulsive potential energy barrier exists that prevents the particles from moving to the minimum energy separation because they do not have enough thermal or kinetic energy to surmount the barrier. (In terms of force, there is not enough force applied to the particles to exceed the repulsive force field.) In this case the particles cannot touch each other and reside at a separation distance greater than the extent of the

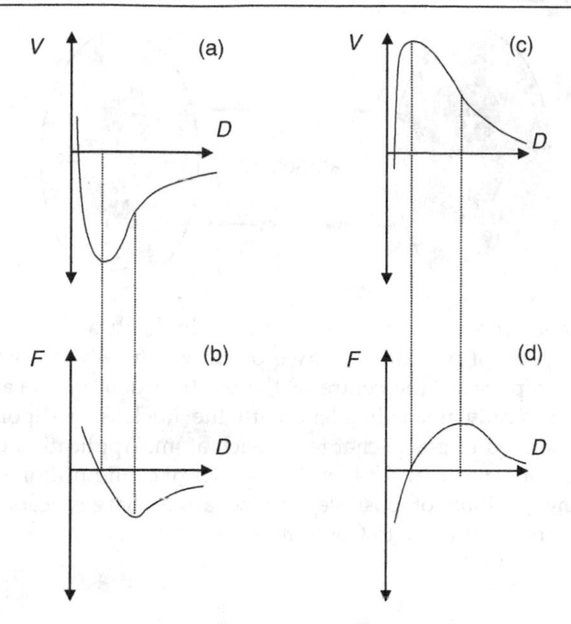

Figure 5.2 Schematic representations of interparticle potential energy (V) and force (F) versus particle surface to surface separation distance (D). (a) Energy versus separation distance curve for an attractive interaction. The particles will reside at the separation distance where the minimum in energy occurs. (b) Force versus separation distance for the attractive potential shown in (a). (The convention used in this book is that positive interparticle forces are repulsive.) The particles feel no force if they are at the equilibrium separation distance. An applied force greater than a maximum is required to pull the particles apart. (c) Energy versus separation distance curve for a repulsive interaction. When the potential energy barrier is greater than the available thermal and kinetic energy the particles cannot come in contact and move away from each other to reduce their energy. (d) Force versus separation distance for the repulsive potential shown in (c). There is no force on the particles when they are very far apart. There is a maximum force that must be exceeded to push the particles into contact

repulsive barrier which is usually at least several nanometres or more [Figure 5.2 (c) and (d)].

The relationships between force and distance as well as the underlying physical and chemical mechanisms responsible for those forces are described below for several of the forces with the most technological significance.

5.3.1 van der Waals Forces

van der Waals forces is the term commonly used to refer to a group of electrodynamic interactions including Keesom, Debye and London dispersion interactions that occur between the atoms in two different particles. The dominant contribution to the van der Waals interaction between two particles is from the dispersion force. The dispersion force is a result of Columbic interactions between correlated fluctuating instantaneous dipole moments within the atoms

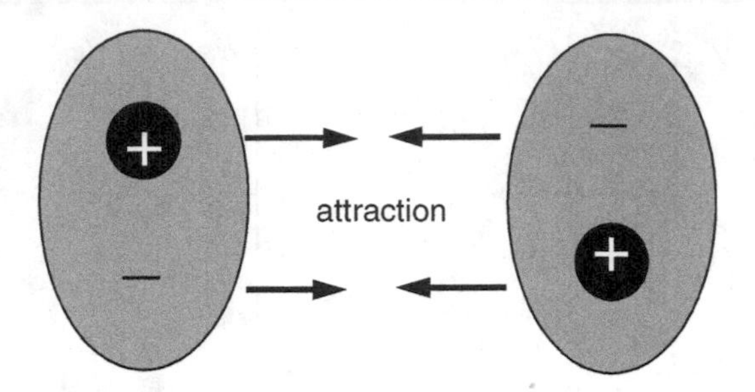

Figure 5.3 Schematic representation of the dipole–dipole attraction that exists between the instantaneous dipoles of two atoms in two particles. The + represents the nucleus of the atom and the − represents the centre of the electron density. Because the centre of electron density is typically not coincident with the nucleus, a dipole moment exists between the two separated opposite charges in each atom. Application of Coulomb's law between the charges indicates that the lowest free energy configuration is as shown in the figure. The resulting position of positive and negative charges leads to an attraction between the two atoms, again due to Coulomb's law

that comprise the two particles. To understand this concept, imagine that each atom in a material contains a positively charged nucleus and orbiting negative electrons. The nucleus and the electrons are separated by a short distance, on the order of an Ångstrom (10^{-10} m). At any instant in time, a dipole moment exists between the nucleus and the centre of electron density. This dipole moment fluctuates very rapidly with time, revolving around the nucleus as the electrons orbit. The dipole moment of each atom creates an electric field that emanates from the atom and is felt by all other atoms in both particles. In order to lower the overall energy of the system, the dipole moments of all the atoms in both particles correlate their dipole moments (i.e. they align themselves like a group of choreographed pairs of dancers in a musical, who remain 'in sync' although they are always moving). When the two particles are composed of the same material, the lowest energy configuration of the correlated dipoles is such that there is attraction between the dipoles as shown in Figure 5.3. The combined attraction between all the dipoles in the two particles results in an overall attraction between the particles. In general the van der Waals interaction can be attractive or repulsive depending on the dielectric properties of the two particles and the medium between the particles.

Pairwise summation of the interactions between all atoms in both particles results in a surprisingly simple equation for the overall interaction between two spherical particles of the same size when the distance between the particles (D) is much less than the diameter of the particles (x).

$$V_{\mathrm{vdW}} = -Ax/24D \tag{5.8a}$$

and

$$F_{\mathrm{vdW}} = -Ax/24D^2 \tag{5.8b}$$

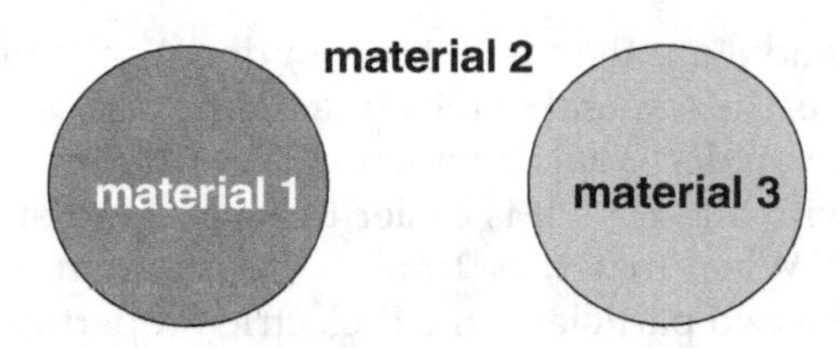

Figure 5.4 Notation used to indicate the type of material for each particle and the intervening medium

where V_{vdW} is the van der Waals interaction energy and F_{vdW} is the van der Waals force. More complicated relationships arise if the particles are not the same size or are small relative to the distance between them. The sign and magnitude of the interaction for a particular pair of particles interacting in a given medium is expressed as the numerical value of the Hamaker constant (A). When the Hamaker constant is greater than zero the interaction is attractive and when the Hamaker constant is less than zero the interaction is repulsive. Figure 5.4 shows the configuration of two particles (materials 1 and 3) and an intervening medium (material 2). The Hamaker constant can be calculated from the dielectric properties of the three materials. Table 5.1 shows the Hamaker constants for several combinations of particles and intervening media. Note that oil droplets in emulsions and bubbles in foams may be considered particles in the sense of understanding interaction forces and the stability of the emulsions and foams.

Table 5.1 Hamaker constants of some common material combinations

Material 1	Material 2	Material 3	Hamaker constant (approximate) (J)	Example
Alumina	Air	Alumina	15×10^{-20}	Oxide minerals in air are strongly attractive and cohesive
Silica	Air	Silica	6.5×10^{-20}	
Zirconia	Air	Zirconia	20×10^{-20}	
Titania	Air	Titania	15×10^{-20}	
Alumina	Water	Alumina	5.0×10^{-20}	Oxide minerals in water are attractive but less so than in air
Silica	Water	Silica	0.7×10^{-20}	
Zirconia	Water	Zirconia	8.0×10^{-20}	
Titania	Water	Titania	5.5×10^{-20}	
Metals	Water	Metals	40×10^{-20}	Conductivity of metals makes them strongly attractive
Air	Water	Air	3.7×10^{-20}	Foams
Octane	Water	Octane	0.4×10^{-20}	Oil in water emulsions
Water	Octane	Water	0.4×10^{-20}	Water in oil emulsions
Silica	Water	Air	-0.9×10^{-20}	Particle bubble attachment in mineral flotation, weak repulsion

When materials 1 and 3 are the same, the van der Waals interaction is always attractive, for example, the mineral oxides interacting across water or air shown in Table 5.1. Note the van der Waals interaction is reduced when the particles are in water compared with air. Thus it is easier to separate (disperse) fine particles in liquids than in air. When materials 1 and 3 are different materials, repulsion will result between the two particles if the dielectric properties of the intervening medium are between that of the two particles, such as for silica particles and air bubbles interacting across water.

5.3.2 Electrical Double Layer Forces

When particles are immersed in a liquid they may develop a surface charge by any one of a number of mechanisms. Here, consider the case of oxide particles immersed in aqueous solutions. The surface of a particle is comprised of atoms that have unsatisfied bonds. In vacuum, these unfulfilled bonds result in an equal number of positively charged metal ions and negatively charged oxygen ions as shown in Figure 5.5(a). When exposed to ambient air (which usually has at least 15 % relative humidity) or immersed in water, the surface reacts with water to produce surface hydroxyl groups (denoted M-OH) as shown in Figure 5.5(b).

The surface hydroxyl groups react with acid and base at low and high pH, respectively, via surface ionization reactions as follows (Hunter, 2001):

$$\text{M-OH} + \text{H}^+ \xrightarrow{K_a} \text{M-OH}_2^+ \tag{5.9a}$$

$$\text{M-OH} + \text{OH}^- \xrightarrow{K_b} \text{M-O}^- + \text{H}_2\text{O} \tag{5.9b}$$

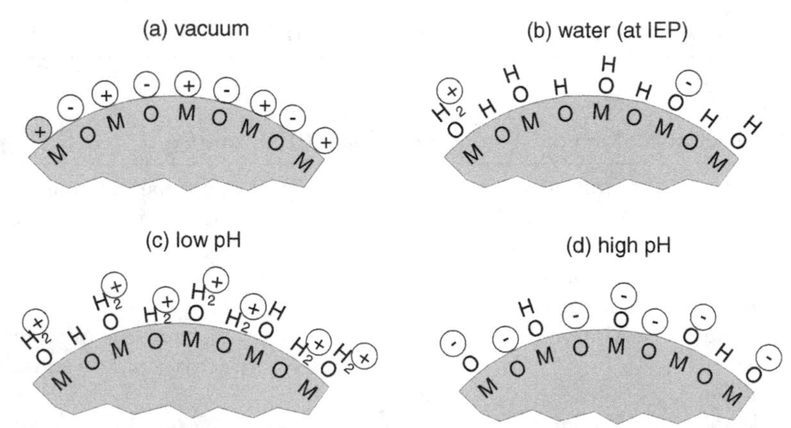

Figure 5.5 Schematic representation of the surface of metal oxides (a) in vacuum. Unsatisfied bonds lead to positive and negative sites associated with metal and oxygen atoms, respectively. (b) The surface sites react with water or water vapour in the environment to form surface hydroxyl groups (M-OH). At the isoelectric point (IEP) the neutral sites dominate, and the few positive and negative sites present exist in equal numbers. (c) At low pH the surface hydroxyl groups react with H^+ in solution to create a positively charged surface composed mainly of (M-OH_2^+) species. (d) At high pH the surface hydroxyl groups react with OH^- in solution to create a negatively charged surface composed mainly of (M-O^-) species

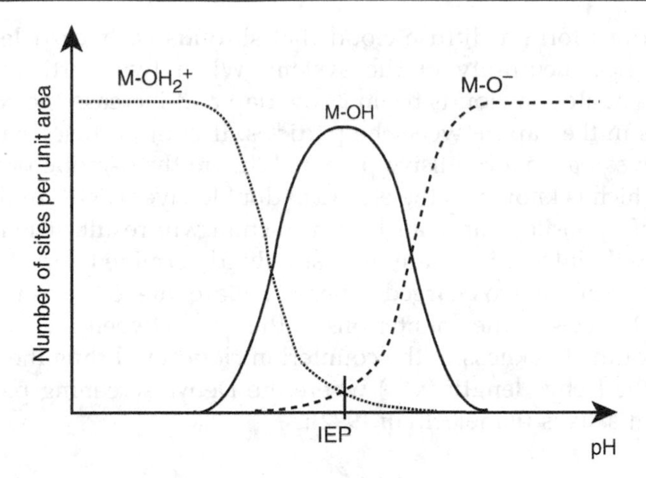

Figure 5.6 Number density per unit area of neutral (M-OH), positive ($M-OH_2^+$) and negative ($M\text{-}O^-$) surface sites as a function of pH

resulting in either a positively charged surface ($M\text{-}OH_2^+$) as in Figure 5.5(c), or a negatively charged surface ($M-O^-$) as in Figure 5.5(d). The value of the surface ionization reaction constants (K_a and K_b) depend upon the particular type of material (for example SiO_2, Al_2O_3 and TiO_2). For each type of material, there is a pH known as the isoelectric point (IEP), where the majority of surface sites are neutral (M-OH) and the net charge on the surface is zero. At a pH below or above the IEP, the particle's surfaces become positively ($M\text{-}OH_2^+$) or negatively ($M\text{-}O^-$) charged due to the addition of either acid (H^+) or base (OH^-), respectively. Figure 5.6 shows how the concentration of surface sites changes with pH. Table 5.2 contains a listing of the IEPs of some common materials.

For each charged surface site there is a counterion of opposite charge in solution. For example, the counterion for a positive surface site is a Cl^- anion in the case when HCl is used to reduce the pH and the counterion for a negative surface site is a Na^+ cation if NaOH is used to increase the pH. The entire system is electrically neutral. The separation of charge between the surface and the bulk solution results in a potential difference known as the surface potential (Ψ_0).

Table 5.2 Isoelectric points of some common materials

Material	pH of IEP
Silica	2–3
Alumina	8.5–9.5
Titania	5–7
Zirconia	7–8
Hematite	7–9
Calcite	8
Oil	3–4
Air	3–4

The counterions form a diffuse cloud that shrouds each particle in order to maintain electrical neutrality of the system. When two particles are forced together their counterion clouds begin to overlap and increase the concentration of counterions in the gap between the particles. If both particles have the same charge, this gives rise to a repulsive potential due to the osmotic pressure of the counterions which is known as the electrical double layer (EDL) repulsion. If the particles are of opposite charge an EDL attraction will result. It is important to realize that EDL interactions are not simply determined by the Columbic interaction between the two charged spheres, but are due to the osmotic pressure (concentration) effects of the counterions in the gap between the particles.

A measure of the thickness of the counterion cloud (and thus the range of the repulsion) is the Debye length (κ^{-1}) where the Debye screening parameter (κ), for monovalent salts is (Israelachvili,1992):

$$\kappa = 3.29\sqrt{[c]}\,(\text{nm}^{-1}) \tag{5.10}$$

where $[c]$ is the molar concentration of monovalent electrolyte. When the Debye length is large (small counterion concentration) the particles are repulsive at large separation distances so that the van der Waals attraction is overwhelmed as in Figure 5.2(c) and (d). The electrical double layer is compressed (Debye length is reduced) by adding a salt, which increases the concentration of the counterions around the particle. When sufficient salt is added, the range of the EDL repulsion is decreased sufficiently to allow the van der Waals attraction to dominate at large separation distances. At this point, an attractive potential energy well as shown in Figure 5.2(a) and (b) results.

An approximate expression for the EDL potential energy (V_{EDL}) versus the surface to surface separation distance (D) between two spherical particles of diameter (x) with the same surface charge is (Israelachvili,1992):

$$V_{\text{EDL}} = \pi\varepsilon\varepsilon_0 x \Psi_0^2 e^{-\kappa D} \tag{5.11}$$

where Ψ_0 is the surface potential (created by the surface charge), ε the relative permittivity of water, not voidage as frequently used in other parts of the book, ε_0 the permittivity of free space which is $8.854 \times 10^{-12}\,\text{C}^2/\text{J/m}$, and κ the inverse Debye length. This expression is valid when the surface potential is constant and below about 25 mV and the separation distance between the particles is small relative to their size (Israelachvili, 1992).

Because a layer of immobile ions and water molecules exists at the surface of the particle it is not easy to directly measure the surface potential of particles. Instead, a closely related potential known as the zeta potential is usually measured. The zeta potential can be determined by measuring the particle's velocity in an electric field. The zeta potential is the potential at the plane of shear between the immobilized surface layer and the bulk solution. This plane is typically located only a few Angstroms from the surface so that there is little difference between the zeta potential and the surface potential. In practice, the zeta potential can be used in place of the surface potential in Equation (5.11) to predict the interparticle forces as a function of separation distance with little

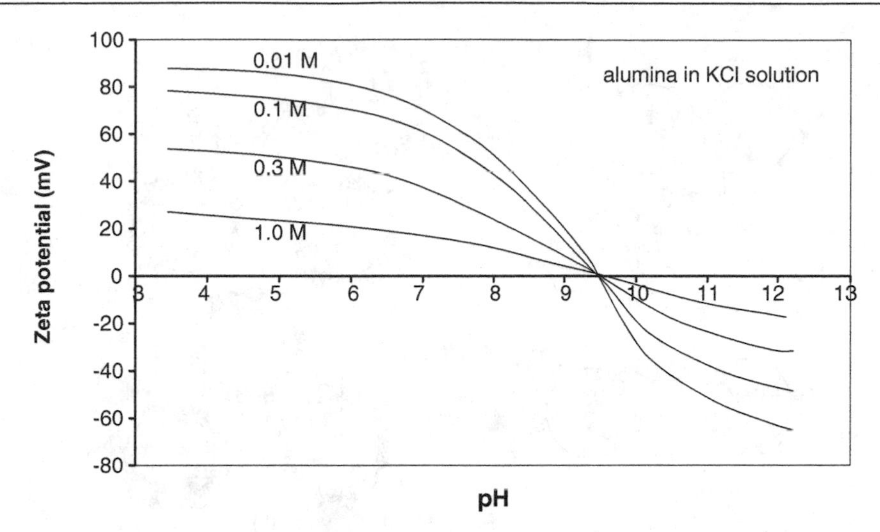

Figure 5.7 Zeta potential of alumina particles as a function of pH and salt concentration. (Data from Johnson *et al.*, 2000)

error. Addition of salt to a suspension reduces the magnitude of the zeta potential as well as compressing the range of the double layer (reducing Debye length) as described above. Figure 5.7 is an example of how pH and salt concentration influence the zeta potential of alumina particles.

5.3.3 Adsorbing Polymers, Bridging and Steric Forces

Another method that is useful in controlling surface forces between particles in suspensions is through the addition of soluble polymer to the solution. Consider the situation where the polymer has affinity for the particle's surface and tends to adsorb on the particles' surfaces. Either attraction by polymer bridging or repulsion due to steric interactions can result depending on the polymer molecular weight and amount adsorbed as shown in Figure 5.8.

Bridging flocculation (Gregory, 2006; Hiemenz and Rajagopolan, 1997) is a method in which polymer that adsorbs onto the particle's surface is added in a quantity that is less than sufficient to fully cover the surface. The polymer chains adsorbed onto one particle's surface can then extend and adsorb on another particle's surface and hold them together. The optimum amount of polymer to add is usually just enough to cover half of the total particle surface area. The best bridging flocculation is usually found with polymers of high to very high molecular weight (typically $1 \times 10^6 - 20 \times 10^6$ g/mol) so that they can easily bridge between particles. Commercial polymeric flocculants are typically charged or nonionic copolymers of polyacrylamide. They are used extensively in the water treatment, waste water, paper and mineral processing industries to aid in solid/liquid separation. They operate by creating attraction between the fine particles resulting in the formation of aggregates known as flocs. The flocs' mass is much greater than the individual particle's mass so that body forces

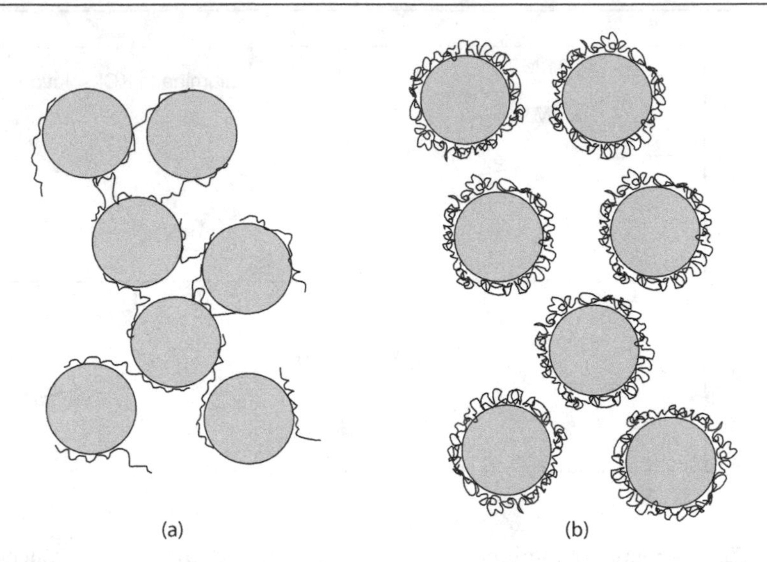

Figure 5.8 Schematic representation of (a) bridging flocculation and (b) steric repulsion

become important in controlling the behaviour of the floc and gravity sedimentation dominates relative to the randomizing effect of Brownian motion that dominates the individual particle's behaviour.

The addition of polymers can also create repulsion between particles by a steric mechanism. Steric repulsion occurs when the particle's surfaces are completely covered with a thick layer of polymer. The polymer must adsorb to the surfaces of the particles and extend out into solution. In a good solvent, as the separation distance becomes less than twice the extent of the adsorbed polymer, the polymer layers begin to overlap and a strong repulsion results as shown in Figure 5.2(c) and (d). The polymers that work best in creating steric repulsion are typically low to moderate molecular weight (typically less than 1×10^6 g/mol) and the surfaces of the particles should be completely covered. When the solvent quality is poor, the steric interaction can be attractive at moderate to long range. This can occur when a poorly soluble polymer is adsorbed to the particle's surface. Electrosteric stabilization occurs when the polymer is a charged polyelectrolyte such that both EDL and steric repulsion are active. Steric and electrosteric stabilization are commonly used in the processing of ceramics to control suspension stability and viscosity.

5.3.4 Other Forces

There are several other mechanisms that may lead to forces between particles. In very dry air static charge may result in Columbic interactions between particles. Columbic interactions are usually of little significance in ambient air which is usually humid enough that the static charge dissipates rapidly. In solution, non-adsorbing polymers can result in another type of weak attraction called depletion attraction. Layers of solvent molecules on particles' surfaces, such as water on strongly hydrated surfaces, can result in short range repulsion. These short ranged repulsions are known as hydration or structural forces. Hydrophobic surfaces (poorly wetted), when immersed in water exhibit a special kind of strong attraction. The so called hydrophobic force causes oil droplets to coalesce and hydrophobic particles to aggregate.

5.3.5 Net Interaction Force

In the 1940s Derjaguin, Landau, Verwey and Overbeek (DLVO), developed the hypothesis that the total particle interaction could be determined by simply summing the contributions from the van der Waals interaction and the EDL interaction. In the meantime, the DLVO theory has been widely verified experimentally. Furthermore, it has been found that many other forces may be combined in the same way to determine the overall interparticle interaction. Examples of some net interparticle interaction forces are shown in Figure 5.9.

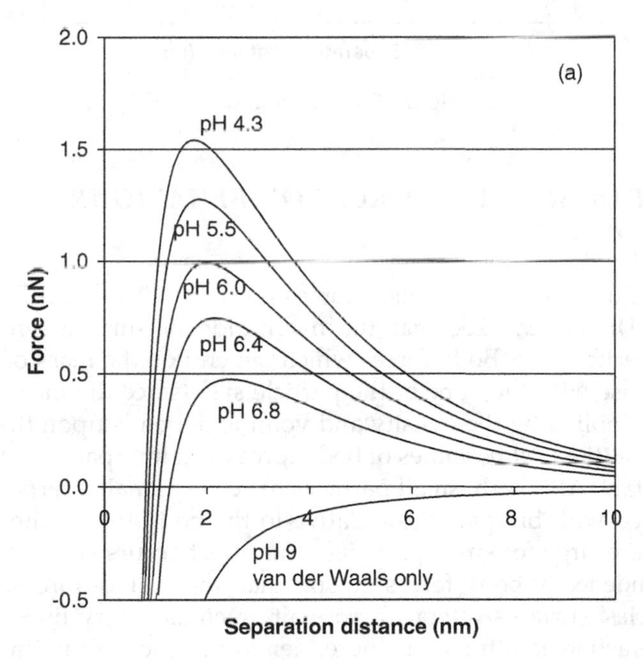

Figure 5.9 (a) Force versus distance curves for alumina at different pH values calculated from Equations (5.8) and (5.11) with parameters as detailed in Franks *et al.* (2000). At pH 9 the van der Waals attraction dominates. As pH is decreased the range and magnitude of the EDL repulsion increases as zeta potential increases (see Figure 5.8). At very small separation distances the van der Waals attraction always dominates the EDL repulsion. (b) Force versus distance curves for silica particles interacting with an adsorbed polymer (Zhou *et al.*, 2008). Upon approach, the adsorbed polymer provides a weak steric repulsion. Upon separation (retraction) the polymer creates a strong long range attraction because chains are adsorbed on both surfaces. The van der Waals only interaction is shown for comparison

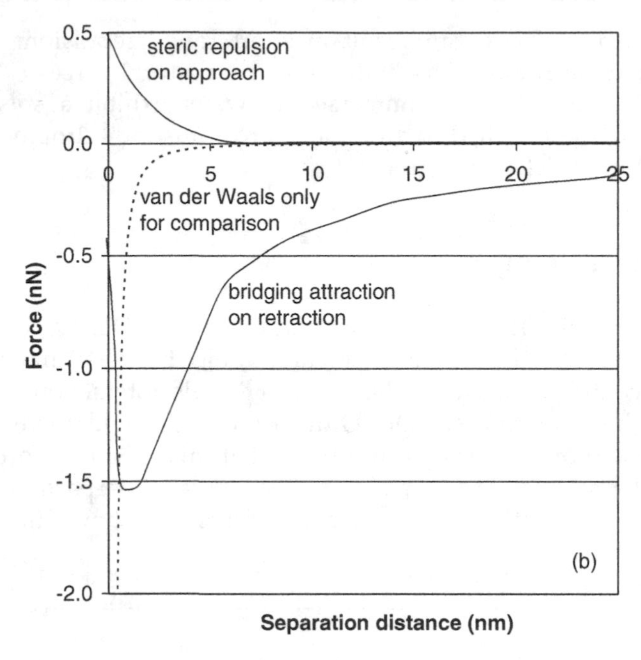

Figure 5.9 *(Continued)*

5.4 RESULT OF SURFACE FORCES ON BEHAVIOUR IN AIR AND WATER

From the equations for van der Waals forces [Equation (5.8)] and EDL repulsion [Equation (5.11)] one can see that the magnitude of surface forces increases linearly with particle size. Body forces which depend on the mass of the particle, however, increase with the cube of the particle size (since the mass is related to the volume multiplied by the density and volume depends upon the cube of the particle size). It is the *relative* values of body forces and interparticle surface forces that are important. Although, small particles have very small interparticle surface forces compared with big particles, relative to the body forces, the interparticle surface forces are large for small particles. This effect occurs because of the much stronger dependence of body forces on size than the surface forces.

When particles' surfaces interact across air, such as in dry fine powders, the dominant interaction is attraction due either to van der Waals interactions or capillary bridges (see Chapter 13). In air, other gases and vacuum the only mechanism which could generate repulsion is electrostatic charging (for example due to friction). If the charge on particles is the same sign, repulsion will result due to Coulomb's law while attraction would result between oppositely charged particles. Electrostatic interaction, although possible is usually not of significance if the relative humidity is greater than about 45 % because the charge rapidly dissipates at room temperature in humid air.

The result is that fine powders in air are cohesive due to van der Waals and capillary attraction. Attraction between particles results in cohesive behaviour of the powder. The strong cohesion of the particles is the reason why fine particles are difficult to fluidize (Geldart's Group C powders described in Chapter 7). The strong cohesion is also the cause of the high unconfined yield stresses of powders described in Chapter 10. The high unconfined yield stress of these powders means that the powders are not free flowing and will require a larger dimension hopper opening relative to free flowing powders of the same bulk density. Either larger primary particles or granules of the fine powder will have greater mass so that body forces (rather than adhesive surface forces) will dominate behaviour of these bigger particles and free flowing powders will result.

The influence of attractive forces between fine dry powders is observed as the effect of particle size on bulk density. As the particle size decreases the loose packed and tapped bulk densities tend to decrease. This is because as the particle size becomes smaller, the influence of the attractive surface forces becomes stronger than the body forces. Consolidation is aided by body forces (such as gravity) which allow the particles to rearrange into denser packing structures. The attractive surface forces between fine particles hinder their rearrangement into dense packing structures. Note this may seem counterintuitive to some readers (who may think that attractive forces would increase packing density). This is not the case because attractive forces create strong bonds that hinder rearrangement of particles into more dense packing structures.

One difference between dry fine powders and colloids in liquids is that the low viscosity of air (and other gases) make hydrodynamic drag forces minimal for dry powders in many instances, except when the particles have very low density (such as dust and smoke) or the gas velocity is very high. However, fine particles in liquids are strongly influenced by hydrodynamic drag forces as described in Chapter 2 because the viscosity of liquids is much greater than that of gases.

When fine particles are suspended or dispersed in liquids, such as water, we are able to control the interaction forces by prudent choice of the solution chemistry. This control of interaction forces is of significant technological importance because we can thus control the suspension behaviour such as stability, sedimentation rate, viscosity, and sediment density. Additives such as acids, bases, polymers and surfactants can easily be used in formulations to develop the range and magnitude of either repulsion or attraction as demonstrated in Figure 5.9. When fine particles are suspended or dispersed in liquids, such as water, there are several mechanisms that can produce repulsive forces between particles that can overwhelm the attractive van der Waals interaction between like particles if we want to keep the particles dispersed. One example of a situation where dispersed particles are desirable is in ceramic processing. In this application, the low viscosity of dispersed particles is desirable as well as the uniform and dense packing of particles in the shaped ceramic component afforded by the repulsion. However, suspended particles with high magnitude zeta potential with strong repulsion between them may be made to aggregate by development of an attractive interaction so that they settle rapidly to increase efficiency of solid/liquid separation. This application of bridging polymers is discussed in Section 5.5.

In general, as shown in Figure 5.10, suspension behaviour depends upon the interparticle forces which in turn depend upon the solution conditions. The

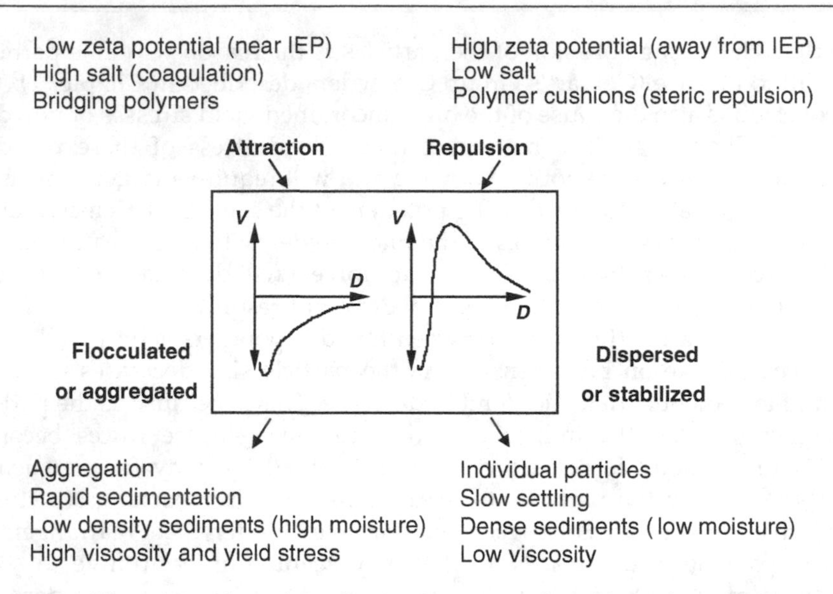

Figure 5.10 The top section of the figure gives examples of how the solution conditions influence the forces between particles. The bottom section shows how attractive and repulsive forces influence some behaviour of suspensions

suspension behaviour of interest such as stability, sedimentation, sediment density, particle packing and rheological (flow) behaviour are discussed in the following sections.

5.5 INFLUENCES OF PARTICLE SIZE AND SURFACE FORCES ON SOLID/LIQUID SEPARATION BY SEDIMENTATION

The two primary factors that influence the efficiency of solid/liquid separation by gravity are the rate of sedimentation and the moisture content (solids concentration) of sediment. The rate of sedimentation should be maximized while the moisture content of the sediment should be minimized.

5.5.1 Sedimentation Rate

The time frame for the stability of a colloidal suspension against gravity depends upon the ratio of the sedimentation flux to the Brownian flux. The sedimentation flux tends to move particles denser than the fluid downward and particles less dense than the fluid upward. The Brownian flux tends to randomize the position of the particles. It is possible to estimate the time frame of stability from the particle and fluid properties assuming that the suspension is stable for the period of time that the average distance travelled by a particle due to Brownian motion is greater than the distance it settles over the same time period. This can be determined by equating Equation (5.6) with a rearranged Stokes' settling law (see Chapter 2) such that:

$$L = \sqrt{\frac{2kT}{3\pi x\mu}t} = \frac{(\rho_p - \rho_f)x^2 g}{18\mu}t \tag{5.12}$$

Solving for (non-zero) time:

$$t = \frac{216kT\mu}{\pi g^2(\rho_p - \rho_f)^2 x^5} \tag{5.13}$$

Because the Brownian distance depends upon the square root of time and the distance settled depends linearly on time, given enough time all suspensions will eventually settle out. The time frame of stability is important when the engineering objective is solid/liquid separation because it is typically only economically viable to conduct solid/liquid separation by sedimentation in a unit operation such as a thickener when the residence times are on the order of hours rather than on the order of weeks or months.

In order to increase the sedimentation rate of colloidal suspensions which would otherwise remain stable for days or weeks, a polymeric flocculant which produces bridging attraction is typically added to the suspension. The attraction between particles results in the formation of aggregates which are larger than the primary particles. The larger aggregates then have sedimentation velocities that exceed the randomizing effect of Brownian motion so that economical solid/liquid separation is possible in conventional thickeners by gravity sedimentation.

5.5.2 Sediment Concentration and Consolidation

The moisture content of the sediment and how the sediment consolidates in response to an applied consolidation pressure (by for instance the weight of the sediment above it or during filtration) depends upon the interparticle forces. During batch sedimentation, given enough time, the sediment/supernatant interface will stop moving downward and a final equilibrium sediment concentration will result. In fact the concentration will vary from the top to the bottom of the sediment due to the local solids pressure because fine particles and colloids typically produce compressible sediments. Although the sedimentation rate is slow when repulsion and Brownian motion dominate, the sediment bed that eventually forms is quite concentrated and approaches a value near random dense packing of monodisperse spheres $[\psi_{max} = (1 - \varepsilon)_{max} = 0.64]$. (Note, here ε refers to the voidage and ϕ is the volume fraction of solids.) This is because the repulsive particles joining the sediment bed are able to rearrange into a lower energy (lower height) position as illustrated in Figure 5.11(a). Attractive particles (and aggregates), however, form sediments that are quite open and contain high levels of residual moisture. This is because the strong attraction between particles creates a strong bond between individual particles that prevents rearrangement into a compact sediment structure as shown in Figure 5.11(b).

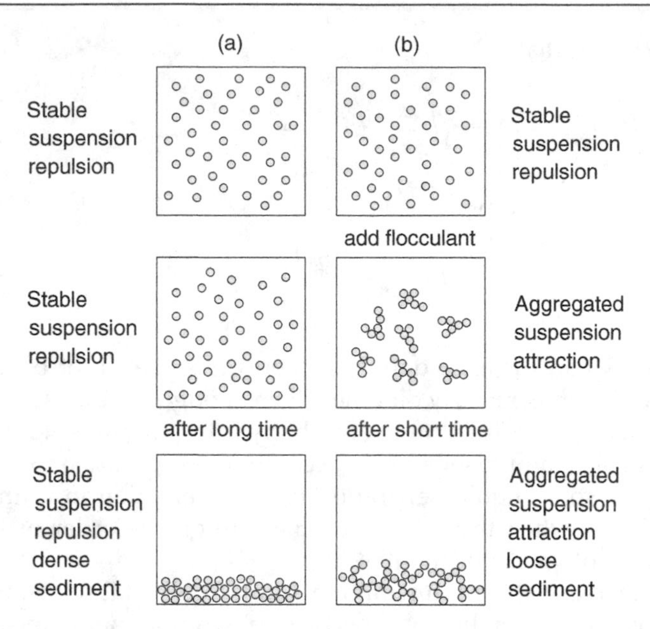

Figure 5.11 (a) Repulsive colloidal particles result in stable dispersions that only form sediments after extended periods. The sediment is quite concentrated. (b) When a flocculant is added to a stable dispersion, the resulting attraction causes aggregation of the particles and rapid sedimentation of the flocs. The sediment in this case is quite open

A pressure may be applied to a particle network in a number of ways including direct application of pressure as in a filter press or centrifuge and by the weight of the particles sitting above a particular level in a sediment. The response of the particle network to the applied pressure depends upon the interparticle force between individual particles. The difference between repulsive particles and attractive particles is demonstrated in Figure 5.12. The repulsive particles (dispersed suspensions) easily pack to near the maximum random close packing limit at all consolidation pressures. Strongly attractive particles have the lowest packing densities at a particular applied pressure and weakly attractive particles have intermediate behaviour.

One can see the dilemma in solid/liquid separation where it appears rapid sedimentation and low sediment moisture are mutually exclusive. Current research in improving solid/liquid separation focuses on controlling the interparticle interaction to be optimized for each step of the separation process; that is, attraction when rapid sedimentation is required and repulsion when consolidation is desired. The approach by the author of this chapter is to use stimulant responsive flocculants to achieve this outcome.

5.6 *SUSPENSION RHEOLOGY*

Rheology is the study of flow and deformation of matter (Barnes *et al.*, 1989). It encompasses a wide range of mechanical behaviour from Hookian elastic

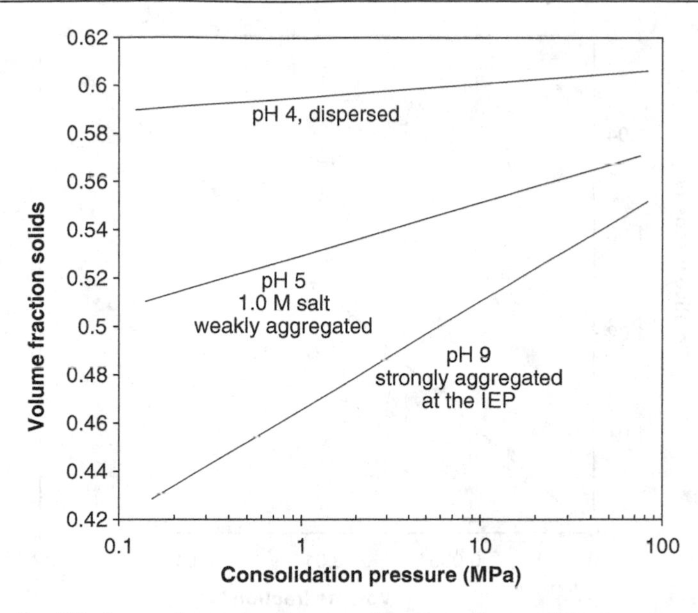

Figure 5.12 Equilibrium volume fraction as a function of consolidation pressure in a filter press (data from Franks and Lange, 1996) for 200 nm diameter alumina. At pH 4 the strong repulsion between particles results in consolidation to high densities over a wide range of pressures. At pH 9, the IEP of the powder, the strong attraction produces difficult to consolidate and pressure-dependent filtration behaviour. The weak attraction at pH 5, with added salt results in intermediate behaviour

behaviour to Newtonian fluid behaviour. Suspensions of particles can exhibit this entire range of behaviour from Newtonian liquids with viscosities near water to high yield stress and high viscosity pastes such as mortar or toothpaste. The parameters that mainly influence the rheological behaviour of suspensions are the volume fraction of solids, the viscosity of the fluid, surface forces between particles, particle size and particle shape.

First, in this section, the influence of volume fraction of particles is discussed in the case where there are no surface forces between particles. Only hydrodynamic forces and Brownian motion are considered in this case, which is known as the non-interacting hard sphere model. The influence of surface forces is considered in the following section.

Consider a molecular liquid with Newtonian behaviour (see Chapter 4) such as water, benzene, alcohol, decane, etc. The addition of a spherical particle to the liquid will increase its viscosity due to the additional energy dissipation related to the hydrodynamic interaction between the liquid and the sphere. Further addition of spherical particles increases the viscosity of the suspension linearly. Einstein developed the relationship between the viscosity of a dilute suspension and the volume fraction of solid spherical particles as follows (Einstein, 1906):

$$\mu_s = \mu_l(1 + 2.5\phi) \tag{5.14}$$

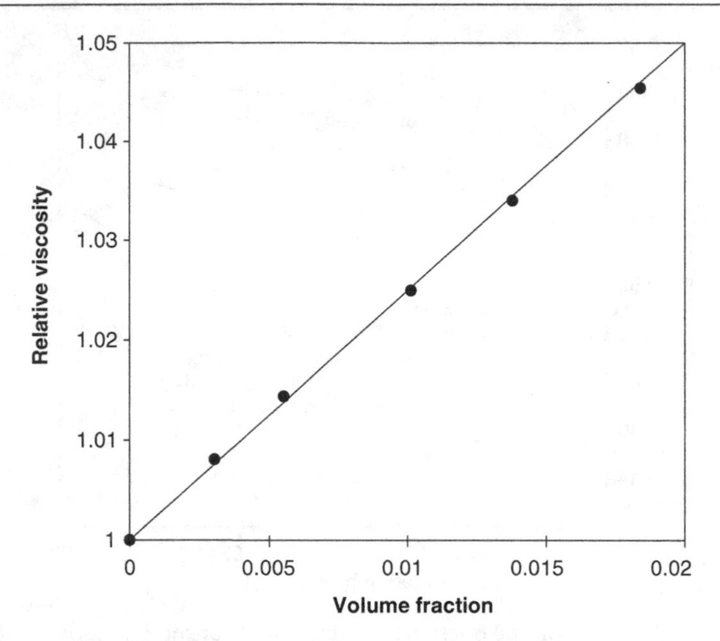

Figure 5.13 Relative viscosity (μ_s/μ_l) of hard sphere silica particle suspensions (black circles) and Einstein's relationship (line). (Data from Jones *et al.*, 1991)

where μ_s is the suspension viscosity, μ_l is the liquid viscosity and ϕ is the volume fraction of solids ($\phi = 1 - \varepsilon$ where ε is the voidage). Note that the viscosity of the suspension remains Newtonian and follows Einstein's prediction when the volume fraction of solids is less than about 7%. This relationship has been verified extensively. Figure 5.13 shows the relationship as measured by Jones *et al.*, (1991) for silica spheres.

Einstein's analysis was based on the assumption that the particles are far enough apart so that they do not influence each other. Once the volume fraction of solids reaches about 10%, the average separation distance between particles is about equal to their diameter. This is when the hydrodynamic disturbance of the liquid by one sphere begins to influence other spheres. In this semi-dilute concentration regime (about 7–15 vol% solids), the hydrodynamic interactions between spheres results in positive deviation for Einstein's relationship. Batchelor (1977) extended the analysis to include higher order terms in volume fraction and found that the suspension viscosities are still Newtonian but increase with volume fraction according to:

$$\mu_s = \mu_l(1 + 2.5\phi + 6.2\phi^2) \tag{5.15}$$

At even higher concentrations of particles, the particle–particle hydrodynamic interactions become even more significant and the suspension viscosity increases even faster than predicted by Batchelor and the suspension rheology becomes shear thinning (see Chapter 4) rather than Newtonian.

Brownian motion dominates the behaviour of concentrated suspensions at rest and at low shear rate such that a random particle structure results that produces a

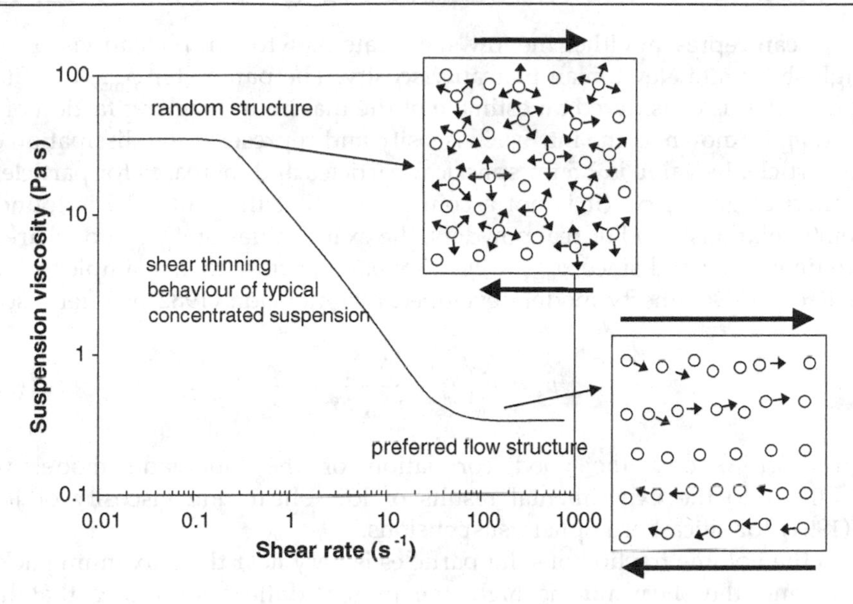

Figure 5.14 The transition from Brownian dominated random structures to preferred flow structures as shear rate is increased is the mechanism for the shear thinning behaviour of concentrated suspensions of hard sphere colloids

viscosity dependent upon the particle volume fraction. As shown in Figure 5.14 there is typically a range of low shear rates over which the viscosity is independent of shear rate. This region is commonly referred to as the low shear rate Newtonian plateau. At high shear rates, hydrodynamic interactions are more significant than Brownian motion and preferred flow structures such as sheets and strings of particles develop as in Figure 5.14. The viscosity of suspensions with such preferred flow structures is much lower than the viscosity of the same volume fraction suspension with randomized structure. The preferred flow structure that minimizes the particle–particle hydrodynamic interaction develops naturally as the shear rate is increased. There is typically a range of high shear rates where the viscosity reaches a plateau. The shear thinning behaviour observed in concentrated hard sphere suspensions is the transition from the randomized structure of the low shear rate Newtonian plateau to the fully developed flow structure of the high shear rate Newtonian plateau as illustrated in Figure 5.14.

As the volume fraction of particles continues to increase, the viscosities of suspensions continue to increase and diverge to infinity as the maximum packing fraction of the powder is approached. Although there is currently no first principles model that is able to predict the rheological behaviour of concentrated suspensions, there are a number of semi-empirical models that are useful in describing concentrated suspension behaviour. The Kreiger–Dougherty model (Kreiger and Dougherty, 1959) takes the form:

$$\mu_s^* = \mu_l \left(1 - \frac{\phi}{\phi_{\max}}\right)^{-[\eta]\phi_{\max}} \tag{5.16}$$

where μ_s^* can represent either the low shear rate Newtonian plateau viscosity or the high shear rate Newtonian plateau viscosity. The parameter ϕ_{max} is a fitting parameter that is considered an estimate of the maximum packing faction of the powder. $[\eta]$ is known as the intrinsic viscosity and represents the dissipation of a single particle. Its value is 2.5 for spherical particles and increases for particles of non-spherical geometry. (It is not a coincidence that the value 2.5 is found in Einstein's relationship.) For real powders, the exact values of ϕ_{max} and $[\eta]$ are not easy to determine and since $\phi_{max} \times [\eta](2.5 \times 0.64)$ is close to 2, a simpler version of the Kreiger–Dougherty model developed by Quemada (1982) is often used.

$$\mu_s^* = \mu_l \left(1 - \frac{\phi}{\phi_{max}}\right)^{-2} \tag{5.17}$$

Figure 5.15 shows the good correlation of the Quemada model with $\phi_{max} = 0.631$ to the experimental results of low shear rate viscosity of Jones *et al.* (1991) for silica hard sphere suspensions.

When the volume fraction of solid particles is very near the maximum packing fraction, and the shear rate is high, the preferred flow structures that have developed become unstable. The large magnitude hydrodynamic interactions push particles together into clusters that do not produce good flow structures. In fact, these hydrodynamic clusters can begin to jam the entire flowing suspension and result in an increase in viscosity. Depending on the conditions, this shear

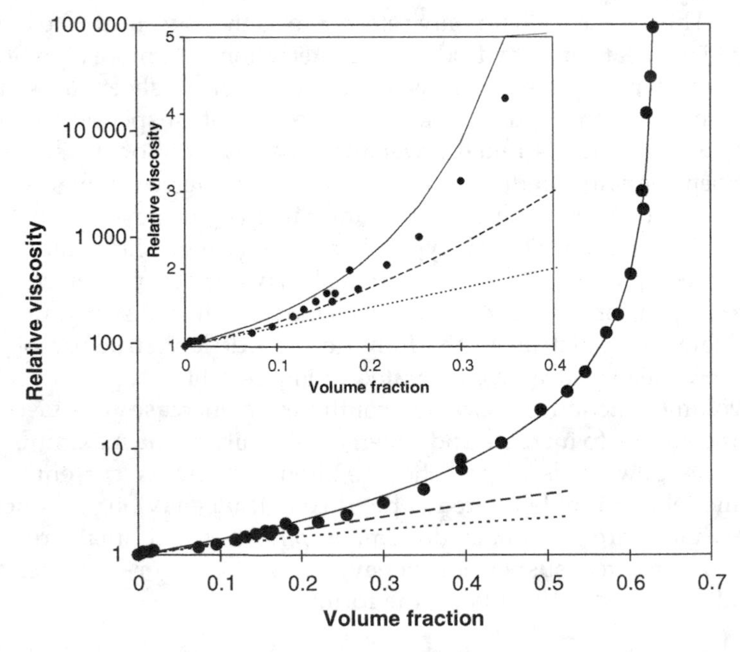

Figure 5.15 Relative viscosity (μ_s/μ_l) at low shear rate of hard sphere silica suspensions (circles). Quemada's model (solid line) with $\phi_{max} = 0.631$; Batchelor's model (dashed line) and Einstein's model (dotted line). (Data from Jones *et al.*, 1991)

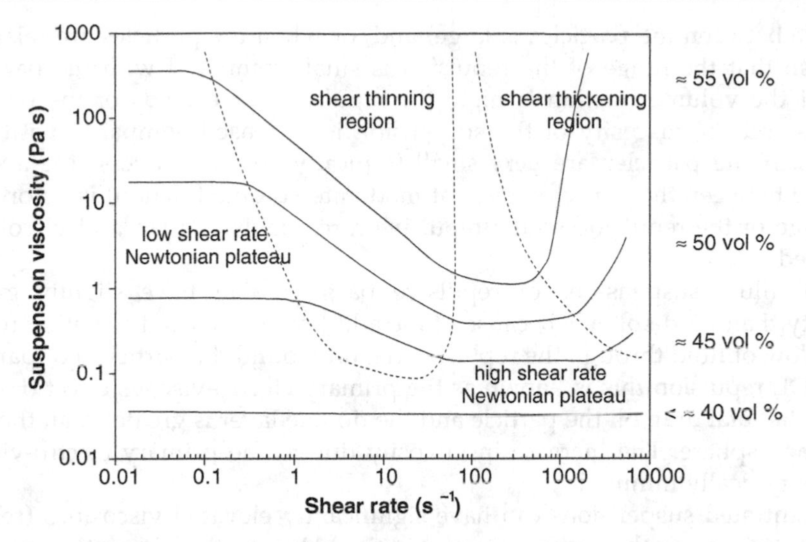

Figure 5.16 Map of typical rheological behaviour of hard sphere suspensions as a function of shear rate for suspensions with volume fractions between about 40 and 55 vol % solid particles. The dashed lines indicate the approximate location of the boundaries between Newtonian and non-Newtonian behaviour

thickening (increase in viscosity with shear rate) or dilatancy may be gradual or abrupt. Strictly speaking, the term dilatancy means that the suspension volume must increase (dilate) for particles to be able to flow past one another, but the term dilatancy is commonly used interchangeably with shear thickening (any increase in viscosity with increasing shear rate). Figure 5.16 shows the behaviour of typical hard sphere suspensions over a wide range of particle concentrations and shear rates.

Note that for hard spheres, surprisingly there is no influence of the particle size on the viscosity. The only concern about particle size in hard sphere suspensions is that if the particles are too big they will settle out. If the particles are neutrally buoyant sedimentation issues are not significant.

5.7 INFLUENCE OF SURFACE FORCES ON SUSPENSION FLOW

The second factor to influence fine particle and colloidal suspension rheology is the interaction force between the particles. (The first factor being the volume fraction of particles.) The sense (attractive or repulsive), range and magnitude of the surface forces all influence the suspension rheological behaviour.

5.7.1 Repulsive Forces

Particles that interact with long range repulsive forces behave much like hard spheres when the distance between the particles is larger than the range of the repulsive force. This is usually the case when the volume fraction is low (average distance between the particles is large) and/or when the particles are relatively large (so that the range of the repulsion is small compared with the particles' size). If the volume fraction is high, the repulsive force fields of the particles overlap and the viscosity of the suspension is increased compared with hard spheres. If the particles are very small (typically 100 nm or less) the average distance between the particles (even at moderate volume fraction) is on order of the range of the repulsion so the repulsive force fields overlap and viscosity is increased.

Even dilute suspensions of repulsive particles will have slightly greater viscosity than hard spheres because of the additional viscous dissipation related to the flow of fluid through the repulsive region around the particle. For particles with EDL repulsion this is known as the primary electro-viscous effect (Hunter, 2001). The total drag on the particle and the double layer is greater than the drag on a hard sphere. The increase in viscosity due to the primary electro-viscous effect is typically minimal.

Concentrated suspensions can have significantly elevated viscosities (relative to hard spheres at the same volume fraction) due to the interaction between overlapping EDLs. For particles to push past each other the double layer must be distorted. This effect is known as the secondary electro-viscous effect (Hunter, 2001). Similar effects occur when the repulsion is by steric mechanism.

The influence of repulsive forces on suspension viscosity is usually handled by considering the effective volume fraction of the particles. The effective volume fraction is the volume fraction of the particles plus the fraction of volume occupied by the repulsive region around the particle.

$$\phi_{\mathrm{eff}} = \frac{\text{volume of solid} + \text{excluded volume}}{\text{total volume}} \tag{5.18}$$

The effective volume fraction accounts for the volume fraction of fluid that cannot be occupied by particles because they are excluded from that region by the repulsive force as illustrated in Figure 5.17. The suspension rheology can be modelled reasonably well using the Kreiger–Dougherty or Quemada model with ϕ_{eff} in place of ϕ.

5.7.2 Attractive Forces

There is a fundamental difference between the rheological behaviour of hard sphere or repulsive particle suspensions and attractive particle suspensions due to the attractive bond between particles. The bonds between particles must be broken in order to pull the particles apart to allow flow to occur. The result of the attractive bonds between particles is that an attractive particle network is formed when the suspension is at rest. The attractive bonding produces material behaviour that is characterized by viscoelasticity, a yield stress (minimum stress required for flow) and shear thinning behaviour.

The shear thinning of an attractive particle network is more pronounced than for hard sphere suspensions of the same particles at the same volume fraction

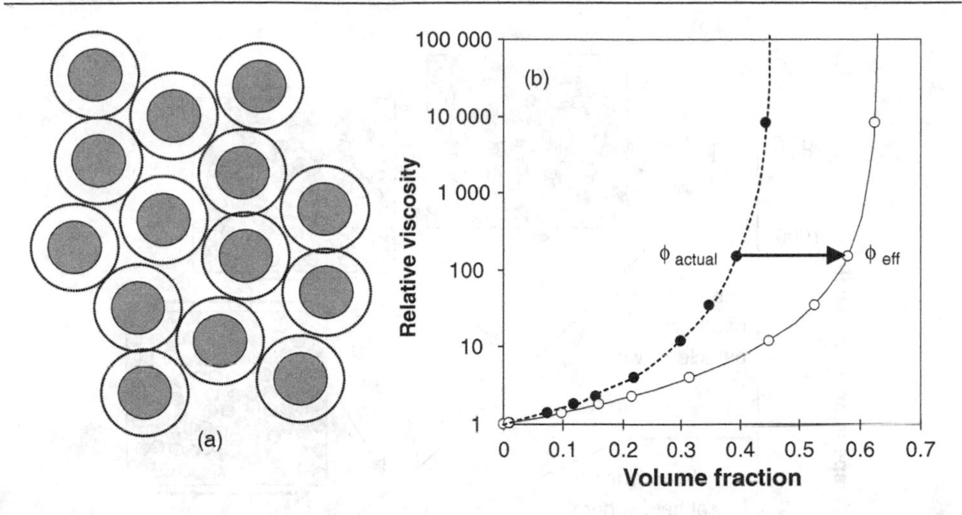

Figure 5.17 (a) Illustration of suspension of particles with volume fraction 0.4 (grey circles) with repulsive interaction extending to the dotted line, resulting in an effective volume fraction of 0.57. (b) Relative viscosity of suspensions of repulsive particles (black dots and dotted line) as a function of actual volume fraction. When the rheological results are plotted as a function of effective volume fraction (open dots) the data maps onto the Quemada model (solid line)

and is caused by a different mechanism. The mechanism for shear thinning is illustrated in Figure 5.18. At rest, the particle network spans the entire volume of the container and resists flow. At low shear rates, the particle network is broken up in to large clusters that flow as units. A large amount of liquid is trapped within the particle clusters and the viscosity is high. As shear rate is increased and hydrodynamic forces overcome interparticle attraction, the particle clusters are broken down into smaller and smaller flow units releasing more and more liquid and reducing the viscosity. At very high shear rates the particle network is completely broken down and the particles flow as individuals again almost as if they were non-interacting.

Greater magnitude interparticle attraction results in increased viscosities at all shear rates. Thus stronger attraction between particles results in higher viscosities. Attractive particle networks also exhibit a yield stress (minimum stress required for flow) because there is an attractive force [as shown in Figure 5.2(a) and (b)] which must be exceeded in order to pull two particles apart. The yield stress of the suspension also depends upon the magnitude of the attraction, with stronger attraction resulting in higher yield stresses. Figure 5.19 shows an example of the yield stress of alumina suspensions as a function of pH. The maximum in yield stress corresponds with the IEP of the powder. At low salt concentration, as the pH is adjusted away from the IEP the EDL repulsion increases as the zeta potential increases (see Figure 5.7) thus reducing the overall attraction and decreasing the yield stress. As the salt concentration is increased at pH away from the IEP, the magnitude of the zeta potential decreases (see Figure 5.7) and the EDL repulsion decreases. As such, the resulting overall interaction is attractive and the attraction increases as salt content is increased.

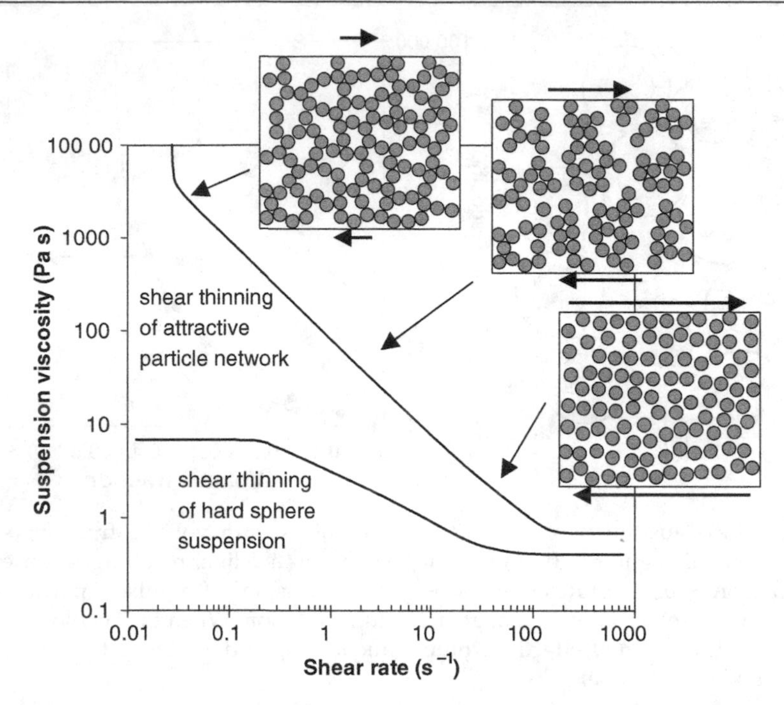

Figure 5.18 Comparison of typical shear thinning behaviour of attractive particle network with less pronounced shear thinning of hard sphere suspensions. The attractive particle network is broken down into smaller flow units as the shear rate is increased

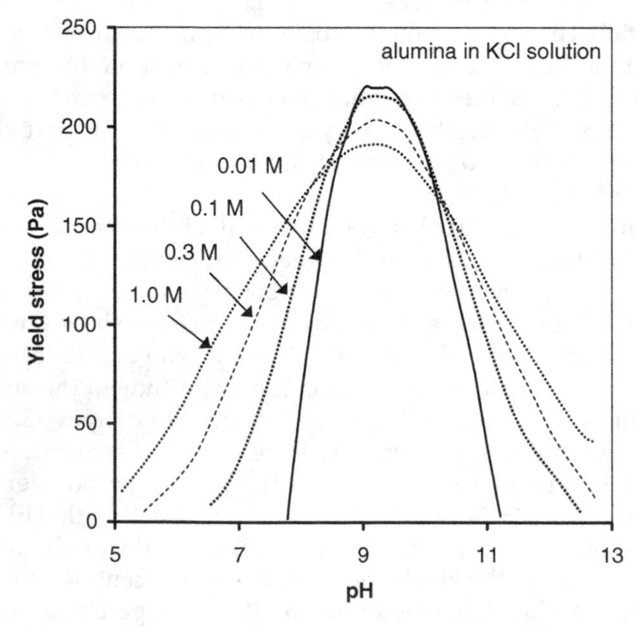

Figure 5.19 Yield stress of 25 vol % alumina suspensions (0.3 μm diameter) as a function of pH and salt concentration. (Data from Johnson *et al.*, 1999)

The result presented in Figure 5.19 indicates that the yield stress increases as salt is increased at pH away from the IEP consistent with the force predictions.

It has been pointed out in the previous section that the rheological behaviour of hard sphere suspensions (non-interacting particles) was not influenced by the size of the particles. This is not true of attractive particle networks. When the particles in a suspension are attractive, smaller particle size results in increased rheological properties such as yield stress, viscosity and elastic modulus (described next). The influence of particle size can be determined by considering that the rheological properties of attractive particle networks depend upon the strength of the bond between particles and the number of bonds per unit volume that need to be broken. For example, consider the shear yield stress:

$$\tau_Y \propto \frac{\text{Number of bonds}}{\text{Unit volume}} \times \text{Strength of bond} \tag{5.19}$$

The strength of the bond of an attractive particle network increases linearly with particle size as indicated by Equations (5.8) and (5.11).

$$\text{Strength of bond} \propto x \tag{5.20}$$

This would make one think that the larger size particles result in suspensions with greater yield stress, viscosity and elastic modulus. It is the influence of the number of bonds per unit volume that produces the opposite result. The number of bonds that need to be broken per unit volume depends upon the structure of the particle network and the size of the particles. In the first instance we assume that the structure of the particle network does not vary with particle size. (Details of the aggregate and particle network structure are beyond the scope of the present text.) Then the number of bonds per unit volume simply varies with the inverse cube of the particle size:

$$\frac{\text{Number of bonds}}{\text{Unit volume}} \propto \frac{1}{x^3} \tag{5.21}$$

When the contributions of the strength of the bond and the number of bonds to be broken are considered one finds that the rheological properties such as yield stress, viscosity and elastic modulus vary inversely with the square of the particle size:

$$\tau_Y \propto \left(\frac{1}{x^3} \times x\right) \propto \frac{1}{x^2} \tag{5.22}$$

Many experimental measurements confirm this result although there is considerable variation from the inverse square dependence in many other cases. One example of well controlled experiments that confirm the inverse square dependence of the yield stress on the particle size is shown in Figure 5.20.

When a stress less than the yield stress is applied to an attractive particle network, the network responds with an elastic-like response. The attractive bonds between the particles are stretched rather than broken and when the stress is removed the particles are pulled back together again by the attractive bonds and the suspension returns to near its original shape. Because the stretching and

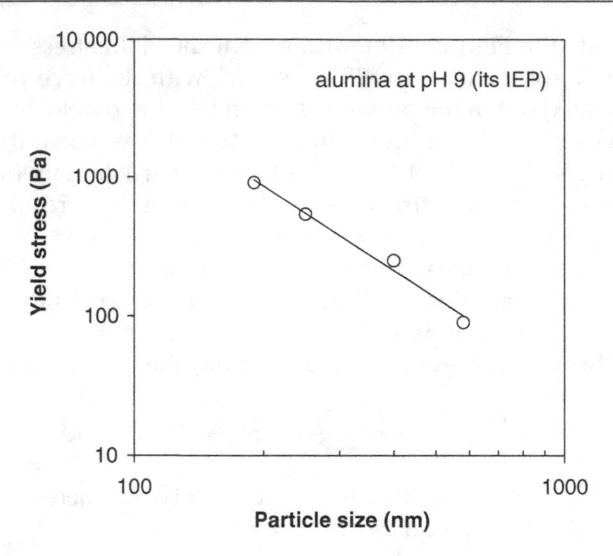

Figure 5.20 Yield stress of alumina suspensions at their IEP as a function of particle size. The best fit line has a slope of −2.01, correlating quite well with the predicted inverse square particle size dependence as per Equation (5.22). (Data from Zhou *et al.*, 2001).

breaking of bonds is a statistical phenomenon, pure elasticity is not usually achieved, rather the attractive particle network is a visco-elastic material displaying behaviour characteristic of both solids and fluids. Some of the energy imparted to deform the material is stored elastically and some of the energy is dissipated by viscous mechanism.

5.8 NANOPARTICLES

Nanoparticles are finding application in many areas of technology due to their unique properties. The very high ratio of surface atoms to bulk atoms is the primary reason for the unique properties of nanoparticles. Due to the large surface area and quantum effects related to very small nanoparticles, the optical, electronic, and other properties are quite different than for larger particles of the same materials. Because of the unusual properties of nanoparticles, there are numerous emerging applications and processes where nanoparticles will be used. Examples of applications of nanoparticles include such diverse topics as:

anti-reflective coatings;

fluorescent labels for biotechnology;

drug delivery systems;

clear inorganic (ZnO) sunscreens;

high performance solar cells;

catalysts;

high density magnetic storage media;

high energy density batteries;

self cleaning glass;

improved LEDs;

high performance fuel cells;

nanostructured materials.

The success of these potential applications for nanoparticles relies heavily on the ability to efficiently produce, transport, separate and safely handle nanoparticles. The concepts presented in this chapter on fine particles and colloids provide a starting point for dealing with these issues.

5.9 *WORKED EXAMPLES*

WORKED EXAMPLE 5.1

Brownian Motion and Settling

Estimate the amount of time that each of the following suspensions will remain stabilized against sedimentation due to Brownian motion at room temperature (300K).

(a) 200 nm diameter alumina ($\rho = 3980\,\mathrm{kg/m^3}$) in water (typical ceramic processing suspension);

(b) 200 nm diameter latex particles ($\rho = 1060\,\mathrm{kg/m^3}$) in water (typical paint formulation);

(c) 150 nm diameter fat globules ($\rho = 780\,\mathrm{kg/m^3}$) in water (homogenized milk);

(d) 1000 nm diameter fat globules ($\rho = 780\,\mathrm{kg/m^3}$) in water (non-homogenized milk).

Solution

The time that the suspension remains stable against gravity can be approximated by equating the average distance moved by a particle due to Brownian motion to the distance settled due to gravity. This time is presented in Equation (5.13) as follows:

$$t = \frac{216kT\mu}{\pi g^2(\rho_p \quad \rho_f)^2 x^5}$$

where $k = 1.381 \times 10^{-23}\,\mathrm{J/K}$, $\mu_{water} = 0.001\,\mathrm{Pa\,s}$, $g = 9.8\,\mathrm{m/s^2}$, $\rho_{water} = 1000\,\mathrm{kg/m^3}$.

then $$t = \frac{216(1.381 \times 10^{-23}\,\mathrm{J/K})300\,\mathrm{K}(0.001\,\mathrm{Pa\,s})}{\pi(9.8\,\mathrm{m/s^2})^2(\rho_p - 1000\,\mathrm{kg/m^3})^2 x^5}$$

$$t = \frac{2.96 \times 10^{-24}\,\mathrm{kg^2\,s/m}}{(\rho_p\,\mathrm{kg/m^3} - 1000\,\mathrm{kg/m^3})^2 x^5 \mathrm{m^5}}$$

(a) For the alumina suspension

$$t = \frac{2.96 \times 10^{-24}\,\text{kg}^2\,\text{s/m}}{(3980\,\text{kg/m}^3 - 1000\,\text{kg/m}^3)^2(200 \times 10^{-9})^5\,\text{m}^5} = 1042\,\text{s} = 17.4\,\text{min}$$

(b) For the latex particles in paint

$$t = \frac{2.96 \times 10^{-24}\,\text{kg}^2\,\text{s/m}}{(1060\,\text{kg/m}^3 - 1000\,\text{kg/m}^3)^2(200 \times 10^{-9})^5\,\text{m}^5} = 2.57 \times 10^6 s = 30\ \text{days}$$

(c) For the homogenized milk

$$t = \frac{2.96 \times 10^{-24}\,\text{kg}^2\,\text{s/m}}{(780\,\text{kg/m}^3 - 1000\,\text{kg/m}^3)^2(150 \times 10^{-9})^5\,\text{m}^5} = 8.06 \times 10^5\,\text{s} = 9.3\ \text{days}$$

(d) For the non-homogenized milk

$$t = \frac{2.96 \times 10^{-24}\,\text{kg}^2\,\text{s/m}}{(780\,\text{kg/m}^3 - 1000\,\text{kg/m}^3)^2(1000 \times 10^{-9})^5\text{m}^5} = 61\,\text{s}$$

These characteristic times correspond best with the time for the first particle to settle out. The time for all the particles to settle out depends upon the height of the container. Nonetheless, one can understand the reasons why the alumina suspension needs to be mixed to keep all of the particles suspended for extended periods of time, why latex paints must be stirred if kept for a month and why milk is homogenized to prevent cream from forming while the milk is in your fridge for a week or two.

WORKED EXAMPLE 5.2

van der Waals and EDL Forces

Use the DLVO equation, $F_T = \pi\varepsilon\varepsilon_o x\Psi_o^2\kappa e^{-\kappa D}(Ax/24\,D^2)$, to plot the total interparticle force (F_T) versus interparticle separation distance (D) for two alumina particles and for two oil droplets under the following conditions. The particles are spherical, 1 μm in diameter and suspended in water that contains 0.01 M NaCl. Plot three conditions for each material: (a) at the IEP; (b) with $\zeta = 30\,\text{mV}$; and (c) with $\zeta = 60\,\text{mV}$. Comment on the differences in the behaviour of the two different materials. Which particles are easier to disperse and why?

Solution

Assume the surface potential equals the zeta potential ($\Psi_o = \zeta$).

Calculate the inverse Debye length (κ) with Equation (5.10).

$$\kappa = 3.29\sqrt{[c]}\,(\text{nm}^{-1}) = \kappa = 3.29\sqrt{0.01}\,(\text{nm}^{-1}) = 0.329\,(\text{nm}^{-1}) = 3.29 \times 10^8\,\text{m}^{-1}$$

The relative permittivity of water (ε) is 80 and the permittivity of free space (ε_o) is $8.854 \times 10^{-12}\,\text{C}^2/\text{J/m}$.

The diameter of the particles is 1×10^{-6} m.

Then

$$F_T = \pi 80(8.854 \times 10^{-12}\,C^2/J/m)(1 \times 10^{-6}\,m)\Psi_o^2(3.29 \times 10^8\,m^{-1})e^{-(3.29\times10^8\,m^{-1})D} - \frac{A(1 \times 10^{-6}\,m)}{24D^2}$$

$$F_T = (7.32 \times 10^{-7}\,C^2/J/m)\Psi_o^2 e^{-(3.2\times\times10^8\,m^{-1})D} - \frac{(1 \times 10^{-6}\,m)A}{24D^2}$$

where Ψ_o is in volts and D in meters.
From Table 5.1 the Hamaker constants (A) are:
For alumina $A = 5.0 \times 10^{-20}$ J
For oil $A = 0.4 \times 10^{-20}$ J
Then for alumina

$$F_T = (7.32 \times 10^{-7}\,C^2/J/m)\Psi_o^2 e^{-(3.29\times10^8\,m^{-1})D} - \frac{(1 \times 10^{-6}\,m)(5 \times 10^{-20}\,J)}{24\,D^2}$$

$$F_T = (7.32 \times 10^{-7}\,C^2/J/m)\Psi_o^2 e^{-(3.29\times10^8 m^{-1})D} - \frac{(5 \times 10^{-26}\,J\,m)}{24\,D^2}$$

and for oil

$$F_T = (7.32 \times 10^{-7}\,C^2/J/m)\Psi_o^2 e^{-(3.29\times10^8\,m^{-1})D} - \frac{(1 \times 10^{-6}\,m)(0.4 \times 10^{-20}\,J)}{24D^2}$$

$$F_T = (7.32 \times 10^{-7}\,C^2/J/m)\Psi_o^2 e^{-(3.29\times10^8 m^{-1})D} - \frac{(0.4 \times 10^{-26}\,J\,m)}{24D^2}$$

These equations can be plotted by a standard data plotting software such as Excel, KG, or Sigmaplot, but first the units and typical values must be checked by hand to insure no errors are made while writing the equation to the spreadsheet.

Unit analysis

$$F_T = C^2/J/m\,V^2 e^{m^{-1}\times m} - \frac{J\,m}{m^2} \quad \text{where } V = J/C$$

$$F_T = C^2/J/m\,J^2/C^2 e^{m^{-1}\times m} - \frac{J\,m}{m^2} \quad \text{so} \quad F_T = m^{-1}J - \frac{J}{m} = \frac{J}{m} \text{ and } J = N\,m \text{ so}$$

F_T is in newtons so the units are OK.

Figures 5W2.1 and 5W2.2 are the plotted results.

The difference between the two materials is that the Hamaker constant for the alumina is much greater than for the oil and thus the attraction between alumina particles is much stronger between alumina than between oil droplets. Thus, it is possible to just stabilize the oil droplets with 30 mV zeta potential, whereas when the alumina has 30 mV zeta potential, there is still attraction between the particles. Hence 60 mV is needed to stabilize the alumina to the same extent that 30 mV was able to stabilize oil droplets.

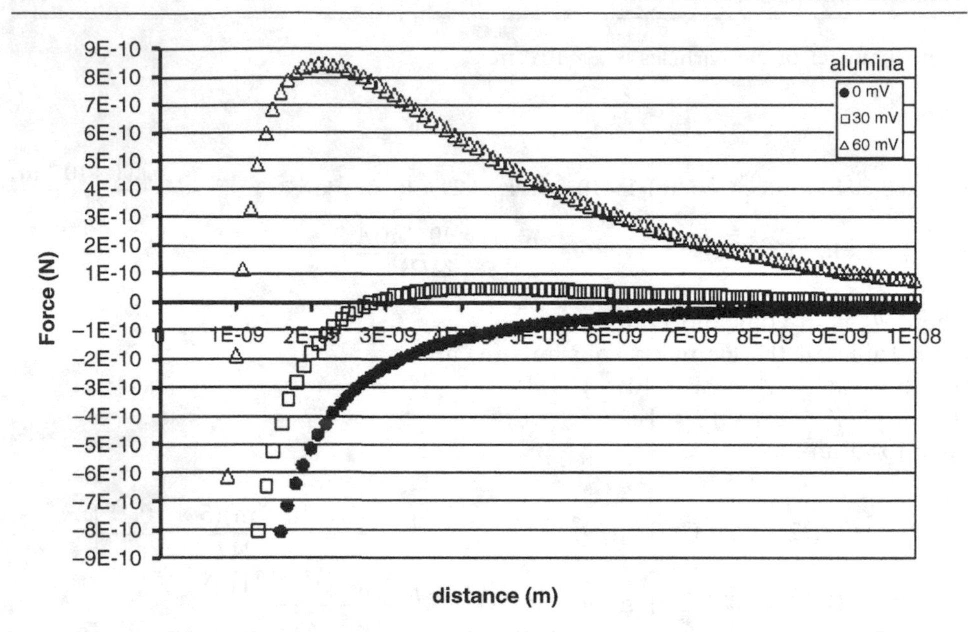

Figure 5W2.1 Force versus separation distance curves for alumina particles

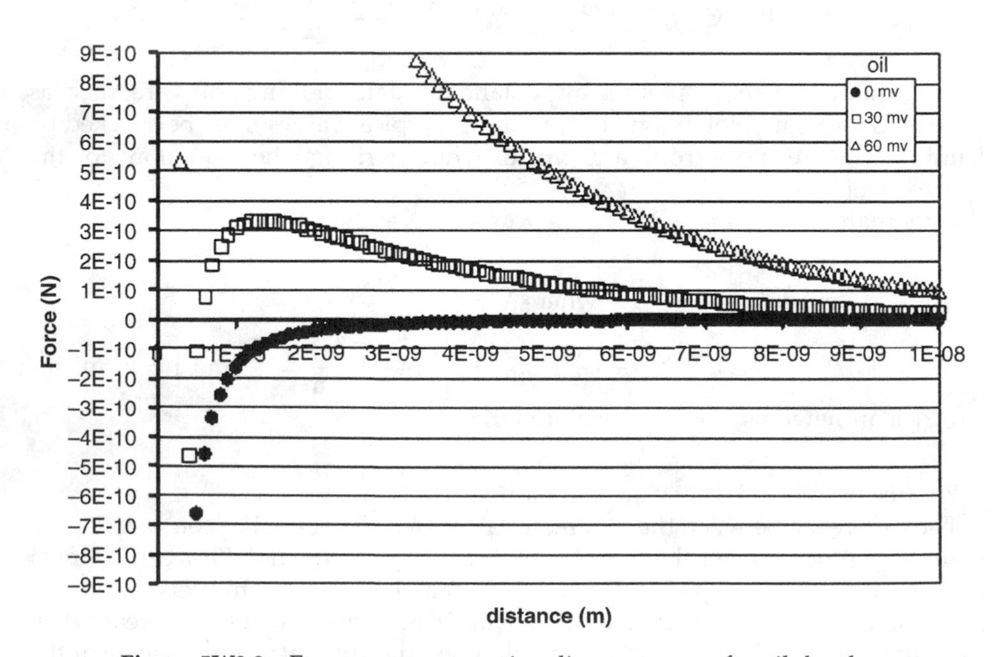

Figure 5W2.2 Force versus separation distance curves for oil droplets

TEST YOURSELF

5.1 What is the typical size range of colloidal particles?

5.2 What two influences are more important for colloidal particles than body forces?

5.3 What is the influence of Brownian motion on a suspension of colloidal particles?

5.4 What is the relationship between surface forces and the potential energy between a pair of particles?

5.5 Under what conditions can you expect van der Waals interactions to be attractive and under what conditions can you expect van der Waals interactions to be repulsive? Which set of conditions is more commonly encountered?

5.6 What are surface hydroxyl groups? What are surface ionization reactions? What is the isoelectric point?

5.7 What is the physical basis for the electrical double layer repulsion between similarly charged particles?

5.8 What is bridging flocculation? What type of polymers are most suitable to induce bridging attraction? What relative surface coverage of the polymer on the particles surface is typically optimum for flocculation?

5.9 What is steric repulsion? What type of polymers are most suitable to induce steric repulsion? What relative surface coverage of the polymer on the particles surface is typically optimum for steric stabilization?

5.10 What is meant by the DLVO theory?

5.11 Why are fine particles in typical atmospheric conditions cohesive? What would happen if all humidity were removed from the air?

5.12 What can be done to particles suspended in a liquid which typically cannot be done to particles suspended in a gas?

5.13 Explain why suspension of repulsive colloidal particles cannot be economically separated from liquid by sedimentation. What is the important parameter that is changed when the particles are flocculated that allows the particles to be ecomomically separated from the liquid by sedimentation?

5.14 How do interparticle interaction forces influence suspension consolidation?

5.15 Why does Einstein's prediction of suspension rheology break down as solids concentration increases above about 7 vol %?

5.16 What is the mechanism for shear thinning of hard sphere suspensions?

5.17 What happens to suspension viscosity as the volume fraction of solids is increased?

5.18 How do repulsive forces influence suspension rheology?

5.19 How do attractive forces influence suspension rheology?

5.20 What three types of rheological behaviour are typical of attractive particle networks?

5.21 What is the mechanism for shear thinning in attractive particle networks?

5.22 Describe the influence of particle size on rheological properties of attractive particle networks.

5.23 How do you think the concepts presented in this chapter will be important for producing products from nano-particles?

EXERCISES

5.1 Colloidal particles may be either 'dispersed' or 'aggregated'.

(a) What causes the difference between these two cases? Answer in terms of inter-particle interactions.

(b) Name and describe at least two methods to create each type of colloidal dispersion.

(c) Describe the differences in the behaviour of the two types of dispersions (including but not limited to rheological behaviour, settling rate, sediment bed properties.)

5.2

(a) What forces are important for colloidal particles? What forces are important for non-colloidal particles?

(b) What is the relationship between interparticle potential energy and interparticle force?

(c) Which three types of rheological behaviour are characteristic of suspensions of attractive particles?

5.3

(a) Describe the mechanism responsible for shear thinning behaviour observed for concentrated suspensions of micrometre sized hard sphere suspensions.

(b) Consider the same suspension as in (a) except instead of hard sphere interactions, the particles are interacting with a strong attraction such as when they are at their isoelectric point. In this case describe the mechanism for the shear thinning behaviour observed.

(c) Draw a schematic plot (log–log) of the relative viscosity as a function of shear rate comparing the behaviour of the two suspensions described in (a) and (b). Be sure to indicate the relative magnitude of the low shear rate viscosites.

(d) Consider two suspensions of particles. All factors are the same except for the particle shape. One suspension has spherical particles and the other rod-shaped particles like grains of rice.

(i) Which suspension will have a higher viscosity?

(ii) What two physical parameters does the shape of the particles influence that affect the suspension viscosity.

5.4

(a) Explain why the permeability of the sediment from a flocculated mineral suspension (less than 5 μm) is greater than the permeability of the sediment of the same mineral suspension that settles while dispersed.

(b) Fine clay particles (approximately 0.15 μm diameter) wash from a farmer's soil into a river due to rain.

(i) Explain why the particles will remain suspended and be carried down stream in the fast flowing fresh water.

(ii) Explain what happens to the clay when the river empties into the ocean.

5.5 Calculate the effective volume fraction for a suspension of 150 nm silica particles at 40 vol % solids in a solution of 0.005 M NaCl.

(Answer: 0.473.)

5.6 You are a sales engineer working for a polymer supply company selling poly acrylic acid (PAA). PAA is a water soluble anionic (negatively charged polymer) that comes in different molecular weights: 10000, 100000, 1 million and 10 million. You have two

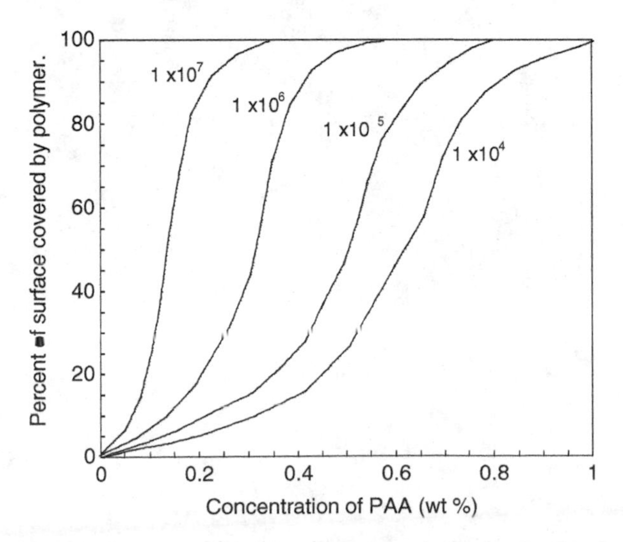

Figure 5E6.1 Adsorption isotherms for PAA of various molecular weights

customers. The first customer is using 0.8 μm alumina to produce ceramics. This customer would like to reduce the viscosity of the suspension of 40 vol % solids suspensions. The second customer is trying to remove 0.8 μm alumina from wastewater. There is about 2 vol % alumina in the water and he wants to remove it by settling. What would you recommend to each customer? Consider if PAA is the right material to use, what molecular weight should be used and how much should be used. Figure 5E6.1 shows the adsorption isotherms for PAA with different molecular weights.

5.7

(a) Draw the typical log μ versus log $\dot{\gamma}$ plot for suspensions of hard spheres of approximately micrometre sized particles at 40, 45, 50 and 55 vol % solids.

(b) Draw the relative viscosity (μ_s/μ_l) versus volume fraction curve for the low shear viscosities of a typical hard sphere suspension.

6

Fluid Flow Through a Packed Bed of Particles

6.1 PRESSURE DROP–FLOW RELATIONSHIP

6.1.1 Laminar Flow

In the nineteenth century Darcy (1856) observed that the flow of water through a packed bed of sand was governed by the relationship:

$$\begin{pmatrix}\text{pressure}\\ \text{gradient}\end{pmatrix} \propto \begin{pmatrix}\text{liquid}\\ \text{velocity}\end{pmatrix} \quad \text{or} \quad \frac{(-\Delta p)}{H} \propto U \tag{6.1}$$

where U is the superficial fluid velocity through the bed and $(-\Delta p)$ is the frictional pressure drop across a bed depth H. (Superficial velocity = fluid volumetric flow rate/cross-sectional area of bed, Q/A.)

The flow of a fluid through a packed bed of solid particles may be analysed in terms of the fluid flow through tubes. The starting point is the Hagen–Poiseuille equation for laminar flow through a tube:

$$\frac{(-\Delta p)}{H} = \frac{32\mu U}{D^2} \tag{6.2}$$

where D is the tube diameter and μ is the fluid viscosity.

Consider the packed bed to be equivalent to many tubes of equivalent diameter D_e following tortuous paths of equivalent length H_e and carrying fluid with a velocity U_i. Then, from Equation (6.2),

$$\frac{(-\Delta p)}{H_e} = K_1 \frac{\mu U_i}{D_e^2} \tag{6.3}$$

U_i is the actual velocity of fluid through the interstices of the packed bed and is related to superficial fluid velocity by:

$$U_i = U/\varepsilon \tag{6.4}$$

where ε is the voidage or void fraction of the packed bed. (Refer to Section 8.1.4 for discussion on actual and superficial velocities.)

Although the paths of the tubes are tortuous, we can assume that their actual length is proportional to the bed depth, that is,

$$H_e = K_2 H \tag{6.5}$$

The tube equivalent diameter is defined as

$$\frac{4 \times \text{flow area}}{\text{wetted perimeter}}$$

where flow area $= \varepsilon A$, where A is the cross-sectional area of the vessel holding the bed; wetted perimeter $= S_B A$, where S_B is the particle surface area per unit volume of the bed.

That this is so may be demonstrated by comparison with pipe flow:

Total particle surface area in the bed $= S_B AH$. For a pipe,

$$\text{wetted perimeter} = \frac{\text{wetted surface}}{\text{length}} = \frac{\pi DL}{L}$$

and so for the packed bed, wetted perimeter $= \dfrac{S_B AH}{H} = S_B A$.

Now if S_v is the surface area per unit volume of particles, then

$$S_v(1 - \varepsilon) = S_B \tag{6.6}$$

since

$$\left(\frac{\text{surface of particles}}{\text{volume of particles}}\right) \times \left(\frac{\text{volume of particles}}{\text{volume of bed}}\right) = \left(\frac{\text{surface of particles}}{\text{volume of bed}}\right)$$

and so

$$\text{equivalent diameter, } D_e = \frac{4\varepsilon A}{S_B} = \frac{4\varepsilon}{S_v(1 - \varepsilon)} \tag{6.7}$$

Substituting Equations (6.4), (6.5) and (6.7) in (6.3):

$$\frac{(-\Delta p)}{H} = K_3 \frac{(1 - \varepsilon)^2}{\varepsilon^3} \mu U S_v^2 \tag{6.8}$$

where $K_3 = K_1 K_2$. Equation (6.8) is known as the Carman–Kozeny equation [after the work of Carman and Kozeny (Carman, 1937; Kozeny, 1927, 1933)] describing laminar flow through randomly packed particles. The constant K_3 depends on particle shape and surface properties and has been found by experiment to have a value of about 5. Taking $K_3 = 5$, for laminar flow through a randomly packed bed of monosized spheres of diameter x (for which $S = 6/x$) the Carman–Kozeny equation becomes:

$$\frac{(-\Delta p)}{H} = 180 \frac{\mu U}{x^2} \frac{(1-\varepsilon)^2}{\varepsilon^3} \tag{6.9}$$

This is the most common form in which the Carman–Kozeny equation is quoted.

6.1.2 Turbulent Flow

For turbulent flow through a randomly packed bed of monosized spheres of diameter x the equivalent equation is:

$$\frac{(-\Delta p)}{H} = 1.75 \frac{\rho_f U^2}{x} \frac{(1-\varepsilon)}{\varepsilon^3} \tag{6.10}$$

6.1.3 General Equation for Turbulent and Laminar Flow

Based on extensive experimental data covering a wide range of size and shape of particles, Ergun (1952) suggested the following general equation for any flow conditions:

$$\frac{(-\Delta p)}{H} = \underbrace{150 \frac{\mu U}{x^2} \frac{(1-\varepsilon)^2}{\varepsilon^3}}_{\begin{pmatrix}\text{laminar}\\ \text{component}\end{pmatrix}} + \underbrace{1.75 \frac{\rho_f U^2}{x} \frac{(1-\varepsilon)}{\varepsilon^3}}_{\begin{pmatrix}\text{turbulent}\\ \text{component}\end{pmatrix}} \tag{6.11}$$

This is known as the Ergun equation for flow through a randomly packed bed of spherical particles of diameter x. Ergun's equation additively combines the laminar and turbulent components of the pressure gradient. Under laminar conditions, the first term dominates and the equation reduces to the Carman–Kozeny equation [Equation (6.9)], but with the constant 150 rather than 180. (The difference in the values of the constants is probably due to differences in shape and packing of the particles.) In laminar flow the pressure gradient increases linearly with superficial fluid velocity and independent of fluid density. Under turbulent flow conditions, the second term dominates; the pressure gradient increases as the square of superficial fluid velocity and is independent of fluid viscosity. In terms of the Reynolds number defined in Equation (6.12), fully laminar condition exist for Re^* less than about 10 and fully turbulent flow exists at Reynolds numbers greater than around 2000.

$$Re^* = \frac{x U \rho_f}{\mu(1-\varepsilon)} \tag{6.12}$$

In practice, the Ergun equation is often used to predict packed bed pressure gradient over the entire range of flow conditions. For simplicity, this practice is followed in the Worked Examples and Exercises in this chapter.

Ergun also expressed flow through a packed bed in terms of a friction factor defined in Equation (6.13):

$$\text{Friction factor, } f^* = \frac{(-\Delta p)}{H} \frac{x}{\rho_f U^2} \frac{\varepsilon^3}{(1-\varepsilon)} \tag{6.13}$$

(Compare the form of this friction factor with the familiar Fanning friction factor for flow through pipes.)

Equation (6.11) then becomes

$$f^* = \frac{150}{Re^*} + 1.75 \tag{6.14}$$

with

$$f^* = \frac{150}{Re^*} \text{ for } Re^* < 10 \quad \text{and} \quad f^* = 1.75 \text{ for } Re^* > 2000$$

(see Figure 6.1).

6.1.4 Non-spherical Particles

The Ergun and Carman–Kozeny equations also accommodate non-spherical particles if x is replaced by x_{sv} the diameter of a sphere having the same surface to volume ratio as the non-spherical particles in question. Use of x_{sv} gives the correct value of specific surface S (surface area of particles per unit volume of particles). The relevance of this will be apparent if Equation (6.8) is recalled.

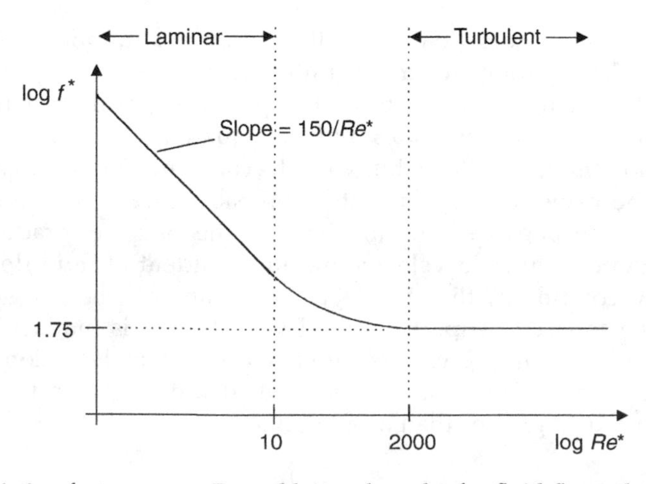

Figure 6.1 Friction factor versus Reynolds number plot for fluid flows through a packed bed of spheres

Thus, in general, the Ergun equation for flow through a randomly packed bed of particles of surface-volume diameter x_{sv} becomes

$$\frac{(-\Delta p)}{H} = 150\frac{\mu U}{x_{sv}^2}\frac{(1-\varepsilon)^2}{\varepsilon^3} + 1.75\frac{\rho_f U^2}{x_{sv}}\frac{(1-\varepsilon)}{\varepsilon^3} \qquad (6.15)$$

and the Carman–Kozeny equation for laminar flow through a randomly packed bed of particles of surface-volume diameter x_{sv} becomes:

$$\frac{(-\Delta p)}{H} = 180\frac{\mu U}{x_{sv}^2}\frac{(1-\varepsilon)^2}{\varepsilon^3} \qquad (6.16)$$

It was shown in Chapter 1 that if the particles in the bed are not mono-sized, then the correct mean size to use in these equations is the surface-volume mean $\bar{x}_{sv}$.

6.2 *FILTRATION*

6.2.1 Introduction

As an example of the application of the above analysis for flow through packed beds of particles, we will briefly consider cake filtration. Cake filtration is widely used in industry to separate solid particles from suspension in liquid. It involves the build up of a bed or 'cake' of particles on a porous surface known as the filter medium, which commonly takes the form of a woven fabric. In cake filtration the pore size of the medium is less than the size of the particles to be filtered. It will be appreciated that this filtration process can be analysed in terms of the flow of fluid through a packed bed of particles, the depth of which is increasing with time. In practice the voidage of the cake may also change with time. However, we will first consider the case where the cake voidage is constant, i.e. an incompressible cake.

6.2.2 Incompressible Cake

First, if we ignore the filter medium and consider only the cake itself, the pressure drop versus liquid flow relationship is described by the Ergun equation [Equation (6.15)]. The particle size and range of liquid flow and properties commonly used in industry give rise to laminar flow and so the second (turbulent) term vanishes. For a given slurry (particle properties fixed) the resulting cake resistance is defined as:

$$\text{cake resistance, } r_c = \frac{150}{x_{sv}^2}\frac{(1-\varepsilon)^2}{\varepsilon^3} \qquad (6.17)$$

and so Equation (6.15) becomes

$$\frac{(-\Delta p)}{H} = r_c \mu U \tag{6.18}$$

If V is the volume of filtrate (liquid) passed in a time t and dV/dt is the instantaneous volumetric flow rate of filtrate at time t, then:

$$\text{superficial filtrate velocity at time } t, \ U = \frac{1}{A}\frac{dV}{dt} \tag{6.19}$$

Each unit volume of filtrate is assumed to deposit a certain mass of particles, which form a certain volume of cake. This is expressed as ϕ, the volume of cake formed by the passage of unit volume of filtrate.

$$\phi = \frac{HA}{V} \tag{6.20}$$

and so Equation (6.18) becomes

$$\frac{dV}{dt} = \frac{A^2(-\Delta p)}{r_c \mu \phi V} \tag{6.21}$$

Constant rate filtration

If the filtration rate dV/dt is constant, then the pressure drop across the filter cake will increase in direct proportion to the volume of filtrate passed V.

Constant pressure drop filtration

If $(-\Delta p)$ is constant then,

$$\frac{dV}{dt} \propto \frac{1}{V}$$

or, integrating Equation (6.21),

$$\frac{t}{V} = C_1 V \tag{6.22}$$

where

$$C_1 = \frac{r_c \mu \phi}{2A^2(-\Delta p)} \tag{6.23}$$

6.2.3 Including the Resistance of the Filter Medium

The total resistance to flow is the sum of the resistance of the cake and the filter medium. Hence,

$$\begin{pmatrix}\text{total pressure}\\ \text{drop}\end{pmatrix} = \begin{pmatrix}\text{pressure drop}\\ \text{across medium}\end{pmatrix} + \begin{pmatrix}\text{pressure drop}\\ \text{across cake}\end{pmatrix}$$

$$(-\Delta p) = (-\Delta p_{\mathrm{m}}) + (-\Delta p_{\mathrm{c}})$$

If the medium is assumed to behave as a packed bed of depth H_{m} and resistance r_{m} obeying the Carman–Kozeny equation, then

$$(-\Delta p) = \frac{1}{A}\frac{\mathrm{d}V}{\mathrm{d}t}(r_{\mathrm{m}}\mu H_{\mathrm{m}} + r_{\mathrm{c}}\mu H_{\mathrm{c}}) \tag{6.24}$$

The medium resistance is usually expressed as the equivalent thickness of cake H_{eq}:

$$r_{\mathrm{m}}H_{\mathrm{m}} = r_{\mathrm{c}}H_{\mathrm{eq}}$$

Hence, combining with Equation (6.20),

$$H_{\mathrm{eq}} = \frac{\phi V_{\mathrm{eq}}}{A} \tag{6.25}$$

where V_{eq} is the volume of filtrate that must pass in order to create a cake of thickness H_{eq}. The volume V_{eq} depends only on the properties of the suspension and the filter medium.

Equation (6.24) becomes

$$\frac{1}{A}\frac{\mathrm{d}V}{\mathrm{d}t} = \frac{(-\Delta p)A}{r_{\mathrm{c}}\mu(V + V_{\mathrm{eq}})\phi} \tag{6.26}$$

Considering operation at constant pressure drop, which is the most common case, integrating Equation (6.26) gives:

$$\frac{t}{V} = \frac{r_{\mathrm{c}}\phi\mu}{2A^2(-\Delta p)}V + \frac{r_{\mathrm{c}}\phi\mu}{A^2(-\Delta p)}V_{\mathrm{eq}} \tag{6.27}$$

6.2.4 Washing the Cake

The solid particles separated by filtration often must be washed to remove filtrate from the pores. There are two processes involved in washing. Much of the filtrate occupying the voids between particles may be removed by displacement as clean solvent is passed through the cake. Removal of filtrate held in less accessible regions of the cake and from pores in the particles takes place by diffusion into

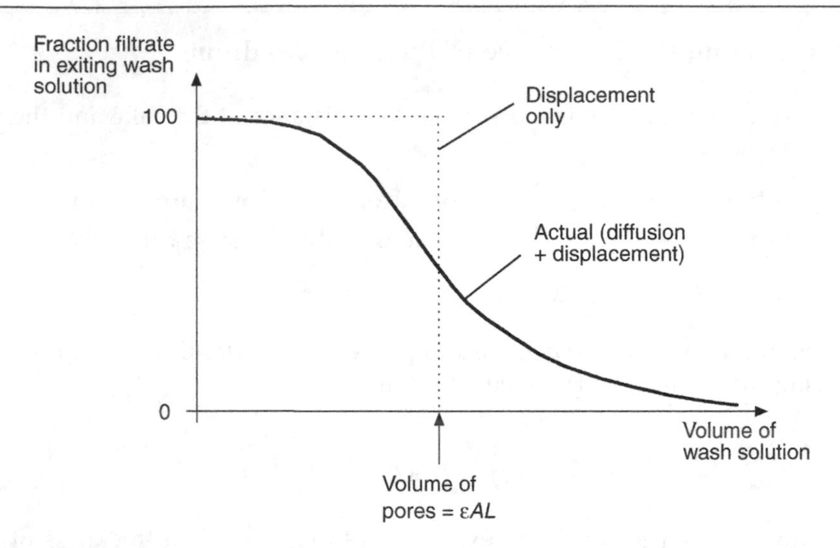

Figure 6.2 Removal of filtrate during washing of the filter cake

the wash water. Figure 6.2 shows how the filtrate concentration in the wash solvent leaving the cake varies typically with volume of wash solvent passed.

6.2.5 Compressible Cake

In practice many materials give rise to compressible filter cakes. A compressible cake is one whose cake resistance r_c increases with applied pressure difference $(-\Delta p)$. Change in r_c is due mainly to the effect on the cake voidage [recall Equation (6.17)]. Fluid drag on the particles in the cake causes a force which is transmitted through the bed. Particles deeper in the bed experience the sum of the forces acting on the particles above. The force on the particles causes the particle packing to become more dense, i.e. cake voidage decreases. In the case of soft particles, the shape or size of the particles may change, adding to the increase in cake resistance.

Referring to Figure 6.3, liquid flows at a superficial velocity U through a filter cake of thickness H. Consider an element of the filter cake of thickness $\mathrm{d}L$ across which the pressure drop is $\mathrm{d}p$. Applying the Carman–Kozeny equation [Equation (6.18)] for flow through this element,

$$-\frac{\mathrm{d}p}{\mathrm{d}L} = r_c \mu U \tag{6.28}$$

where r_c is the resistance of this element of the cake. For a compressible cake, r_c is a function of the pressure difference between the upstream surface of the cake and the element (i.e. referring to Figure 6.3, $p_1 - p$).

$$\text{Letting} \quad p_s = p_1 - p \tag{6.29}$$

$$\text{then} \quad -\mathrm{d}p = \mathrm{d}p_s \tag{6.30}$$

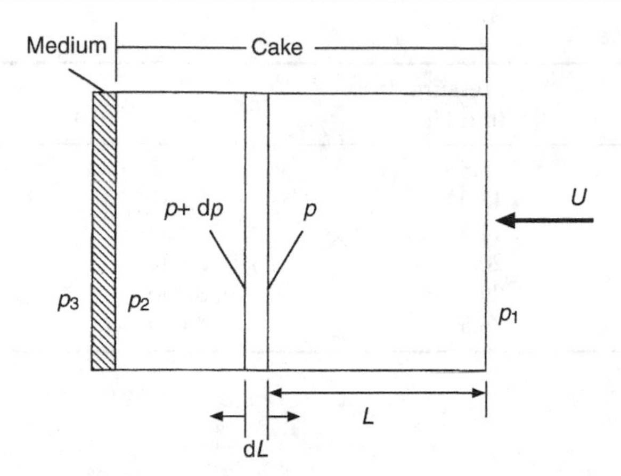

Figure 6.3 Analysis of the pressure drop – flow relationship for a compressible cake

and Equation (6.28) becomes

$$\frac{dp_s}{dL} = r_c \mu U \tag{6.31}$$

In practice the relationship between r_c and p_s must be found from laboratory experiments before Equation (6.29) can be used in design.

6.3 *FURTHER READING*

For further information on fluid flow through packed beds and on filtration the reader is referred to the following:

Coulson, J. M. and Richardson, J. R. (1991) *Chemical Engineering*, Vol. 2, *Particle Technology and Separation Processes*, 5th Edition, Pergamon, Oxford.
Perry, R. H. and Green, D. (eds) (1984) *Perry's Chemical Engineering Handbook*, 6th or later editions, McGraw-Hill, New York.

6.4 *WORKED EXAMPLES*

WORKED EXAMPLE 6.1

Water flows through 3.6 kg of glass particles of density 2590 kg/m^3 forming a packed bed of depth 0.475 m and diameter 0.0757 m. The variation in frictional pressure drop across the bed with water flow rate in the range 200–1200 cm^3/min is shown in columns one and two in Table 6.W1.1.

(a) Demonstrate that the flow is laminar.

(b) Estimate the mean surface-volume diameter of the particles.

(c) Calculate the relevant Reynolds number.

Table 6W1.1

Water flow rate (cm^3/min)	Pressure drop (mmHg)	U (m/s $\times 10^4$)	Pressure drop (Pa)
200	5.5	7.41	734
400	12.0	14.81	1600
500	14.5	18.52	1935
700	20.5	25.92	2735
1000	29.5	37.00	3936
1200	36.5	44.40	4870

Solution

(a) First, convert the volumetric water flow rate values into superficial velocities and the pressure drop in millimetres of mercury into pascal. These values are shown in columns 3 and 4 of Table 6W1.1.

If the flow is laminar then the pressure gradient across the packed bed should increase linearly with superficial fluid velocity, assuming constant bed voidage and fluid viscosity. Under laminar conditions, the Ergun equation [Equation (6.15)] reduces to:

$$\frac{(-\Delta p)}{H} = 150\frac{\mu U}{x_{sv}^2}\frac{(1-\varepsilon)^2}{\varepsilon^3}$$

Hence, since the bed depth H, the water viscosity μ and the packed bed voidage ε may be assumed constant, then $(-\Delta p)$ plotted against U should give a straight line of gradient

$$150\frac{\mu H}{x_{sv}^2}\frac{(1-\varepsilon)^2}{\varepsilon^3}$$

This plot is shown in Figure 6W1.1. The data points fall reasonably on a straight line confirming laminar flow. The gradient of the straight line is 1.12×10^6 Pa s/m and so:

$$150\frac{\mu H}{x_{sv}^2}\frac{(1-\varepsilon)^2}{\varepsilon^3} = 1.12 \times 10^6 \text{ Pa} \cdot \text{s/m}$$

(b) Knowing the mass of particles in the bed, the density of the particles and the volume of the bed, the voidage may be calculated:

$$\text{mass of bed} = AH(1-\varepsilon)\rho_p$$
$$\text{giving } \varepsilon = 0.3497$$

Substituting $\varepsilon = 0.3497$, $H = 0.475$ m and $\mu = 0.001$ Pa s in the expression for the gradient of the straight line, we have

$$x_{sv} = 792\,\mu\text{m}$$

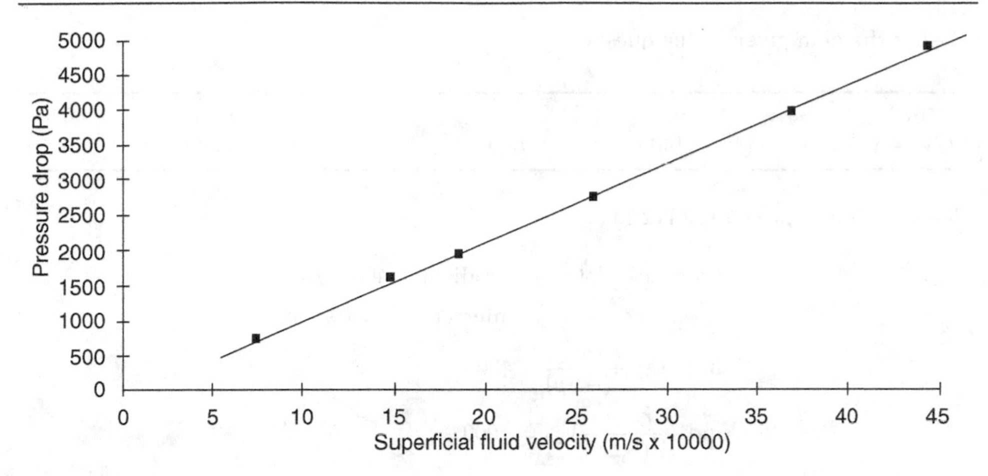

Figure 6W1.1 Plot of packed bed pressure drop versus superficial fluid velocity

(c) The relevant Reynolds number is $Re^* = \dfrac{xU\rho_f}{\mu(1-\varepsilon)}$ [Equation (6.12)] giving $Re^* = 5.4$ (for the maximum velocity used). This is less than the limiting value for laminar flow (10); a further confirmation of laminar flow.

WORKED EXAMPLE 6.2

A leaf filter has an area of 0.5 m^2 and operates at a constant pressure drop of 500 kPa. The following test results were obtained for a slurry in water which gave rise to a filter cake regarded as incompressible:

Volume of filtrate collected (m^3)	0.1	0.2	0.3	0.4	0.5
Time (s)	140	360	660	1040	1500

Calculate:

(a) the time need to collect 0.8 m^3 of filtrate at a constant pressure drop of 700 kPa;

(b) the time required to wash the resulting cake with 0.3 m^3 of water at a pressure drop of 400 kPa.

Solution

For filtration at constant pressure drop we use Equation (6.27), which indicates that if we plot t/V versus V a straight line will have a gradient

$$\frac{r_c\phi\mu}{2A^2(-\Delta p)}$$

and an intercept $\dfrac{r_c\psi\mu}{A^2(-\Delta p)}V_{eq}$ on the t/V axis.

Using the data given in the question:

$V(\mathrm{m^3})$	0.1	0.2	0.3	0.4	0.5
$t/V\ (\mathrm{s/m^3})$	1400	1800	2200	2600	3000

This is plotted in Figure 6W2.1.

$$\text{From the plot:} \quad \text{gradient} = 4000\,\mathrm{s/m^6}$$
$$\text{intercept} = 1000\,\mathrm{s/m^3}$$

$$\text{hence } \frac{r_c \phi \mu}{2A^2(-\Delta p)} = 4000$$

$$\text{and } \frac{r_c \phi \mu}{A^2(-\Delta p)} V_{eq} = 1000$$

which, with $A = 0.5\,\mathrm{m^2}$ and $(-\Delta p) = 500 \times 10^3\,\mathrm{Pa}$, gives

$$r_c \phi \mu = 1 \times 10^9\,\mathrm{Pa\,s/m^2}$$

and $V_{eq} = 0.125\,\mathrm{m^3}$

Substituting in Equation (6.27):

$$\frac{t}{V} = \frac{0.5 \times 10^9}{(-\Delta p)}(4V + 1)$$

which applies to the filtration of the same slurry in the same filter at any pressure drop.

(a) To calculate the time required to pass $0.8\,\mathrm{m^3}$ of filtrate at a pressure drop of 700 kPa, we substitute $V = 0.8\,\mathrm{m^3}$ and $(-\Delta p) = 700 \times 10^3\,\mathrm{Pa}$ in the above equation, giving

$$t = 2400\,\mathrm{s}\ (\text{or } 40\ \text{min})$$

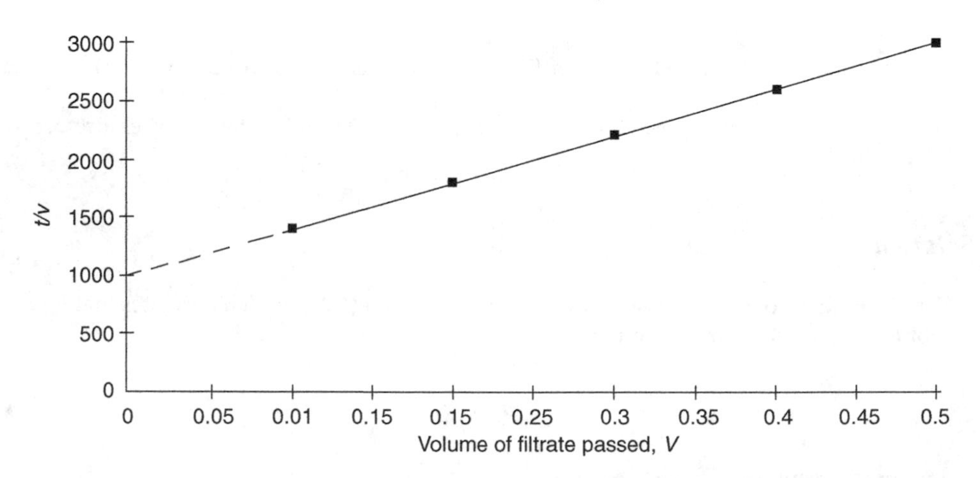

Figure 6W2.1 Plot of t/V versus V

(b) During the filtration the cake thickness is continuously increasing and, since the pressure drop is constant, the volume flow rate of filtrate will continuously decrease. The filtration rate is given by Equation (6.26). Substituting the volume of filtrate passed at the end of the filtration period ($V = 0.8\,\text{m}^3$), $r_c\phi\mu = 1 \times 10^9\,\text{Pa s/m}^2$, $V_{eq} = 0.125\,\text{m}^3$ and $(-\Delta p) = 700 \times 10^3\,\text{Pa}$, we find the filtration rate at the end of the filtration period is $dV/dt = 1.89 \times 10^{-4}\,\text{m}^3/\text{s}$.

If we assume that the wash water has the same physical properties as the filtrate, then during a wash period at a pressure drop of 700 kPa the wash rate is also $1.89 \times 10^{-4}\,\text{m}^3/\text{s}$. However, the applied pressure drop during the wash cycle is 400 kPa. According to Equation (6.26) the liquid flow rate is directly proportional to the applied pressure drop, and so

$$\text{flow rate of wash water (at 400 kPa)} = 1.89 \times 10^{-4} \times \left(\frac{400 \times 10^3}{700 \times 10^3}\right)$$
$$= 1.08 \times 10^{-4}\,\text{m}^3/\text{s}$$

Hence, the time needed to pass $0.3\,\text{m}^3$ of wash water at this rate is 2778 s (or 46.3 min).

TEST YOURSELF

6.1 For low Reynolds number (<10) flow of a fluid through a packed bed of particles how does the frictional pressure drop across the bed depend on (a) superficial fluid velocity, (b) particle size, (c) fluid density, (d) fluid viscosity and (e) voidage?

6.2 For high Reynolds number (>500) flow of a fluid through a packed bed of particles how does the frictional pressure drop across the bed depend on (a) superficial fluid velocity, (b) particle size, (c) fluid density, (d) fluid viscosity and (e) voidage?

6.3 What is the correct mean particle diameter to be used in the Ergun equation? How can this diameter be derived from a volume distribution?

6.4 During constant pressure drop filtration of an incompressible cake, how does filtrate flow rate vary with time?

EXERCISES

6.1 A packed bed of solid particles of density $2500\,\text{kg/m}^3$ occupies a depth of 1 m in a vessel of cross-sectional area $0.04\,\text{m}^2$. The mass of solids in the bed is 50 kg and the surface-volume mean diameter of the particles is 1 mm. A liquid of density $800\,\text{kg/m}^3$ and viscosity 0.002 Pa s flows upwards through the bed, which is restrained at its upper surface.

(a) Calculate the voidage (volume fraction occupied by voids) of the bed.

(b) Calculate the frictional pressure drop across the bed when the volume flow rate of liquid is $1.44\,\text{m}^3/\text{h}$.

[Answer: (a) 0.50; (b) 6560 Pa (Ergun).]

6.2 A packed bed of solids of density 2000 kg/m^3 occupies a depth of 0.6 m in a cylindrical vessel of inside diameter 0.1 m. The mass of solids in the bed is 5 kg and the surface-volume mean diameter of the particles is 300 μm. Water (density 1000 kg/m^3 and viscosity 0.001 Pa s) flows upwards through the bed.

(a) What is the voidage of the packed bed?

(b) Calculate the frictional superficial liquid velocity at which the pressure drop across the bed is 4130 Pa.

[Answer: (a) 0.4692; (b) 1.5 mm/s (Ergun).]

6.3 A gas absorption tower of diameter 2 m contains ceramic Raschig rings randomly packed to a height of 5 m. Air containing a small proportion of sulfur dioxide passes upwards through the absorption tower at a flow rate of 6 m^3/s. The viscosity and density of the gas may be taken as 1.80×10^{-5} Pa s and 1.2 kg/m^3, respectively. Details of the packing are given below:

Ceramic Raschig rings

surface area per unit volume of packed bed, $S_B = 190\,\text{m}^2/\text{m}^3$

voidage of randomly packed bed = 0.71

(a) Calculate the diameter, d_{sv}, of a sphere with the same surface-volume ratio as the Raschig rings.

(b) Calculate the frictional pressure drop across the packing in the tower.

(c) *Discuss* how this pressure drop will vary with flow rate of the gas within ±10% of the quoted flow rate.

(d) *Discuss* how the pressure drop across the packing would vary with gas pressure and temperature.

[Answer: (a) 9.16 mm; (b) 3460 Pa; for (c), (d) use the hint that turbulence dominates.]

6.4 A solution of density 1100 kg/m^3 and viscosity 2×10^{-3} Pa s is flowing under gravity at a rate of 0.24 kg/s through a bed of catalyst particles. The bed diameter is 0.2 m and the depth is 0.5 m. The particles are cylindrical, with a diameter of 1 mm and length of 2 mm. They are loosely packed to give a voidage of 0.3. Calculate the depth of liquid above the top of the bed. (*Hint*: apply the mechanical energy equation between the bottom of the bed and the surface of the liquid.)

[Answer: 0.716 m.]

6.5 In the regeneration of an ion exchange resin, hydrochloric acid of density 1200 kg/m^3 and viscosity 2×10^{-3} Pa s flows upwards through a bed of resin particles of density 2500 kg/m^3 resting on a porous support in a tube 4 cm in diameter. The particles are spherical, have a diameter 0.2 mm and form a bed of void fraction 0.5. The bed is 60 cm deep and is unrestrained at its upper surface. Plot the frictional pressure drop across the bed as function of acid flow rate up to a value of 0.1 litres/min.

[Answer: Pressure drop increases linearly up to a value of 3826 Pa beyond which point the bed will fluidize and maintain this pressure drop (see Chapter 7).]

6.6 The reactor of a catalytic reformer contains spherical catalyst particles of diameter 1.46 mm. The packed volume of the reactor is to be 3.4 m^3 and the void fraction is 0.45. The reactor feed is a gas of density 30 kg/m^3 and viscosity 2×10^{-5} Pa s flowing at a rate of 11 320 m^3/h. The gas properties may be assumed constant. The pressure loss through the reactor is restricted to 68.95 kPa. Calculate the cross-sectional area for flow and the bed depth required.

[Answer: area = 4.78 m^2; depth = 0.711 m.]

6.7 A leaf filter has an area of 2 m^2 and operates at a constant pressure drop of 250 kPa. The following results were obtained during a test with an incompressible cake:

Volume of filtrate collected (litre)	280	430	540	680	800
Time (min)	10	20	30	45	60

Calculate:

(a) the time required to collect 1200 litre of filtrate at a constant pressure drop of 400 kPa with the same feed slurry;

(b) the time required to wash the resulting filter cake with 500 litre of water (same properties as the filtrate) at a pressure drop of 200 kPa.

[Answer: (a) 79.4 min; (b) 124 min.]

6.8 A laboratory leaf filter has an area of 0.1 m^2, operates at a constant pressure drop of 400 kPa and produces the following results during a test on filtration of a slurry:

Volume of filtrate collected (litre)	19	31	41	49	56	63
Time (s)	300	600	900	1200	1500	1800

(a) Calculate the time required to collect 1.5 m^3 of filtrate during filtration of the same slurry at a constant pressure drop of 300 kPa on a similar full-scale filter with an area of 2 m^2.

(b) Calculate the rate of passage of filtrate at the end of the filtration in (a).

(c) Calculate the time required to wash the resulting filter cake with 0.5 m^3 of water at a constant pressure drop of 200 kPa.

(Assume the cake is incompressible and that the flow properties of the filtrate are the same as those of the wash solution.)

[Answer: (a) 37.2 min; (b) 20.4 litre/min; (c) 36.7 min.]

6.9 A leaf filter has an area of 1.73 m^2, operates at a constant pressure drop of 300 kPa and produces the following results during a test on filtration of a slurry:

Volume of filtrate collected (m^3)	0.19	0.31	0.41	0.49	0.56	0.63
Time (s)	300	600	900	1200	1500	1800

(a) Calculate the time required to collect 1 m^3 of filtrate during filtration of the same slurry at a constant pressure drop of 400 kPa.

(b) Calculate the time required to wash the resulting filter cake with 0.8 m^3 of water at a constant pressure drop of 250 kPa.

(Assume the cake is incompressible and that the flow properties of the filtrate are the same as those of the wash solution.)

[Answer: (a) 49.5 min; (b) 110.9 min.]

7

Fluidization

7.1 FUNDAMENTALS

When a fluid is passed upwards through a bed of particles the pressure loss in the fluid due to frictional resistance increases with increasing fluid flow. A point is reached when the upward drag force exerted by the fluid on the particles is equal to the apparent weight of particles in the bed. At this point the particles are lifted by the fluid, the separation of the particles increases, and the bed becomes fluidized. The force balance across the fluidized bed dictates that the fluid pressure loss across the bed of particles is equal to the apparent weight of the particles per unit area of the bed. Thus:

$$\text{pressure drop} = \frac{\text{weight of particles} - \text{upthrust on particle}}{\text{bed cross-sectional area}}$$

For a bed of particles of density ρ_p, fluidized by a fluid of density ρ_f to form a bed of depth H and voidage ε in a vessel of cross-sectional area A:

$$\Delta p = \frac{HA(1-\varepsilon)(\rho_p - \rho_f)g}{A} \tag{7.1}$$

or

$$\Delta p = H(1-\varepsilon)(\rho_p - \rho_f)g \tag{7.2}$$

A plot of fluid pressure loss across the bed versus superficial fluid velocity through the bed would have the appearance of Figure 7.1. Referring to Figure 7.1, the straight line region OA is the packed bed region. Here the solid particles do not move relative to one another and their separation is constant. The pressure loss

Introduction to Particle Technology - 2nd Edition Martin Rhodes

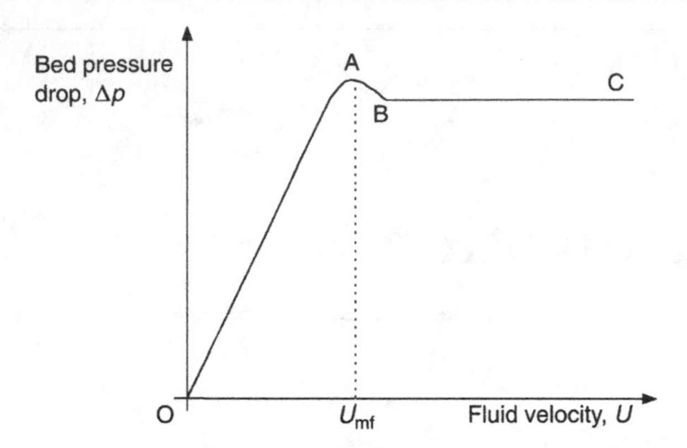

Figure 7.1 Pressure drop versus fluid velocity for packed and fluidized beds

versus fluid velocity relationship in this region is described by the Carman–Kozeny equation [Equation (6.9)] in the laminar flow regime and the Ergun equation in general [Equation (6.11)]. (See Chapter 6 for a detailed analysis of packed bed flow.)

The region BC is the fluidized bed region where Equation (7.1) applies. At point A it will be noticed that the pressure loss rises above the value predicted by Equation (7.1). This rise is more marked in small vessels and in powders which have been compacted to some extent before the test and is associated with the extra force required to overcome wall friction and adhesive forces between bed and the distributor.

The superficial fluid velocity at which the packed bed becomes a fluidized bed is known as the minimum fluidization velocity, U_{mf}. This is also sometimes referred to as the velocity at incipient fluidization (incipient meaning beginning). U_{mf} increases with particle size and particle density and is affected by fluid properties. It is possible to derive an expression for U_{mf} by equating the expression for pressure loss in a fluidized bed [Equation (7.2)] with the expression for pressure loss across a packed bed. Thus recalling the Ergun equation [Equation (6.11)]:

$$\frac{(-\Delta p)}{H} = 150\frac{(1-\varepsilon)^2}{\varepsilon^3}\frac{\mu U}{x_{sv}^2} + 1.75\frac{(1-\varepsilon)}{\varepsilon^3}\frac{\rho_f U^2}{x_{sv}} \tag{7.3}$$

substituting the expression for $(-\Delta p)$ from Equation (7.2):

$$(1-\varepsilon)(\rho_p - \rho_f)g = 150\frac{(1-\varepsilon)^2}{\varepsilon^3}\frac{\mu U_{mf}}{x_{sv}^2} + 1.75\frac{(1-\varepsilon)}{\varepsilon^3}\frac{\rho_f U_{mf}^2}{x_{sv}} \tag{7.4}$$

Rearranging,

$$\begin{aligned}(1-\varepsilon)(\rho_p - \rho_f)g &= 150\frac{(1-\varepsilon)^2}{\varepsilon^3}\left(\frac{\mu^2}{\rho_f x_{sv}^3}\right)\left(\frac{U_{mf} x_{sv}\rho_f}{\mu}\right) \\ &\quad + 1.75\frac{(1-\varepsilon)}{\varepsilon^3}\left(\frac{\mu^2}{\rho_f x_{sv}^3}\right)\left(\frac{U_{mf}^2 x_{sv}^2 \rho_f^2}{\mu^2}\right)\end{aligned} \tag{7.5}$$

and so

$$(1-\varepsilon)(\rho_p-\rho_f)g\left(\frac{\rho_f x_{sv}^3}{\mu^2}\right)=150\frac{(1-\varepsilon)^2}{\varepsilon^3}Re_{mf}+1.75\frac{(1-\varepsilon)}{\varepsilon^3}Re_{mf}^2 \quad (7.6)$$

or

$$Ar=150\frac{(1-\varepsilon)}{\varepsilon^3}Re_{mf}+1.75\frac{1}{\varepsilon^3}Re_{mf}^2 \quad (7.7)$$

where Ar is the dimensionless number known as the Archimedes number,

$$Ar=\frac{\rho_f(\rho_p-\rho_f)gx_{sv}^3}{\mu^2}$$

and Re_{mf} is the Reynolds number at incipient fluidization,

$$Re_{mf}=\left(\frac{U_{mf}x_{sv}\rho_f}{\mu}\right)$$

In order to obtain a value of U_{mf} from Equation (7.7) we need to know the voidage of the bed at incipient fluidization, $\varepsilon=\varepsilon_{mf}$. Taking ε_{mf} as the voidage of the packed bed, we can obtain a crude U_{mf}. However, in practice voidage at the onset of fluidization may be considerably greater than the packed bed voidage. A typical often used value of ε_{mf} is 0.4. Using this value, Equation (7.7) becomes

$$Ar=1406\,Re_{mf}+27.3\,Re_{mf}^2 \quad (7.8)$$

Wen and Yu (1966) produced an empirical correlation for U_{mf} with a form similar to Equation (7.8):

$$Ar=1652Re_{mf}+24.51Re_{mf}^2 \quad (7.9)$$

The Wen and Yu correlation is often expressed in the form:

$$Re_{mf}=33.7[(1+3.59\times10^{-5}Ar)^{0.5}-1] \quad (7.10)$$

and is valid for spheres in the range $0.01 < Re_{mf} < 1000$.

For gas fluidization the Wen and Yu correlation is often taken as being most suitable for particles larger than 100 μm, whereas the correlation of Baeyens and Geldart (1974), shown in Equation (7.11), is best for particles less than 100 μm.

$$U_{mf}=\frac{(\rho_p-\rho_f)^{0.934}g^{0.934}x_p^{1.8}}{1110\mu^{0.87}\rho_f^{0.066}} \quad (7.11)$$

7.2 RELEVANT POWDER AND PARTICLE PROPERTIES

The correct density for use in fluidization equations is the particle density, defined as the mass of a particle divided by its hydrodynamic volume. This is the volume 'seen' by the fluid in its fluid dynamic interaction with the particle and includes the volume of all the open and closed pores (see Figure 7.2):

$$\text{particle density} = \frac{\text{mass of particle}}{\text{hydrodynamic volume of particle}}$$

For non-porous solids, this is easily measured by a gas pycnometer or specific gravity bottle, but these devices should not be used for porous solids since they give the true or absolute density ρ_{abs} of the material of which the particle is made and this is not appropriate where interaction with fluid flow is concerned:

$$\text{absolute density} = \frac{\text{mass of particle}}{\text{volume of solids material making up the particle}}$$

For porous particles, the particle density ρ_p (also called apparent or envelope density) is not easy to measure directly although several methods are given in Geldart (1990). Bed density is another term used in connection with fluidized beds; bed density is defined as

$$\text{bed density} = \frac{\text{mass of particles in a bed}}{\text{volume occupied by particles and voids between them}}$$

For example, 600 kg of powder is fluidized in a vessel of cross-sectional area 1 m^2 and achieves a bed height of 0.5 m. What is the bed density?

Mass of particles in the bed = 600 kg

Volume occupied by particles and voids $= 1 \times 0.5 = 0.5\,\text{m}^3$

Hence, bed density $= 600/0.5 = 1200\,\text{kg/m}^3$.

If the particle density of these solids is $2700\,\text{kg/m}^3$, what is the bed voidage?

Bed density ρ_B is related to particle density ρ_p and bed voidage ε by Equation (7.12):

$$\rho_B = (1 - \varepsilon)\rho_p \tag{7.12}$$

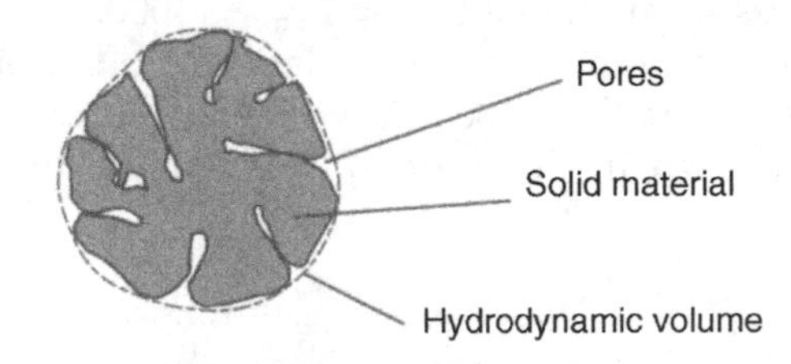

Figure 7.2 Hydrodynamic volume of a particle

Hence, voidage $= 1 - \dfrac{1200}{2700} = 0.555$.

Another density often used when dealing with powders is the bulk density. It is defined in a similar way to fluid bed density:

$$\text{bulk density} = \frac{\text{mass of particles}}{\text{volume occupied by particles and voids between them}}$$

The most appropriate particle size to use in equations relating to fluid–particle interactions is a hydrodynamic diameter, i.e. an equivalent sphere diameter derived from a measurement technique involving hydrodynamic interaction between the particle and fluid. In practice, however, in most industrial applications sizing is done using sieving and correlations use either sieve diameter, x_p or volume diameter, x_v. For spherical or near spherical particles x_v is equal to x_p. For angular particles, $x_v \approx 1.13x_p$.

For use in fluidization applications, starting from a sieve analysis the mean size of the powder is often calculated from

$$\text{mean } x_p = \frac{1}{\sum m_i/x_i} \tag{7.13}$$

where x_i is the arithmetic mean of adjacent sieves between which a mass fraction m_i is collected. This is the harmonic mean of the mass distribution, which was shown in Chapter 1 to be equivalent to the arithmetic mean of a surface distribution.

7.3 BUBBLING AND NON-BUBBLING FLUIDIZATION

Beyond the minimum fluidization velocity bubbles or particle-free voids may appear in the fluidized bed. Figure 7.3 shows bubbles in a gas fluidized bed. The

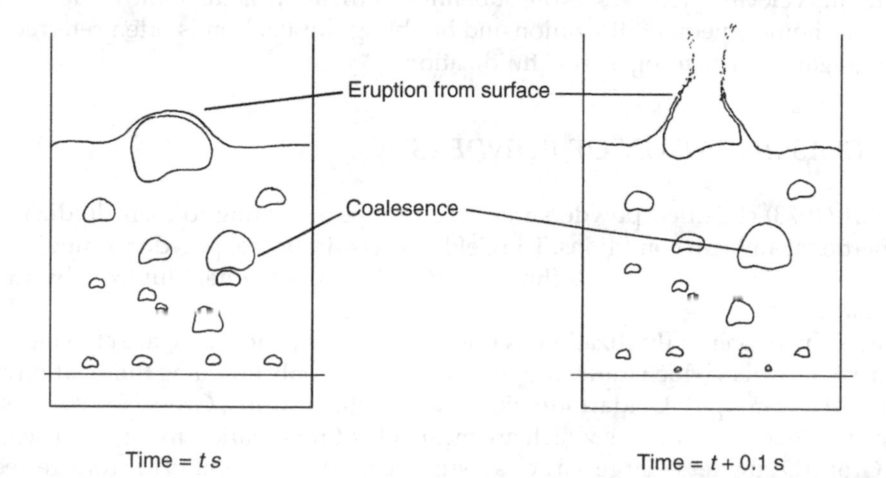

Figure 7.3 Sequence showing bubbles in a 'two-dimensional' fluidized bed of Group B powder. Sketches taken from video

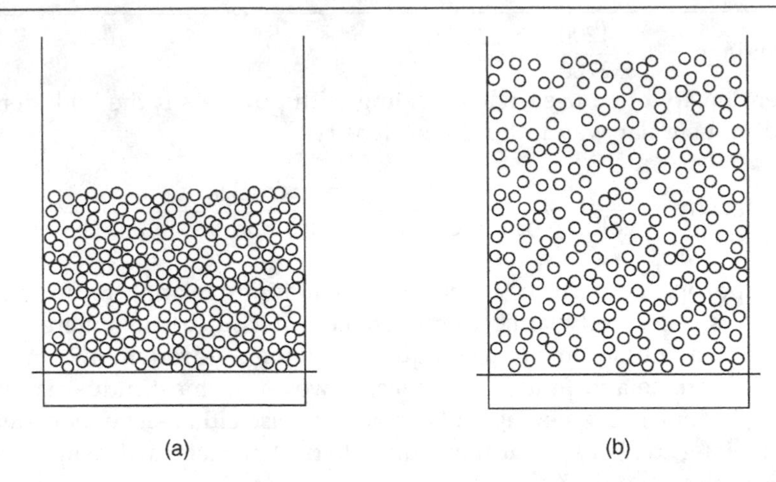

Figure 7.4 Expansion of a liquid fluidized bed: (a) just above U_{mf}; (b) liquid velocity several times U_{mf}. Note uniform increase in void fraciton. Sketches taken from video

equipment used Figure 7.3 is a so-called 'two-dimensional fluidized bed'. A favourite tool of researchers looking at bubble behaviour, this is actually a vessel of a rectangular cross-section, whose shortest dimension (into the page) is usually only 1 cm or so.

At superficial velocities above the minimum fluidization velocity, fluidization may in general be either bubbling or non-bubbling. Some combinations of fluid and particles give rise to *only bubbling* fluidization and some combinations give *only non-bubbling* fluidization. Most liquid fluidized systems, except those involving very dense particles, do not give rise to bubbling. Figure 7.4 shows a bed of glass spheres fluidized by water exhibiting non-bubbling fluidized bed behaviour. Gas fluidized systems, however, give either only bubbling fluidization or non-bubbling fluidization beginning at U_{mf}, followed by bubbling fluidization as fluidizing velocity increases. Non-bubbling fluidization is also known as particulate or homogeneous fluidization and bubbling fluidization is often referred to as aggregative or heterogeneous fluidization.

7.4 CLASSIFICATION OF POWDERS

Geldart (1973) classified powders into four groups according to their fluidization properties at ambient conditions. The Geldart classification of powders is now used widely in all fields of powder technology. Powders which when fluidized by air at ambient conditions give a region of non-bubbling fluidization beginning at U_{mf}, followed by bubbling fluidization as fluidizing velocity increases, are classified as Group A. Powders which under these conditions give only bubbling fluidization are classified as Group B. Geldart identified two further groups: Group C powders – very fine, cohesive powders which are incapable of fluidization in the strict sense, and Group D powders – large particles distinguished by their ability to produce deep spouting beds (see Figure 7.5). Figure 7.6 shows how the group classifications are related to the particle and gas properties.

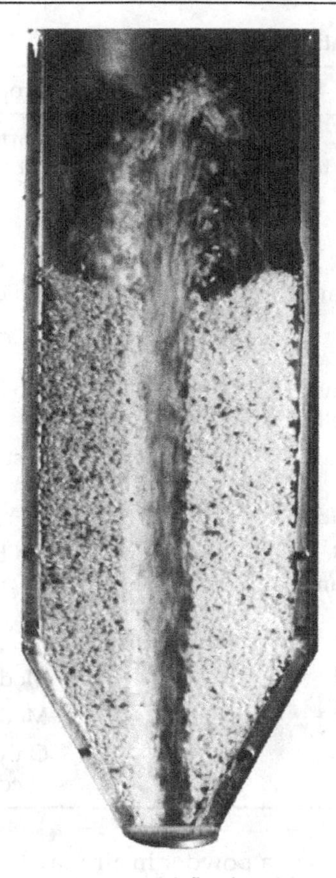

Figure 7.5 A spouted fluidized bed of rice

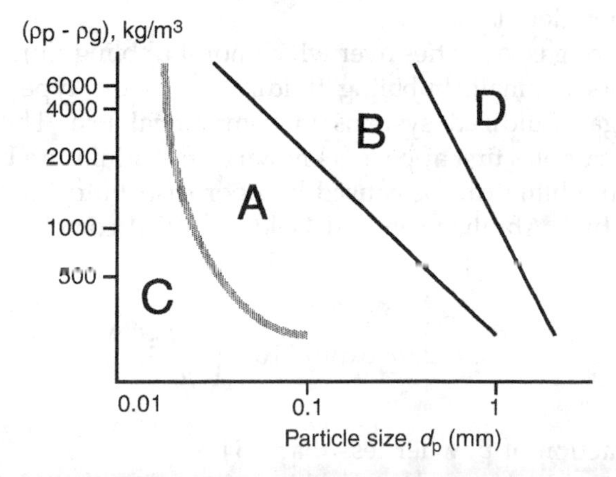

Figure 7.6 Simplified diagram showing Geldart's classification of powders according to their fluidization behaviour in air under ambient conditions (Geldart, 1973)

Table 7.1 Geldart's classification of powders

	Group C	Group A	Group B	Group D
Most obvious characteristic	Cohesive, difficult to fluidize	Ideal for fluidization. Exhibits range of non-bubbling fluidization	Starts bubbling at U_{mf}	Coarse solids
Typical solids	Flour, cement	Cracking catalyst	Building sand	Gravel, coffee beans
Property				
Bed expansion	Low because of chanelling	High	Moderate	Low
De-aeration rate	Initially fast, then exponential	Slow, linear	Fast	Fast
Bubble properties	No bubbles– only channels	Bubbles split and coalesce. Maximum bubble size	No limit to size	No limit to size
Solids mixing	Very low	High	Moderate	Low
Gas backmixing	Very low	High	Moderate	Low
Spouting	No	No	Only in shallow beds	Yes, even in deep beds

The fluidization properties of a powder in air may be predicted by establishing in which group it lies. It is important to note that at operating temperatures and pressures above ambient a powder may appear in a different group from that which it occupies at ambient conditions. This is due to the effect of gas properties on the grouping and may have serious implications as far as the operation of the fluidized bed is concerned. Table 7.1 presents a summary of the typical properties of the different powder classes.

Since the range of gas velocities over which non-bubbling fluidization occurs in Group A powders is small, bubbling fluidization is the type most commonly encountered in gas fluidized systems in commercial use. The superficial gas velocity at which bubbles first appear is known as the minimum bubbling velocity U_{mb}. Premature bubbling can be caused by poor distributor design or protuberances inside the bed. Abrahamsen and Geldart (1980) correlated the maximum values of U_{mb} with gas and particle properties using the following correlation:

$$U_{mb} = 2.07\ \exp(0.716F)\left(\frac{x_p \rho_g^{0.06}}{\mu^{0.347}}\right) \qquad (7.14)$$

where F is the fraction of powder less than 45 μm.

In Group A powders $U_{mb} > U_{mf}$, bubbles are constantly splitting and coalescing, and a maximum stable bubble size is achieved. This makes for good quality,

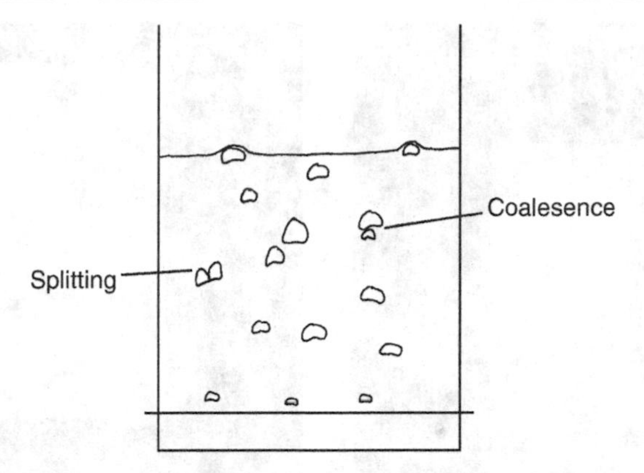

Figure 7.7 Bubbles in a 'two-dimensional' fluidized bed of Group A powder. Sketch taken from video

smooth fluidization. Figure 7.7 shows bubbles in a Group A powder in a two-dimensional fluidized bed.

In Groups B and D powders $U_{mb} = U_{mf}$, bubbles continue to grow, never achieving a maximum size (see Figure 7.3). This makes for rather poor quality fluidization associated with large pressure fluctuations.

In Group C powders the interparticle forces are large compared with the inertial forces on the particles. As a result, the particles are unable to achieve the separation they require to be totally supported by drag and buoyancy forces and true fluidization does not occur. Bubbles, as such, do not appear; instead the gas flow forms channels through the powder (see Figure 7.8). Since the particles are not fully supported by the gas, the pressure loss across the bed is always less than apparent weight of the bed per unit cross-sectional area. Consequently, measurement of bed pressure drop is one means of detecting this Group C behaviour if visual observation is inconclusive. Fluidization, of sorts, can be achieved with the assistance of a mechanical stirrer or vibration.

When the size of the bubbles is greater than about one-third of the diameter of the equipment their rise velocity is controlled by the equipment and they become slugs of gas. Slugging is attended by large pressure fluctuations and so it is generally avoided in large units since it can cause vibration to the plant. Slugging is unlikely to occur at any velocity if the bed is sufficiently shallow. According to Yagi and Muchi (1952), slugging will not occur provided the following criterion is satisfied:

$$\left(\frac{H_{mf}}{D}\right) \leq \frac{1.9}{(\rho_p x_p)^{0.3}} \tag{7.15}$$

This criterion works well for most powders. If the bed is deeper than this critical height then slugging will occur when the gas velocity exceeds U_{ms} as given by (Baeyens and Geldart, 1974):

$$U_{ms} = U_{mf} + 0.16(1.34D^{0.175} - H_{mf})^2 + 0.07(gD)^{0.5} \tag{7.16}$$

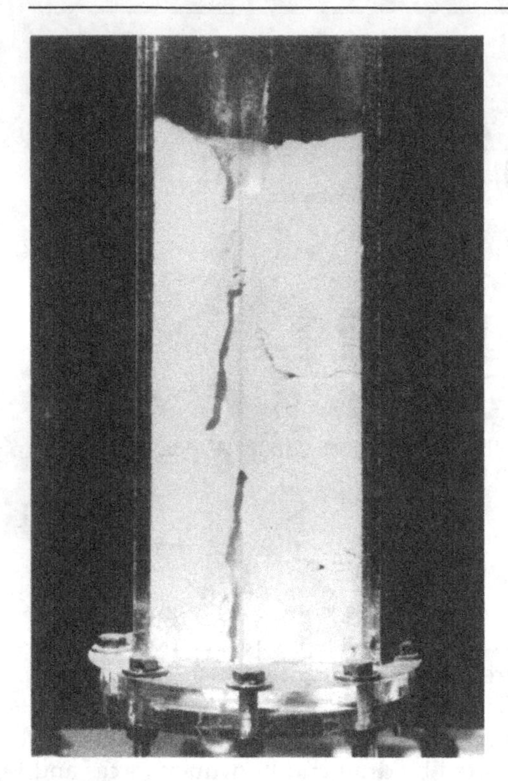

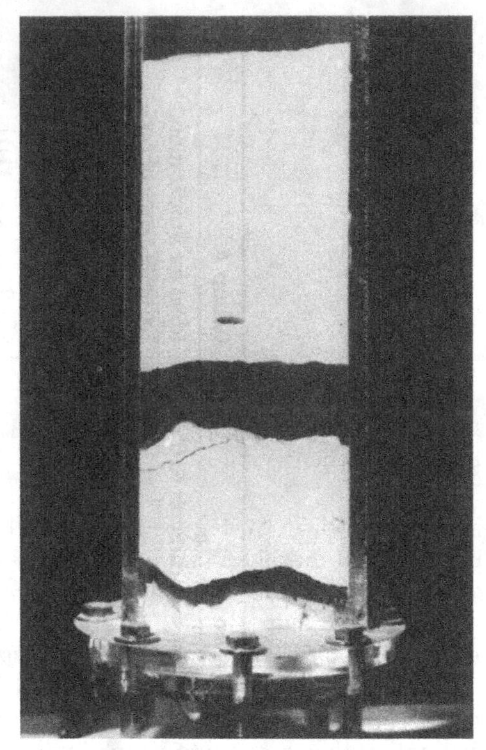

Figure 7.8 Attempts to fluidize Group C powder producing cracks and channels or discrete solid plugs

7.5 EXPANSION OF A FLUIDIZED BED

7.5.1 Non-bubbling Fluidization

In a non-bubbling fluidized bed beyond U_{mf} the particle separation increases with increasing fluid superficial velocity whilst the pressure loss across the bed remains constant. This increase in bed voidage with fluidizing velocity is referred to as bed expansion (see Figure 7.4). The relationship between fluid velocity and bed voidage may be determined by recalling the analysis of multiple particle systems (Chapter 3). For a particle suspension settling in a fluid under force balance conditions the relative velocity U_{rel} between particles and fluid is given by:

$$U_{rel} = U_p - U_f = U_T \varepsilon f(\varepsilon) \tag{7.17}$$

where U_p and U_f are the actual downward vertical velocities of the particles and the fluid, and U_T is the single particle terminal velocity in the fluid. In the case of a fluidized bed the time-averaged actual vertical particle velocity is zero ($U_p = 0$) and so

$$U_f = -U_T \varepsilon f(\varepsilon) \tag{7.18}$$

or

$$U_{fs} = -U_T \varepsilon^2 f(\varepsilon) \tag{7.19}$$

where U_{fs} is the downward volumetric fluid flux. In common with fluidization practice, we will use the term superficial velocity (U) rather than volumetric fluid flux. Since the upward superficial fluid velocity (U) is equal to the upward volumetric fluid flux ($-U_{fs}$), and $U_{fs} = U_f \varepsilon$, then:

$$U = U_T \varepsilon^2 f(\varepsilon) \tag{7.20}$$

Richardson and Zaki (1954) found the function $f(\varepsilon)$ which applied to both hindered settling and to non-bubbling fluidization. They found that in general, $f(\varepsilon) = \varepsilon^n$, where the exponent n was independent of particle Reynolds number at very low Reynolds numbers, when the drag force is independent of fluid density, and at high Reynolds number, when the drag force is independent of fluid viscosity, i.e.

$$\text{In general}: \ U = U_T \varepsilon^n \tag{7.21}$$

$$\text{For } Re_p \leq 0.3;\ f(\varepsilon) = \varepsilon^{2.65} \Rightarrow U = U_T \varepsilon^{4.65} \tag{7.22}$$

$$\text{For } Re_p \geq 500;\ f(\varepsilon) = \varepsilon^{0.4} \Rightarrow U = U_T \varepsilon^{2.4} \tag{7.23}$$

where Re_p is calculated at U_T.

Khan and Richardson (1989) suggested the correlation given in Equation (3.25) (Chapter 3) which permits the determination of the exponent n at intermediate values of Reynolds number (although it is expressed in terms of the Archimedes number Ar there is a direct relationship between Re_p and Ar). This correlation also incorporates the effect of the vessel diameter on the exponent. Thus Equations (7.21), (7.22) and (7.23) in conjunction with Equation (3.25) permit calculation of the variation in bed voidage with fluid velocity beyond U_{mf}. Knowledge of the bed voidage allows calculation of the fluidized bed height as illustrated below:

$$\text{mass of particles in the bed} = M_B = (1 - \varepsilon)\rho_p A H \tag{7.24}$$

If packed bed depth (H_1) and voidage (ε_1) are known, then if the mass remains constant the bed depth at any voidage can be determined:

$$(1 - \varepsilon_2)\rho_p A H_2 = (1 - \varepsilon_1)\rho_p A H_1 \tag{7.25}$$

hence

$$H_2 = \frac{(1 - \varepsilon_1)}{(1 - \varepsilon_2)} H_1$$

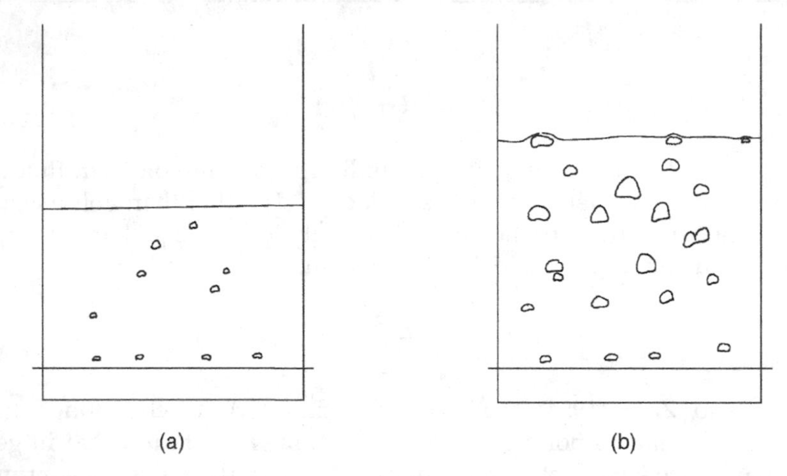

Figure 7.9 Bed expansion in a 'two-dimensional' fluidized bed of Group A powder: (a) just above U_{mb}; (b) fluidized at several times U_{mb}. Sketches taken from video

7.5.2 Bubbling Fluidization

The simplest description of the expansion of a bubbling fluidized bed is derived from the two-phase theory of fluidization of Toomey and Johnstone (1952). This theory considers the bubbling fluidized bed to be composed of two phases: the bubbling phase (the gas bubbles); and the particulate phase (the fluidized solids around the bubbles). The particulate phase is also referred to as the emulsion phase. The theory states that any gas in excess of that required at incipient fluidization will pass through the bed as bubbles. Figure 7.9 shows the effect of fluidizing gas velocity on bed expansion of a Group A powder fluidized by air. Thus, referring to Figure 7.10, Q is the actual gas flow rate to the fluid bed and Q_{mf} is the gas flow rate at incipient fluidization, then

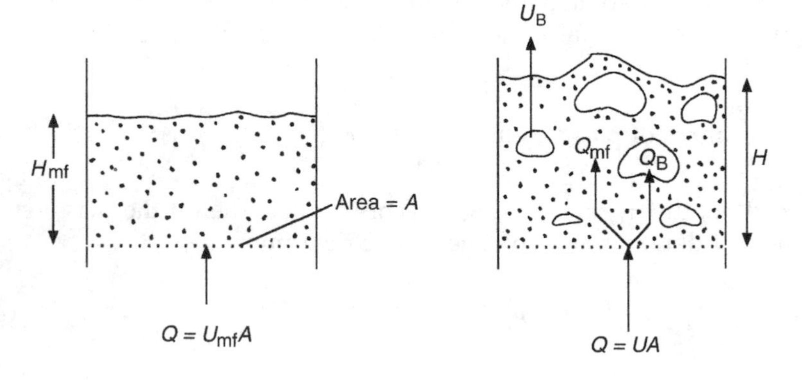

Figure 7.10 Gas flows in a fluidized bed according to the two-phase theory

$$\text{gas passing through the bed as bubbles} = Q - Q_{mf} = (U - U_{mf})A \tag{7.26}$$
$$\text{gas passing through the emulsion phase} = Q_{mf} = U_{mf}A \tag{7.27}$$

Expressing the bed expansion in terms of the fraction of the bed occupied by bubbles, ε_B:

$$\varepsilon_B = \frac{H - H_{mf}}{H} = \frac{Q - Q_{mf}}{AU_B} = \frac{(U - U_{mf})}{U_B} \tag{7.28}$$

where H is the bed height at U and H_{mf} is the bed height at U_{mf} and U_B is the mean rise velocity of a bubble in the bed (obtained from correlations; see below). The voidage of the emulsion phase is taken to be that at minimum fluidization ε_{mf}. The mean bed voidage is then given by:

$$(1 - \varepsilon) = (1 - \varepsilon_B)(1 - \varepsilon_{mf}) \tag{7.29}$$

In practice, the elegant two-phase theory overestimates the volume of gas passing through the bed as bubbles (the visible bubble flow rate) and better estimates of bed expansion may be obtained by replacing $(Q - Q_{mf})$ in Equation (7.28) with

$$\begin{aligned} &\text{visible bubble flow rate}, Q_B = YA(U - U_{mf}) \\ &\text{where } 0.8 < Y < 1.0 \text{ for Group A powders} \\ &\quad 0.6 < Y < 0.8 \text{ for Group B powders} \\ &\quad 0.25 < Y < 0.6 \text{ for Group D powders} \end{aligned} \tag{7.30}$$

Strictly the equations should be written in terms of U_{mb} rather than U_{mf} and Q_{mb} rather than Q_{mf}, so that they are valid for both Group A and Group B powders. Here they have been written in their original form. In practice, however, it makes little difference, since both U_{mb} and U_{mf} are usually much smaller than the superficial fluidizing velocity, U [so $(U - U_{mf}) \cong (U - U_{mb})$]. In rare cases where the operating velocity is not much greater than U_{mb}, then U_{mb} should be used in place of U_{mf} in the equations.
The above analysis requires a knowledge of the bubble rise velocity U_B, which depends on the bubble size d_{Bv} and bed diameter D. The bubble diameter at a given height above the distributor depends on the orifice density in the distributor N, the distance above the distributor L and the excess gas velocity $(U - U_{mf})$.

For Group B powders

$$d_{Bv} = \frac{0.54}{g^{0.2}}(U - U_{mf})^{0.4}(L + 4N^{-0.5})^{0.8} \quad \text{(Darton } et\ al.\text{, 1977)} \tag{7.31}$$
$$U_B = \Phi_B(gd_{Bv})^{0.5} \quad \text{(Werther, 1983)} \tag{7.32}$$

where

$$\left\{\begin{array}{lll}\Phi_B = 0.64 & \text{for} & D \leq 0.1\,\text{m} \\ \Phi_B = 1.6D^{0.4} & \text{for} & 0.1 < D \leq 1\,\text{m} \\ \Phi_B = 1.6 & \text{for} & D > 1\,\text{m}\end{array}\right\} \tag{7.33}$$

For Group A powders

Bubbles reach a maximum stable size which may be estimated from

$$d_{Bv\,max} = 2(U_{T2.7})^2/g \text{ (Geldart, 1992)} \tag{7.34}$$

where $U_{T2.7}$ is the terminal free fall velocity for particles of diameter 2.7 times the actual mean particle diameter.

Bubble velocity for Group A powders is given by:

$$U_B = \Phi_A(gd_{Bv})^{0.5} \quad \text{(Werther, 1983)} \tag{7.35}$$

where

$$\left\{\begin{array}{lll}\Phi_A = 1 & \text{for} & D \leq 0.1\,\text{m} \\ \Phi_A = 2.5D^{0.4} & \text{for} & 0.1 < D \leq 1\,\text{m} \\ \Phi_A = 2.5 & \text{for} & D > 1\,\text{m}\end{array}\right\} \tag{7.36}$$

7.6 *ENTRAINMENT*

The term entrainment will be used here to describe the ejection of particles from the surface of a bubbling bed and their removal from the vessel in the fluidizing gas. In the literature on the subject other terms such as 'carryover' and 'elutriation' are often used to describe the same process. In this section we will study the factors affecting the rate of entrainment of solids from a fluidized bed and develop a simple approach to the estimation of the entrainment rate and the size distribution of entrained solids.

Consider a single particle falling under gravity in a static gas in the absence of any solids boundaries. We know that this particle will reach a terminal velocity when the forces of gravity, buoyancy and drag are balanced (see Chapter 2). If the gas of infinite extent is now considered to be moving upwards at a velocity equal to the terminal velocity of the particle, the particle will be stationary. If the gas is moving upwards in a pipe at a superficial velocity equal to the particle's terminal velocity, then:

(a) in laminar flow: the particle may move up or down depending on its radial position because of the parabolic velocity profile of the gas in the pipe.

(b) in turbulent flow: the particle may move up or down depending on its radial position. In addition the random velocity fluctuations superimposed on the time-averaged velocity profile make the actual particle motion less predictable.

If we now introduce into the moving gas stream a number of particles with a range of particle size some particles may fall and some may rise depending on their size and their radial position. Thus the entrainment of particles in an upward-flowing gas stream is a complex process. We can see that the rate of entrainment and the size distribution of entrained particles will in general depend on particle size and density, gas properties, gas velocity, gas flow regime–radial velocity profile and fluctuations and vessel diameter. In addition (i) the mechanisms by which the particles are ejected into the gas stream from the fluidized bed are dependent on the characteristics of the bed – in particular bubble size and velocity at the surface, and (ii) the gas velocity profile immediately above the bed surface is distorted by the bursting bubbles. It is not surprising then that prediction of entrainment from first principles is not possible and in practice an empirical approach must be adopted.

This empirical approach defines coarse particles as particles whose terminal velocities are greater than the superficial gas velocity ($U_T > U$) and fine particles as those for which $U_T < U$, and considers the region above the fluidized bed surface to be composed of several zones shown in Figure 7.11:

- *Freeboard*. Region between the bed surface and the gas outlet.

- *Splash zone*. Region just above the bed surface in which coarse particles fall back down.

- *Disengagement zone*. Region above the splash zone in which the upward flux and suspension concentration of fine particles decreases with increasing height.

- *Dilute-phase transport zone*. Region above the disengagement zone in which all particles are carried upwards; particle flux and suspension concentration are constant with height.

Note that, although in general fine particles will be entrained and leave the system and coarse particles will remain, in practice fine particles may stay in the system at velocities several times their terminal velocity and coarse particles may be entrained.

The height from the bed surface to the top of the disengagement zone is known as the transport disengagement height (TDH). Above TDH the entrainment flux and concentration of particles is constant. Thus, from the design point of view, in order to gain maximum benefit from the effect of gravity in the freeboard, the gas exit should be placed above the TDH. Many empirical correlations for TDH are available in the literature; those of Horio *et al.* (1980) presented in Equation (7.37)

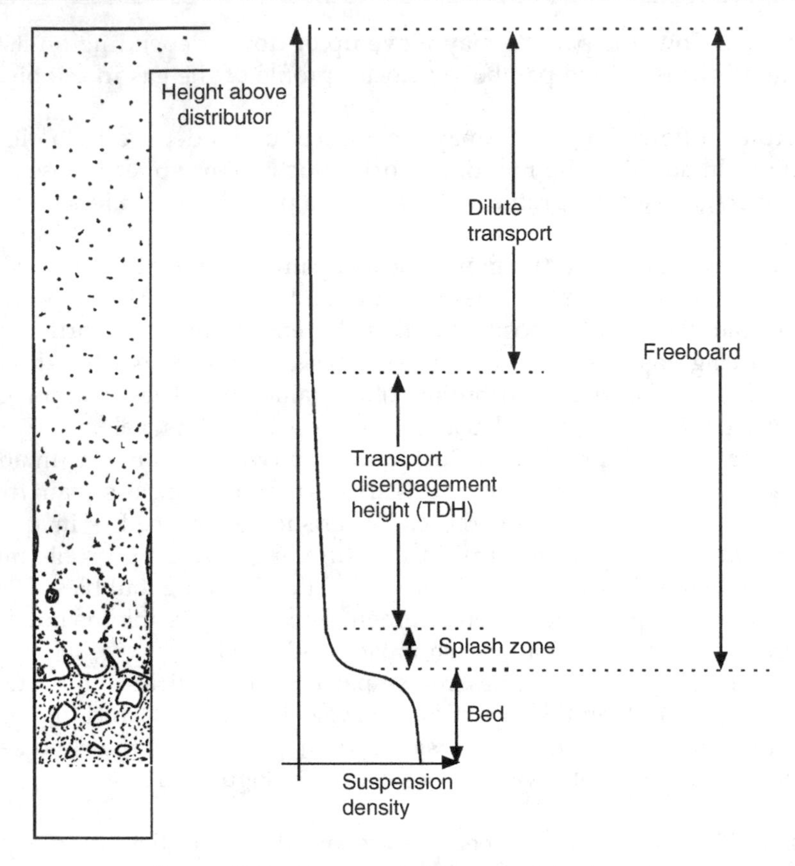

Figure 7.11 Zones in fluidized bed freeboard

and Zenz (1983) presented graphically in Figure 7.12 are two of the more reliable ones.

$$\mathrm{TDH} = 4.47 d_{\mathrm{bvs}}^{0.5} \tag{7.37}$$

Where d_{bvs} is the equivalent volume diameter of a bubble at the surface.

The empirical estimation of entrainment rates from fluidized beds is based on the following rather intuitive equation:

$$\begin{pmatrix}\text{instantaneous rate of loss} \\ \text{of solids of size } x_i\end{pmatrix} \propto \text{bed area} \times \begin{pmatrix}\text{fraction of bed with} \\ \text{size } x_i \text{ at time} t\end{pmatrix}$$
$$\text{i.e. } R_i = -\frac{\mathrm{d}}{\mathrm{d}t}(M_{\mathrm{B}} m_{\mathrm{B}i}) = K_{i\mathrm{h}}^{*} A m_{\mathrm{B}i} \tag{7.38}$$

where $K_{i\mathrm{h}}^{*}$ is the elutriation rate constant (the entrainment flux at height h above the bed surface for the solids of size x_i, when $m_{\mathrm{B}i} = 1.0$), M_{B} is the total mass of solids in the bed, A is the area of bed surface and $m_{\mathrm{B}i}$ is the fraction of the bed mass with size x_i at time t.

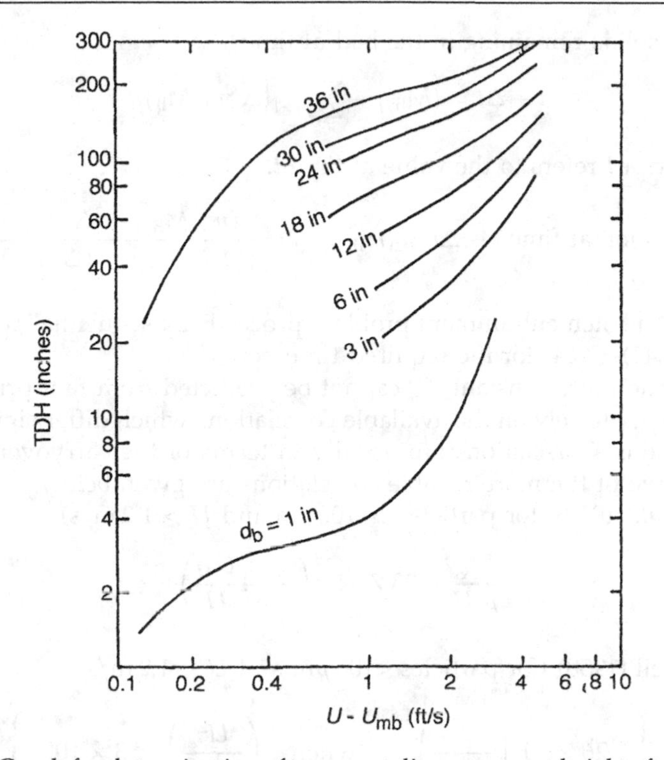

Figure 7.12 Graph for determination of transport disengagement height after the method of Zenz (1983). Reproduced by permission. [Note: TDH and the bubble diameter at the bubble surface d_b are given in inches (1 in. = 27.4 mm)]

For continuous operation, m_{Bi} and M_B are constant and so

$$R_i = K^*_{ih} A m_{Bi} \tag{7.39}$$

and

$$\text{total rate of entrainment, } R_T = \sum R_i = \sum K^*_{ih} A m_{Bi} \tag{7.40}$$

The solids loading of size x_i in the off-gases is $\rho_i = R_i/UA$ and the total solids loading of the gas leaving the freeboard is $\rho_T = \sum \rho_i$.

For batch operation, the rates of entrainment of each size range, the total entrainment rate and the particle size distribution of the bed change with time. The problem can best be solved by writing Equation (7.38) in finite increment form:

$$-\Delta(m_{Bi} M_B) = K^*_{ih} A m_{Bi} \Delta t \tag{7.41}$$

where $\Delta(m_{Bi} M_B)$ is the mass of solids in size range i entrained in time increment Δt.

$$\text{Total mass entrained in time } \Delta t = \sum_{i=1}^{k} [\Delta(m_{Bi} M_B)] \tag{7.42}$$

and mass of solids remaining in the bed at time

$$t + \Delta t = (M_B)_t - \sum_{i=1}^{k} [\Delta(m_{Bi} M_B)_t] \tag{7.43}$$

where subscript t refers to the value at time t.

$$\text{Bed composition at time } t + \Delta t = (m_{Bi})_{t+\Delta t} = \frac{(m_{Bi} M_B)_t - [\Delta(m_{Bi} M_B)_t]}{(M_B)_t - \sum_{i=1}^{k} \{\Delta(m_{Bi} M_B)_t\}} \tag{7.44}$$

Solution of a batch entrainment problem proceeds by sequential application of Equations (7.41)–(7.44) for the required time period.

The elutriation rate constant K_{ih}^* cannot be predicted from first principles and so it is necessary to rely on the available correlations which differ significantly in their predictions. Correlations are usually in terms of the carryover rate above TDH, $K_{i\infty}^*$. Two of the more reliable correlations are given below.

Geldart *et al.* (1979) (for particles $> 100\,\mu$m and $U > 1.2\,$m/s)

$$\frac{K_{i\infty}^*}{\rho_g U} = 23.7 \exp\left(-5.4 \frac{U_{Ti}}{U}\right) \tag{7.45}$$

Zenz and Weil (1958) (for particles $<100\,\mu$m and $U < 1.2\,$m/s)

$$\frac{K_{i\infty}^*}{\rho_g U} = \left\{ \begin{array}{lll} 1.26 \times 10^7 \left(\dfrac{U^2}{g x_i \rho_p^2}\right)^{1.88} & \text{when} & \left(\dfrac{U^2}{g x_i \rho_p^2}\right) < 3 \times 10^{-4} \\ 4.31 \times 10^4 \left(\dfrac{U^2}{g x_i \rho_p^2}\right)^{1.18} & \text{when} & \left(\dfrac{U^2}{g x_i \rho_p^2}\right) > 3 \times 10^{-4} \end{array} \right\} \tag{7.46}$$

7.7 *HEAT TRANSFER IN FLUIDIZED BEDS*

The transfer of heat between fluidized solids, gas and internal surfaces of equipment is very good. This makes for uniform temperatures and ease of control of bed temperature.

7.7.1 Gas–Particle Heat Transfer

Gas to particle heat transfer coefficients are typically small, of the order of 5–20 $\text{W} \cdot \text{m}^2\text{K}$. However, because of the very large heat transfer surface area provided by a mass of small particles (1 m^3 of 100 μm particles has a surface area of 60 000 m^2), the heat transfer between gas and particles is rarely limiting in fluid bed heat transfer. One of the most commonly used correlations for gas–particle heat transfer coefficient is that of Kunii and Levenspiel (1969):

$$Nu = 0.03 Re_p^{1.3} \quad (Re_p < 50) \tag{7.47}$$

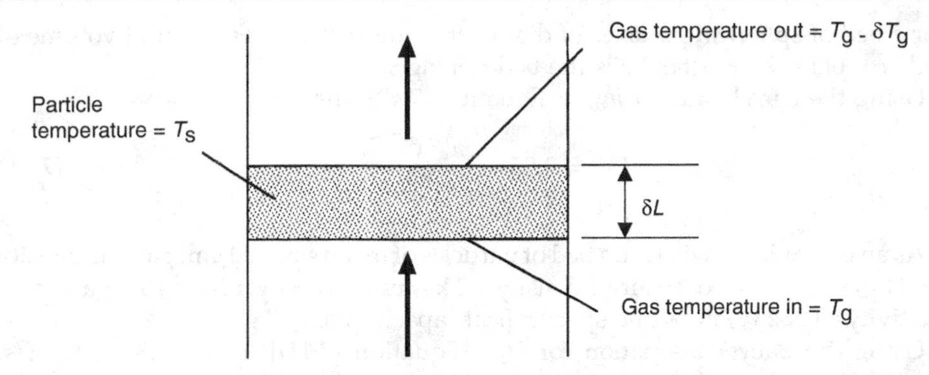

Figure 7.13 Analysis of gas–particle heat transfer in an element of a fluidized bed

where Nu is the Nusselt number $[h_{gp}x/k_g]$ and the single particle Reynolds number is based on the relative velocity between fluid and particle as usual.

Gas–particle heat transfer is relevant where a hot fluidized bed is fluidized by cold gas. The fact that particle–gas heat transfer presents little resistance in bubbling fluidized beds can be demonstrated by the following example:

Consider a fluidized bed of solids held at a constant temperature T_s. Hot fluidizing gas at temperature T_{g0} enters the bed. At what distance above the distributor is the difference between the inlet gas temperature and the bed solids temperature reduced to half its original value?

Consider an element of the bed of height δL at a distance L above the distributor (Figure 7.13). Let the temperature of the gas entering this element be T_g and the change in gas temperature across the element be δT_g. The particle temperature in the element is T_s.

The energy balance across the element gives

rate of heat loss by the gas = rate of heat transfer to the solids

that is

$$-(C_g U \rho_g)\mathrm{d}T_g = h_{gp}a(T_g - T_s)\mathrm{d}L \tag{7.48}$$

where a is the surface area of solids per unit volume of bed, C_g is the specific heat capacity of the gas, ρ_p is particle density, h_{gp} is the particle-to-gas heat transfer coefficient and U is superficial gas velocity.

Integrating with the boundary condition $T_g = T_{g0}$ at $L = 0$,

$$\ln\left(\frac{T_g - T_s}{T_{g0} - T_s}\right) = -\left(\frac{h_{gp}a}{U_{rel}\rho_g C_g}\right)L \tag{7.49}$$

The distance over which the temperature difference is reduced to half its initial value, $L_{0.5}$, is then

$$L_{0.5} = -\ln(0.5)\frac{C_g U_{rel}\rho_g}{h_{gp}a} = 0.693\frac{C_g U_{rel}\rho_g}{h_{gp}a} \tag{7.50}$$

For a bed of spherical particles of diameter x, the surface area per unit volume of bed, $a = 6(1 - \varepsilon)/x$, where ε is the bed voidage.

Using the correlation for h_{gp} in Equation (7.47), then

$$L_{0.5} = 3.85 \frac{\mu^{1.3} x^{0.7} C_g}{U_{rel}^{0.3} \rho_g^{0.3} (1 - \varepsilon) k_g} \tag{7.51}$$

As an example we will take a bed of particles of mean size 100 μm, particle density 2500 kg/m^3, fluidized by air of density 1.2 kg/m^3, viscosity 1.84×10^{-5} Pa s, conductivity 0.0262 W/m/K and specific heat capacity 1005 J/kg/K.

Using the Baeyens equation for U_{mf} [Equation (7.11)], $U_{mf} = 9.3 \times 10^{-3}$ m/s. The relative velocity between particles and gas under fluidized conditions can be approximated as U_{mf}/ε under these conditions.

Hence, assuming a fluidized bed voidage of 0.47, $U_{rel} = 0.02$ m/s.

Substituting these values in Equation (7.51), we find $L_{0.5} = 0.95$ mm. So, within 1 mm of entering the bed the difference in temperature between the gas and the bed will be reduced by half. Typically for particles less than 1 mm in diameter the temperature difference between hot bed and cold fluidizing gas would be reduced by half within the first 5 mm of the bed depth.

7.7.2 Bed–Surface Heat Transfer

In a bubbling fluidized bed the coefficient of heat transfer between bed and immersed surfaces (vertical bed walls or tubes) can be considered to be made up of three components which are approximately additive (Botterill, 1975).

$$\text{bed–surface heat transfer coefficient}, h = h_{pc} + h_{gc} + h_r$$

where h_{pc} is the particle convective heat transfer coefficient and describes the heat transfer due to the motion of packets of solids carrying heat to and from the surface, h_{gc} is the gas convective heat transfer coefficient describing the transfer of heat by motion of the gas between the particles and h_r is the radiant heat transfer coefficient. Figure 7.14, after Botterill (1986), gives an indication of the range of bed–surface heat transfer coefficients and the effect of particle size on the dominant heat transfer mechanism.

Particle convective heat transfer: On a volumetric basis the solids in the fluidized bed have about one thousand times the heat capacity of the gas and so, since the solids are continuously circulating within the bed, they transport the heat around the bed. For heat transfer between the bed and a surface the limiting factor is the gas conductivity, since all the heat must be transferred through a gas film between the particles and the surface (Figure 7.15). The particle–surface contact area is too small to allow significant heat transfer. Factors affecting the gas film thickness or the gas conductivity will therefore influence the heat transfer under particle convective conditions. Decreasing particle size, for example, decreases the mean gas film thickness and so improves h_{pc}. However, reducing particle size into the Group C range will reduce particle mobility and so reduce particle convective heat transfer. Increasing gas temperature increases gas conductivity and so improves h_{pc}.

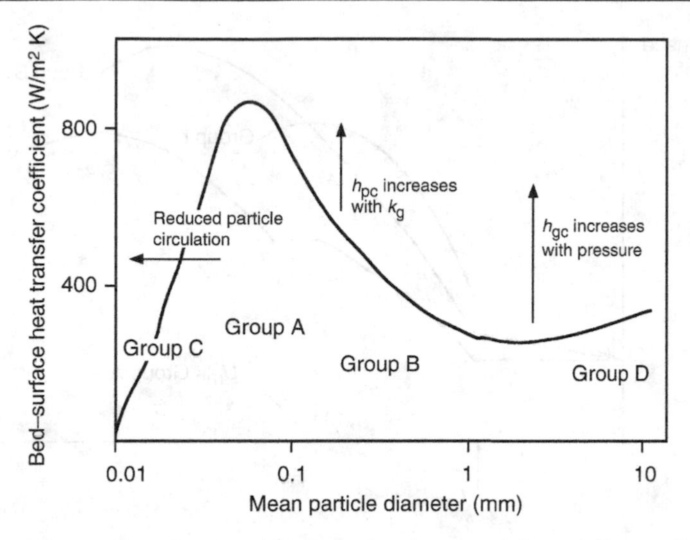

Figure 7.14 Range of bed–surface heat transfer coefficients

Particle convective heat transfer is dominant in Group A and B powders. Increasing gas velocity beyond minimum fluidization improves particle circulation and so increases particle convective heat transfer. The heat transfer coefficient increases with fluidizing velocity up to a broad maximum h_{max} and then declines as the heat transfer surface becomes blanketed by bubbles. This is shown in Figure 7.16 for powders in Groups A, B and D. The maximum in h_{pc} occurs relatively closer to U_{mf} for Group B and D powders since these powders give rise to bubbles at U_{mf} and the size of these bubbles increases with increasing gas velocity. Group A powders exhibit a non-bubbling fluidization between U_{mf} and U_{mb} and achieve a maximum stable bubble size.

Botterill (1986) recommends the Zabrodsky (1966) correlation for h_{max} for Group B powders:

$$h_{max} = 35.8 \frac{k_g^{0.6} \rho_p^{0.2}}{x^{0.36}} \quad \text{W/m}^2\text{/K} \tag{7.52}$$

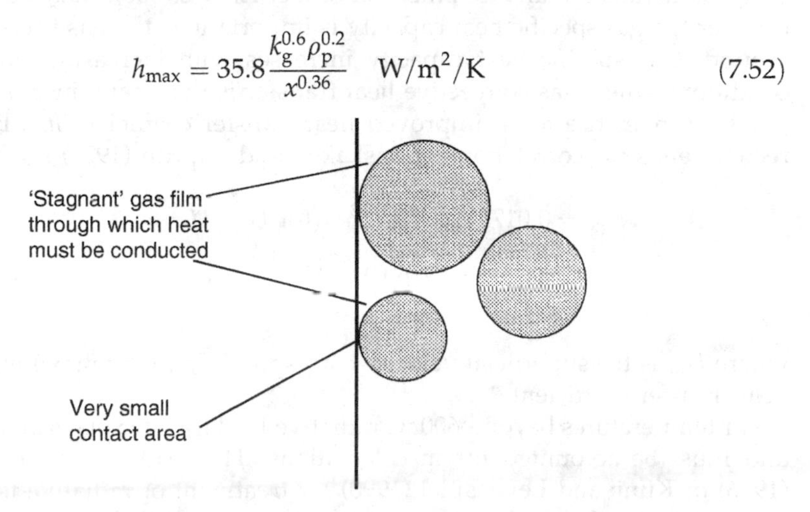

Figure 7.15 Heat transfer from bed particles to an immersed surface

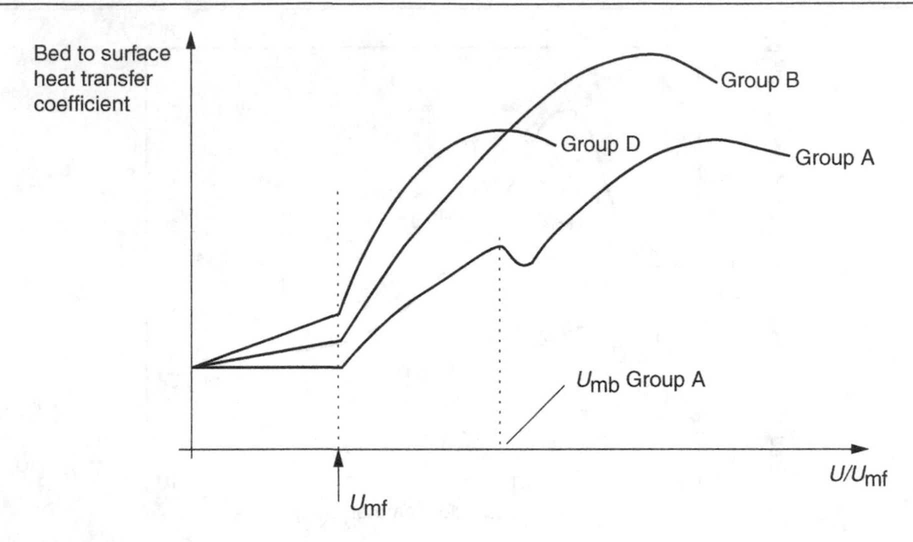

Figure 7.16 Effect of fluidizing gas velocity on bed–surface heat transfer coefficient in a fluidized bed

and the correlation of Khan *et al.* (1978) for Group A powders:

$$Nu_{\max} = 0.157\,Ar^{0.475} \tag{7.53}$$

Gas convective heat transfer is not important in Group A and B powders where the flow of interstitial gas is laminar but becomes significant in Group D powders, which fluidize at higher velocities and give rise to transitional or turbulent flow of interstitial gas. Botterill suggests that the gas convective mechanism takes over from particle convective heat transfer as the dominant mechanism at $Re_{\mathrm{mf}} \approx 12.5$ (Re_{mf} is the Reynolds number at minimum fluidization and is equivalent to an Archimedes number $Ar \approx 26\,000$). In gas convective heat transfer the gas specific heat capacity is important as the gas transports the heat around. Gas specific heat capacity increases with increasing pressure and in conditions where gas convective heat transfer is dominant, increasing operating pressure gives rise to an improved heat transfer coefficient h_{gc}. Botterill (1986) recommends the correlations of Baskakov and Suprun (1972) for h_{gc}.

$$Nu_{\mathrm{gc}} = 0.0175\,Ar^{0.46}Pr^{0.33} \quad (\text{for } U > U_{\mathrm{m}}) \tag{7.54}$$

$$Nu_{\mathrm{gc}} = 0.0175Ar^{0.46}Pr^{0.33}\left(\frac{U}{U_{\mathrm{m}}}\right)^{0.3} \quad (\text{for } U_{\mathrm{mf}} < U < U_{\mathrm{m}}) \tag{7.55}$$

where U_{m} is the superficial velocity corresponding to the maximum overall bed heat transfer coefficient.

For temperatures beyond 600 °C radiative heat transfer plays an increasing role and must be accounted for in calculations. The reader is referred to Botterill (1986) or Kunii and Levenspiel (1990) for treatment of radiative heat transfer or for a more detailed look at heat transfer in fluidized beds.

7.8 APPLICATIONS OF FLUIDIZED BEDS

7.8.1 Physical Processes

Physical processes which use fluidized beds include drying, mixing, granulation, coating, heating and cooling. All these processes take advantage of the excellent mixing capabilities of the fluid bed. Good solids mixing gives rise to good heat transfer, temperature uniformity and ease of process control. One of the most important applications of the fluidized bed is to the drying of solids. Fluidized beds are currently used commercially for drying such materials as crushed minerals, sand, polymers, pharmaceuticals, fertilizers and crystalline products. The reasons for the popularity of fluidized bed drying are:

- The dryers are compact, simple in construction and of relatively low capital cost.
- The absence of moving parts, other than the feeding and discharge devices, leads to reliable operation and low maintenance.
- The thermal efficiency of these dryers is relatively high.
- Fluidized bed dryers are gentle in the handling of powders and this is useful when dealing with friable materials.

Fluidized bed granulation is dealt with in Chapter 13 and mixing is covered in Chapter 11. Fluidized beds are often used to cool particulate solids following a reaction. Cooling may be by fluidizing air alone or by the use of cooling water passing through tubes immersed in the bed (see Figure 7.17 for example). Fluidized beds are used for coating particles in the pharmaceutical and agricultural industries. Metal components may be plastic coated by dipping them hot into an air-fluidized bed of powdered thermosetting plastic.

7.8.2 Chemical Processes

The gas fluidized bed is a good medium in which to carry out a chemical reaction involving a gas and a solid. Advantages of the fluidized bed for chemical reaction include:

- The gas–solid contacting is generally good.
- The excellent solids circulation within the bed promotes good heat transfer between bed particles and the fluidizing gas and between the bed and heat transfer surfaces immersed in the bed.

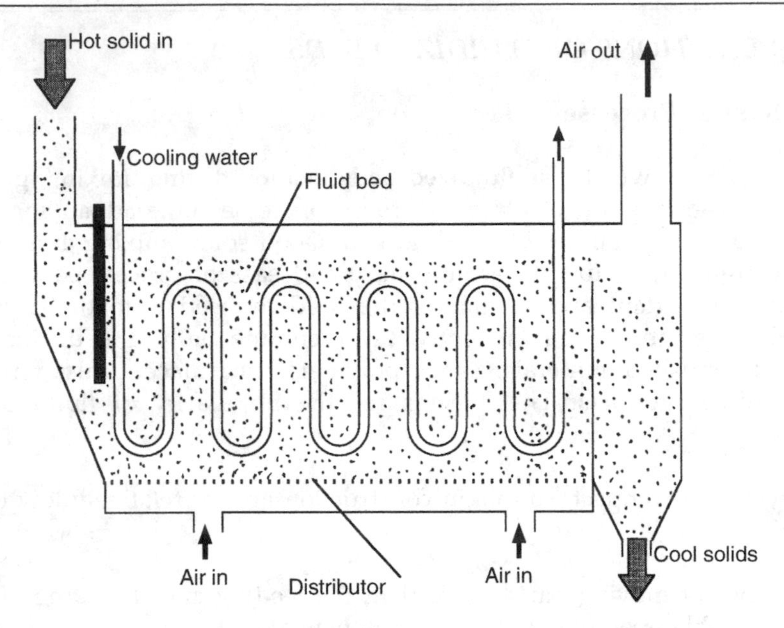

Figure 7.17 Schematic diagram of a fluidized bed solid cooler

- This gives rise to near isothermal conditions even when reactions are strongly exothermic or endothermic.
- The good heat transfer also gives rise to ease of control of the reaction.
- The fluidity of the bed makes for ease of removal of solids from the reactor.

However, it is far from ideal; the main problems arise from the two phase (bubbles and fluidized solids) nature of such systems. This problem is particularly acute where the bed solids are the catalyst for a gas-phase reaction. In such a case the ideal fluidized bed chemical reactor would have excellent gas–solid contacting, no gas by-passing and no back-mixing of the gas against the main direction of flow. In a bubbling fluidized bed the gas bypasses the solids by passing through the bed as bubbles. This means that unreacted reactants appear in the product. Also, gas circulation patterns within a bubbling fluidized bed are such that products are back-mixed and may undergo undesirable secondary reactions. These problems lead to serious practical difficulties particularly in the scaling-up of a new fluidized bed process from pilot plant to full industrial scale. This subject is dealt with in more detail in Kunii and Levenspiel (1990), Geldart (1986) and Davidson and Harrison (1971).

Figure 7.18 is a schematic diagram of one type of fluid catalytic cracking (FCC) unit, a celebrated example of fluidized bed technology for breaking down large molecules in crude oil to small molecules suitable for gasoline, etc. Other examples of the application of fluidized bed technology to different kinds of chemical reaction are shown in Table 7.2.

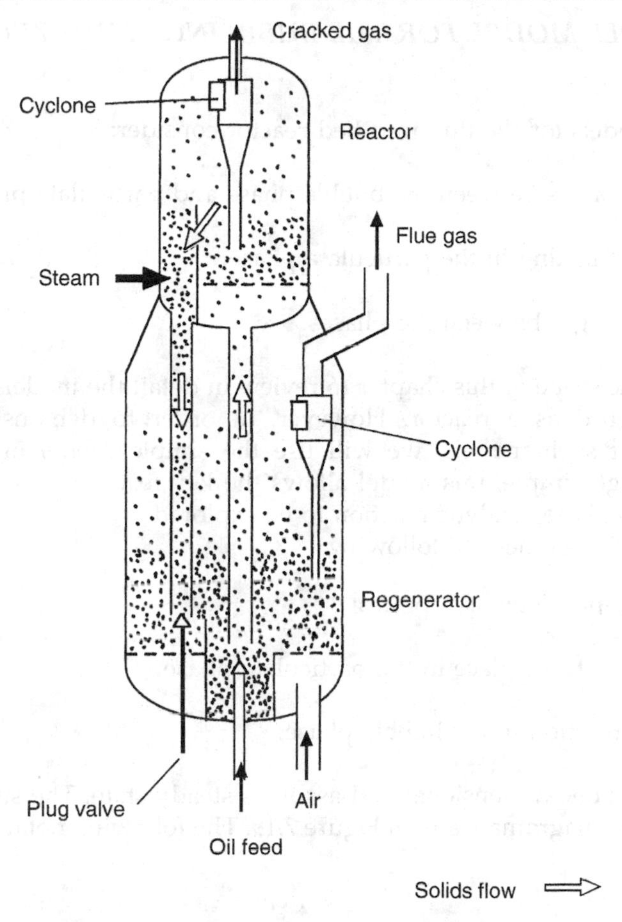

Figure 7.18 Kellogg's Model A Orthoflow FCC unit

Table 7.2 Summary of the types of gas–solid chemical reactions employing fluidization

Type	Example	Reasons for using a fluidized bed
Homogeneous gas-phase reactions	Ethylene hydrogenation	Rapid heating of entering gas. Uniform controllable temperature
Heterogeneous non-catalytic reactions	Sulfide ore roasting, combustion	Ease of solids handling. Temperature uniformity. Good heat transfer
Heterogeneous catalytic reactions	Hydrocarbon cracking, phthalic anhydride, acrylonitrile	Ease of solids handling. Temperature uniformity. Good heat transfer

7.9 A SIMPLE MODEL FOR THE BUBBLING FLUIDIZED BED REACTOR

In general, models for the fluidized bed reactor consider:

- the division of gas between the bubble phase and particulate phase;
- the degree of mixing in the particulate phase;
- the transfer of gas between the phases.

It is outside the scope of this chapter to review in detail the models available for the fluidized bed as a reactor. However, in order to demonstrate the key components of such models, we will use the simple model of Orcutt *et al.* (1962). Although simple, this model allows the key features of a fluidized bed reactor for gas-phase catalytic reaction to be explored.

The approach assumes the following:

- Original two-phase theory applies;
- Perfect mixing takes place in the particulate phase;
- There is no reaction in the bubble phase;

The model is one-dimensional and assumes steady state. The structure of the model is shown diagramatically in Figure 7.19. The following notation is used: C_0

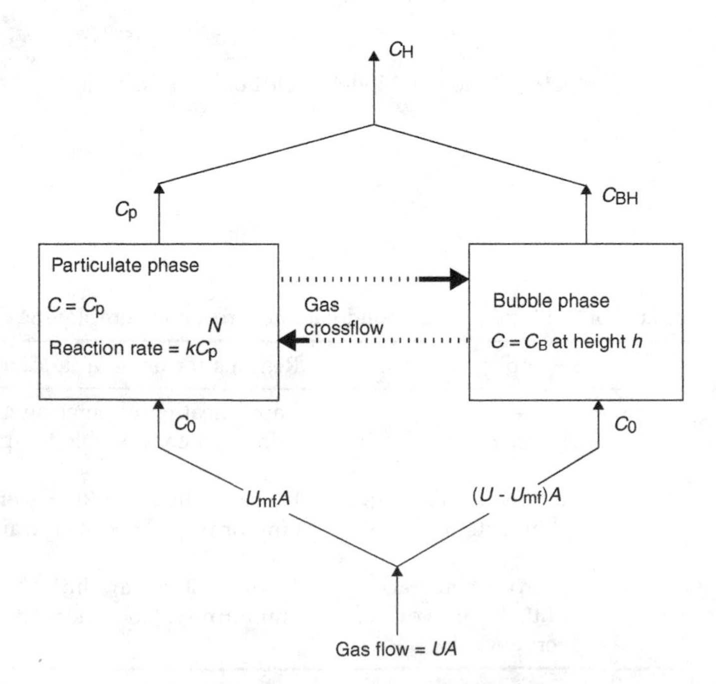

Figure 7.19 Schematic diagram of the Orcutt fluidized bed reactor model

is the concentration of reactant at distributor; C_p is the concentration of reactant in the particulate phase; C_B is the concentration of the reactant in the bubble phase at height h above the distributor; C_{BH} is the concentration of reactant leaving the bubble phase; and C_H is the concentration of reactant leaving the reactor.

In steady state, the concentration of reactant in the particulate phase is constant throughout the particulate phase because of the assumption of perfect mixing in the particulate phase. Throughout the bed, gaseous reactant is assumed to pass between particulate phase and bubble phase.

The overall mass balance on the reactant is:

$$\begin{pmatrix} \text{molar flow of} \\ \text{reactant into} \\ \text{reactor} \\ (1) \end{pmatrix} = \begin{pmatrix} \text{molar flow out} \\ \text{in the bubble phase} \\ (2) \end{pmatrix} + \begin{pmatrix} \text{molar flow out in} \\ \text{the particulate phase} \\ (3) \end{pmatrix} + \begin{pmatrix} \text{rate of} \\ \text{conversion} \\ (4) \end{pmatrix} \quad (7.56)$$

Term (1) $= UAC_0$

Term (2): molar reactant flow in bubble phase changes with height L above the distributor as gas is exchanged with the particulate phase. Consider an element of bed of thickness δL at a height L above the distributor. In this element:

$$\begin{pmatrix} \text{rate of increase of} \\ \text{reactant in bubble phase} \end{pmatrix} = \begin{pmatrix} \text{rate of transfer of} \\ \text{reactant from particulate phase} \end{pmatrix}$$

$$\text{i.e.} \quad (U - U_{mf})A\delta C_B = -K_C(\varepsilon_B A \delta L)(C_B - C_p) \quad (7.57)$$

in the limit as $\delta L \to 0$,

$$\frac{dC_B}{dL} = -\frac{K_C \varepsilon_B (C_B - C_p)}{(U - U_{mf})} \quad (7.58)$$

where K_C is the mass transfer coefficient per unit bubble volume and ε_B is the bubble fraction. Integrating with the boundary condition that $C_B = C_0$ at $L_0 = 0$:

$$C_B = C_p + (C_0 - C_p)\exp\left(-\frac{K_C L}{U_B}\right) \quad (7.59)$$

since $\varepsilon_B = (U - U_{mf})/U_B$ [Equation (7.28)].

At the surface of the bed, $L = H$ and so the reactant concentration in the bubble phase at the bed surface is given by:

$$C_{BH} = C_p + (C_0 - C_p)\exp\left(-\frac{K_C H}{U_B}\right) \quad (7.60)$$

Term (2) $= C_{BH}(U - U_{mf})A$

Term (3) $= U_{mf}AC_p$

Term (4): For a reaction which is jth order in the reactant under consideration,

$$\begin{pmatrix}\text{molar rate of conversion}\\ \text{per unit volume of solids}\end{pmatrix} = kC_p^j$$

where k is the reaction rate constant per unit volume of solids.

Therefore,

$$\begin{pmatrix}\text{molar rate of}\\ \text{conversion in bed}\end{pmatrix} = \begin{pmatrix}\text{molar rate of}\\ \text{conversion per unit}\\ \text{volume of solids}\end{pmatrix} \times \begin{pmatrix}\text{volume of solids}\\ \text{per unit volume of}\\ \text{particulate phase}\end{pmatrix} \times \begin{pmatrix}\text{volume of particulate}\\ \text{phase per unit}\\ \text{volume of bed}\end{pmatrix} \times \begin{pmatrix}\text{volume}\\ \text{of bed}\end{pmatrix}$$

hence, term (4),

$$\begin{pmatrix}\text{molar rate of}\\ \text{conversion in bed}\end{pmatrix} = kC_p^j(1-\varepsilon_p)(1-\varepsilon_B)AH \tag{7.61}$$

where ε_p is the particulate phase voidage.

Substituting these expressions for terms (1)–(4), the mass balance becomes:

$$UAC_0 = \left[C_p + (C_0 - C_p)\exp\left(-\frac{K_C H}{U_B}\right)\right](U - U_{mf})A + U_{mf}AC_p + kC_p^j(1-\varepsilon_p)(1-\varepsilon_B)AH \tag{7.62}$$

From this mass balance C_p may be found. The reactant concentration leaving the reactor C_H is then calculated from the reactant concentrations and gas flows through the bubble and particulate phases:

$$C_H = \frac{U_{mf}C_p + (U - U_{mf})C_{BH}}{U} \tag{7.63}$$

In the case of a first-order reaction ($j = 1$), solving the mass balance for C_p gives:

$$C_p = \frac{C_0[U - (U - U_{mf})e^{-\chi}]}{kH_{mf}(1-\varepsilon_p) + [U - (U - U_{mf})e^{-\chi}]} \tag{7.64}$$

where $\chi = K_C H/U_B$, equivalent to a number of mass transfer units for gas exchange between the phases. χ is related to bubble size and correlations are available. Generally χ decreases as bubble size increases and so small bubbles are preferred.

Thus from Equations (7.63) and (7.64), we obtain an expression for the conversion in the reactor:

$$1 - \frac{C_H}{C_0} = (1 - \beta e^{-\chi}) - \frac{(1 - \beta e^{-\chi})^2}{\frac{kH_{mf}(1 - \varepsilon_p)}{U} + (1 - \beta e^{-\chi})} \tag{7.65}$$

where $\beta = (U - U_{mf})/U$, the fraction of gas passing through the bed as bubbles. It is interesting to note that although the two-phase theory does not always hold, Equation (7.65) often holds with β still the fraction of gas passing through the bed as bubbles, but not equal to $(U - U_{mf})/U$.

Readers interested in reactions of order different from unity, solids reactions and more complex reactor models for the fluidized bed, are referred to Kunii and Levenspiel (1990).

Although the Orcutt model is simple, it does allow us to explore the effects of operating conditions, reaction rate and degree of interphase mass transfer on performance of a fluidized bed as a gas-phase catalytic reactor. Figure 7.20 shows the variation of conversion with reaction rate (expressed as $kH_{mf}(1 - \varepsilon_p/U)$) with excess gas velocity (expressed as β) calculated using Equation (7.65) for a first-order reaction.

Noting that the value of χ is dictated mainly by the bed hydrodynamics, we see that:

- *For slow reactions*, overall conversion is insensitive to bed hydrodynamics and so reaction rate k is the rate controlling factor.

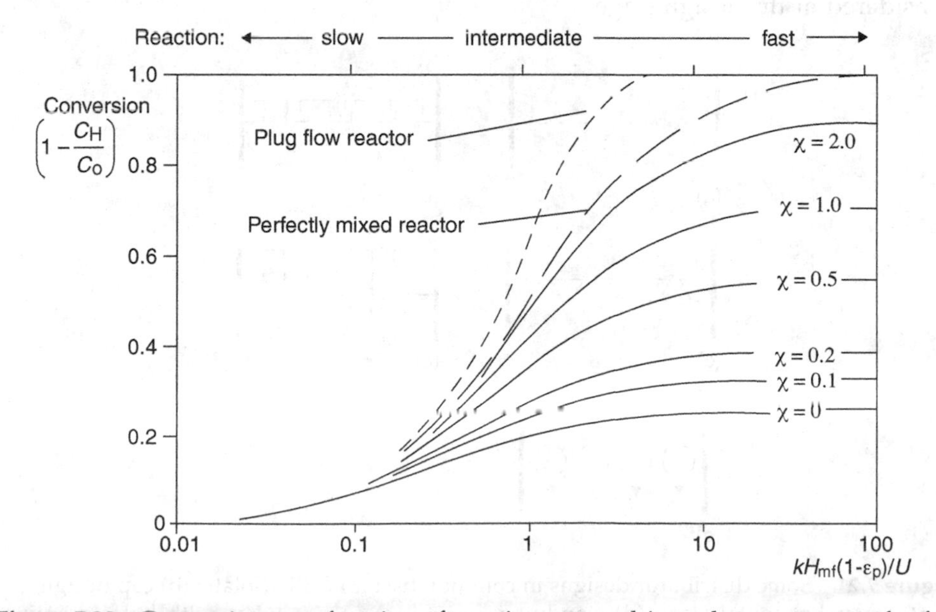

Figure 7.20 Conversion as a function of reaction rate and interphase mass transfer for $\beta = 0.75$ for a first-order gas phase catalytic reaction [based on Equation (7.65)]

- *For intermediate reactions,* both reaction rate and bed hydrodynamics affect the conversion.

- *For fast reactions,* the conversion is determined by the bed hydrodynamics.

These results are typical for a gas-phase catalytic reaction in a fluidized bed.

7.10 SOME PRACTICAL CONSIDERATIONS

7.10.1 Gas Distributor

The distributor is a device designed to ensure that the fluidizing gas is always evenly distributed across the cross-section of the bed. It is a critical part of the design of a fluidized bed system. Good design is based on achieving a pressure drop which is a sufficient fraction of the bed pressure drop. Readers are referred to Geldart (1986) for guidelines on distributor design. Many operating problems can be traced back to poor distributor design. Some distributor designs in common use are shown in Figure 7.21.

7.10.2 Loss of Fluidizing Gas

Loss of fluidizing gas will lead to collapse of the fluidized bed. If the process involves the emission of heat then this heat will not be dissipated as well from the packed bed as it was from the fluidized bed. The consequences of this should be considered at the design stage.

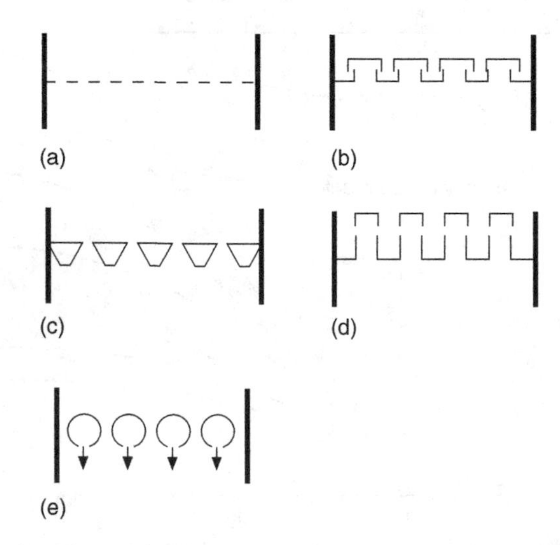

Figure 7.21 Some distributor designs in common use: (a) drilled plate; (b) cap design; (c) continuous horizontal slots; (d) standpipe design; (e) sparge tubes with holes pointing downwards

7.10.3 Erosion

All parts of the fluidized bed unit are subject to erosion by the solid particles. Heat transfer tubes within the bed or freeboard are particularly at risk and erosion here may lead to tube failure. Erosion of the distributor may lead to poor fluidization and areas of the bed becoming deaerated.

7.10.4 Loss of Fines

Loss of fine solids from the bed reduces the quality of fluidization and reduces the area of contact between the solids and the gas in the process. In a catalytic process this means lower conversion.

7.10.5 Cyclones

Cyclone separators are often used in fluidized beds for separating entrained solids from the gas stream (see Chapter 9). Cyclones installed within the fluidized bed vessel would be fitted with a dip-leg and seal in order to prevent gas entering the solids exit. Fluidized systems may have two or more stages of cyclone in series in order to improve separation efficiency. Cyclones are also the subject of erosion and must be designed to cope with this.

7.10.6 Solids Feeders

Various devices are available for feeding solids into the fluidized bed. The choice of device depends largely on the nature of the solids feed. Screw conveyors, spray feeders and pneumatic conveying are in common use.

7.11 *WORKED EXAMPLES*

WORKED EXAMPLE 7.1

3.6 kg of solid particles of density 2590 kg/m^3 and surface-volume mean size 748 μm form a packed bed of height 0.475 m in a circular vessel of diameter 0.0757 m. Water of density 1000 kg/m^3 and viscosity 0.001 Pa s is passed upwards through the bed. Calculate (a) the bed pressure drop at incipient fluidization, (b) the superficial liquid velocity at incipient fluidization, (c) the mean bed voidage at a superficial liquid velocity of 1.0 cm/s, (d) the bed height at this velocity and (e) the pressure drop across the bed at this velocity.

Solution

(a) Applying Equation (7.24) to the packed bed, we find the packed bed voidage:

$$\text{mass of solids} = 3.6 = (1 - \varepsilon) \times 2590 \times \frac{\pi(0.0757)^2}{4} \times 0.475$$

hence, $\varepsilon = 0.3498$

Frictional pressure drop across the bed when fluidized:

$$(-\Delta p) = \frac{\text{weight of particles} - \text{upthrust on particles}}{\text{cross-sectional area}}$$

$$(-\Delta p) = \frac{Mg - Mg(\rho_f/\rho_p)}{A} \quad \text{(since upthrust = weight of fluid displaced by particles)}$$

$$\text{Hence, } (-\Delta p) = \frac{Mg}{A}\left(1 - \frac{\rho_f}{\rho_p}\right) = \frac{3.6 \times 9.81}{4.50 \times 10^{-3}}\left(1 - \frac{1000}{2590}\right) = 4817\,\text{Pa}$$

(b) Assuming that the voidage at the onset of fluidization is equal to the voidage of the packed bed, we use the Ergun equation to express the relationship between packed pressure drop and superficial liquid velocity:

$$\frac{(-\Delta p)}{H} = 3.55 \times 10^7 U^2 + 2.648 \times 10^6 U$$

Equating this expression for pressure drop across the packed bed to the fluidized bed pressure drop, we determine superficial fluid velocity at incipient fluidization, U_{mf}.

$$U_{mf} = 0.365 \text{ cm/s}$$

(c) The Richardson–Zaki equation [Equation (7.21)] allows us to estimate the expansion of a liquid fluidized bed.

$$U = U_T \varepsilon^n \qquad \text{[Equation(7.21)]}$$

Using the method given in Chapter 2, we determine the single particle terminal velocity, U_T.

$$Ar = 6527.9; \quad C_D Re_p^2 = 8704; \quad Re_p = 90; \quad U_T = 0.120 \text{ m/s}$$

Note that Re_p is calculated at U_T. At this value of Reynolds number, the flow is intermediate between viscous and inertial, and so we must use the correlation of Khan and Richardson [Equation (3.25)] to determine the exponent n in Equation (7.21):

$$\text{With } Ar = 6527.9, \ n = 3.202$$

Hence from Equation (7.21), $\varepsilon = 0.460$ when $U = 0.01$ m/s.

Mean bed voidage is 0.460 when the superficial liquid velocity is 1 cm/s.

(d) From Equation (7.25), we now determine the mean bed height at this velocity:

$$\text{Bed height(at } U = 0.01 \text{ m/s)} = \frac{(1 - 0.3498)}{(1 - 0.460)} 0.475 = 0.572\,\text{m}$$

(e) The frictional pressure drop across the bed remains essentially constant once the bed is fluidized. Hence at a superficial liquid velocity of 1 cm/s the frictional pressure drop across the bed is 4817 Pa.

However, the measured pressure drop across the bed will include the hydrostatic head of the liquid in the bed. Applying the mechanical energy equation between the bottom (1) and the top (2) of the fluidized bed:

$$\frac{p_1 - p_2}{\rho_f g} + \frac{U_1^2 - U_2^2}{2g} + (z_1 - z_2) = \text{friction head loss} = \frac{4817}{\rho_f g}$$

$$U_1 = U_2; \qquad z_1 - z_2 = -H = -0.572\,\text{m}$$

Hence, $p_1 - p_2 = 10\,428\,\text{Pa}$.

WORKED EXAMPLE 7.2

A powder having the size distribution given below and a particle density of 2500 kg/m^3 is fed into a fluidized bed of cross-sectional area 4 m^2 at a rate of 1.0 kg/s.

Size range number (i)	Size range (μm)	Mass fraction in feed
1	10–30	0.20
2	30–50	0.65
3	50–70	0.15

The bed is fluidized using air of density 1.2 kg/m^3 at a superficial velocity of 0.25 m/s. Processed solids are continuously withdrawn from the base of the fluidized bed in order to maintain a constant bed mass. Solids carried over with the gas leaving the vessel are collected by a bag filter operating at 100% total efficiency. None of the solids caught by the filter are returned to the bed. Assuming that the fluidized bed is well-mixed and that the freeboard height is greater than the transport disengagement height under these conditions, calculate at equilibrium:

(a) the flow rate of solids entering the filter bag;

(b) the size distribution of the solids in the bed;

(c) the size distribution of the solids entering the filter bag;

(d) the rate of withdrawal of processed solids from the base of the bed;

(e) the solids loading in the gas entering the filter.

Solution

First calculate the elutriation rate constants for the three size ranges under these conditions from the Zenz and Weil correlation [Equation (7.46)]. The value of particle size x used in the correlation is the arithmetic mean of each size range:

$$x_1 = 20 \times 10^{-6}\,\text{m}; \quad x_2 = 40 \times 10^{-6}\,\text{m}; \quad x_3 = 60 \times 10^{-6}\,\text{m}$$

$$\text{With } U = 0.25\,\text{m/s}, \rho_p = 2500\,\text{kg/m}^3 \quad \text{and} \quad \rho_f = 1.2\,\text{kg/m}^3$$
$$K_{1\infty}^* = 3.21 \times 10^{-2}\text{kg/m}^2\,\text{s}$$
$$K_{2\infty}^* = 8.74 \times 10^{-3}\text{kg/m}^2\,\text{s}$$
$$K_{3\infty}^* = 4.08 \times 10^{-3}\text{kg/m}^2\,\text{s}$$

Referring to Figure 7.W2.1 the overall and component material balances over the fluidized bed system are:

$$\text{Overall balance}: F = Q + R \tag{7W2.1}$$
$$\text{Component balance}: Fm_{F_i} = Qm_{Q_i} + Rm_{R_i} \tag{7W2.2}$$

where F, Q and R are the mass flow rates of solids in the feed, withdrawal and filter discharge, respectively, and m_{F_i}, m_{Q_i} and m_{R_i} are the mass fractions of solids in size range i in the feed, withdrawal and filter discharge, respectively.

From Equation (7.39) the entrainment rate of size range i at the gas exit from the freeboard is given by:

$$R_i = Rm_{R_i} = K_{i\infty}^* A m_{B_i} \tag{7W2.3}$$

and

$$R = \sum R_i = \sum Rm_{R_i} \tag{7W2.4}$$

Combining these equations with the assumption that the bed is well mixed ($m_{Q_i} = m_{B_i}$),

$$m_{B_i} = \frac{Fm_{F_i}}{F - R + K_{i\infty}^* A} \tag{7W2.5}$$

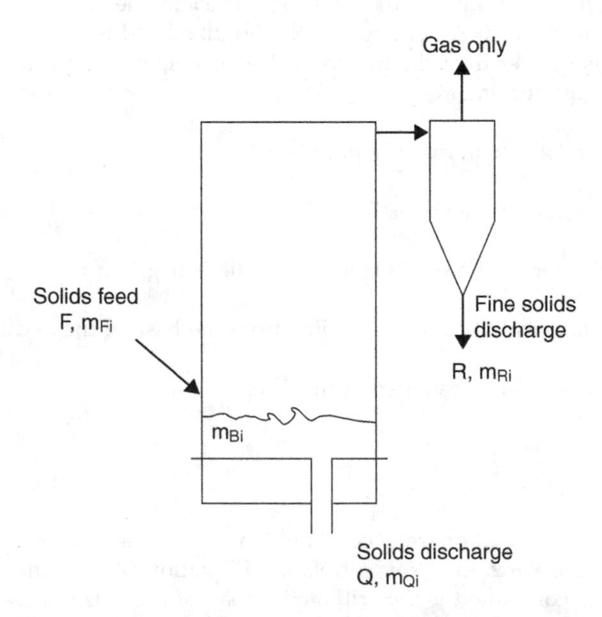

Figure 7W2.1 Schematic diagram showing solids flows and size distributions for the fluidized bed

Now both m_{B_i} and R are unknown. However, noting that $\sum m_{B_i} = 1$, we have

$$\frac{1.0 \times 0.2}{1.0 - R + (3.21 \times 10^{-2} \times 4)} + \frac{1.0 \times 0.65}{1.0 - R + (8.74 \times 10^{-3} \times 4)} + \frac{1.0 \times 0.15}{1.0 - R + (4.08 \times 10^{-3} \times 4)} = 1.0$$

Solving for R by trial and error, $R = 0.05\,\text{kg/s}$

(b) Substituting $R = 0.05\,\text{kg/s}$ in Equation (7W2.5), $m_{B_1} = 0.1855$, $m_{B_2} = 0.6599$ and $m_{B_3} = 0.1552$

Therefore size distribution of bed:

Size range number (*i*)	Size range (μm)	Mass fraction in bed
1	10–30	0.1855
2	30–50	0.6599
3	50–70	0.1552

(c) From Equation (7W2.3), knowing R and m_{B_i}, we can calculate m_{R_i}:

$$m_{R_i} = \frac{K^*_{1\infty} A m_{B_1}}{R} = \frac{3.21 \times 10^{-2} \times 4 \times 0.1855}{0.05} = 0.476$$

similarly, $m_{R_2} = 0.4614$, $m_{R_3} = 0.0506$

Therefore size distribution of solids entering filter:

Size range number (*i*)	Size range (μm)	Mass fraction entering filter
1	10–30	0.476
2	30–50	0.4614
3	50–70	0.0506

(d) From Equation (7W2.1), the rate of withdrawal of solids from the bed, $Q = 0.95\,\text{kg/s}$

(e) Solids loading for gas entering the filter,

$$\frac{\text{mass flow of solids}}{\text{volume flow of gas}} = \frac{R}{UA} = 0.05\,\text{kg/m}^3$$

WORKED EXAMPLE 7.3

A gas phase catalytic reaction is performed in a fluidized bed operating at a superficial gas velocity of 0.3 m/s. For this reaction under these conditions it is known that the reaction is first order in reactant A. The following information is given:

- bed height at incipient fluidization = 1.5 m;
- operating mean bed height = 1.65 m;
- voidage at incipient fluidization = 0.47;
- reaction rate constant = 75.47 (per unit volume of solids);
- $U_{mf} = 0.033$ m/s;
- mean bubble rise velocity = 0.111 m/s;
- mass transfer coefficient between bubbles and emulsion = 0.1009 (based on unit bubble volume) minimum fluidization velocity = 0.033 m/s.

Use the reactor model of Orcutt *et al.* to determine:

(a) the conversion of reactant A;

(b) the effect on the conversion found in (a) of reducing the inventory of catalyst by one half;

(c) the effect on the conversion found in (a) of halving the bubble size (assuming the interphase mass transfer coefficient is inversely proportional to the square root of the bubble diameter).

Discuss your answers to (b) and (c) and state which mechanism is controlling conversion in the reactor.

Solution

(a) From Section 7.9 the model of Orcutt *et al.* gives for a first order reaction:

$$\text{conversion, } 1 - \frac{C_H}{C_0} = (1 - \beta e^{-\chi}) - \frac{(1 - \beta e^{-\chi})^2}{\dfrac{kH_{mf}(1 - \varepsilon_p)}{U} + (1 - \beta e^{-\chi})} \qquad \text{[Equation (7.65)]}$$

where

$$\chi = \frac{K_C H}{U_B} \text{ and } \quad \beta = (U - U_{mf})/U$$

From the information given,

$$K_C = 0.1009,\ U_B = 0.111 \text{ m/s},\ U = 0.3 \text{ m/s},\ U_{mf} = 0.033 \text{ m/s},$$
$$H = 1.65 \text{ m},\ H_{mf} = 1.5 \text{ m},\ k = 75.47.$$

Hence, $\chi = 1.5$, $\beta = 0.89$ and $kH_{mf}(1 - \varepsilon_p)/U = 200$ (assuming $\varepsilon_p = \varepsilon_{mf}$)

So, from Equation (7.65), conversion = 0.798.

(b) If the inventory of solids in the bed is halved, both the operating bed height H and the height at incipient fluidization H_{mf} are halved. Thus, assuming all else remains constant, under the new conditions

$$\chi = 0.75,\ \beta = 0.89 \text{ and } kH_{mf}(1 - \varepsilon_p)/U = 100$$

and so the new conversion = 0.576.

(c) If the bubble size is halved and K_C is proportional to $1/\sqrt{(\text{double diameter})}$,

$$\text{new } K_C = 1.414 \times 0.1009 = 0.1427$$

Hence, $\chi = 2.121$, giving conversion = 0.889.

(d) Comparing the conversion achieved in (c) with that achieved in (a), we see that improving interphase mass transfer has a significant effect on the conversion. We may also note that doubling the reaction rate (say by increasing the reactor temperature) and keeping everything else constant has a negligible effect on the conversion achieved in (a). We conclude, therefore, that under these conditions the transfer of gas between bubble phase and emulsion phase controls the conversion.

TEST YOURSELF

7.1 Write down the equation for the force balance across a fluidized bed and use it to come up with an expression for the pressure drop across a fluidized bed.

7.2 15 kg of particles of particle density 2000 kg/m^3 are fluidized in a vessel of cross-sectional area 0.03 m^2 by a fluid of density 900 kg/m^3. (a) What is the frictional pressure drop across the bed? (b) If the bed height is 0.6 m, what is the bed voidage?

7.3 Sketch a plot of pressure drop across a bed of powder versus velocity of the fluid flowing upwards through it. Include packed bed and fluidized bed regions. Mark on the incipient fluidization velocity.

7.4 What are the chief behaviour characteristics of the four Geldart powder groups?

7.5 What differentiates a Geldart Group A powder from a Geldart Group B powder?

7.6 According to Richardson and Zaki, how does bed voidage in a liquid-fluidized bed vary with fluidizing velocity at Reynolds numbers less than 0.3?

7.7 What is the basic assumption of the two-phase theory? Write down an equation that describes bed expansion as a function of superficial fluidizing velocity according to the two-phase theory.

7.8 Explain what is meant by particle convective heat transfer in a fluidized bed. In which Geldart group is particle convective heat transfer dominant?

7.9 Under what conditions does gas convective heat transfer play a significant role?

7.10 A fast gas phase catalytic reaction is performed in a fluidized bed using a particulate catalyst. Would conversion be increased by improving conditions for mass transfer between particulate phase and bubble phase?

EXERCISES

7.1 A packed bed of solid particles of density 2500 kg/m^3, occupies a depth of 1 m in a vessel of cross-sectional area 0.04 m^2. The mass of solids in the bed is 50 kg and the surface-volume mean diameter of the particles is 1 mm. A liquid of density 800 kg/m^2 and viscosity 0.002 Pa s flows upwards through the bed.

(a) Calculate the voidage (volume fraction occupied by voids) of the bed.

(b) Calculate the pressure drop across the bed when the volume flow rate of liquid is 1.44 m^3/h.

(c) Calculate the pressure drop across, the bed when it becomes fluidized.

[Answer: (a) 0.5; (b) 6560 Pa; (c) 8338 Pa.]

7.2 130 kg of uniform spherical particles with a diameter of 50 μm and particle density 1500 kg/m^3 are fluidized by water (density 1000 kg/m^3, viscosity 0.001 Pa s) in a circular bed of cross-sectional area 0.2 m^2. The single particle terminal velocity of the particles is 0.68 mm/s and the voidage at incipient fluidization is known to be 0.47.

(a) Calculate the bed height at incipient fluidization.

(b) Calculate the mean bed voidage when the liquid flow rate is 2×10^{-5} m^3/s.

[Answer: (a) 0.818 m; (b) 0.6622.]

7.3 130 kg of uniform spherical particles with a diameter of 60 μm and particle density 1500 kg/m^3 are fluidized by water (density 1000 kg/m^3, viscosity 0.001 Pa s) in a circular bed of cross-sectional area 0.2 m^2. The single particle terminal velocity of the particles is 0.98 mm/s and the voidage at incipient fluidization is known to be 0.47.

(a) Calculate the bed height at incipient fluidization.

(b) Calculate the mean fluidized bed voidage when the liquid flow rate is 2×10^{-5} m^3/s.

[Answer: (a) 0.818 m; (b) 0.6121.]

7.4 A packed bed of solid particles of density 2500 kg/m^3, occupies a depth of 1 m in a vessel of cross-sectional area 0.04 m^2. The mass of solids in the bed is 59 kg and the surface-volume mean diameter of the particles is 1 mm. A liquid of density 800 kg/m^3 and viscosity 0.002 Pa s flows upwards through the bed.

(a) Calculate the voidage (volume fraction occupied by voids) of the bed.

(b) Calculate the pressure drop across the bed when the volume flow rate of liquid is $0.72\,m^3/h$.

(c) Calculate the pressure drop across the bed when it becomes fluidized.

[Answer: (a) 0.41; (b) 7876 Pa; (c) 9839 Pa.]

7.5 12 kg of spherical resin particles of density $1200\,kg/m^3$ and uniform diameter 70 μm are fluidized by water (density $1000\,kg/m^3$ and viscosity 0.001 Pa s) in a vessel of diameter 0.3 m and form an expanded bed of height 0.25 m.

(a) Calculate the difference in pressure between the base and the top of the bed.

(b) If the flow rate of water is increased to $7\,cm^3/s$, what will be the resultant bed height and bed voidage (liquid volume fraction)?

State and justify the major assumptions.

[Answer: (a) Frictional pressure drop = 277.5 Pa, pressure difference = 2730 Pa; (b) height = 0.465 m; voidage = 0.696.]

7.6 A packed bed of solids of density $2000\,kg/m^3$ occupies a depth of 0.6 m in a cylindrical vessel of inside diameter 0.1 m. The mass of solids in the bed is 5 kg and the surface-volume mean diameter of the particles is 300 μm. Water (density $1000\,kg/m^3$ and viscosity 0.001 Pa s) flows upwards through the bed.

(a) What is the voidage of the packed bed?

(b) Use a force balance over the bed to determine the bed pressure drop when fluidized.

(c) Hence, assuming laminar flow and that the voidage at incipient fluidization is the same as the packed bed voidage, determine the minimum fluidization velocity. Verify the assumption of laminar flow.

[Answer: (a) 0.4692; (b) 3124 Pa; (c) 1.145 mm/s.]

7.7 A packed bed of solids of density $2000\,kg/m^3$ occupies a depth of 0.5 m in a cylindrical vessel of inside diameter 0.1 m. The mass of solids in the bed is 4 kg and the surface-volume mean diameter of the particles is 400 μm. Water (density 1000 kg and viscosity 0.001 Pa s) flows upwards through the bed.

(a) What is the voidage of the packed bed?

(b) Use a force balance over the bed to determine the bed pressure drop when fluidized.

(c) Hence, assuming laminar flow and that the voidage at incipient fluidization is the same as the packed bed voidage, determine the minimum fluidization velocity. Verify the assumption of laminar flow.

[Answer: (a) 0.4907; (b) 2498 Pa; (c) 2.43 mm/s.]

7.8 By applying a force balance, calculate the incipient fluidizing velocity for a system with particles of particle density 5000 kg/m^3 and mean volume diameter 100 μm and a fluid of density 1.2 kg/m^3 and viscosity 1.8×10^{-5} Pa s. Assume that the voidage at incipient fluidization is 0.5.

If in the above example the particle size is changed to 2 mm, what is U_{mf}?

[Answer: 0.045 m/s; 2.26 m/s.]

7.9 A powder of mean sieve size 60 μm and particle density 1800 kg/m^3 is fluidized by air of density 1.2 kg/m^3 and viscosity 1.84×10^{-5} Pa s in a circular vessel of diameter 0.5 m. The mass of powder charged to the bed is 240 kg and the volume flow rate of air to the bed is 140 m^3/h. It is known that the average bed voidage at incipient fluidization is 0.45 and correlation reveals that the average bubble rise velocity under the conditions in question is 0.8 m/s. Estimate:

(a) the minimum fluidization velocity, U_{mf};

(b) the bed height at incipient fluidization;

(c) the visible bubble flow rate;

(d) the bubble fraction;

(e) the particulate phase voidage;

(f) the mean bed height;

(g) the mean bed voidage.

[Answer: (a) Baeyens and Geldart correlation [Equation (7.11)], 0.0027 m/s; (b) 1.24 m; (c) 0.038 m^3/s (assumes $U_{mf} \cong U_{mb}$); (d) 0.245; (e) 0.45; (f) 1.64 m; (g) 0.585.]

7.10 A batch fluidized bed process has an initial charge of 2000 kg of solids of particle density 1800 kg/s^3 and with the size distribution shown below:

Size range number (i)	Size range (μm)	Mass fraction in feed
1	15–30	0.10
2	30–50	0.20
3	50–70	0.30
4	70–100	0.40

The bed is fluidized by a gas of density 1.2 kg/m^3 Pa s at a superficial gas velocity of 0.4 m/s.

The fluid bed vessel has a cross-sectional area of 1 m^2.

Using a discrete time interval calculation with a time increment of 5 min, calculate:

(a) the size distribution of the bed after 50 min;

(b) the total mass of solids lost from the bed in that time;

(c) the maximum solids loading at the process exit;

(d) the entrainment flux above the transport disengagement height of solids in size range 1 (15–30 μm) after 50 min.

Assume that the process exit is positioned above TDH and that none of the entrained solids are returned to the bed.

[Answer: (a) (range 1) 0.029, (2) 0.165, (3) 0.324, (4) 0.482; (b) 527 kg; (c) 0.514 kg/m^3 s; (d) 0.024 kg/m^2 s.]

7.11 A powder having a particle density of 1800 kg/m^3 and the following size distribution:

Size range number (*i*)	Size range (μm)	Mass fraction in feed
1	20–40	0.10
2	40–60	0.35
3	60–80	0.40
4	80–100	0.15

is fed into a fluidized bed 2 m in diameter at a rate of 0.2 kg/s. The cyclone inlet is 4 m above the distributor and the mass of solids in the bed is held constant at 4000 kg by withdrawing solids continuously from the bed. The bed is fluidized using dry air at 700 K (density 0.504 kg/m^3 and viscosity 3.33×10^{-5} Pa s) giving a superficial gas velocity of 0.3 m/s. Under these conditions the mean bed voidage is 0.55 and the mean bubble size at the bed surface is 5 cm. For this powder, under these conditions, $U_{mb} = 0.155$ cm/s.

Assuming that none of the entrained solids are returned to the bed, estimate:

(a) the flow rate and size distribution of the entrained solids entering the cyclone;

(b) the equilibrium size distribution of solids in the bed;

(c) the solids loading of the gas entering the cyclone;

(d) the rate at which solids are withdrawn from the bed.

[Answer: (a) 0.0485 kg/s, (range 1) 0.213, (2) 0.420, (3) 0.295, (4) 0.074; (b) (range 1) 0.0638, (2) 0.328, (3) 0.433, (4) 0.174; (c) 51.5 g/m^3; (d) 0.152 kg/s.]

7.12 A gas phase catalytic reaction is performed in a fluidized bed operating at a superficial gas velocity equivalent to $10 \times U_{mf}$. For this reaction under these conditions it is known that the reaction is first order in reactant A. Given the following information:

$$kH_{mf}(1 - \varepsilon_p)/U = 100; \ \chi = \frac{K_C H}{U_B} = 1.0$$

use the reactor model of Orcutt *et al.* to determine:

(a) the conversion of reactant A;

(b) the effect on the conversion found in (a) of doubling the inventory of catalyst;

(c) the effect on the conversion found in (a) of halving the bubble size by using suitable baffles (assuming the interphase mass transfer coefficient is inversely proportional to the bubble diameter).

If the reaction rate were two orders of magnitude smaller, comment on the wisdom of installing baffles in the bed with a view to improving conversion.

[Answer: (a) 0.6645; (b) 0.8744; (c) 0.8706.]

8

Pneumatic Transport and Standpipes

In this chapter we deal with two examples of the transport of particulate solids in the presence of a gas. The first example is pneumatic transport (sometimes referred to as pneumatic conveying), which is the use of a gas to transport a particulate solid through a pipeline. The second example is the standpipe, which has been used for many years, particularly in the oil industry, for transferring solids downwards from a vessel at low pressure to a vessel at a higher pressure.

8.1 PNEUMATIC TRANSPORT

For many years gases have been used successfully in industry to transport a wide range of particulate solids – from wheat flour to wheat grain and plastic chips to coal. Until quite recently most pneumatic transport was done in dilute suspension using large volumes of air at high velocity. Since the mid-1960s, however, there has been increasing interest in the so-called 'dense phase' mode of transport in which the solid particles are not fully suspended. The attractions of dense phase transport lie in its low air requirements. Thus, in dense phase transport, a minimum amount of air is delivered to the process with the solids (a particular attraction in feeding solids into fluidized bed reactors, for example). A low air requirement also generally means a lower energy requirement (in spite of the higher pressures needed). The resulting low solids velocities mean that in dense phase transport product degradation by attrition, and pipeline erosion are not the major problems they are in dilute phase pneumatic transport.

In this section we will look at the distinguishing characteristics of dense and dilute phase transport and the types of equipment and systems used with each. The design of dilute phase systems is dealt with in detail and the approach to design of dense phase systems is summarized.

Introduction to Particle Technology - 2nd Edition Martin Rhodes

8.1.1 Dilute Phase and Dense Phase Transport

The pneumatic transport of particulate solids is broadly classified into two flow regimes: dilute (or lean) phase flow; and dense phase flow. Dilute phase flow in its most recognizable form is characterized by high gas velocities (greater than 20 m/s), low solids concentrations (less than 1% by volume) and low pressure drops per unit length of transport line (typically less than 5 mbar/m). Dilute phase pneumatic transport is limited to short route, continuous transport of solids at rates of less than 10 t/h and is the only system capable of operation under negative pressure. Under these dilute flow conditions the solid particles behave as individuals, fully suspended in the gas, and fluid-particle forces dominate. At the opposite end of the scale is dense phase flow, characterized by low gas velocities (1–5 m/s, high solids concentrations (greater than 30% by volume) and high pressure drops per unit length of pipe (typically greater than 20 mbar/m). In dense phase transport particles are not fully suspended and there is much interaction between the particles.

The boundary between dilute phase flow and dense phase flow, however, is not clear cut and there are as yet no universally accepted definitions of dense phase and dilute phase transport.

Konrad (1986) lists four alternative means of distinguishing dense phase flow from dilute phase flow:

(a) on the basis of solids/air mass flow rates;

(b) on the basis of solids concentration;

(c) dense phase flow exists where the solids completely fill the cross–section of the pipe at some point;

(d) dense phase flow exists when, for horizontal flow, the gas velocity is insufficient to support all particles in suspension, and, for vertical flow, where reverse flow of solids occurs.

In all these cases different authors claim different values and apply different interpretations.

In this chapter the 'choking' and 'saltation' velocities will be used to mark the boundaries between dilute phase transport and dense phase transport in vertical and horizontal pipelines, respectively. These terms are defined below in considering the relationships between gas velocity, solids mass flow rate and pressure drop per unit length of transport line in both horizontal and vertical transport.

8.1.2 The Choking Velocity in Vertical Transport

We will see in Section 8.1.4 that the pressure drop across a length of transport line has in general six components:

- pressure drop due to gas acceleration;

- pressure drop due to particle acceleration;
- pressure drop due to gas-to-pipe friction;
- pressure drop related to solid-to-pipe friction;
- pressure drop due to the static head of the solids;
- pressure drop due to the static head of the gas.

The general relationship between gas velocity and pressure gradient $\Delta p/\Delta L$ for a vertical transport line is shown in Figure 8.1. Line AB represents the frictional pressure loss due to gas only in a vertical transport line. Curve CDE is for a solids flux of G_1 and curve FG is for a higher feed rate G_2. At point C the gas velocity is high, the concentration is low, and frictional resistance between gas and pipe wall predominates. As the gas velocity is decreased the frictional resistance decreases but, since the concentration of the suspension increases, the static head required to support these solids increases. If the gas velocity is decreased below point D then the increase in static head outweighs the decrease in frictional resistance and $\Delta p/\Delta L$ rises again. In the region DE the decreasing velocity causes a rapid increase in solids concentration and a point is reached when the gas can no longer entrain all the solids. At this point a flowing, slugging fluidized bed (see Chapter 7) is formed in the transport line. The phenomenon is known as 'choking' and is usually attended by large pressure fluctuations. The choking velocity, U_{CH}, is the lowest velocity at which this dilute phase transport line can be operated at the solids feed rate G_1. At the higher solids feed rate, G_2, the choking velocity is higher. The choking velocity marks the boundary between dilute phase and dense phase vertical pneumatic

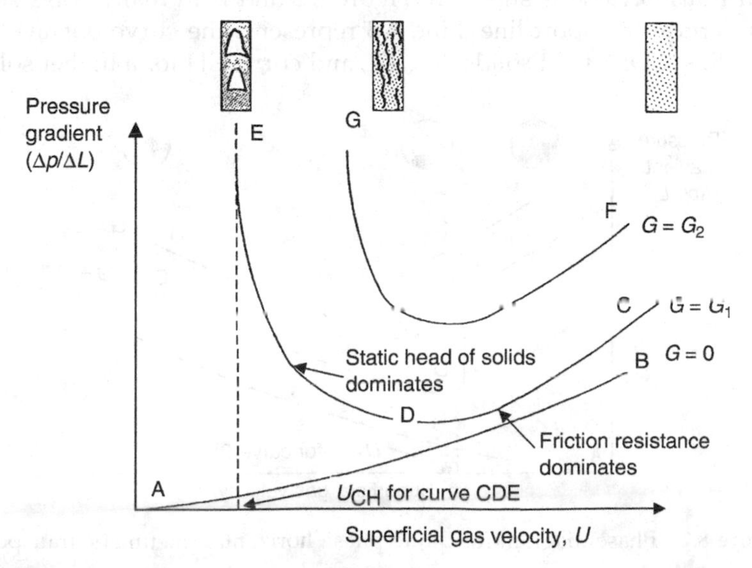

Figure 8.1 Phase diagram for dilute-phase vertical pneumatic transport

transport. Note that choking can be reached by decreasing the gas velocity at a constant solids flow rate, or by increasing the solids flow rate at a constant gas velocity.

It is not possible to theoretically predict the conditions for choking to occur. However, many correlations for predicting choking velocities are available in the literature. Knowlton (1986) recommends the correlation of Punwani *et al.* (1976), which takes account of the considerable effect of gas density. This correlation is presented below:

$$\frac{U_{\mathrm{CH}}}{\varepsilon_{\mathrm{CH}}} - U_{\mathrm{T}} = \frac{G}{\rho_{\mathrm{p}}(1 - \varepsilon_{\mathrm{CH}})} \tag{8.1}$$

$$\rho_{\mathrm{f}}^{0.77} = \frac{2250D(\varepsilon_{\mathrm{CH}}^{-4.7} - 1)}{\left(\frac{U_{\mathrm{CH}}}{\varepsilon_{\mathrm{CH}}} - U_{T}\right)^{2}} \tag{8.2}$$

where $\varepsilon_{\mathrm{CH}}$ is the voidage in the pipe at the choking velocity U_{CH}, ρ_{p} is the particle density, ρ_{f} is the gas density, G is the mass flux of solids ($= M_{\mathrm{p}}/A$) and U_{T} is the free fall, or terminal velocity, of a single particle in the gas. (Note that the constant is dimensional and that SI units must be used.)

Equation (8.1) represents the solids velocity at choking and includes the assumption that the slip velocity U_{slip} is equal to U_{T} (see Section 8.1.4 below for definition of slip velocity). Equations (8.1) and (8.2) must be solved simultaneously by trial and error to give $\varepsilon_{\mathrm{CH}}$ and U_{CH}.

8.1.3 The Saltation Velocity in Horizontal Transport

The general relationship between gas velocity and pressure gradient $\Delta p/\Delta L$ for a horizontal transport line is shown in Figure 8.2 and is in many ways similar to that for a vertical transport line. Line AB represents the curve obtained for gas only in the line, CDEF for a solids flux, G_1, and curve GH for a higher solids feed

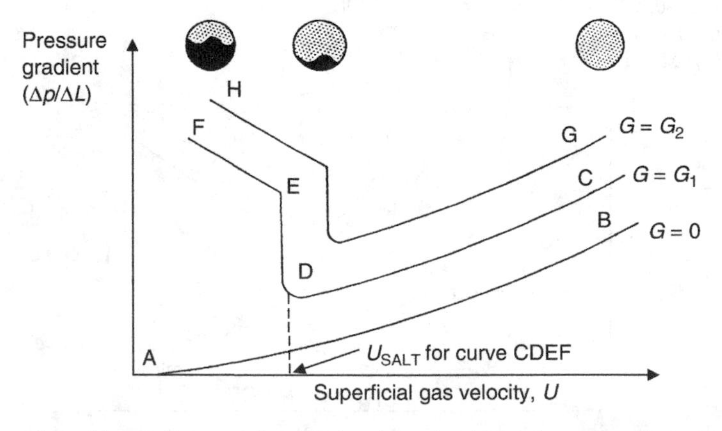

Figure 8.2 Phase diagram for dilute-phase horizontal pneumatic transport

rate, G_2. At point C, the gas velocity is sufficiently high to carry all the solids in very dilute suspension. The solid particles are prevented from settling to the walls of the pipe by the turbulent eddies generated in the flowing gas. If the gas velocity is reduced whilst solids feed rate is kept constant, the frictional resistance and $\Delta p/\Delta L$ decrease. The solids move more slowly and the solids concentration increases. At point D the gas velocity is insufficient to maintain the solids in suspension and the solids begin to settle out in the bottom of the pipe. The gas velocity at which this occurs is termed the 'saltation velocity'. Further decrease in gas velocity results in rapid 'salting out' of solids and rapid increase in $\Delta p/\Delta L$ as the area available for flow of gas is restricted by settled solids.

In the region E and F some solids may move in dense phase flow along the bottom of the pipe whilst others travel in dilute phase flow in the gas in the upper part of the pipe. The saltation velocity marks the boundary between dilute phase flow and dense phase flow in horizontal pneumatic transport.

Once again, it is not possible to theoretically predict the conditions under which saltation will occur. However, many correlations for predicting saltation velocity are available in the literature. The correlation by Zenz (1964) is frequently used but is entirely empirical and requires the use of a graph. It is reported by Leung and Jones (1978) to have an average error of ±54 %. The correlation of Rizk (1973), based on a semi-theoretical approach, is considerably simpler to use, and has a similar error range. It is most unambiguously expressed as:

$$\frac{M_p}{\rho_f U_{salt} A} = \left(\frac{1}{10^{(1440x+1.96)}}\right)\left(\frac{U_{salt}}{\sqrt{gD}}\right)^{(1100x+2.5)} \tag{8.3}$$

where $\dfrac{M_p}{\rho_f U_{salt} A}$ is the solids loading $\left(\dfrac{\text{mass flow rate of solids}}{\text{mass flow rate of gas}}\right)$

$\dfrac{U_{salt}}{\sqrt{gD}}$ is the Froude number at saltation

U_{salt} is the superficial gas velocity (see Section 8.1.4 for the definition of superficial velocity) at saltation when the mass flow rate of solids is M_p, the pipe diameter is D and the particle size is x. (The units are SI.)

8.1.4 Fundamentals

In this section we generate some basic relationships governing the flow of gas and particles in a pipe.

Gas and particle velocities

We have to be careful in the definition of gas and particle velocities and in the relative velocity between them, the slip velocity. The terms are often used loosely in the literature and are defined below.

The term 'superficial velocity' is also commonly used. Superficial gas and solids (particles) velocities are defined as:

$$\text{superficial gas velocity, } U_{fs} = \frac{\text{volume flow of gas}}{\text{cross-sectional area of pipe}} = \frac{Q_f}{A} \quad (8.4)$$

$$\text{superficial solids velocity, } U_{ps} = \frac{\text{volume flow of solids}}{\text{cross-sectional area of pipe}} = \frac{Q_p}{A} \quad (8.5)$$

where subscript 's' denotes superficial and subscripts 'f' and 'p' refer to the fluid and particles, respectively.

The fraction of pipe cross-sectional area available for the flow of gas is usually assumed to be equal to the volume fraction occupied by gas, i.e. the voidage or void fraction ε. The fraction of pipe area available for the flow of solids is therefore $(1 - \varepsilon)$.

And so, actual gas velocity,

$$U_f = \frac{Q_f}{A\varepsilon} \quad (8.6)$$

and actual particle velocity,

$$U_p = \frac{Q_p}{A(1 - \varepsilon)} \quad (8.7)$$

Thus superficial velocities are related to actual velocities by the equations:

$$U_f = \frac{U_{fs}}{\varepsilon} \quad (8.8)$$

$$U_p = \frac{U_{ps}}{1 - \varepsilon} \quad (8.9)$$

It is common practice in dealing with fluidization and pneumatic transport to simply use the symbol U to denote superficial fluid velocity. This practice will be followed in this chapter. Also, in line with common practice, the symbol G will be used to denote the mass flux of solids, i.e. $G = M_p/A$, where M_p is the mass flow rate of solids.

The relative velocity between particle and fluid U_{rel} is defined as:

$$U_{rel} = U_f - U_p \quad (8.10)$$

This velocity is often also referred to as the 'slip velocity' U_{slip}.
It is often assumed that in vertical dilute phase flow the slip velocity is equal to the single particle terminal velocity U_T.

Continuity

Consider a length of transport pipe into which are fed particles and gas at mass flow rates of M_p and M_f, respectively. The continuity equations for particles and gas are:

for the particles

$$M_p = AU_p(1 - \varepsilon)\rho_p \tag{8.11}$$

for the gas

$$M_f = AU_f\varepsilon\rho_f \tag{8.12}$$

Combining these continuity equations gives an expression for the ratio of mass flow rates. This ratio is known as the solids loading:

$$\text{Solids loading, } \frac{M_p}{M_f} = \frac{U_p(1 - \varepsilon)\rho_p}{U_f\varepsilon\rho_f} \tag{8.13}$$

This shows us that the average voidage ε, at a particular position along the length of the pipe, is a function of the solids loading and the magnitudes of the gas and solids velocities for given gas and particle density.

Pressure drop

In order to obtain an expression for the total pressure drop along a section of transport line we will write down the momentum equation for a section of pipe. Consider a section of pipe of cross-sectional area A and length δL inclined to the horizontal at an angle θ and carrying a suspension of voidage ε (see Figure 8.3).

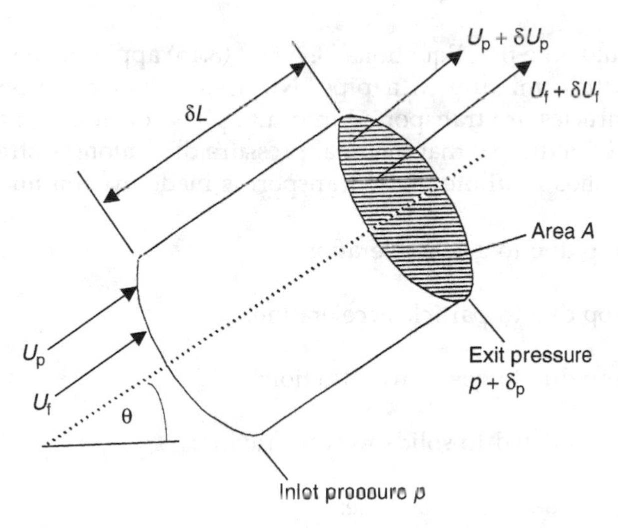

Figure 8.3 Section of conveying pipe: basis for momentum equation

The momentum balance equation is:

$$\begin{pmatrix}\text{net force acting}\\ \text{on pipe contents}\end{pmatrix} = \begin{pmatrix}\text{rate of increase in}\\ \text{momentum of contents}\end{pmatrix}$$

Therefore,

$$\begin{pmatrix}\text{pressure}\\ \text{force}\end{pmatrix} - \begin{pmatrix}\text{gas–wall}\\ \text{friction force}\end{pmatrix} - \begin{pmatrix}\text{solids–wall}\\ \text{friction force}\end{pmatrix} - \begin{pmatrix}\text{gravitational}\\ \text{force}\end{pmatrix}$$
$$= \begin{pmatrix}\text{rate of increase}\\ \text{in momentum}\\ \text{of the gas}\end{pmatrix} + \begin{pmatrix}\text{rate of increase}\\ \text{in momentum}\\ \text{of the solids}\end{pmatrix}$$

or

$$\begin{aligned}-A\delta p - F_{\text{fw}}A\,\delta L - F_{\text{pw}}A\,\delta L - [A(1-\varepsilon)\rho_{\text{p}}\delta L]g\sin\theta - (A\varepsilon\rho_{\text{f}}\delta L)g\sin\theta \\ = \rho_{\text{f}}A\varepsilon U_{\text{f}}\delta U_{\text{f}} + \rho_{\text{p}}A(1-\varepsilon)U_{\text{p}}\delta U_{\text{p}}\end{aligned} \tag{8.14}$$

where F_{fw} and F_{pw} are the gas-to-wall friction force and solids-to-wall friction force per unit volume of pipe, respectively.

Rearranging Equation (8.14) and integrating assuming constant gas density and voidage:

$$\begin{aligned}p_1 - p_2 = \underset{(1)}{\frac{1}{2}\varepsilon\rho_{\text{f}}U_{\text{f}}^2} + \underset{(2)}{\frac{1}{2}(1-\varepsilon)\rho_{\text{p}}U_{\text{p}}^2} + \underset{(3)}{F_{\text{fw}}L} + \underset{(4)}{F_{\text{pw}}L} \\ + \underset{(5)}{\rho_{\text{p}}L(1-\varepsilon)g\sin\theta} + \underset{(6)}{\rho_{\text{f}}L\varepsilon g\sin\theta}\end{aligned} \tag{8.15}$$

Readers should note that Equations (8.4) and (8.15) apply in general to the flow of any gas–particle mixture in a pipe. No assumption has been made as to whether the particles are transported in dilute phase or dense phase.

Equation (8.15) indicates that the total pressure drop along a straight length of pipe carrying solids in dilute phase transport is made up of a number of terms:

(1) pressure drop due to gas acceleration;

(2) pressure drop due to particle acceleration;

(3) pressure drop due to gas-to-wall friction;

(4) pressure drop related to solids-to-wall friction;

(5) pressure drop due to the static head of the solids;

(6) pressure drop due to the static head of the gas.

Some of these terms may be ignored depending on circumstance. If the gas and the solids are already accelerated in the line, then the first two terms should be omitted from the calculation of the pressure drop; if the pipe is horizontal, terms (5) and (6) can be omitted. The main difficulties are in knowing what the solids-to-wall friction is, and whether the gas-to-wall friction can be assumed independent of the presence of the solids; these will be covered in Section 8.1.5.

8.1.5 Design for Dilute Phase Transport

Design of a dilute phase transport system involves selection of a combination of pipe size and gas velocity to ensure dilute flow, calculation of the resulting pipeline pressure drop and selection of appropriate equipment for moving the gas and separating the solids from the gas at the end of the line.

Gas velocity

In both horizontal and vertical dilute phase transport it is desirable to operate at the lowest possible velocity in order to minimize frictional pressure loss, reduce attrition and reduce running costs. For a particular pipe size and solids flow rate, the saltation velocity is always higher than the choking velocity. Therefore, in a transport system comprising both vertical and horizontal lines, the gas velocity must be selected to avoid saltation. In this way choking will also be avoided. These systems would ideally operate at a gas velocity slightly to the right of point D in Figure 8.2. In practice, however, U_{salt} is not known with great confidence and so conservative design leads to operation well to the right of point D with the consequent increase in frictional losses. Another factor encouraging caution in selecting the design velocity is the fact that the region near to point D is unstable; slight perturbations in the system may bring about saltation.

If the system consists only of a lift line, then the choking velocity becomes the important criterion. Here again, since U_{CH} cannot be predicted with confidence, conservative design is necessary. In systems using a centrifugal blower, characterized by reduced capacity at increased pressure, choking can almost be self-induced. For example, if a small perturbation in the system gives rise to an increase in solids feed rate, the pressure gradient in the vertical line will increase (Figure 8.1). This results in a higher back pressure at the blower giving rise to reduced volume flow of gas. Less gas means higher pressure gradient and the system soon reaches the condition of choking. The system fills with solids and can only be restarted by draining off the solids.

Bearing in mind the uncertainty in the correlations for predicting choking and saltation velocities, safety margins of 50 % and greater are recommended when selecting the operating gas velocity.

Pipeline pressure drop

Equation (8.15) applies in general to the flow of any gas–particle mixture in a pipe. In order to make the equation specific to dilute phase transport, we must find expressions for terms 3 (gas-to-wall friction) and 4 (solids-to-wall friction).

In dilute transport the gas-to-wall friction is often assumed independent of the presence of the solids and so the friction factor for the gas may be used (e.g. Fanning friction factor – see worked example on dilute pneumatic transport).

Several approaches to estimating solids-to-wall friction are presented in the literature. Here we will use the modified Konno and Saito (1969) correlation for estimating the pressure loss due to solid-to-pipe friction in vertical transport and the Hinkle (1953) correlation for estimating this pressure loss in horizontal transport. Thus for vertical transport (Konno and Saito, 1969):

$$F_{pw}L = 0.057GL\sqrt{\frac{g}{D}} \tag{8.16}$$

and for horizontal transport:

$$F_{pw}L = \frac{2f_p(1-\varepsilon)\rho_p U_p^2 L}{D} \tag{8.17a}$$

or

$$F_{pw}L = \frac{2f_p G U_p L}{D} \tag{8.17b}$$

where

$$U_p = U(1 - 0.0638x^{0.3}\rho_p^{0.5}) \tag{8.18}$$

and (Hinkle, 1953)

$$f_p = \frac{3}{8}\frac{\rho_f}{\rho_p}\frac{D}{x}C_D\left(\frac{U_f - U_p}{U_p}\right)^2 \tag{8.19}$$

where C_D is the drag coefficient between the particle and gas (see Chapter 2).

Note:

Hinkle's analysis assumes that particles lose momentum by collision with the pipe walls. The pressure loss due to solids–wall friction is the gas pressure loss as a result of re-accelerating the solids. Thus, from Chapter 2, the drag force on a single particle is given by:

$$F_D = \frac{\pi x^2}{4}\rho_f C_D \frac{(U_f - U_p)^2}{2} \tag{8.20}$$

If the void fraction is ε, then the number of particles per unit volume of pipe N_v is

$$N_v = \frac{(1-\varepsilon)}{\pi x^3/6} \tag{8.21}$$

Therefore the force exerted by the gas on the particles in unit volume of pipe F_v is

$$F_v = F_D \frac{(1-\varepsilon)}{\pi x^3/6} \tag{8.22}$$

Based on Hinkle's assumption, this is equal to the solids–wall friction force per unit volume of pipe, F_{pw}. Hence,

$$F_{pw}L = \frac{3}{4}\rho_f C_D \frac{L}{x}(1-\varepsilon)(U_f - U_p)^2 \tag{8.23}$$

Expressing this in terms of a friction factor, f_p we obtain Equations (8.17) and (8.19).

Equation (8.15) relates to pressure losses along lengths of straight pipe. Pressure losses are also associated with bends in pipelines and estimations of the value of these losses will be covered in the next section.

Bends

Bends complicate the design of pneumatic dilute phase transport systems and when designing a transport system it is best to use as few bends as possible. Bends increase the pressure drop in a line, and also are the points of most serious erosion and particle attrition.

Solids normally in suspension in straight, horizontal or vertical pipes tend to salt out at bends due to the centrifugal force encountered while travelling around the bend. Because of this operation, the particles slow down and are then re entrained and re-accelerated after they pass through the bend, resulting in the higher pressure drops associated with bends.

There is a greater tendency for particles to salt out in a horizontal pipe which is preceded by a downflowing vertical to horizontal bend than in any other configuration. If this type of bend is present in a system, it is possible for solids to remain on the bottom of the pipe for very long distances following the bend before they redisperse. Therefore, it is recommended that downflowing vertical to horizontal bends be avoided if at all possible in dilute phase pneumatic transport systems.

In the past, designers of dilute phase pneumatic transport systems intuitively thought that gradually sloped, long radius elbows would reduce the erosion and increase bend service life relative to 90° elbows. Zenz (1964), however, recommended that blinded tees (Figure 8.4) be used in place of elbows in pneumatic transport systems. The theory behind the use of the blinded tee is that a cushion of stagnant particles collects in the blinded or unused branch of the tee, and the conveyed particles then impinge upon the stagnant particles in the tee rather

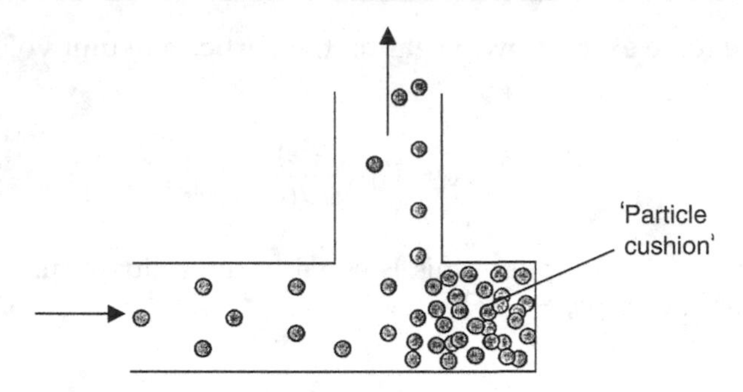

Figure 8.4 Blinded tee bend

than on the metal surface, as in a long radius or short radius elbow. Bodner (1982) determined the service life and pressure drop of various bend configurations. He found that the service life of the blinded tee configuration was far better than any other configuration tested and that it gave a service life 15 times greater than that of radius bends or elbows. This was due to the cushioning accumulation of particles in the blinded branch of the tee which he observed in glass bend models. Bodner also reported that pressure drops and solid attrition rates for the blinded tee were approximately the same as those observed for radius bends.

In spite of a considerable amount of research into bend pressure drop, there is no reliable method of predicting accurate bend pressure drops other than by experiment for the actual conditions expected. In industrial practice, bend pressure drop is often approximated by assuming that it is equivalent to approximately 7.5 m of vertical section pressure drop. In the absence of any reliable correlation to predict bend pressure drop, this crude method is probably as reliable and as conservative as any.

Equipment

Dilute phase transport is carried in systems in which the solids are fed into the air stream. Solids are fed from a hopper at a controlled rate through a rotary air lock into the air stream. The system may be positive pressure, negative pressure or employ a combination of both. Positive pressure systems are usually limited to a maximum pressure of 1 bar gauge and negative pressure systems to a vacuum of about 0.4 bar by the types of blowers and exhausters used.

Typical dilute phase systems are shown in Figures 8.5 and 8.6. Blowers are normally of the positive displacement type which may or may not have speed control in order to vary volume flow rate. Rotary airlocks enable solids to be fed at a controlled rate into the air stream against the air pressure. Screw feeders are frequently used to transfer solids. Cyclone separators (see Chapter 9) are used to recover the solids from the gas stream at the receiving end of the transport line. Filters of various types and with various methods of solids recovery are used to clean up the transport gas before discharge or recycle.

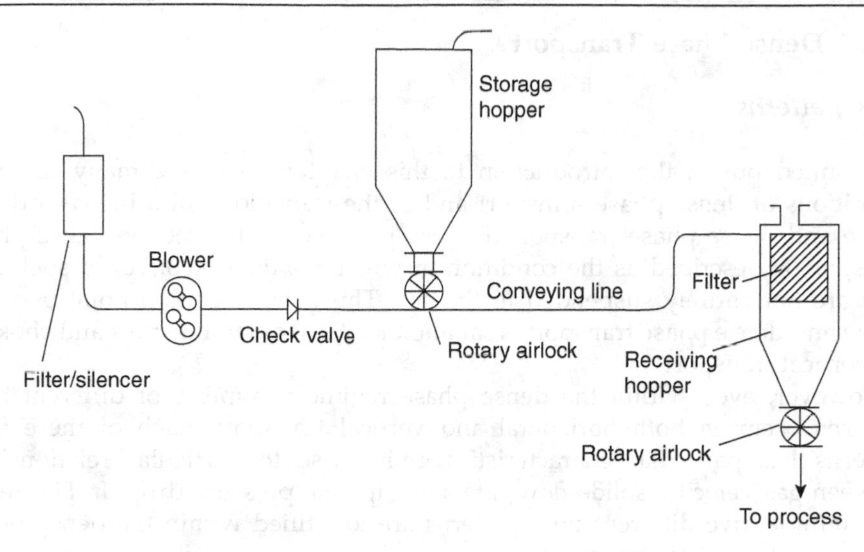

Figure 8.5 Dilute-phase transport: positive pressure system

In some circumstances it may not be desirable to use once-through air as the transport gas (e.g. for the risk of contamination of the factory with toxic or radioactive substances; for risk of explosion an inert gas may be used; in order to control humidity when the solids are moisture sensitive). In these cases a closed loop system is used. If a rotary positive displacement blower is used then the solids must be separated from the gas by cyclone separator and by inline fabric filter. If lower system pressures are acceptable (0.2 bar gauge) then a centrifugal blower may be used in conjunction with only a cyclone separator. The centrifugal fan is able to pass small quantities of solids without damage, whereas the positive displacement blower will not pass dust.

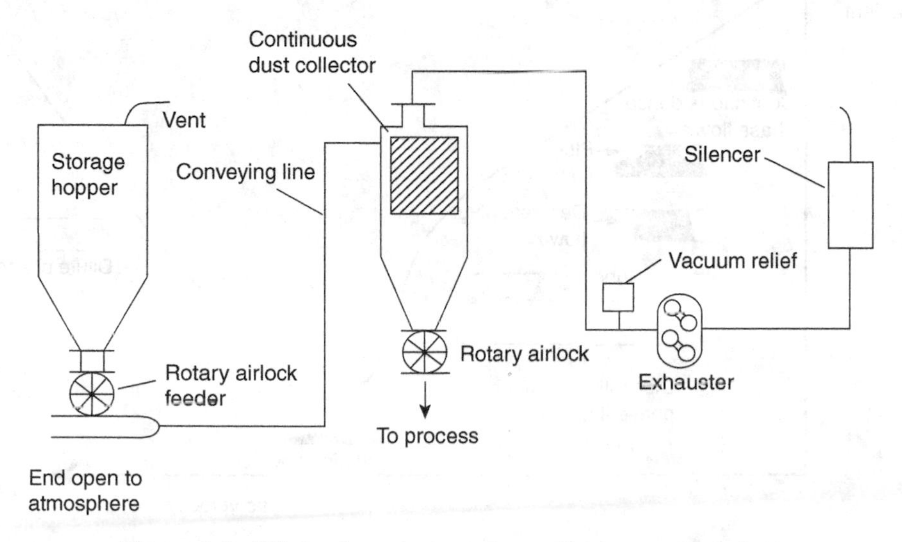

Figure 8.6 Dilute-phase transport: negative pressure system

8.1.6 Dense Phase Transport

Flow patterns

As pointed out in the introduction to this chapter, there are many different definitions of dense phase transport and of the transition point between dilute phase and dense phase transport. For the purpose of this section dense phase transport is described as the condition in which solids are conveyed such that they are not entirely suspended in the gas. Thus, the transition point between dilute and dense phase transport is saltation for horizontal transport and choking for vertical transport.

However, even within the dense phase regime a number of different flow patterns occur in both horizontal and vertical transport. Each of these flow patterns has particular characteristics giving rise to particular relationships between gas velocity, solids flow rate and pipeline pressure drop. In Figure 8.7 for example, five different flow patterns are identified within the dense phase regime for horizontal transport.

The continuous dense phase flow pattern, in which the solids occupy the entire pipe, is virtually extrusion. Transport in this form requires very high gas pressures and is limited to short straight pipe lengths and granular materials (which have a high permeability).

Discontinuous dense phase flow can be divided into three fairly distinct flow patterns: 'discrete plug flow' in which discrete plugs of solids occupy the full pipe cross-section; 'dune flow' in which a layer of solids settled at the bottom of

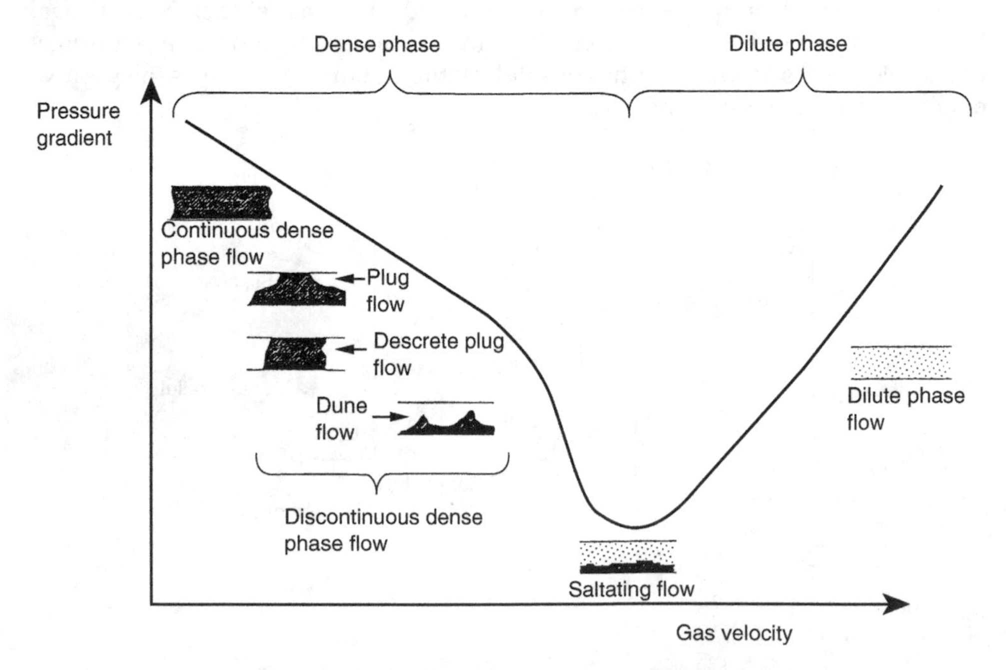

Figure 8.7 Flow patterns in horizontal pneumatic conveying

the pipe move along in the form of rolling dunes; a hybrid of discrete plug flow and dune flow in which the rolling dunes completely fill the pipe cross-section but in which there are no discrete plugs (also known as 'plug flow').

Saltating flow is encountered at gas velocities just below the saltation velocity. Particles are conveyed in suspension above a layer of settled solids. Particles may be deposited and re-entrained from this layer. As the gas velocity is decreased the thickness of the layer of settled solids increases and eventually we have dune flow.

It should be noted, first, that not all powders exhibit all these flow patterns and, secondly, that within any transport line it is possible to encounter more than one regime.

The main advantages of dense phase transport arise from the low gas requirements and low solids velocities. Low gas volume requirements generally mean low energy requirements per kilogram of product conveyed, and also mean that smaller pipelines and recovery and solids–gas separation are required. Indeed in some cases, since the solids are not suspended in the transport gas, it may be possible to operate without a filter at the receiving end of the pipeline. Low solids velocities means that abrasive and friable materials may be conveyed without major pipeline erosion or product degradation.

It is interesting to look at the characteristics of the different dense phase flow patterns with a view to selecting the optimum for a dense phase transport system. The continuous dense phase flow pattern is the most attractive from the point of view of low gas requirements and solid velocities, but has the serious drawback that it is limited to use in the transport of granular materials along short straight pipes and requires very high pressures. Saltating flow occurs at a velocity too close to the saltation velocity and is therefore unstable. In addition this flow pattern offers little advantage in the area of gas and solids velocity. We are then left with the so-called discontinuous dense phase flow pattern with its plugs and dunes. However, performance in this area is unpredictable, can give rise to complete pipeline blockages and requires high pressures. Most commercial dense phase transport systems operate in this flow pattern and incorporate some means of controlling plug length in order to increase predictability and reduce the chance of blockages.

It is therefore necessary to consider how the pressure drop across a plug of solids depends on its length. Unfortunately contradictory experimental evidence is reported in the literature. Konrad (1986) points out that the pressure drop across a moving plug has been reported to increase (a) linearly with plug length, (b) as the square of the plug length and (c) exponentially with plug length. A possible explanation of these apparent contradictions is reported by Klintworth and Marcus (1985) who cite the work of Wilson (1981) on the effect of stress on the deformation within the plug. Large cohesionless particles [typically Geldart Group D particles (Geldart's classification of powders – see Chapter 7 on Fluidization)] give rise to a permeable plug permitting the passage of a significant gas flow at low pressure drops. In this case the stress developed in the plug would be low and a linear dependence of pressure drop on plug length would result. Plugs of fine cohesive particles (typically Geldart Group C) would be virtually impermeable to gas flow at the pressures usually encountered. In this case, the plug moves as a piston in a cylinder by purely mechanical means. The stress developed within the plug is high. The high stress translates to a high wall shear stress which gives rise to an exponential increase in pressure drop with plug length. Thus it is the degree of permeability of the plug which determines the relationship between plug length and pressure drop: the pressure drop across a plug can vary between a linear and an exponential function of the plug length depending on the permeability of the plug.

Large cohesionless particles form permeable plugs and are therefore suitable for discontinuous dense phase transport. In other materials, where interaction under the action of stress and interparticle forces give rise to low permeability plugs, discontinuous dense phase transport is only possible if some mechanism is used to limit plug length, avoiding blockages.

Equipment

In commercial systems, the problem of plug formation is tackled in three ways:

1. Detect the plug at its formation and take appropriate action to either:

(a) use a bypass system in which the pressure build-up behind a plug causes more air to flow around the bypass line and break up the plug from its front end (Figure 8.8);

(b) detect the pressure build-up using pressure actuated valves which divert auxiliary air to break up the plugs into smaller lengths (Figure 8.9).

2. Form stable plugs – stable plugs of granular material do form naturally under certain conditions. However, to form stable plugs of manageable length of other materials, it is generally necessary to induce them artificially by one of the following means:

(a) use an air knife to chop up solids fed in continuous dense phase flow from a blow tank (Figure 8.10);

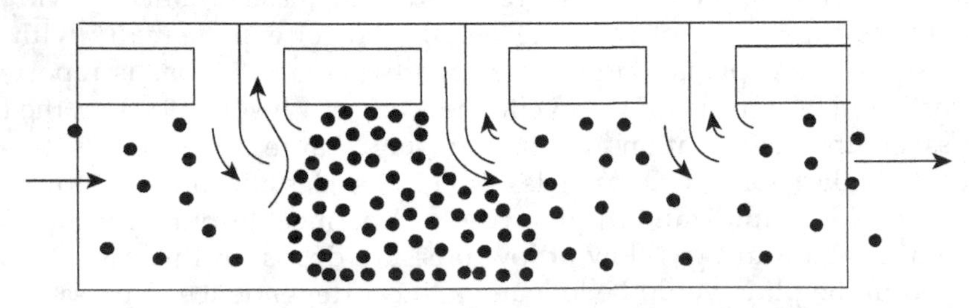

Figure 8.8 Dense phase conveying system using a bypass line to break up plugs of solids

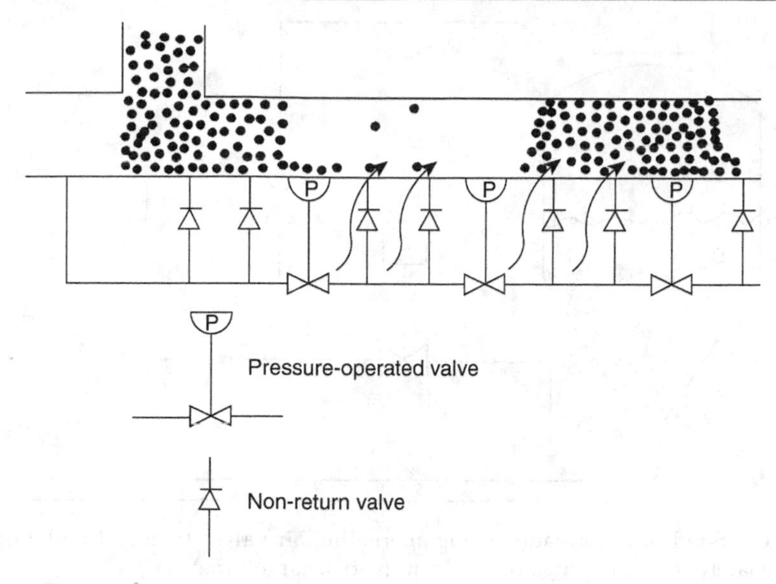

Figure 8.9 Dense phase conveying system using pressure-actuated valves to direct gas

(b) use an alternating valves system (Figure 8.11) in order to cut up the continuous dense phase flow from the blow tank;

(c) for free-flowing materials it is possible to use an air-operated diaphragm in the blow tank to create plugs (Figure 8.12);

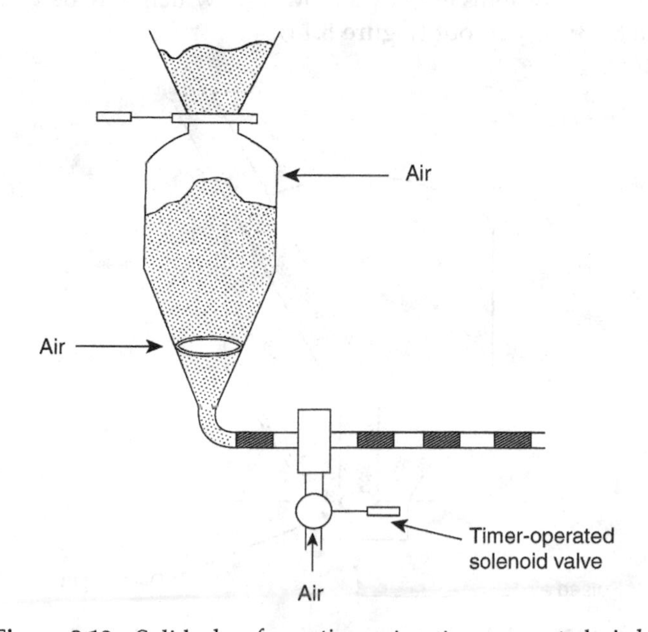

Figure 8.10 Solid plug formation using timer-operated air knife

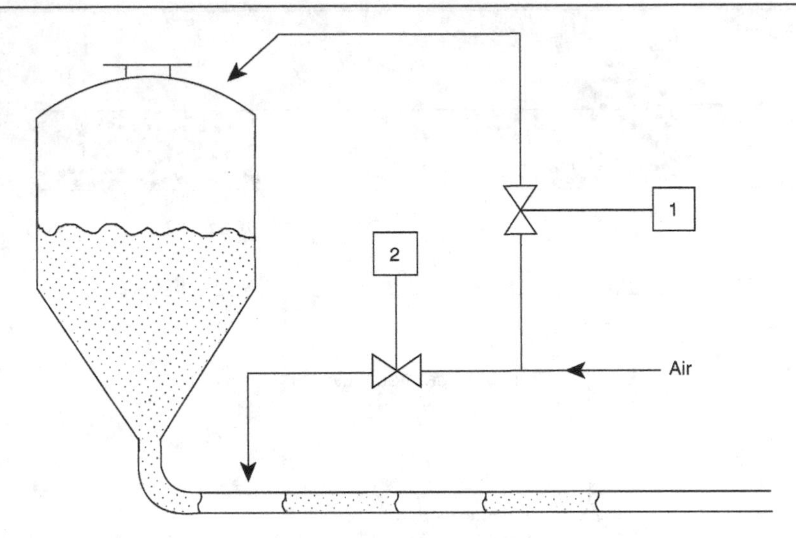

Figure 8.11 Solid plug formation using alternating air valves (valves 1 and 2 open and close alternately to create plugs of solids in the discharge pipe)

(d) a novel idea reported by Tsuji (1983) uses table tennis balls to separate solids into plugs.

3. Fluidization–add extra air along the transport line in order to maintain the aeration of the solids and hence avoid the formation of blockages.

Whatever the mechanism used to tackle the plug problem, all commercial dense phase transport systems employ a blow tank which may be with fluidizing element (Figure 8.13) or without (Figure 8.14).

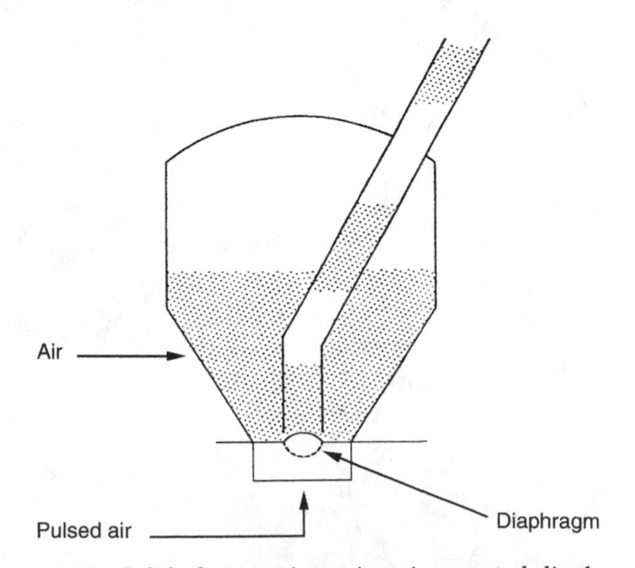

Figure 8.12 Solid plug creation using air-operated diaphragm

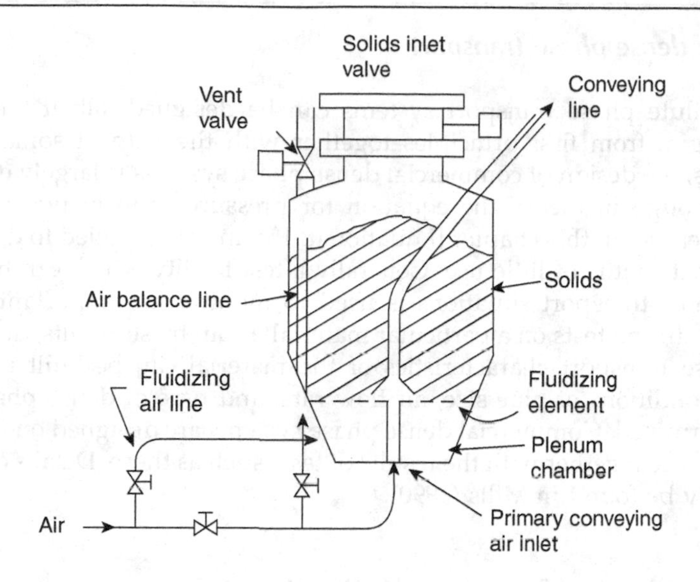

Figure 8.13 Dense phase transport blow tank with fluidizing element

The blow tank is automatically taken through repeated cycles of filling, pressurizing and discharging. Since one third of the cycle time is used for filling the blow tank, a system required to give a mean delivery rate of 20 t/h must be able to deliver a peak rate of over 30 t/h. Dense phase transport is thus a batch operation because of the high pressures involved, whereas dilute phase transport can be continuous because of the relatively low pressures and the use of rotary valves. The dense phase system can be made to operate in semi-continuous mode by the use of two blow tanks in parallel.

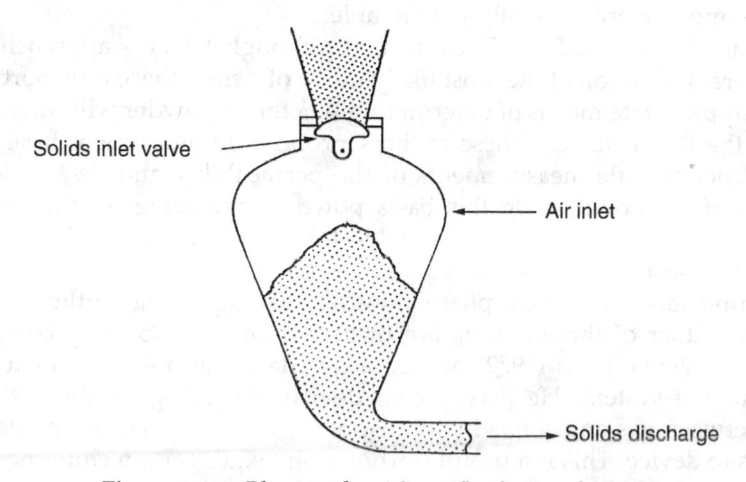

Figure 8.14 Blow tank without fluidizing element

Design for dense phase transport

Whereas dilute phase transport systems can be designed, albeit with a large safety margin, from first principles together with the help of some empirical correlations, the design of commercial dense phase systems is largely empirically based. Although in theory the equation for pressure drop in two phase flow developed earlier in this chapter [Equation (8.15)] may be applied to dense phase flow, in practice it is of little use. Generally a test facility which can be made to simulate most transport situations is used to monitor the important transport parameters during tests on a particular material. From these results, details of the dense phase transport characteristics of the material can be built up and the optimum conditions of pipe size, air flow rate, and type of dense phase system can be determined. Commercial dense phase systems are designed on the basis of past experience together with the results of tests such as these. Details of how this is done may be found in Mills (1990).

8.1.7 Matching the System to the Powder

Generally speaking it is possible to convey any powder in the dilute phase mode, but because of the attractions of dense phase transport, there is great interest in assessing the suitability of a powder for transport in this mode. The most commonly used procedure is to undertake a series of tests on a sample of the powder in a pilot plant. This is obviously expensive. An alternative approach offered by Dixon (1979) is available. Dixon recognized the similarities between gas fluidization and dense phase transport and proposed a method of assessing the suitability of a powder for transport in the dense phase based on Geldart's (1973) classification of powders (see Chapter 7 on Fluidization). Dixon proposed a 'slugging diagram' which allows prediction of the possible dense phase flow patterns from a knowledge of particle size and density. Dixon concluded that Geldart's Groups A and D were suitable for dense phase transport whereas Groups B and C were generally not suitable.

Mainwaring and Reed (1987) claim that although Dixon's approach gives a good general indication of the most likely mode of dense phase transport, it is not the most appropriate means of determining whether a powder will convey in this mode in the first instance. These authors propose an assessment based on the results of bench-scale measurements of the permeability and de-aeration characteristics of the powder. On this basis powders achieving a sufficiently high permeability in the test would be suitable for plug type dense phase transport and powders scoring high on air retention would be suitable for transport in the rolling dune mode of dense phase flow. According to the authors, powders satisfying neither of these criteria are unsuitable for transport by conventional blow tank systems. Flain (1972) offered a qualitative approach to matching the powder to the system. He lists twelve devices for bringing about the initial contact between gas and solids in a transport system and matches powder characteristics to device. This is a useful starting point since certain equipment can be excluded for use with a particular powder.

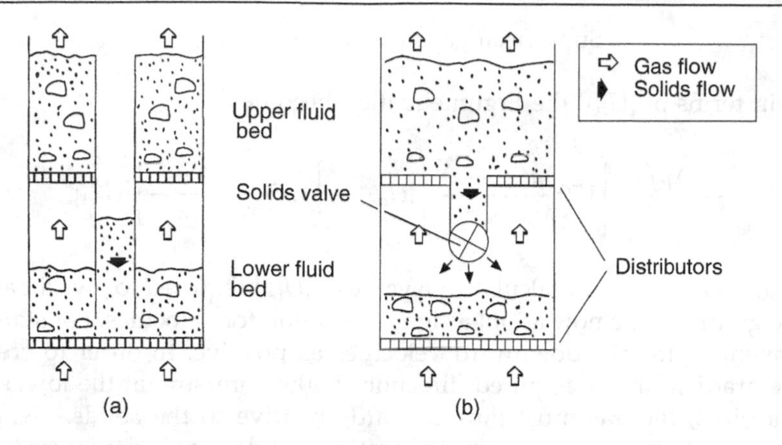

Figure 8.15 (a) Overflow and (b) underflow type standpipes transporting solids from low pressure fluidized bed to bed at higher pressure

8.2 *STANDPIPES*

Standpipes have been used for many years, particularly in the oil industry, for transferring solids downwards from a region of low pressure to a region of higher pressure. The overview of standpipe operation given here is based largely on the work of Knowlton (1997).

Typical overflow and underflow standpipes are shown in Figure 8.15, where they are used to continuously transfer solids from an upper fluidized bed to a lower fluidized bed. For solids to be transferred downwards against the pressure gradient gas must flow upward relative to the solids. The friction losses developed by the flow of the gas through the packed or fluidized bed of solids in the standpipe generates the required pressure gradient. If the gas must flow upwards relative to the downflowing solids there are two possible cases: (i) the gas flows upward relative to the standpipe wall; and (ii) the gas flows downwards relative to the standpipe wall; but at a lower velocity than the solids.

A standpipe may operate in two basic flow regimes depending on the relative velocity of the gas to the solids; packed bed flow and fluidized bed flow.

8.2.1 Standpipes in Packed Bed Flow

If the relative upward velocity of the gas $(U_f - U_p)$ is less than the relative velocity at incipient fluidization $(U_f - U_p)_{mf}$, then packed bed flow results and the relationship between gas velocity and pressure gradient is in general determined by the Ergun equation [see Chapter 6, Equation (6.11)].

The Ergun equation is usually expressed in terms of the superficial gas velocity through the packed bed. However, for the purposes of standpipe calculations it is useful to write the Ergun equation in terms of the magnitude of the velocity of the gas relative to the velocity of the solids $|U_{rel}|(= |U_f - U_p|)$. (Refer to Section 8.1.4 for clarification of relationships between superficial and actual velocities.)

$$\text{Superficial gas velocity, } U = \varepsilon|U_{rel}| \tag{8.24}$$

And so in terms of $|U_{rel}|$ the Ergun equation becomes:

$$\frac{(-\Delta p)}{H} = \left[150\frac{\mu}{x_{sv}^2}\frac{(1-\varepsilon)^2}{\varepsilon^2}\right]|U_{rel}| + \left[1.75\frac{\rho_f}{x_{sv}}\frac{(1-\varepsilon)}{\varepsilon}\right]|U_{rel}|^2 \tag{8.25}$$

The equation allows us to calculate the value of $|U_{rel}|$ required to give a particular pressure gradient. We now adopt a sign convention for velocities. For standpipes it is convenient to take downward velocities as positive. In order to create the pressure gradient in the required direction (higher pressure at the lower end of the standpipe), the gas must flow upwards relative to the solids. Hence, U_{rel} should always be negative in normal operation. Solids flow is downwards, so U_p, the actual velocity of the solids (relative to the pipe wall), is always positive.

Knowing the magnitude and direction of U_p and U_{rel}, the magnitude and direction of the actual gas velocity (relative to the pipe wall) may be found from $U_{rel} = U_f - U_p$. In this way the quantity of gas passing up or down the standpipe may be estimated.

8.2.2 Standpipes in Fluidized Bed Flow

If the relative upward velocity of the gas $(U_f - U_p)$ is greater than the relative velocity at incipient fluidization $(U_f - U_p)_{mf}$, then fluidized bed flow will result. In fluidized bed flow the pressure gradient is independent of relative gas velocity. Assuming that in fluidized bed flow the entire apparent weight of the particles is supported by the gas flow, then the pressure gradient is given by (see Chapter 7):

$$\frac{(-\Delta p)}{H} = (1-\varepsilon)(\rho_p - \rho_f)g \tag{8.26}$$

where $(-\Delta p)$ is the pressure drop across a height H of solids in the standpipe, ε is the voidage and ρ_p is the particle density.

Fluidized bed flow may be non-bubbling flow or bubbling flow. Non-bubbling flow occurs only with Geldart Group A solids (described in Chapter 7) when the relative gas velocity lies between the relative velocity for incipient fluidization and the relative velocity for minimum bubbling $(U_f - U_p)_{mb}$. For Geldart Group B materials (Chapter 7) with $(U_f - U_p) > (U_f - U_p)_{mf}$ and for Group A solids with $(U_f - U_p) > (U_f - U_p)_{mb}$ bubbling fluidized flow results.

Four types of bubbling fluidized bed flow in standpipes are possible depending on the direction of motion of the gas in the bubble phase and emulsion phases relative to the standpipe walls. These are depicted in Figure 8.16. In practice, bubbles are undesirable in a standpipe. The presence of rising bubbles hinders the flow of solids and reduces the pressure gradient developed in the standpipe. If the bubble rise velocity is greater than the solids velocity, then the bubbles will

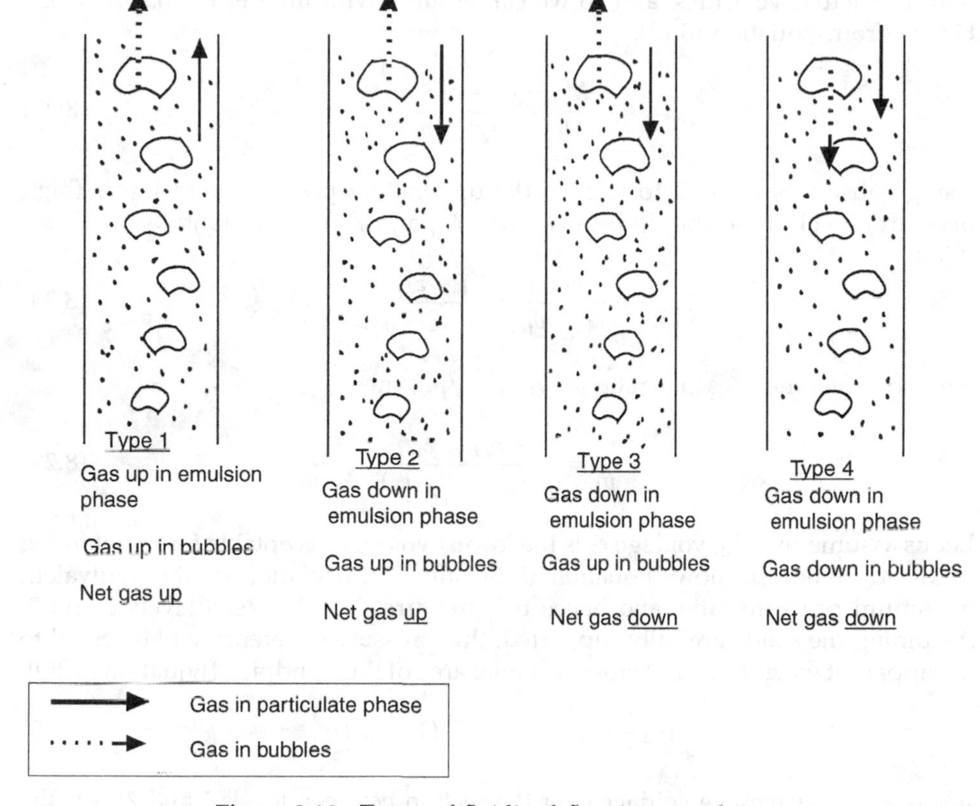

Figure 8.16 Types of fluidized flow in a standpipe

rise and grow by coalescence. Larger standpipes are easier to operate since they can tolerate larger bubbles than smaller standpipes. For optimum standpipe operation, when using Group B solids, the relative gas velocity should be slightly greater than relative velocity for incipient fluidization. For Group A solids, relative gas velocity should lie between $(U_f - U_p)_{mf}$ and $(U_f - U_p)_{mb}$.

In practice, aeration is often added along the length of a standpipe in order to maintain the solids in a fluidized state just above minimum fluidization velocity. If this were not done then, with a constant mass flow of gas, relative velocities would decrease towards the high-pressure end of the standpipe. The lower velocities would result in lower mean voidages and the possibility of an unfluidized region at the bottom of the standpipe. Aeration gas is added in stages along the length of the standpipe and only the minimum requirement is added at any level. If too much is added, bubbles are created which may hinder solids flow. The analysis below, based on that of Kunii and Levenspiel (1990), enables calculation of the position and quantity of aeration gas to be added.

The starting point is Equation (8.13), the equation derived from the continuity equations for gas and solids flow in a pipe. For fine Group A solids in question the relative velocity between gas and particles will be very small in comparison with the actual velocities, and so we can assume with little error that $U_p = U_f$. Hence, from Equation (8.13):

$$\frac{M_p}{M_f} = \frac{(1-\varepsilon)}{\varepsilon}\frac{\rho_p}{\rho_f} \qquad (8.27)$$

Using subscripts 1 and 2 to refer to the upper (low pressure) and lower (high pressure) level in the standpipe, since M_p, M_f and ρ_p are constant:

$$\frac{(1-\varepsilon_1)}{\varepsilon_1}\frac{1}{\rho_{f_1}} = \frac{(1-\varepsilon_2)}{\varepsilon_2}\frac{1}{\rho_{f_2}} \qquad (8.28)$$

And so, since the pressure ratio $p_2/p_1 = \rho_{f_2}/\rho_{f_1}$, then

$$\frac{p_2}{p_1} = \frac{(1-\varepsilon_2)}{\varepsilon_2}\frac{\varepsilon_1}{(1-\varepsilon_1)} \qquad (8.29)$$

Let us assume that the voidage ε_2 is the lowest voidage acceptable for maintaining fluidized standpipe flow. Equation (8.29) allows calculation of the equivalent maximum pressure ratio, and hence the pressure drop between levels 1 and 2. Assuming the solids are fully supported, this pressure difference will be equal to the apparent weight per unit cross-sectional area of the standpipe [Equation (8.26)].

$$(p_2 - p_1) = (\rho_p - \rho_f)(1-\varepsilon_a)Hg \qquad (8.30)$$

where ε_a is the average voidage over the section between levels 1 and 2, H is the distance between the levels and g is the acceleration due to gravity.

If ε_1 and ε_2 are known and gas density is regarded as negligible compared to particle density, H may be calculated from Equation (8.30).

The objective of adding aeration gas is to raise the voidage at the lower level to equal that at the upper level. Applying Equation (8.27),

$$\frac{(1-\varepsilon_2)}{\varepsilon_2} = \frac{M_p}{(M_f + M_{f_2})}\frac{\rho_{f_2}}{\rho_p} = \frac{M_p}{M_f}\frac{\rho_{f_1}}{\rho_p} \qquad (8.31)$$

where M_{f_2} is the mass flow of aeration air added at level 2.

Then rearranging,

$$M_{f_2} = M_f\left(\frac{\rho_{f_2}}{\rho_{f_1}} - 1\right) \qquad (8.32)$$

and, since from Equation (8.27), $M_f = M_p \dfrac{\varepsilon_1}{(1-\varepsilon_1)}\dfrac{\rho_{f_1}}{\rho_p}$

$$M_{f_2} = M_p \frac{\varepsilon_1}{(1-\varepsilon_1)}\frac{\rho_{f_1}}{\rho_p}\left(\frac{\rho_{f_2}}{\rho_{f_1}} - 1\right) \qquad (8.33)$$

and so mass flow of aeration air to be added,

$$M_{f_2} = \frac{\varepsilon_1}{(1-\varepsilon_1)}\frac{M_p}{\rho_p}(\rho_{f_2} - \rho_{f_1}) \tag{8.34}$$

from which, it can also be shown that

$$Q_{f_2} = Q_p \frac{\varepsilon_1}{(1-\varepsilon_1)}\left(1 - \frac{\rho_{f_1}}{\rho_{f_2}}\right) \tag{8.35}$$

where Q_{f_2} is the volume flow rate of gas to be added at pressure p_2 and Q_p is the volume flow rate of solids down the standpipe.

For long standpipes aeration gas will need to be added at several levels in order to keep the voidage within the required range (see the worked example on standpipe aeration).

8.2.3 Pressure Balance During Standpipe Operation

As an example of the operation of a standpipe, we will consider how an overflow standpipe operating in fluidized bed flow reacts to a change in gas flow rate. Figure 8.17(a) shows the pressure profile over such a system. The pressure balance equation over this system is:

$$\Delta p_{SP} = \Delta p_{LB} + \Delta p_{UB} + \Delta p_d \tag{8.36}$$

where Δp_{SP}, Δp_{LB}, Δp_{UB} and Δp_d are the pressure drops across the standpipe, the lower fluidized bed, the upper fluidized bed and the distributor of the upper fluidized bed, respectively.

Let us consider a disturbance in the system such that the gas flow through the fluidized beds increases [Figure 8.17(b)]. If the gas flow through the lower bed increases, although the pressure drops across the lower and upper beds will remain constant, the pressure drop across the upper distributor will increase $\Delta p_{d(new)}$. To match this increase, the pressure across the standpipe must rise to $\Delta p_{SP(new)}$ [Figure 8.17(b)]. In the case of an overflow standpipe operating in fluidized flow the increase in standpipe pressure drop results from a rise in the height of solids in the standpipe to $H_{SP(new)}$.

Consider now the case of an underflow standpipe operating in packed bed flow (Figure 8.18), the pressure balance across the system is given by:

$$\Delta p_{SP} = \Delta p_d + \Delta p_V \tag{8.37}$$

where Δp_{SP}, Δp_d and Δp_v are the pressure drops across the standpipe, the distributor of the upper fluidized bed and the standpipe valve, respectively.

If the gas flow from the lower bed increases, the pressure drop across the upper bed distributor increases to $\Delta p_{d(new)}$. The pressure balance then calls for an increase in standpipe pressure drop. Since in this case the standpipe length is

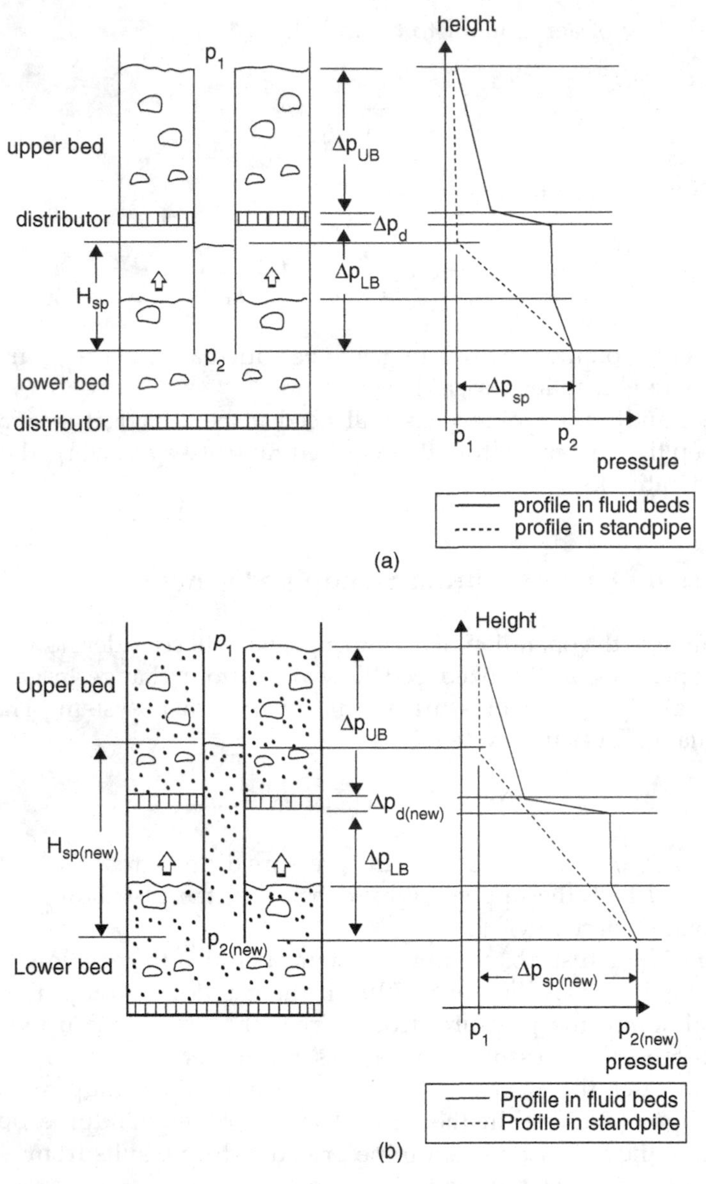

Figure 8.17 Operation of an overflow standpipe: (a) before increasing gas flow; (b) change in pressure profile due to increased gas flow through the fluid beds

fixed, in packed bed flow this increase in pressure drop is achieved by an increase in the magnitude of the relative velocity $|U_{\mathrm{rel}}|$. The standpipe pressure drop will increase to $\Delta p_{\mathrm{SP(new)}}$ and the valve pressure drop, which depends on the solids flow, will remain essentially constant. Once the standpipe pressure gradient reaches that required for fluidized bed flow, its pressure drop will remain constant so it will not be able to adjust to system changes.

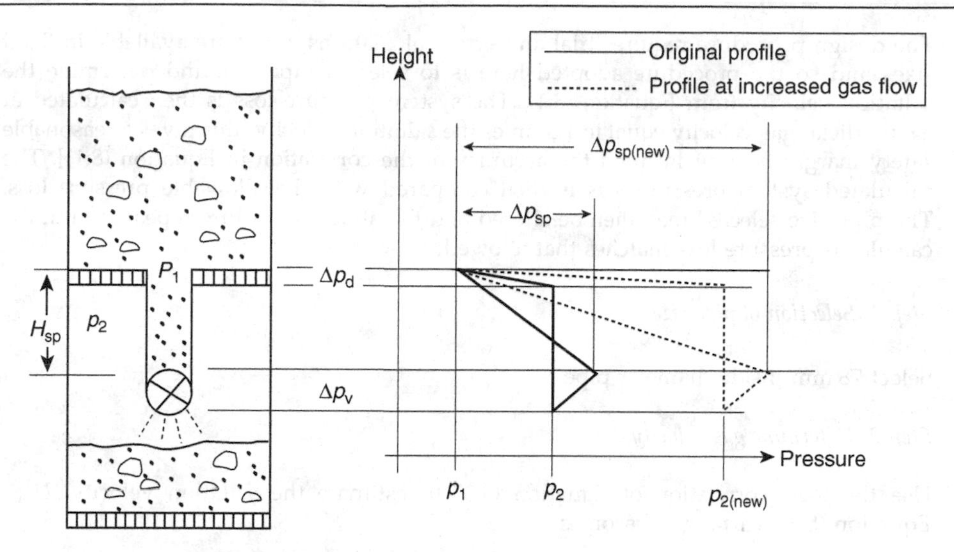

Figure 8.18 Pressure balance during operation of an underflow standpipe: effect of increasing gas flow through fluid beds

A standpipe commonly used in the petroleum industry is the underflow vertical standpipe with slide valve at the lower end. In this case the standpipe generates more head than is required and the excess is used across the slide valve in controlling the solids flow. Such a standpipe is used in the fluid catalytic cracking (FCC) unit to transfer solids from the reactor to the regenerator.

8.3 FURTHER READING

Readers wishing to learn more about solids circulation systems, standpipe flow and non-mechanical valves are referred to Kunii and Levenspiel (1990) or the chapters by Knowlton in either Geldart (1986) or Grace *et al.* (1997).

8.4 WORKED EXAMPLES

WORKED EXAMPLE 8.1

Design a positive pressure dilute-phase pneumatic transport system to transport 900 kg/h of sand of particle density 2500 kg/m^3 and mean particle size 100 μm between two points in a plant separated by 10 m vertical distance and 30 m horizontal distance using ambient air. Assume that six 90° bends are required and that the allowable pressure loss is 0.55 bar.

Solution

In this case, to design the system means to determine the pipe size and air flow rate which would give a total system pressure loss near to the allowable pressure loss.

The design procedure requires trial and error calculations. Pipes are available in fixed sizes and so the procedure adopted here is to select a pipe size and determine the saltation velocity from Equation (8.1). The system pressure loss is then calculated at a superficial gas velocity equal to 1.5 times the saltation velocity [this gives a reasonable safety margin bearing in mind the accuracy of the correlation in Equation (8.1)]. The calculated system pressure loss is then compared with the allowable pressure loss. The pipe size selected may then be altered and the above procedure repeated until the calculated pressure loss matches that allowed.

Step 1. Selection of pipe size

Select 78 mm inside diameter pipe.

Step 2. Determine gas velocity

Use the Rizk correlation of Equation (8.3) to estimate the saltation velocity, U_{salt}. Equation (8.3) rearranged becomes

$$U_{salt} = \left(\frac{4M_p 10^{\alpha} g^{\beta/2} D^{(\beta/2)-2}}{\pi \rho_f} \right)^{1/(\beta+1)}$$

where $\alpha = 1440x + 1.96$ and $\beta = 1100x + 2.5$.

In the present case $\alpha = 2.104$, $\beta = 2.61$ and $U_{salt} = 9.88$.

Therefore, superficial gas velocity, $U = 1.5 \times 9.88\,\text{m/s} = 14.82\,\text{m/s}$.

Step 3. Pressure loss calculations

(a) *Horizontal sections.* Starting with Equation (8.15) an expression for the total pressure loss in the horizontal sections of the transport line may be generated. We will assume that all the initial acceleration of the solids and the gas take place in the horizontal sections and so terms (1) and (2) are required. For term (3) the Fanning friction equation is used assuming that the pressure loss due to gas-to-wall friction is independent of the presence of solids. For term (4) we employ the Hinkle correlation [Equation (8.17)]. Terms (5) and (6) become zero as $\theta = 0$ for horizontal pipes. Thus, the pressure loss, Δp_H, in the horizontal sections of the transport line is given by:

$$\Delta p_H = \frac{\rho_f \varepsilon_H U_{fH}^2}{2} + \frac{\rho_p (1 - \varepsilon_H) U_{pH}^2}{2} + \frac{2 f_g \rho_f U^2 L_H}{D} + \frac{2 f_p \rho_p (1 - \varepsilon_H) U_{pH}^2 L_H}{D}$$

where the subscript H refers to the values specific to the horizontal sections.

To use this equation we need to know ε_H, U_{f_H} and U_{pH}. Hinkle's correlation gives us U_{p_H}:

$$U_{pH} = U(1 - 0.0638 x^{0.3} \rho_p^{0.5}) = 11.84\,\text{m/s}$$

From continuity, $G = \rho_p(1 - \varepsilon_H)U_{pH}$

$$\text{thus } \varepsilon_H = 1 - \frac{G}{\rho_p U_{pH}} = 0.9982$$

$$\text{and } U_{fH} = \frac{U}{\varepsilon_H} = \frac{14.82}{0.9982} = 14.85\,\text{m/s}$$

Friction factor f_p is found from Equation (8.19) with C_D estimated at the relative velocity $(U_f - U_p)$, using the approximate correlations given below [or by using an appropriate C_D versus Re chart (see Chapter 2)]:

$$Re_p < 1 \qquad C_D = 24/Re_p$$
$$1 < Re_p < 500 \qquad C_D = 18.5Re_p^{-0.6}$$
$$500 < Re_p < 2 \times 10^5 \qquad C_D = 0.44$$

Thus, for flow in the horizontal sections,

$$Re_p = \frac{\rho_f(U_{fH} - U_{pH})x}{\mu}$$

for ambient air $\rho_f = 1.2\,\text{kg/m}^3$ and $\mu = 18.4 \times 10^{-6}$ Pa s, giving

$$Re_p = 19.63$$

and so, using the approximate correlations above,

$$C_D = 18.5Re_p^{-0.6} = 3.1$$

Substituting $C_D = 3.1$ in Equation (8.19) we have:

$$f_p = \frac{3}{8} \times \frac{1.2}{2500} \times 3.1 \times \frac{0.078}{100 \times 10^{-6}} \left(\frac{14.85 - 11.84}{11.84}\right)^2$$

The gas friction factor is taken as $f_g = 0.005$. This gives $\Delta p_H = 14.864$ Pa.

(b) *Vertical sections.* Starting again with Equation (8.15), the general pressure loss equation, an expression for the total pressure loss in the vertical section may be derived. Since the initial acceleration of solids and gas was assumed to take place in the horizontal sections, terms (1) and (2) become zero. The Fanning friction equation is used to estimate the pressure loss due to gas-to-wall friction [term (3)] assuming solids have negligible effect on this pressure loss. For term (4) the modified Konno and Saito correlation [Equation (8.16)] is used. For vertical transport θ becomes equal to 90° in terms (5) and (6).

Thus, the pressure loss, Δp_v, in the vertical sections of the transport line is given by:

$$\Delta p_v = \frac{2f_g\rho_f U^2 L_v}{D} + 0.057GL_v\sqrt{\frac{g}{D}} + \rho_p(1 - \varepsilon_v)gL_v + \rho_f\varepsilon_v gL_v$$

where subscript v refers to values specific to the vertical sections.

To use this equation we need to calculate the voidage of the suspension in the vertical pipe line ε_v.

Assuming particles behave as individuals, then slip velocity is equal to single particle terminal velocity, U_T (also noting that the superficial gas velocity in both horizontal and vertical sections is the same and equal to U), i.e.

$$U_{pv} = \frac{U}{\varepsilon_v} - U_T$$

continuity gives particle mass flux, $G = \rho_p(1 - \varepsilon_v)U_{pv}$.

Combining these equations gives a quadratic in ε_v which has only one possible root.

$$\varepsilon_v^2 U_T - \left(U_T + U + \frac{G}{\rho_p}\right)\varepsilon_v + U = 0$$

The single particle terminal velocity, U_T may be estimated as shown in Chapter 2, giving $U_T = 0.52\,\text{m/s}$ assuming the particles are spherical.

And so, solving the quadratic equation, $\varepsilon_v = 0.9985$ and thus $\Delta p_v = 1148\,\text{Pa}$.

(c) *Bends*. The pressure loss across each 90° bend is taken to be equivalent to that across 7.5 m of vertical pipe.

$$\text{Pressure loss per metre of vertical pipe} = \frac{\Delta p_v}{L_v} = 114.8\,\text{Pa/m}$$

Therefore, pressure loss across six 90° bends

$$\begin{aligned} &= 6 \times 7.5 \times 114.8\,\text{Pa} \\ &= 5166\,\text{Pa} \end{aligned}$$

And so,

$$\begin{pmatrix}\text{total pressure}\\ \text{loss}\end{pmatrix} = \begin{pmatrix}\text{loss across}\\ \text{vertical sections}\end{pmatrix} + \begin{pmatrix}\text{loss across}\\ \text{horizontal}\\ \text{sections}\end{pmatrix} + \begin{pmatrix}\text{loss across}\\ \text{bends}\end{pmatrix}$$

$$\begin{aligned} &= 11.48 + 14\,864 + 5166\,\text{Pa} \\ &= 0.212\,\text{bar} \end{aligned}$$

Step 4. Compare calculated and allowable pressure losses

The allowable system pressure loss is 0.55 bar and so we may select a smaller pipe size and repeat the above calculation procedure. The table below gives the results for a range of pipe sizes.

Pipe inside diameter (mm)	Total system pressure loss (bar)
78	0.212
63	0.322
50	0.512
40	0.809

In this case we would select 50 mm pipe which gives a total system pressure loss of 0.512 bar. (An economic option could be found if capital and running cost were incorporated.) The design details for this selection are given below:

$$\begin{aligned} \text{pipe size} &= 50\,\text{mm inside diameter} \\ \text{air flow rate} &= 0.0317\,\text{m}^3/\text{s} \\ \text{air superficial velocity} &= 16.15\,\text{m/s} \\ \text{saltation velocity} &= 10.77\,\text{m/s} \\ \text{solids loading} &= 6.57\,\text{kg solid/kg air} \\ \text{total system pressure loss} &= 0.512\,\text{bar} \end{aligned}$$

WORKED EXAMPLE 8.2

A 20 m long standpipe carrying a Group A solids at a rate of 80 kg/s is to be aerated in order to maintain fluidized flow with a voidage in the range 0.5–0.53. Solids enter the top of the standpipe at a voidage of 0.53. The pressure and gas density at the top of the standpipe are 1.3 bar (abs) and 1.0 kg/m^3, respectively.

The particle density of the solids is 1200 kg/m^3.

Determine the aeration positions and rates.

Solution

From Equation (8.29), pressure ratio,

$$\frac{p_2}{p_1} = \frac{(1-0.50)}{0.50}\frac{0.53}{(1-0.53)} = 1.128$$

Therefore, $p_2 = 1.466\,\text{bar(abs)}$

Pressure difference, $p_2 - p_1 = 0.166 \times 10^5$ Pa.

Hence, from Equation (8.30) [with $\varepsilon_a = (0.5 + 0.53)/2 = 0.515$],

$$\text{length to first aeration point, } H = \frac{0.166 \times 10^5}{1200 \times (1 - 0.515) \times 9.81} = 2.91\,\text{m}$$

Assuming ideal gas behaviour, density at level 2, $\rho_{f_2} = \rho_{f_1}\left(\frac{p_2}{p_1}\right) = 1.128\,\text{kg/m}^3$

Applying Equation (8.34), aeration gas mass flow at first aeration point,

$$M_{f_2} = \frac{0.53}{(1-0.53)}\frac{80}{1200}(1.128 - 1.0) = 0.0096\,\text{kg/s}$$

The above calculation is repeated in order to determine the position and rates of subsequent aeration points. The results are summarized below:

	First point	Second point	Third point	Fourth point	Fifth point
Distance from top of standpipe (m)	2.91	6.18	9.88	14.04	18.75
Aeration rate (kg/s)	0.0096	0.0108	0.0122	0.0138	0.0155
Pressure at aeration point (bar)	1.47	1.65	1.86	2.10	2.37

WORKED EXAMPLE 8.3

A 10 m long vertical standpipe of inside diameter 0.1 m transports solids at a flux of 100 kg/m^{2}s from an upper vessel which is held at a pressure 1.0 bar to a lower vessel held at 1.5 bar. The particle density of the solids is 2500 kg/m^3 and the surface-volume mean particle size is 250 μm. Assuming that the voidage is constant along the standpipe and equal to 0.50, and that the effect of pressure change may be ignored, determine the direction and flow rate of gas passing between the vessels. (Properties of gas in the system: density, 1 kg/m^3; viscosity, 2×10^{-5} Pa s.)

Solution

First check that the solids are moving in packed bed flow. We do this by comparing the actual pressure gradient with the pressure gradient for fluidization.

Assuming that in fluidized flow the apparent weight of the solids will be supported by the gas flow, Equation (8.26) gives the pressure gradient for fluidized bed flow:

$$\frac{(-\Delta p)}{H} = (1 - 0.5) \times (2500 - 1) \times 9.81 = 12258\,\text{Pa/m}$$

$$\text{Actual pressure gradient} = \frac{(1.5 - 1.0) \times 10^5}{10} = 5000\,\text{Pa/m}$$

Since the actual pressure gradient is well below that for fluidized flow, the standpipe is operating in packed bed flow.

The pressure gradient in packed bed flow is generated by the upward flow of gas relative to the solids in the standpipe. The Ergun equation [Equation (8.25)] provides the relationship between gas flow and pressure gradient in a packed bed.

Knowing the required pressure gradient, the packed bed voidage and the particle and gas properties, Equation (8.25) can be solved for $|U_{rel}|$, the magnitude of the relative gas velocity:

Ignoring the negative root of the quadratic, $|U_{rel}| = 0.1026\,\text{m/s}$

We now adopt a sign convention for velocities. For standpipes it is convenient to take downward velocities as positive. In order to create the pressure gradient in the required direction, the gas must flow upwards relative to the solids. Hence, U_{rel} is negative:

$$U_{rel} = -0.1026\,\mathrm{m/s}$$

From the continuity for the solids [Equation (8.11)],

$$\text{solids flux, } \frac{M_p}{A} = U_p(1-\varepsilon)\rho_p$$

The solids flux is given as 100 kg/m^{2}s and so

$$U_p = \frac{100}{(1-0.5)\times 2500} = 0.08\,\mathrm{m/s}$$

Solids flow is downwards, so $U_p = +0.08\,\mathrm{m/s}$

The relative velocity, $U_{rel} = U_f - U_p$

hence, actual gas velocity, $U_f = -0.1026 + 0.08 = -0.0226\,\mathrm{m/s}$ (upwards)

Therefore the gas flows *upwards* at a velocity of 0.0226 m/s relative to the standpipe walls. The superficial gas velocity is therefore:

$$U = \varepsilon U_f = -0.0113\,\mathrm{m/s}$$

From the continuity for the gas [Equation (8.12)] mass flow rate of gas,

$$M_f = \varepsilon U_f \rho_f A$$
$$= -8.9\times 10^{-5}\,\mathrm{kg/s}$$

So for the standpipe to operate as required, 8.9×10^{-5} kg/s of gas must flow from the lower vessel to the upper vessel.

TEST YOURSELF

8.1 In horizontal pneumatic transport of particulate solids, what is meant by the term saltation velocity?

8.2 In vertical pneumatic transport of particulate solids, what is meant by the term choking velocity?

8.3 In horizontal pneumatic transport of particulate solids, why is there a minimum in the pressure drop versus gas velocity plot?

8.4 In vertical pneumatic transport of particulate solids, why is there a minimum in the pressure drop versus gas velocity plot?

8.5 There are six components in the equation describing the pressure drop across a pipe carrying solids by pneumatic transport. Write down these six components, in words.

8.6 In a dilute phase pneumatic transport system, what are the two main reasons for using a rotary airlock?

8.7 In a dense phase pneumatic transport system, why is it necessary to limit plug length in some cases? Describe three ways in which the plug length might be limited in practice?

8.8 How do we determine whether a standpipe is operating in packed bed flow or fluidized bed flow?

8.9 For a standpipe operating in packed bed flow, how do we determine the quantity of gas flow and whether the gas is flowing upwards or downwards?

8.10 For a standpipe operating in fluidized bed flow, why is it often necessary to add aeration gas at several points along the standpipe? What approach is taken to calculation of the quantities of gas to be added and the positions of the aeration points?

EXERCISES

8.1 Design a positive pressure dilute-phase pneumatic transport system to carry 500 kg/h of a powder of particle density 1800 kg/m^3 and mean particle size 150 μm across a horizontal distance of 100 m and a vertical distance of 20 m using ambient air. Assume that the pipe is smooth, that four 90° bends are required and that the allowable pressure loss is 0.7 bar. See below for Blasius correlation for the gas-wall friction factor for smooth pipes.

(Answer: 50 mm diameter pipe gives total pressure drop of 0.55 bar; superficial gas velocity 13.8 m/s.)

8.2 It is required to use an existing 50 mm inside diameter vertical smooth pipe as lift line to transfer 2000 kg/h of sand of mean particle size 270 μm and particle density 2500 kg/m^3 to a process 50 m above the solids feed point. A blower is available which is capable of delivering 60 m^3/h of ambient air at a pressure of 0.3 bar. Will the system operate as required?

(Answer: Using a superficial gas velocity of 8.49 m/s($= 1.55 \times U_{CH}$) the total pressure drop is 0.344 bar. System will not operate as required since allowable $\Delta p = 0.3$ bar)

8.3 Design a negative pressure dilute-phase pneumatic transport system to carry 700 kg/h of plastic spheres of particle density 1000 kg/m^3 and mean particle size 1 mm between two points in a factory separated by a vertical distance of 15 m and a horizontal distance of 80 m using ambient air. Assume that the pipe is smooth, that five 90° bends are required and that the allowable pressure loss is 0.4 bar.

(Answer: Using a superficial gas velocity of 16.4 m/s in a pipe of inside diameter 40 mm, the total pressure drop is 0.38 bar.)

8.4 A 25 m long standpipe carrying Group A solids at a rate of 75 kg/s is to be aerated in order to maintain fluidized flow with a voidage in the range 0.50–0.55. Solids enter the top of the standpipe at a voidage of 0.55. The pressure and gas density at the top of the standpipe are 1.4 bar (abs) and 1.1 kg/m^3, respectively. The particle density of the solids is 1050 kg/m^3.

Determine the aeration positions and rates.

(Answer: Positions: 6.36 m, 14.13 m, 23.6 m. Rates: 0.0213 kg/s, 0.0261 kg/s, 0.0319 kg/s.)

8.5 A 15 m long standpipe carrying Group A solids at a rate of 120 kg/s is to be aerated in order to maintain fluidized flow with a voidage in the range 0.50–0.54. Solids enter the top of the standpipe at a voidage of 0.54. The pressure and gas density at the top of the standpipe are 1.2 bar (abs) and 0.9 kg/m^3, respectively. The particle density of the solids is 1100 kg/m^3.

Determine the aeration positions and rates. What is the pressure at the lowest aeration point?

(Answer: Positions: 4.03 m, 6.76 m, 14.3 m. Rates: 0.0200 kg/s, 0.0235 kg/s, 0.0276 kg/s. Pressure: 1.94 bar.)

8.6 A 5 m long vertical standpipe of inside diameter 0.3 m transports solids at flux of 500 kg/m^{2}s from an upper vessel which is held at a pressure of 1.25 bar to a lower vessel held at 1.6 bar. The particle density of the solids is 1800 kg/m^3 and the surface-volume mean particle size is 200 μm. Assuming that the voidage is 0.48 and is constant along the standpipe, determine the direction and flow rate of gas passing between the vessels. (Properties of gas in the system: density, 1.5 kg/m^3; viscosity, 1.9×10^{-5} Pa s.)

(Answer: 0.023 kg/s downwards.)

8.7 A vertical standpipe of inside diameter 0.3 m transports solids at a flux of 300 kg/m^{2}s from an upper vessel which is held at a pressure 2.0 bar to a lower vessel held at 2.72 bar. The particle density of the solids is 2000 kg/m^3 and the surface-volume mean particle size is 220 μm. The density and viscosity of the gas in the system are 2.0 kg/m^3 and 2×10^{-5} Pa s, respectively.

Assuming that the voidage is 0.47 and is constant along the standpipe.

(a) Determine the minimum standpipe length required to avoid fluidized flow.

(b) Given that the actual standpipe is 8 m long, determine the direction and flow rate of gas passing between the vessels.

(Answer: (a) 6.92 m; (b) 0.0114 kg/s downwards.)

Blasius correlation for the gas-wall friction factor for smooth pipes: $f_g = 0.079\ Re^{-0.25}$

9

Separation of Particles from a Gas: Gas Cyclones

There are many cases during the processing and handling of particulate solids when particles are required to be separated from suspension in a gas. We saw in Chapter 7 that in fluidized bed processes the passage of gas through the bed entrains fine particles. These particles must be removed from the gas and returned to the bed before the gas can be discharged or sent to the next stage in the process. Keeping the very small particles in the fluid bed may be crucial to the successful operation of the process, as is the case in fluid catalytic cracking of oil.

In Chapter 8, we saw how a gas may be used to transport powders within a process. The efficient separation of the product from the gas at the end of the transport line plays an important part in the successful application of this method of powder transportation. In the combustion of solid fuels, fine particles of fuel ash become suspended in the combustion gases and must be removed before the gases can be discharged to the environment.

In any application, the size of the particles to be removed from the gas determine, to a large extent, the method to be used for their separation. Generally speaking, particles larger than about 100 μm can be separated easily by gravity settling. For particles less than 10 μm more energy intensive methods such as filtration, wet scrubbing and electrostatic precipitation must be used. Figure 9.1 shows typical grade efficiency curves for gas–particle separation devices. The grade efficiency curve describes how the separation efficiency of the device varies with particle size. In this chapter we will focus on the device known as the cyclone separator or cyclone. Gas cyclones are generally not suitable for separation involving suspensions with a large proportion of particles less than 10 μm. They are best suited as primary separation devices and for relatively coarse particles, with an electrostatic precipitator or fabric filter being used downstream to remove very fine particles.

Introduction to Particle Technology - 2nd Edition Martin Rhodes

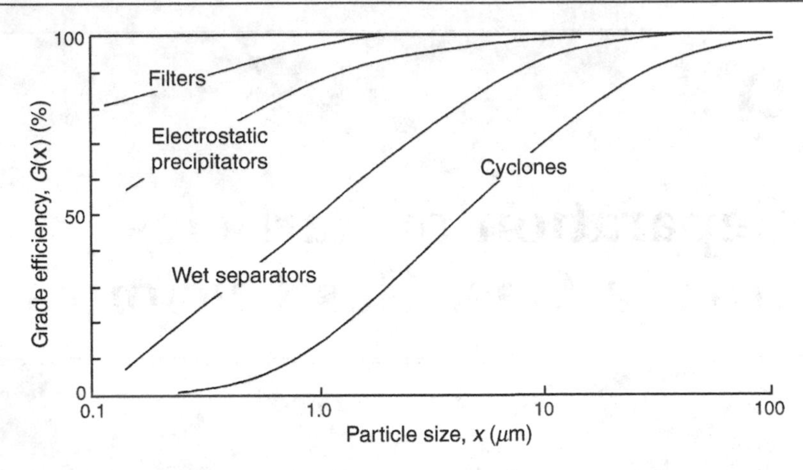

Figure 9.1 Typical grade efficiency curves for gas–particle separators

Readers wishing to know more about other methods of gas–particle separation and about the choice between methods are referred to Svarovsky (1981, 1990) and Perry and Green (1984).

9.1 GAS CYCLONES–DESCRIPTION

The most common type of cyclone is known as the reverse flow type (Figure 9.2). Inlet gas is brought tangentially into the cylindrical section and a strong vortex is thus created inside the cyclone body. Particles in the gas are subjected to centrifugal forces which move them radially outwards, against the inward flow of gas and towards the inside surface of the cyclone on which the solids separate. The direction of flow of the vortex reverses near the bottom of the cylindrical

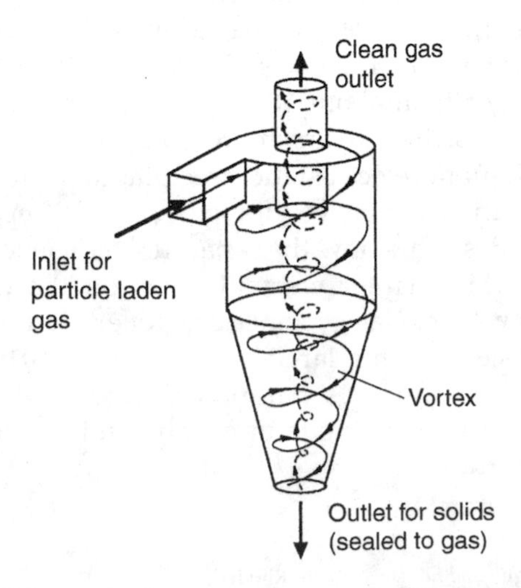

Figure 9.2 Schematic diagram of a reverse flow cyclone separator

section and the gas leaves the cyclone via the outlet in the top (the solids outlet is sealed to gas). The solids at the wall of the cyclone are pushed downwards by the outer vortex and out of the solids exit. Gravity has been shown to have little effect on the operation of the cyclone.

9.2 *FLOW CHARACTERISTICS*

Rotational flow in the forced vortex within the cyclone body gives rise to a radial pressure gradient. This pressure gradient, combined with the frictional pressure losses at the gas inlet and outlet and losses due to changes in flow direction, make up the total pressure drop. This pressure drop, measured between the inlet and the gas outlet, is usually proportional to the square of gas flow rate through the cyclone. A resistance coefficient, the Euler number *Eu*, relates the cyclone pressure drop Δp to a characteristic velocity v:

$$Eu = \Delta p/(\rho_f v^2/2) \tag{9.1}$$

where ρ_f is the gas density.

The characteristic velocity v can be defined for gas cyclones in various ways but the simplest and most appropriate definition is based on the cross-section of the cylindrical body of the cyclone, so that:

$$v = 4q/(\pi D^2) \tag{9.2}$$

where q is the gas flow rate and D is the cyclone inside diameter.

The Euler number represents the ratio of pressure forces to the inertial forces acting on a fluid element. Its value is practically constant for a given cyclone geometry, independent of the cyclone body diameter (see Section 9.4).

9.3 *EFFICIENCY OF SEPARATION*

9.3.1 Total Efficiency and Grade Efficiency

Consider a cyclone to which the solids mass flow rate is M, the mass flow discharged from the solids exit orifice is M_c (known as the coarse product) and the solids mass flow rate leaving with the gas is M_f (known as the fine product). The total material balance on the solids over this cyclone may be written:

$$\text{Total} : M = M_f + M_c \tag{9.3}$$

and the 'component' material balance for each particle size x (assuming no breakage or growth of particles within the cyclone) is

$$\text{Component} : M(\mathrm{d}F/\mathrm{d}x) = M_f(\mathrm{d}F_f/\mathrm{d}x) + M_c(\mathrm{d}F_c/\mathrm{d}x) \tag{9.4}$$

where, $\mathrm{d}F/\mathrm{d}x, \mathrm{d}F_f/\mathrm{d}x$ and $\mathrm{d}F_c/\mathrm{d}x$ are the differential frequency size distributions by mass (i.e. mass fraction of size x) for the feed, fine product and coarse product respectively. F, F_f and F_c are the cumulative frequency size distributions by mass (mass fraction less than size x) for the feed, fine product and coarse product, respectively. Refer to Chapter 1 for further details on representations of particle size distributions.

The total efficiency of separation of particles from gas, E_T, is defined as the fraction of the total feed which appears in the coarse product collected, i.e.

$$E_T = M_c/M \tag{9.5}$$

The efficiency with which the cyclone collects particles of a certain size is described by the grade efficiency, $G(x)$, which is defined as:

$$G(x) = \frac{\text{mass of solids of size } x \text{ in coarse product}}{\text{mass of solids of size } x \text{ in feed}} \tag{9.6}$$

Using the notation for size distribution described above:

$$G(x) = \frac{M_c(\mathrm{d}F_c/\mathrm{d}x)}{M(\mathrm{d}F/\mathrm{d}x)} \tag{9.7}$$

Combining with Equation (9.5), we find an expression linking grade efficiency with total efficiency of separation:

$$G(x) = E_T \frac{(\mathrm{d}F_c/\mathrm{d}x)}{(\mathrm{d}F/\mathrm{d}x)} \tag{9.8}$$

From Equations (9.3)–(9.5), we have

$$(\mathrm{d}F/\mathrm{d}x) = E_T(\mathrm{d}F_c/\mathrm{d}x) + (1 - E_T)(\mathrm{d}F_f/\mathrm{d}x) \tag{9.9}$$

Equation (9.9) relates the size distributions of the feed (no subscript), the coarse product (subscript c) and the fine product (subscript f). In cumulative form this becomes

$$F = E_T F_c + (1 - E_T)F_f \tag{9.10}$$

9.3.2 Simple Theoretical Analysis for the Gas Cyclone Separator

Referring to Figure 9.3, consider a reverse flow cyclone with a cylindrical section of radius R. Particles entering the cyclone with the gas stream are forced into circular motion. The net flow of gas is radially inwards towards the central gas outlet. The forces acting on a particle following a circular path are drag, buoyancy and centrifugal force. The balance between these forces determines the equilibrium orbit adopted by the particle. The drag force is caused by the inward flow of gas past the particle and acts radially inwards. Consider a particle of diameter

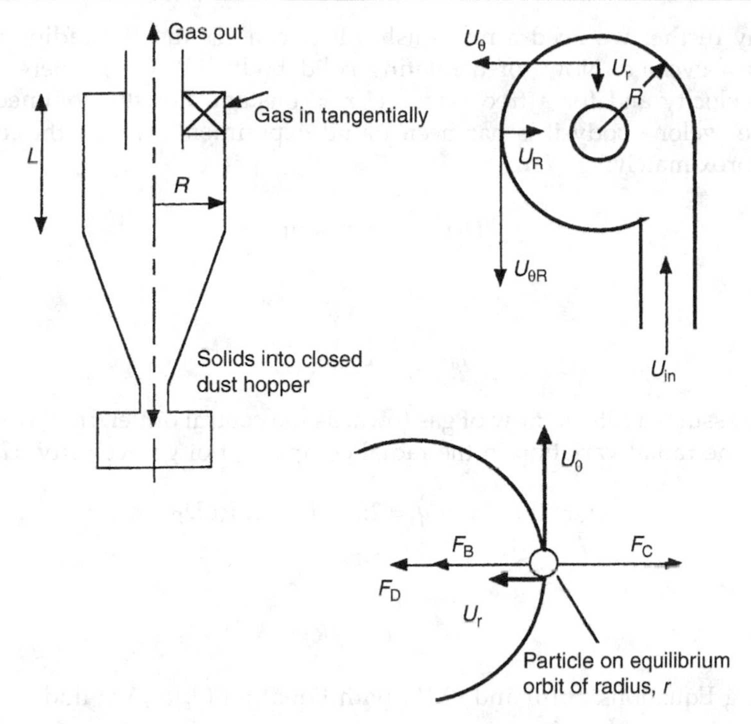

Figure 9.3 Reverse flow cyclone – a simple theory for separation efficiency

x and density ρ_p following an orbit of radius r in a gas of density ρ_f and viscosity μ. Let the tangential velocity of the particle be U_θ and the radial inward velocity of the gas be U_r. If we assume that Stokes' law applies under these conditions then the drag force is given by:

$$F_D = 3\pi x \mu U_r \tag{9.11}$$

The centrifugal and buoyancy forces acting on the particle moving with a tangential velocity component U_θ at radius r are, respectively:

$$F_C = \frac{\pi x^3}{6}\rho_p \frac{U_\theta^2}{r} \tag{9.12}$$

$$F_B = \frac{\pi x^3}{6}\rho_f \frac{U_\theta^2}{r} \tag{9.13}$$

Under the action of these forces the particle moves inwards or outwards until the forces are balanced and the particle assumes its equilibrium orbit. At this point,

$$F_C = F_D + F_B \tag{9.14}$$

and so

$$x^2 = \frac{18\mu}{(\rho_p - \rho_f)}\left(\frac{r}{U_\theta^2}\right)U_r \tag{9.15}$$

To go any further we need a relationship between U_θ and the radius r for the vortex in a cyclone. Now for a rotating solid body, $U_\theta = r\omega$, where ω is the angular velocity and for a free vortex $U_\theta r = \text{constant}$. For the confined vortex inside the cyclone body it is has been found experimentally that the following holds approximately:

$$U_\theta r^{1/2} = \text{constant}$$

hence

$$U_\theta r^{1/2} = U_{\theta R} R^{1/2} \tag{9.16}$$

If we also assume uniform flow of gas towards the central outlet, then we are able to derive the radial variation in the radial component of gas velocity, U_r:

$$\text{gas flow rate}, q = 2\pi r L U_r = 2\pi R L U_R \tag{9.17}$$

hence

$$U_R = U_r(r/R) \tag{9.18}$$

Combining Equations (9.16) and (9.18) with Equation (9.15), we find

$$x^2 = \frac{18\mu}{(\rho_p - \rho_f)} \frac{U_R}{U_{\theta R}^2} r \tag{9.19}$$

where r is the radius of the equilibrium orbit for a particle of diameter x.

If we assume that all particles with an equilibrium orbit radius greater than or equal to the cyclone body radius will be collected, then substituting $r = R$ in Equation (9.19) we derive the expression below for the critical particle diameter for separation, x_{crit}:

$$x_{crit}^2 = \frac{18\mu}{(\rho_p - \rho_f)} \frac{U_R}{U_{\theta R}^2} R \tag{9.20}$$

The values of the radial and tangential velocity components at the cyclone wall, U_R and $U_{\theta R}$, in Equation (9.20) may be found from a knowledge of the cyclone geometry and the gas flow rate.

This analysis predicts an ideal grade efficiency curve shown in Figure 9.4. All particles of diameter x_{crit} and greater are collected and all particles of size less than x_{crit} are not collected.

9.3.3 Cyclone Grade Efficiency in Practice

In practice, gas velocity fluctuations and particle–particle interactions result in some particles larger than x_{crit} being lost and some particles smaller than x_{crit}

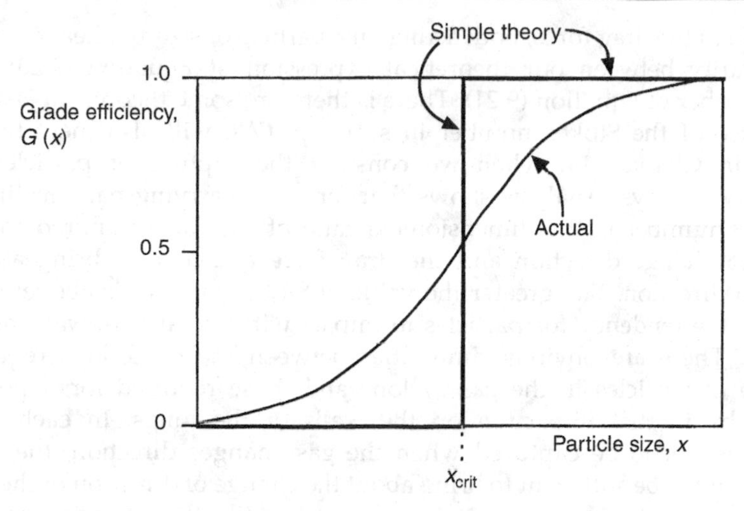

Figure 9.4 Theoretical and actual grade efficiency curves

being collected. Consequently, in practice the cyclone does not achieve such a sharp cut-off as predicted by the theoretical analysis above. In common with other separation devices in which body forces are opposed by drag forces, the grade efficiency curve for gas cyclones is usually S-shaped.

For such a curve, the particle size for which the grade efficiency is 50%, x_{50}, is often used as a single number measurement of the efficiency of the cyclone. x_{50} is also know as the equiprobable size since it is that size of particle which has a 50% probability of appearing in the coarse product. This also means that, in a large population of particles, 50% of the particles of this size will appear in the coarse product. x_{50} is sometimes simply referred to as the cut size of the cyclone (or other separation device).

The concept of x_{50} cut size is useful where the efficiency of a cyclone is to be expressed as a single number independent of the feed solid size distribution, such as in scale-up calculation.

9.4 SCALE-UP OF CYCLONES

The scale-up of cyclones is based on a dimensionless group, the Stokes number, which characterizes the separation performance of a family of geometrically similar cyclones. The Stokes number Stk_{50} is defined as:

$$Stk_{50} = \frac{x_{50}^2 \rho_p v}{18 \mu D} \qquad (9.21)$$

where μ is gas viscosity, ρ_p is solids density, v is the characteristic velocity defined by Equation (9.2) and D is the diameter of the cyclone body. The physical significance of the Stokes number is that it is a ratio of the centrifugal force (less buoyancy) to the drag force, both acting on a particle of size x_{50}. Readers will note the similarity between our theoretical expression of Equation (9.20) and the Stokes number of Equation (9.21). There is therefore some theoretical justification for the use of the Stokes number in scale-up. (We will also meet the Stokes number in Chapter 14, when we consider the capture of particles in the respiratory airways. Analysis shows that for a gas carrying particles in a duct, the Stokes number is the dimensionless ratio of the force required to cause a particle to change direction and the drag force available to bring about that change in direction. The greater the value of Stokes number is above unity, the greater is the tendency for particles to impact with the airway walls and so be captured. There are obvious similarities between the conditions required for collection of particles in the gas cyclone and those required for deposition of particles by inertial impaction on the walls of the lungs. In each case, for the particle not to be captured when the gas changes direction, the available drag force must be sufficient to bring about the change of direction of the particle.

For large industrial cyclones the Stokes number, like the Euler number defined previously, is independent of Reynolds number. For suspensions of concentration less than about 5 g/m^3, the Stokes and Euler numbers are usually constant for a given cyclone geometry (i.e. a set of geometric proportions relative to cyclone diameter D). The geometries and values of Eu and Stk_{50} for two common industrial cyclones, the Stairmand high efficiency (HE) and the Stairmand high rate (HR) are given in Figure 9.5.

The use of the two dimensionless groups Eu and Stk_{50} in cyclone scale-up and design is demonstrated in the worked examples at the end of this chapter.

As can be seen from Equation (9.21), the separation efficiency is described only by the cut size x_{50} and no regard is given to the shape of the grade efficiency curve. If the whole grade efficiency curve is required in performance calculations, it may be generated around the given cut size using plots or analytical functions of a generalized grade efficiency function available from the literature or from previously measured data. For example, Perry and Green (1984) give the grade efficiency expression:

$$\text{grade efficiency} = \frac{(x/x_{50})^2}{[1 + (x/x_{50})^2]} \qquad (9.22)$$

for a reverse flow cyclone with the geometry:

A	B	C	E	J	K	N
4.0	2.0	2.0	0.25	0.625	0.5	0.5

(Letters refer to the cyclone geometry diagram shown in Figure 9.5.)

This expression gives rise to the grade efficiency curve shown in Figure 9.6 for an x_{50} cut size of 5μm. Very little is known how the shape of the grade efficiency curve is affected by operating pressure drop, cyclone size or design, and feed solids concentration.

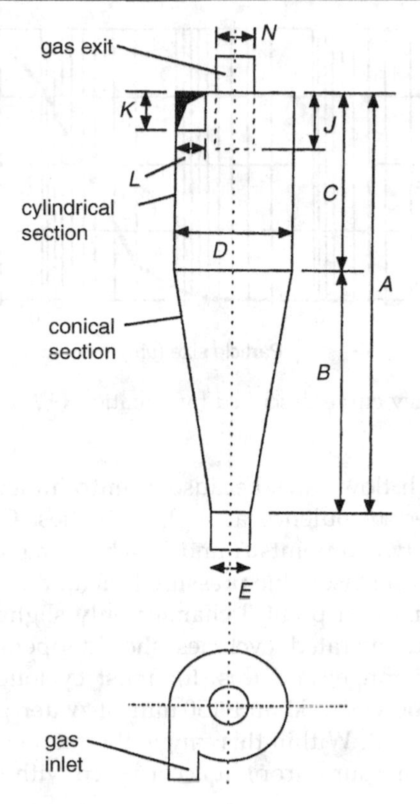

HE – high efficiency Stairmand cyclone
$Stk_{50} = 1.4 \times 10^{-4}$
Eu = 320

HR – high flowrate Stairmand cyclone
$Stk_{50} = 6 \times 10^{-3}$
Eu = 46

	Dimension relative to diameter D							
Cyclone type	A	B	C	E	J	L	K	N
Stairmand, H.E.	4.0	2.5	1.5	0.375	0.5	0.2	0.5	0.5
Stairmand, H.R.	4.0	2.5	1.5	0.575	0.875	0.375	0.75	0.75

Figure 9.5 Geometries and Euler and Stokes numbers for two common cyclones.

9.5 *RANGE OF OPERATION*

One of the most important characteristics of gas cyclones is the way in which their efficiency is affected by pressure drop (or flow rate). For a particular cyclone and inlet particle concentration, total efficiency of separation and pressure drop vary with gas flow rate as shown in Figure 9.7. Theory predicts that efficiency increases with increasing gas flow rate. However, in practice, the total efficiency

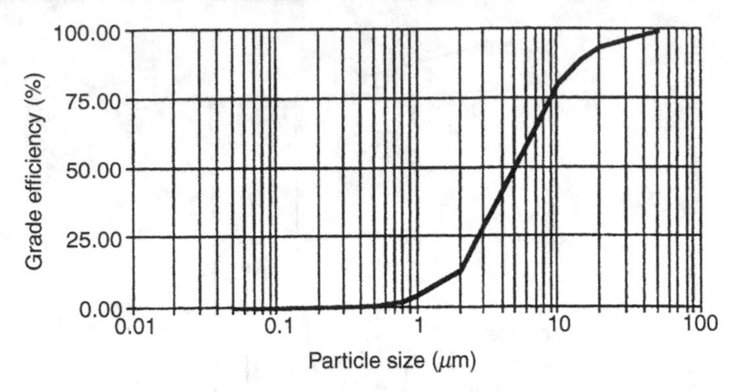

Figure 9.6 Grade efficiency curve described by Equation (9.22) for a cut size $x_{50} = 5\,\mu m$

curve falls away at high flow rates because re-entrainment of separated solids increases with increased turbulence at high velocities. Optimum operation is achieved somewhere between points A and B, where maximum total separation efficiency is achieved with reasonable pressure loss (and hence reasonable power consumption). The position of point B changes only slightly for different dusts. Correctly designed and operated cyclones should operate at pressure drops within a recommended range; and this, for most cyclone designs operated at ambient conditions, is between 50 and 150 mm of water gauge (WG) (approximately from 500 to 1500 Pa). Within this range, the total separation efficiency E_T increases with applied pressure drop, in accordance with the inertial separation theory shown above.

Above the top limit the total efficiency no longer increases with increasing pressure drop and it may actually decline due to re-entrainment of dust from the dust outlet orifice. It is, therefore, wasteful of energy to operate cyclones above the limit. At pressure drops below the bottom limit, the cyclone represents

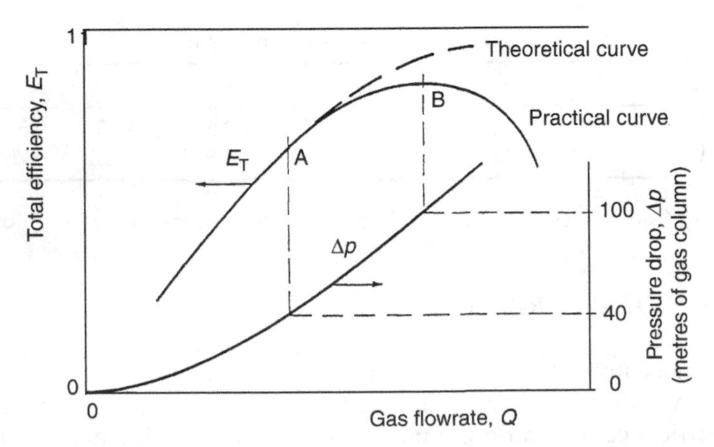

Figure 9.7 Total separation efficiency and pressure drop versus gas flow rate through a reverse flow cyclone

little more than a settling chamber, giving low efficiency due to low velocities within it which may not be capable of generating a stable vortex.

9.6 SOME PRACTICAL DESIGN AND OPERATION DETAILS

The following practical considerations for design and operation of reverse flow gas cyclones are among those listed by Svarovsky (1986).

9.6.1 Effect of Dust Loading on Efficiency

One of the important operating variables affecting total efficiency is the concentration of particles in the suspension (known as the dust loading). Generally, high dust loadings (above about $5\,g/m^3$) lead to higher total separation efficiencies due to particle enlargement through agglomeration of particles (caused, for example, by the effect of humidity).

9.6.2 Cyclone Types

The many reverse flow cyclone designs available today may be divided into two main groups: *high efficiency* designs (e.g. Stairmand HE) and the *high rate* designs (e.g. Stairmand HR). High efficiency cyclones give high recoveries and are characterized by relatively small inlet and gas outlet orifices. The high rate designs have lower total efficiencies, but offer low resistance to flow so that a unit of a given size will give much higher gas capacity than a high efficiency design of the same body diameter. The high rate cyclones have large inlets and gas outlets, and are usually shorter. The geometries and values of Eu and Stk_{50} for two common cyclones, the Stairmand HE and the Stairmand HR are given in Figure 9.5.

For well-designed cyclones there is a direct correlation between Eu and Stk_{50}. High values of the resistance coefficient usually lead to low values of Stk_{50} (therefore low cut sizes and high efficiencies), and vice versa. The general trend can be described by the following *approximate* empirical correlation:

$$Eu = \sqrt{\frac{12}{Stk_{50}}} \tag{9.23}$$

9.6.3 Abrasion

Abrasion in gas cyclones is an important aspect of cyclone performance and it is affected by the way cyclones are installed and operated as much as by the material construction and design. Materials of construction are usually steels of different grades, sometimes lined with rubber, refractory lining or other material. Within the cyclone body there are two critical zones for abrasion: in the cylindrical part just beyond the inlet opening and in the conical part near the dust discharge.

9.6.4 Attrition of Solids

Attrition or break-up of solids is known to take place on collection in gas cyclones but little is known about how it is related to particle properties, although large particles are more likely to be affected by attrition than finer fractions. Attrition is most detectable in recirculating systems such as fluidized beds where cyclones are used to return the carry-over material back to the bed (see Chapter 7). The complete inventory of the bed may pass through the cyclones many times per hour and the effect of attrition is thus exaggerated.

9.6.5 Blockages

Blockages, usually caused by overloading of the solids outlet orifice, is one of the most common causes of failure in cyclone operation. The cyclone cone rapidly fills up with dust, the pressure drop increases and efficiency falls dramatically. Blockages arise due to mechanical defects in the cyclone body (bumps on the cyclone cone, protruding welds or gasket) or changes in chemical or physical properties of the solids (e.g. condensation of water vapour from the gas onto the surface of particles).

9.6.6 Discharge Hoppers and Diplegs

The design of the solids discharge is important for correct functioning of a gas cyclone. If the cyclone operates under vacuum, any inward leakages of air at the discharge end cause particles to be re-entrained and this leads to a sharp decrease in separation efficiency. If the cyclone is under pressure, outward leakages may cause a slight increase in separation efficiency, but also results in loss of product and pollution of the local environment. It is therefore best to keep the solids discharge as gas-tight as possible.

The strong vortex inside a cyclone reaches into the space underneath the solids outlet and it is important that no powder surface is allowed to build up to within at least one cyclone diameter below the underflow orifice. A conical vortex breaker positioned just under the dust discharge orifice may be used to prevent the vortex from intruding into the discharge hopper below. Some cyclone manufacturers use a 'stepped' cone to counter the effects of re-entrainment and abrasion, and Svarovsky (1981) demonstrated the value of this design feature.

In fluidized beds with internal cyclones, 'diplegs' are used to return the collected entrained particles into the fluidized bed. Diplegs are vertical pipes connected directly to the solids discharge orifice of the cyclone extending down to below the fluidized bed surface. Particles discharged from the cyclone collect as a moving settled suspension in the lower part of the dipleg before it enters the bed. The level of the settled suspension in the dipleg is always higher than the fluidized bed surface and it provides a necessary resistance to minimize both the flow of gas up the dipleg and the consequent reduction of cyclone efficiency.

9.6.7 Cyclones in Series

Connecting cyclones in series is often done in practice to increase recovery. Usually the primary cyclone would be of medium or low efficiency design and the secondary and subsequent cyclones of progressively more efficient design or smaller diameter.

9.6.8 Cyclones in Parallel

The x_{50} cut size achievable for a given cyclone geometry and operating pressure drop decreases with decreasing cyclone size [see Equation (9.21)]. The size a single cyclone for treating a given volume flow rate of gas is determined by that gas flow rate [Equations (9.1) and (9.2)]. For large gas flow rates the resulting cyclone may be so large that the x_{50} cut size is unacceptably high. The solution is to split the gas flow into several smaller cyclones operating in parallel. In this way, both the operating pressure drop and x_{50} cut size requirements can be achieved. The worked examples at the end of the chapter demonstrate how the number and diameter of cyclones in parallel are estimated.

9.7 *WORKED EXAMPLES*

WORKED EXAMPLE 9.1 – DESIGN OF A CYCLONE

Determine the diameter and number of gas cyclones required to treat 2 m^3/s of ambient air (viscosity, 18.25×10^{-6} Pa s; density, 1.2 kg/m^3) laden with solids of density 1000 kg/m^3 at a suitable pressure drop and with a cut size of 4 μm. Use a Stairmand HE (high efficiency) cyclone for which $Eu = 320$ and $Stk_{50} = 1.4 \times 10^{-4}$.

$$\text{Optimum pressure drop} = 100\,\text{m gas}$$
$$= 100 \times 1.2 \times 9.81\,\text{Pa}$$
$$= 1177\,\text{Pa}$$

Solution

From Equation (9.1),

$$\text{characteristic velocity}, v = 2.476\,\text{m/s}$$

Hence, from Equation (9.2), diameter of cyclone, $D = 1.014$ m

With this cyclone, using Equation (9.21), cut size, $x_{50} = 4.34\,\mu m$

This is too high and we must therefore opt for passing the gas through several smaller cyclones in parallel.

Assuming that n cyclones in parallel are required and that the total flow is evenly split, then for each cyclone the flow rate will be $q = 2/n$.

Therefore from Equations (9.1) and (9.2), new cyclone diameter, $D = 1.014/n^{0.5}$. Substituting in Equation (9.21) for D, the required cut size and v (2.476 m/s, as originally calculated, since this is determined solely by the pressure drop requirement), we find that

$$n = 1.386$$

We will therefore need two cyclones. Now with $n = 2$, we recalculate the cyclone diameter from $D = 1.014/n^{0.5}$ and the actual achieved cut size from Equation (9.21).

Thus, $D = 0.717$ m, and using this value for D in Equation (9.21) together with required cut size and $v = 2.476$ m/s, we find that the actual cut size is 3.65 μm.

Therefore, two 0.717 m diameter Stairmand HE cyclones in parallel will give a cut size of 3.65 μm using a pressure drop of 1177 Pa.

WORKED EXAMPLE 9.2

Tests on a reverse flow gas cyclone give the results shown in the table below:

Size range (μm)	0–5	5–10	10–15	15–20	20–25	25–30
Feed size analysis, m (g)	10	15	25	30	15	5
Course product size analysis, m_c (g)	0.1	3.53	18.0	27.3	14.63	5.0

(a) From these results determine the total efficiency of the cyclone.

(b) Plot the grade efficiency curve and hence show that the x_{50} cut size is 10 μm.

(c) The dimensionless constants describing this cyclone are: $Eu = 384$ and $Stk_{50} = 1 \times 10^{-3}$. Determine the diameter and number of cyclones to be operated in parallel to achieve this cut size when handling 10 m^3/s of a gas of density 1.2 kg/m^3 and viscosity 18.4×10^{-6} Pa s, laden with dust of particle density 2500 kg/m^3. The available pressure drop is 1200 Pa.

(d) What is the actual cut size of your design?

Solution

(a) From the test results:

Mass of feed, $M = 10 + 15 + 25 + 30 + 15 + 5 = 100$ g

Mass of coarse product, $M_c = 0.1 + 3.53 + 18.0 + 27.3 + 14.63 + 5.0 = 68.56$ g

Therefore, from Equation (9.5), total efficiency,

$$E_T = \frac{M_c}{M} = 0.6856 \text{ (or } 68.56\%)$$

(b) From Equation (9.7), grade efficiency,

$$G(x) = \frac{M_c}{M}\frac{dF_c dx}{dF/dx} = E_T\frac{dF_c/dx}{dF/dx}$$

In this case, $G(x)$ may be obtained directly from the results table as

$$G(x) = \frac{m_c}{m}$$

And so, the grade efficiency curve data becomes:

Size range (μm)	0–5	5–10	10–15	15–20	20–25	25–30
$G(x)$	0.01	0.235	0.721	0.909	0.975	1.00

Plotting these data gives $x_{50} = 10\,\mu$m, as may be seen from Figure 9W.2.1.

For interest, we can calculate the size distributions of the feed, dF/dx and the coarse product, dF_c/dx:

Size range (μm)	0–5	5–10	10–15	15–20	20–25	25–30
dF_c/dx	0.00146	0.0515	0.263	0.398	0.2134	0.0729
dF/dx	0.1	0.15	0.25	0.30	0.15	0.05

We can then verify the calculated $G(x)$ values. For example, in the size range 10–15:

$$G(x) = E_T\frac{dF_c/dx}{dF/dx} = 0.6856 \times \frac{0.263}{0.25} = 0.721$$

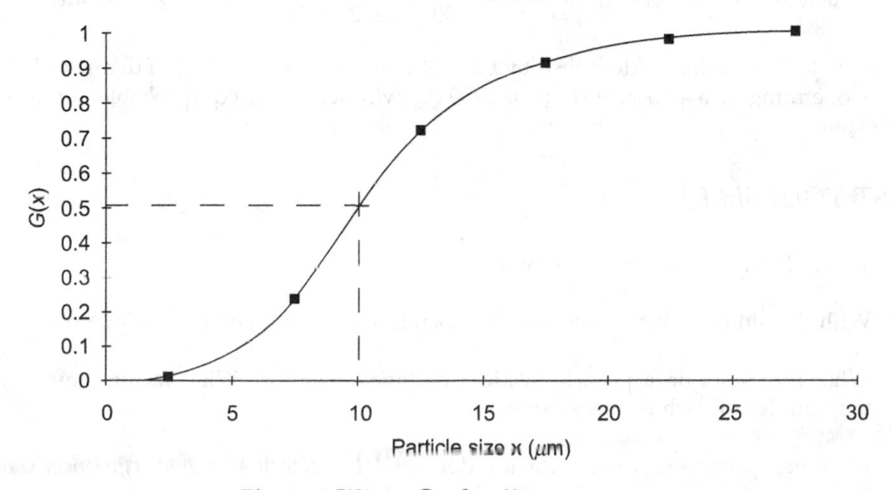

Figure 9.W2.1 Grade efficiency curve

(c) Using Equation (9.1), noting that the allowable pressure is 1200 Pa, we calculate the characteristic velocity, v:

$$v = \sqrt{\frac{2\Delta p}{Eu\rho_f}} = \sqrt{\frac{2 \times 1200}{384 \times 1.2}} = 2.282\,\text{m/s}$$

If we have n cyclones in parallel then assuming even distribution of the gas between the cyclones, flow rate to each cyclone, $q = Q/n$ and from Equation (9.2),

$$D = \sqrt{\frac{4Q}{n\pi v}} = \sqrt{\frac{4 \times 10}{n\pi \times 2.282}} = \frac{2.362}{\sqrt{n}}$$

Now substitute this expression for D and the required cut size x_{50} in Equation (9.21) for Stk_{50}:

$$Stk_{50} = \frac{x_{50}^2 \rho_p v}{18\,\mu D}$$

$$1 \times 10^{-3} = \frac{(10 \times 10^{-6})^2 \times 2500 \times 2.282}{18 \times 18.4 \times 10^{-6} \times (2.362/\sqrt{n})}$$

giving $n = 1.88$.

We therefore require two cyclones. With two cyclones, using all of the allowable pressure drop the characteristic velocity will be the same (2.282 m/s) and the required cyclone diameter may be calculated from the expression derived above:

$$D = \frac{2.362}{\sqrt{n}}$$

giving $D = 1.67\,\text{m}$.

(d) The actual cut size achieved with two cyclones is calculated from Equation (9.21) with $D = 1.67\,\text{m}$ and $v = 2.282\,\text{m/s}$:

$$\text{actual cut size, } x_{50} = \sqrt{\frac{1 \times 10^{-3} \times 18 \times 18.4 \times 10^{-6} \times 1.67}{2500 \times 2.282}} = 9.85 \times 10^{-6}\,\text{m}$$

Summary. Two cyclones (described by $Eu = 384$ and $Stk_{50} = 1 \times 10^{-3}$) of diameter 1.67 m and operating at a pressure drop of 1200 Pa, will achieve a equiprobable cut size of 9.85 μm.

TEST YOURSELF

9.1 Typically in what particle size range are industrial cyclone separators useful?

9.2 With the aid of a sketch, describe the operation of a reverse flow cyclone separator.

9.3 What forces act on a particle inside a cyclone separator? What factors govern the magnitudes of each of these forces?

9.4 For a gas–particle separation device, define *total efficiency* and *grade efficiency*. Using these definitions and the mass balance, derive an expression relating the size distributions of the feed, coarse product and fine product for a gas–particle separation device.

9.5 What is meant by the x_{50} cut size?

9.6 Define the two dimensionless numbers which are used in the scale up of cyclone separators.

9.7 Theory suggests that the total efficiency of a cyclone separator will increase with increasing gas flow rate. Explain why, in practice, cyclone separators are operated within a certain range of pressure drops.

9.8 Under what conditions might we choose to operate cyclone separators in parallel?

EXERCISES

9.1 A gas–particle separation device is tested and gives the results shown in the table below:

Size range (μm)	0–10	10–20	20–30	30–40	40–50
Range mean (μm)	5	15	25	35	45
Feed mass (kg)	45	69	120	45	21
Coarse product mass (kg)	1.35	19.32	99.0	44.33	21.0

(a) Find the total efficiency of the device.

(b) Produce a plot of the grade efficiency for this device and determine the equiprobable cut size.

[Answer: (a) 61.7%; (b) 19.4 μm.]

9.2 A gas–particle separation device is tested and gives the results shown in the table below:

Size range (μm)	6.6–9.4	9.4–13.3	13.3–18.7	18.7–27.0	27.0–37.0	37.0–53.0
Feed size distribution	0.05	0.2	0.35	0.25	0.1	0.05
Coarse product size distribution	0.016	0.139	0.366	0.30	0.12	0.06

Given that the total mass of feed is 200 kg and the total mass of coarse product collected is 166.5 kg:

(a) Find the total efficiency of the device.

(b) Determine the size distribution of the fine product.

(c) Plot the grade efficiency curve for this device and determine the equiprobable size.

(d) If this same device were fed with a material with the size distribution below, what would be the resulting coarse product size distribution?

Size range (μm)	6.6–9.4	9.4–13.3	13.3–18.7	18.7–27.0	27.0–37.0	37.0–53.0
Feed size distribution	0.08	0.13	0.27	0.36	0.14	0.02

[Answer: (a) 83.25%; (b) 0.219, 0.503, 0.271, 0.0015, 0.0006, 0.0003; (c) 10.5 μm; (d) 0.025, 0.089, 0.276, 0.422, 0.165, 0.024].

9.3

(a) Explain what a 'grade efficiency curve' is with reference to a gas–solids separation device and sketch an example of such a curve for a gas cyclone separator.

(b) Determine the diameter and number of Stairmand HR gas cyclones to be operated in parallel to treat 3 m^3/s of gas of density 0.5 kg/m^3 and viscosity 2×10^{-5} Pa s carrying a dust of density 2000 kg/m^3. A x_{50} cut size of at most 7 μm is to be achieved at a pressure drop of 1200 Pa.

(For a Stairmand HR cyclone: $Eu = 46$ and $Stk_{50} = 6 \times 10^{-3}$.)

(c) Give the actual cut size achieved by your design.

(d) A change in process conditions requirements necessitates a 50% drop in gas flow rate. What effect will this have on the cut size achieved by your design?

[Answer: (a) Two cyclones 0.43 m in diameter; (b) $x_{50} = 6.8$ μm; (c) new $x_{50} = 9.6$ μm.]

9.4

(a) Determine the diameter and number of Stairmand HE gas cyclones to be operated in parallel to treat 1 m^3/s of gas of density 1.2 kg/m^3 and viscosity 18.5×10^{-6} Pa s carrying a dust of density 1000 kg/m^3. An x_{50} cut size of at most 5 μm is to be achieved at a pressure drop of 1200 Pa.

(For a Stairmand HE cyclone: $Eu = 320$ and $Stk_{50} = 1.4 \times 10^{-4}$.)

(b) Give the actual cut size achieved by your design.

[Answer: (a) One cyclone, 0.714 m in diameter; (b) $x_{50} = 3.6$ μm.]

9.5 Stairmand HR cyclones are to be used to clean up 2.5 m^3/s of ambient air (density, 1.2 kg/m^3; viscosity, 18.5×10^{-6} Pa s) laden with dust of particle density 2600 kg/m^3. The available pressure drop is 1200 Pa and the required cut size is to be not more than 6 μm.

(a) What size of cyclones are required?

(b) How many cyclones are needed and in what arrangement?

(c) What is the actual cut size achieved?

[Answer: (a) Diameter = 0.311 m; (b) five cyclones in parallel; (c) actual cut size = 6 μm.]

10
Storage and Flow of Powders – Hopper Design

10.1 INTRODUCTION

The short-term storage of raw materials, intermediates and products in the form of particulate solids in process plants presents problems which are often underestimated and which, as was pointed out in the introduction of this text, may frequently be responsible for production stoppages.

One common problem in such plants is the interruption of flow from the discharge orifice in the hopper, or converging section beneath a storage vessel for powders. However, a technology is available which will allow us to design such storage vessels to ensure flow of the powders when desired. Within the bounds of a single chapter it is not possible to cover all aspects of the gravity flow of unaerated powders, and so here we will confine ourselves to a study of the design philosophy to ensure flow from conical hoppers when required. The approach used is that first proposed by Jenike (1964).

10.2 MASS FLOW AND CORE FLOW

Mass flow. In perfect mass flow, all the powder in a silo is in motion whenever any of it is drawn from the outlet as shown in Figure 10.1(b). The flowing channel coincides with the walls of the silo. Mass flow hoppers are smooth and steep. Figure 10.2(a–d) shows sketches taken from a sequence of photographs of a hopper operating in mass flow. The use of alternate layers of coloured powder in this sequence clearly shows the key features of the flow pattern. Note how the powder surface remains level until it reaches the sloping section.

Introduction to Particle Technology - 2nd Edition Martin Rhodes

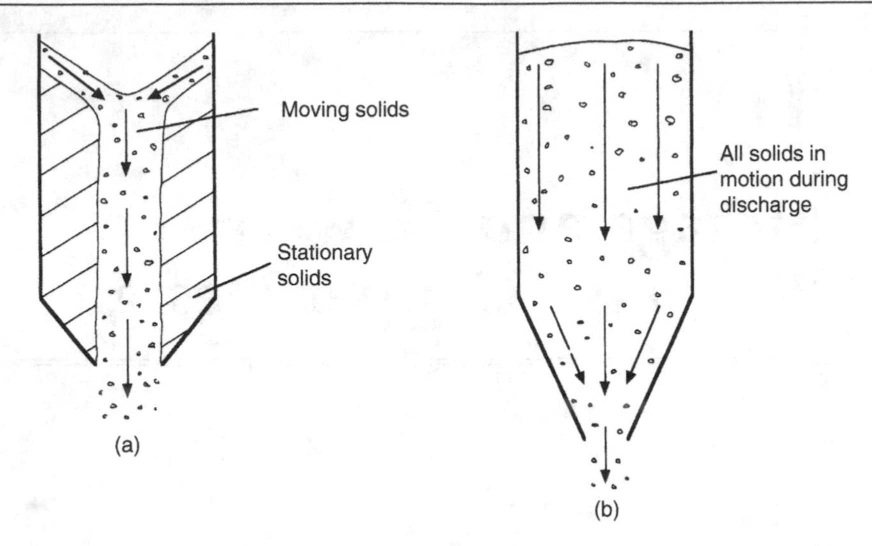

Figure 10.1 Mass flow and core flow in hoppers: (a) core flow; (b) mass flow

Core flow. This occurs when the powder flows towards the outlet of a silo in a channel formed within the powder itself [Figure 10.1(a)]. We will not concern ourselves with core flow silo design. Figure 10.3 (a–d) shows sketches taken from a sequence of photographs of a hopper operating in core flow. Note the regions of

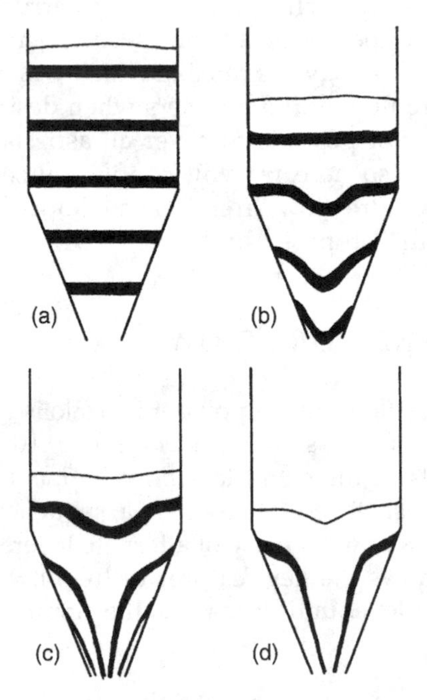

Figure 10.2 Sequence of sketches taken from photographs showing a mass flow pattern as a hopper empties. (The black bands are layers of coloured tracer particles)

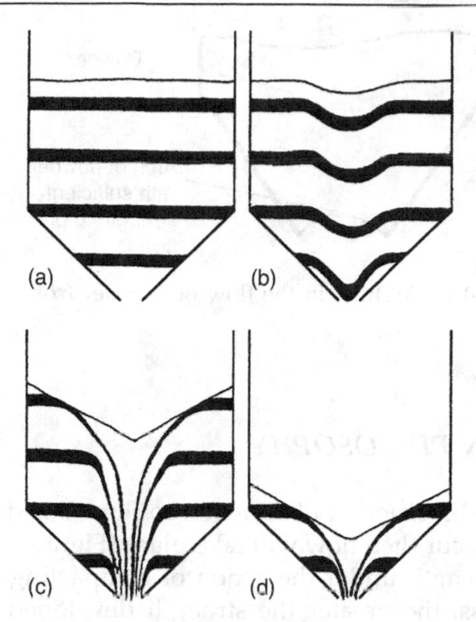

Figure 10.3 Sequence of sketches taken from photographs showing a core flow pattern as a hopper empties. (The black bands are layers of coloured tracer particles)

powder lower down in the hopper are stagnant until the hopper is almost empty. The inclined surface of the powder gives rise to size segregation (see Chapter 11).

Mass flow has many advantages over core flow. In mass flow, the motion of the powder is uniform and steady state can be closely approximated. The bulk density of the discharged powder is constant and practically independent of the height in the silo. In mass flow stresses are generally low throughout the mass of solids, giving low compaction of the powder. There are no stagnant regions in the mass flow hopper. Thus the risk of product degradation is small compared with the case of the core flow hopper. The first-in–first-out flow pattern of the mass flow hopper ensures a narrow range of residence times for solids in the silo. Also, segregation of particles according to size is far less of a problem in mass flow than in core flow. Mass flow has one disadvantage which may be overriding in certain cases. Friction between the moving solids and the silo and hopper walls result in erosion of the wall, which gives rise to contamination of the solids by the material of the hopper wall. If either contamination of the solids or serious erosion of the wall material are unacceptable, then a core flow hopper should be considered.

For conical hoppers the slope angle required to ensure mass flow depends on the powder/powder friction and the powder/wall friction. Later we will see how these are quantified and how it is possible to determine the conditions which give rise to mass flow. Note that there is no such thing as a mass flow hopper; a hopper which gives mass flow with one powder may give core flow with another.

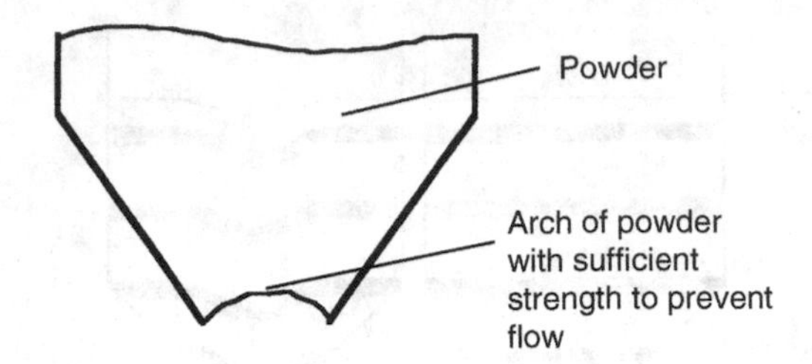

Figure 10.4 Arching in the flow of powder from a hopper

10.3 THE DESIGN PHILOSOPHY

We will consider the blockage or obstruction to flow called arching and assume that if this does not occur then flow will take place (Figure 10.4). Now, in general, powders develop strength under the action of compacting stresses. The greater the compacting stress, the greater the strength developed (Figure 10.5). (Free-flowing solids such as dry coarse sand do not develop strength as the result of compacting stresses and will always flow.)

10.3.1 Flow–No Flow Criterion

Gravity flow of a solid in a channel will take place provided the strength developed by the solids under the action of consolidating pressures is insufficient to support an obstruction to flow. An arch occurs when the strength developed by the solids is greater than the stresses acting within the surface of the arch.

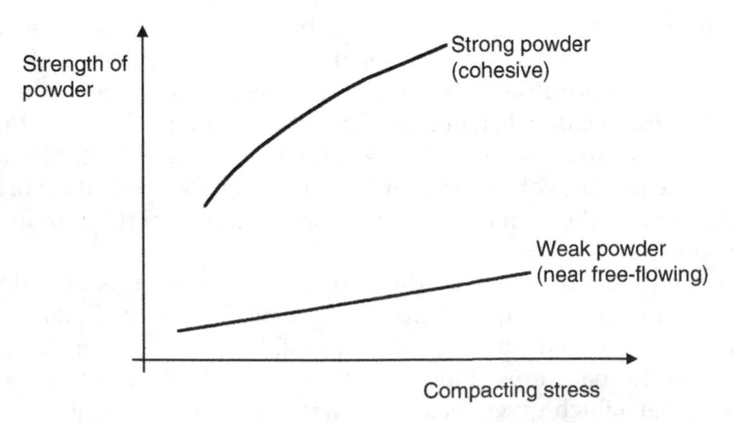

Figure 10.5 Variation of strength of powder with compacting stress for cohesive and free-flowing powders

10.3.2 The Hopper Flow Factor, *ff*

The hopper flow factor, *ff*, relates the stress developed in a particulate solid with the compacting stress acting in a particular hopper. The hopper flow factor is defined as:

$$ff = \frac{\sigma_C}{\sigma_D} = \frac{\text{compacting stress in the hopper}}{\text{stress developed in the powder}} \tag{10.1}$$

A high value of *ff* means low flowability since high σ_C means greater compaction, and a low value of σ_D means more chance of an arch forming.

The hopper flow factor depends on:

- the nature of the solid;
- the nature of the wall material;
- the slope of the hopper wall.

These relationships will be quantified later.

10.3.3 Unconfined Yield Stress, σ_y

We are interested in the strength developed by the powder in the arch surface. Suppose that the yield stress (i.e. the stress which causes flow) of the powder in the exposed surface of the arch is σ_y. The stress σ_y is known as the unconfined yield stress of the powder. Then if the stresses developed in the powder forming the arch are greater than the unconfined yield stress of the powder in the arch, flow will occur. That is, for flow:

$$\sigma_D > \sigma_y \tag{10.2}$$

Incorporating Equation (10.1), this criterion may be rewritten as:

$$\frac{\sigma_C}{ff} > \sigma_y \tag{10.3}$$

10.3.4 Powder Flow Function

Obviously, the unconfined yield stress, σ_y, of the solids varies with compacting stress, σ_C:

$$\sigma_y = \text{fn}(\sigma_C)$$

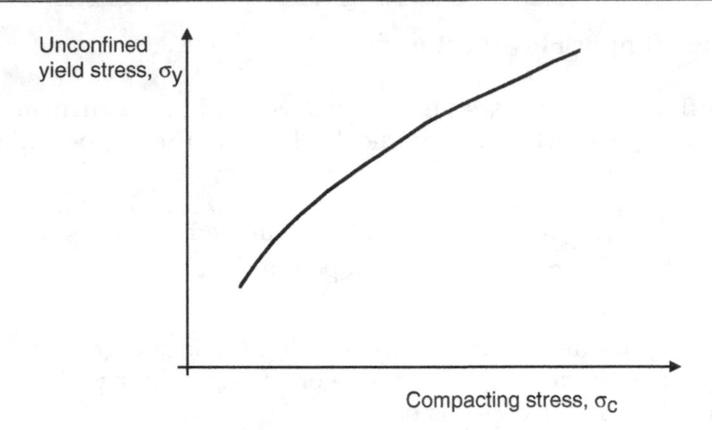

Figure 10.6 Powder flow function (a property of the solids only)

This relationship is determined experimentally and is usually presented graphically (Figure 10.6). This relationship has several different names, some of which are misleading. Here we will call it the *powder flow function*. Note that it is a function *only* of the powder properties.

10.3.5 Critical Conditions for Flow

From Equation (10.3), the limiting condition for flow is:

$$\frac{\sigma_C}{ff} = \sigma_y$$

This may be plotted on the same axes as the powder flow function (unconfined yield stress, σ_y and compacting stress, σ_C) in order to reveal the conditions under which flow will occur for this powder in the hopper. The limiting condition gives a straight line of slope $1/ff$. Figure 10.7 shows such a plot.

Where the powder has a yield stress greater than σ_C/ff, no flow occurs [powder flow function (a)]. Where the powder has a yield stress less than σ_C/ff flow occurs [powder flow function (c)]. For powder flow function (b) there is a critical condition, where unconfined yield stress, σ_y, is equal to stress developed in the powder, σ_C/ff. This gives rise to a critical value of stress, σ_{crit}, which is the critical stress developed in the surface of the arch:

$$\text{If actual stress developed} < \sigma_{crit} \Rightarrow \text{no flow}$$
$$\text{If actual stress developed} > \sigma_{crit} \Rightarrow \text{flow}$$

10.3.6 Critical Outlet Dimension

Intuitively, for a given hopper geometry, one would expect the stress developed in the arch to increase with the span of the arch and the weight of solids in the arch. In practice this is the case and the stress developed in the arch is related to

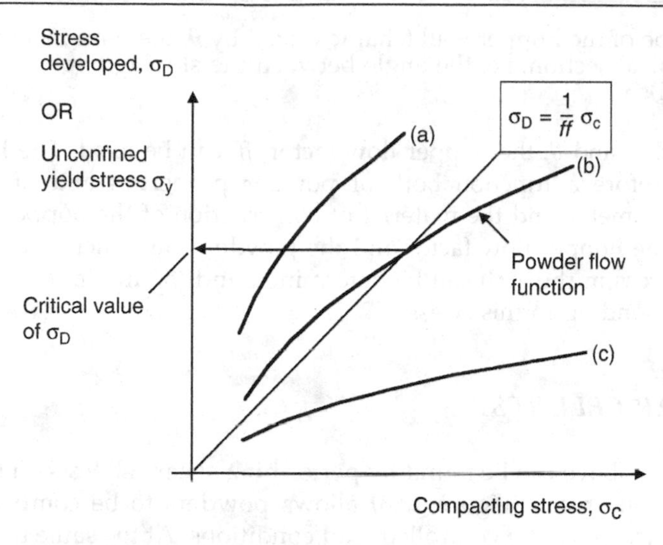

Figure 10.7 Determination of critical conditions for flow

the size of the hopper outlet, B, and the bulk density, ρ_B, of the material by the relationship:

$$\text{minimum outlet dimension}, B = \frac{H(\theta)\sigma_{crit}}{\rho_B g} \tag{10.4}$$

where $H(\theta)$ is a factor determined by the slope of the hopper wall and g is the acceleration due to gravity. An approximate expression for $H(\theta)$ for conical hoppers is

$$H(\theta) = 2.0 + \frac{\theta}{60} \tag{10.5}$$

10.3.7 Summary

From the above discussion of the design philosophy for ensuring mass flow from a conical hopper, we see that the following are required:

(1) the relationship between the strength of the powder in the arch, σ_y (unconfined yield stress) with the compacting stress acting on the powder, σ_C;

(2) the variation of hopper flow factor, *ff*, with:

(a) the nature of the powder (characterized by the effective angle of internal friction, δ);

(b) the nature of the hopper wall (characterized by the angle of wall friction, Φ_W);

(c) the slope of the hopper wall (characterized by θ, the semi-included angle of the conical section, i.e. the angle between the sloping hopper wall and the vertical).

Knowing δ, Φ_W, and θ, the hopper flow factor, *ff*, can be fixed. The hopper flow factor is therefore a function both of powder properties and of the hopper properties (geometry and the material of construction of the hopper walls).

Knowing the hopper flow factor and the powder flow function (σ_y versus σ_C) the critical stress in the arch can be determined and the minimum size of outlet found corresponding to this stress.

10.4 SHEAR CELL TEST

The data listed above can be found by performing shear cell tests on the powder.

The Jenike shear cell (Figure 10.8) allows powders to be compacted to any degree and sheared under controlled load conditions. At the same time the shear force (and hence stress) can be measured.

Generally powders change bulk density under shear. Under the action of shear, for a specific normal load:

- a loosely packed powder would contract (increase bulk density);
- a very tightly packed powder would expand (decrease bulk density);
- a critically packed powder would not change in volume.

For a particular bulk density there is a critical normal load which gives failure (yield) without volume change. A powder flowing in a hopper is in this critical condition. Yield without volume change is therefore of particular interest to us in design.

Using a standardized test procedure five or six samples of powder are prepared all having the same bulk density. Referring to the diagram of the Jenike shear cell shown in Figure 10.8 a normal load is applied to the lid of the cell and the horizontal force applied to the sample via the bracket and loading pin

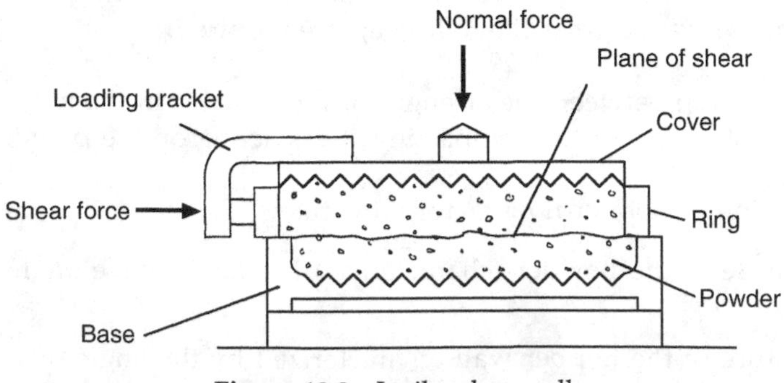

Figure 10.8 Jenike shear cell

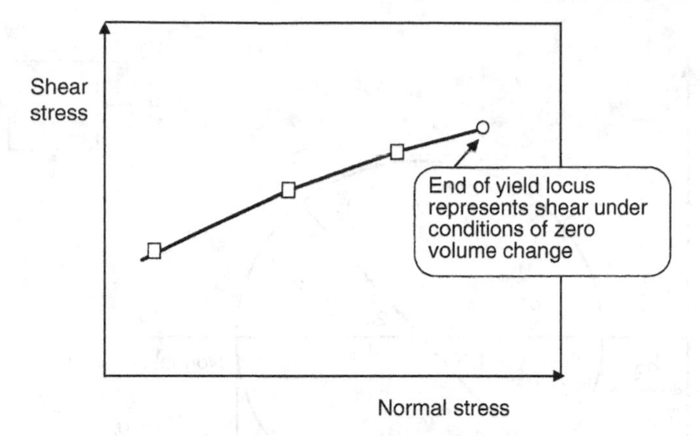

Figure 10.9 A single yield locus

is recorded. That horizontal force necessary to initiate shear of flow of the powder sample is noted. This procedure is repeated for each identical powder sample but with a greater normal load applied to the lid each time. This test thus generates a set of five or six pairs of values for normal load and shear force and hence pairs of values of compacting stress and shear stress for a powder of a particular bulk density. The pairs of values are plotted to give a yield locus (Figure 10.9). The end point of the yield locus corresponds to critical flow conditions where initiation of flow is not accompanied by a change in bulk density. Experience with the procedure permits the operator to select combinations of normal and shear force which achieve the critical conditions. This entire test procedure is repeated two or three times with samples prepared to different bulk densities. In this way a family of yield loci is generated (Figure 10.10).

These yield loci characterize the flow properties of the unaerated powder. The following section deals with the generation of the powder flow function from this family of yield loci.

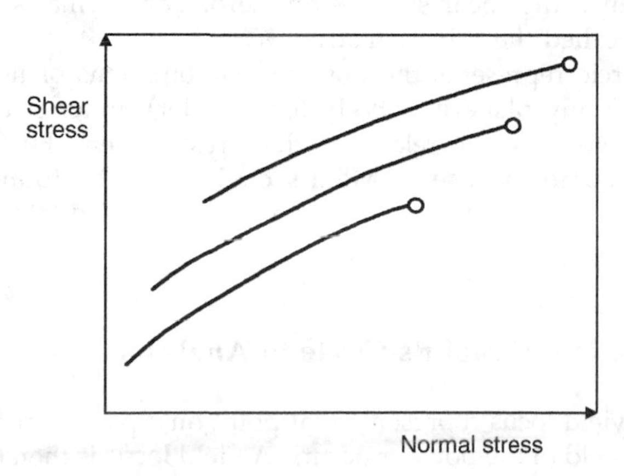

Figure 10.10 A family of yield loci

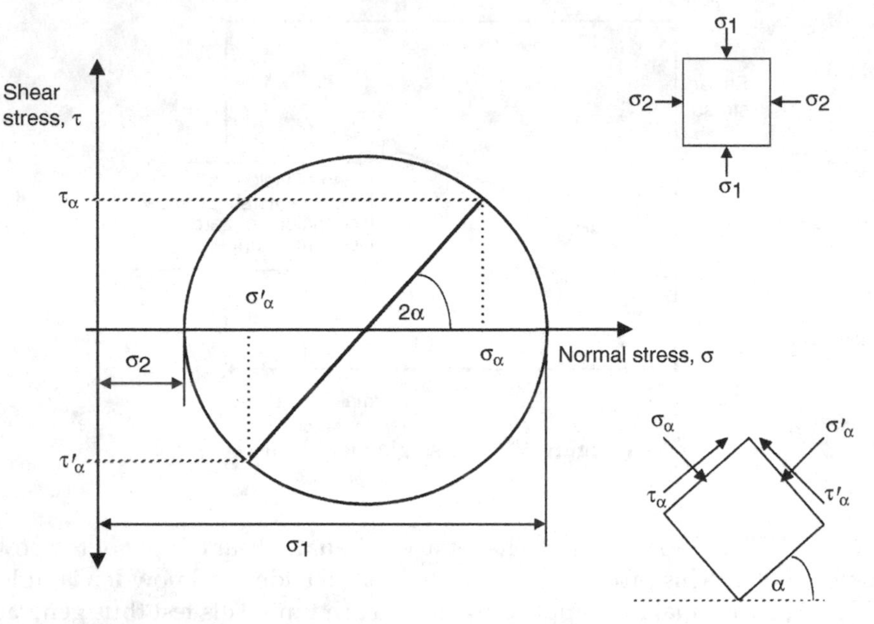

Figure 10.11 Mohr's circle construction

10.5 ANALYSIS OF SHEAR CELL TEST RESULTS

The mathematical stress analysis of the flow of unaerated powders in a hopper requires the use of principal stresses. We therefore need to use the Mohr's stress circle in order to determine principal stresses from the results of the shear tests.

10.5.1 Mohr's Circle – in Brief

Principal stresses – in any stress system there are two planes at right angles to each other in which the shear stresses are zero. The normal stresses acting on these planes are called the principal stresses.

The Mohr's circle represents the possible combinations of normal and shear stresses acting on any plane in a body (or powder) under stress. Figure 10.11 shows how the Mohr's circle relates to the stress system. Further information on the background to the use of Mohr's circles may be found in most texts dealing with the strength of materials and the analysis of stress and strain in solids.

10.5.2 Application of Mohr's Circle to Analysis of the Yield Locus

Each point on a yield locus represents that point on a particular Mohr's circle for which failure or yield of the powder occurs. A yield locus is then tangent to all the Mohr's circles representing stress systems under which the powder will fail (flow).

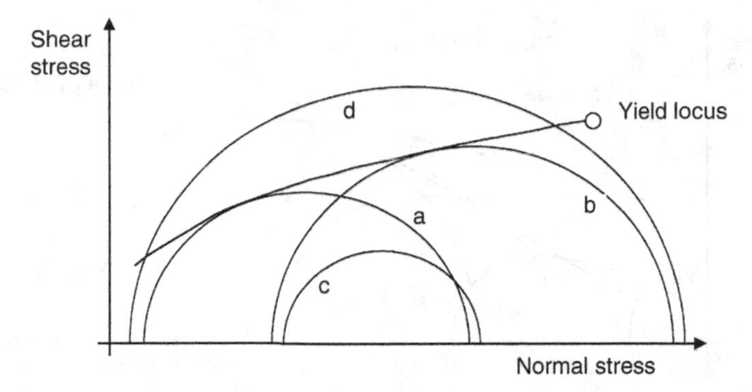

Figure 10.12 Identification of the applicable Mohr's circle

For example, in Figure 10.12 Mohr's circles (a) and (b) represent stress systems under which the powder would fail. In circle (c) the stresses are insufficient to cause flow. Circle (d) is not relevant since the system under consideration cannot support stress combinations above the yield locus. It is therefore Mohr's circles which are tangential to yield loci that are important to our analysis.

10.5.3 Determination of σ_y and σ_c

Two tangential Mohr's circles are of particular interest. Referring to Figure 10.13, the smaller Mohr's circle represents conditions at the free surface of the arch: this free surface is a plane in which there is zero shear and zero normal stress and so the Mohr's circle which represents flow (failure) under these conditions must pass through the origin of the shear stress versus normal stress plot. This Mohr's circle gives the (major principal) unconfined yield stress, and this is the value we use for σ_y. The larger Mohr's circle is tangent to the yield locus at its end point and therefore represents conditions for critical failure. The major principal stress from this Mohr's circle is taken as our value of compacting stress, σ_C.

Pairs of values of σ_y and σ_C are found from each yield locus and plotted against each other to give the powder flow function (Figure 10.6).

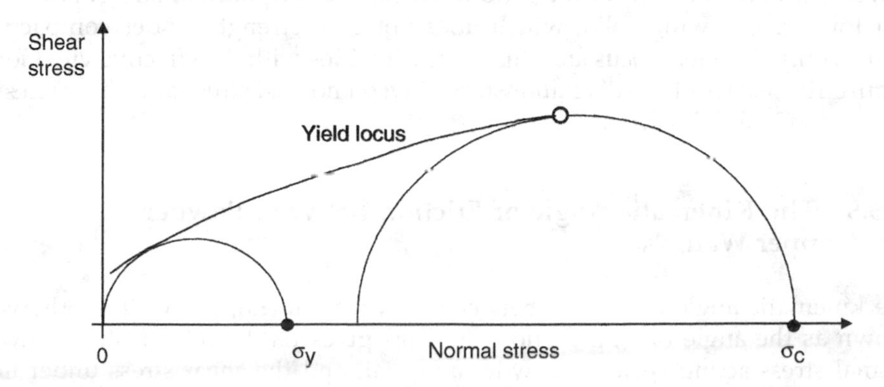

Figure 10.13 Determination of unconfined yield stress, σ_y and compacting stress, σ_C

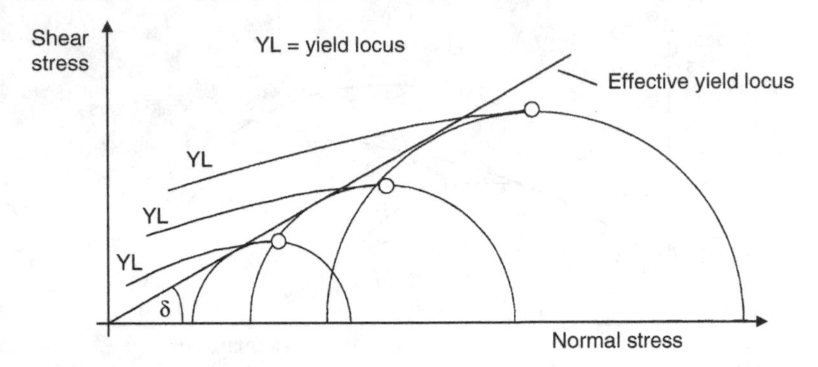

Figure 10.14 Definition of effective yield locus and effective angle of internal friction, δ

10.5.4 Determination of δ from Shear Cell Tests

Experiments carried out on hundreds of bulk solids (Jenike, 1964) have demonstrated that for an element of powder flowing in a hopper:

$$\frac{\sigma_1}{\sigma_2} = \frac{\text{major principal stress on the element}}{\text{minor principal stress on the element}} = \text{a constant}$$

This property of bulk solids is expressed by the relationship:

$$\frac{\sigma_1}{\sigma_2} = \frac{1 + \sin\delta}{1 - \sin\delta} \tag{10.6}$$

where δ is the effective angle of internal friction of the solid. In terms of the Mohr's stress circle this means that Mohr's circles for the critical failure are all tangent to a straight line through the origin, the slope of the line being tan δ (Figure 10.14).

This straight line is called the effective yield locus of the powder. By drawing in this line, δ can be determined. Note that δ is not a real physical angle within the powder; it is the tangent of the ratio of shear stress to normal stress. Note also that for a free-flowing solid, which does not gain strength under compaction, there is only one yield locus and this locus coincides with the effective yield locus (Figure 10.15). (This type of relationship between normal stress and shear stress is known as Coulomb friction.)

10.5.5 The Kinematic Angle of Friction between Powder and Hopper Wall, Φ_W

The kinematic angle of friction between powder and hopper wall is otherwise known as the angle of wall friction, Φ_W This gives us the relationship between normal stress acting between powder and wall and the shear stress under flow conditions. To determine Φ_W it is necessary to first construct the wall yield locus

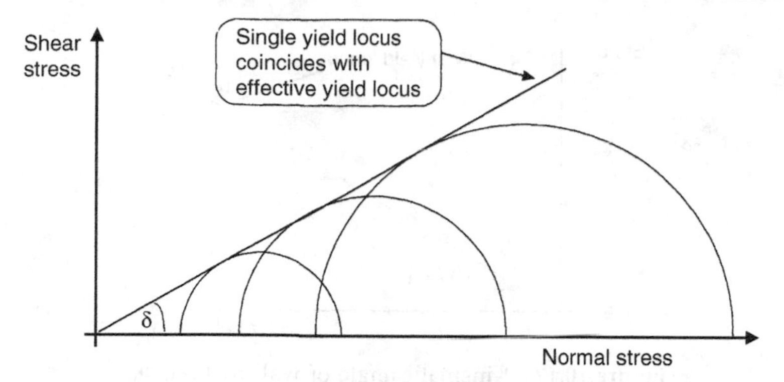

Figure 10.15 Yield locus for a free-flowing powder

from shear cell tests. The wall yield locus is determined by shearing the powder against a sample of the wall material under various normal loads. The apparatus used is shown in Figure 10.16, and a typical wall yield locus is shown in Figure 10.17.

The kinematic angle of wall friction is given by the gradient of the wall yield locus (Figure 10.17), i.e.

$$\tan \Phi_W = \frac{\text{shear stress at the wall}}{\text{normal stress at the wall}}$$

10.5.6 Determination of the Hopper Flow Factor, *ff*

The hopper flow factor, *ff*, is a function of δ, Φ_W, and θ and can be calculated from first principles. However, Jenike (1964) obtained values for a conical hopper and for a wedge-shaped hopper with a slot outlet for values of δ of 30°, 40°, 50°, 60° and 70°. Examples of the 'flow factor charts' for conical hoppers are shown in

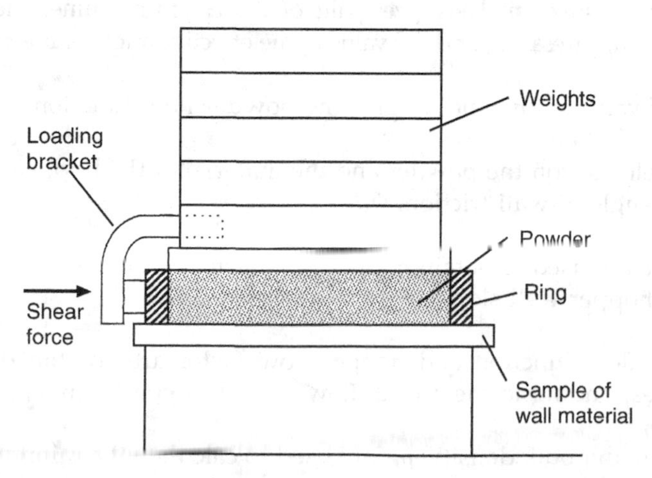

Figure 10.16 Apparatus for the measurement of kinematic angle of wall friction, Φ_W

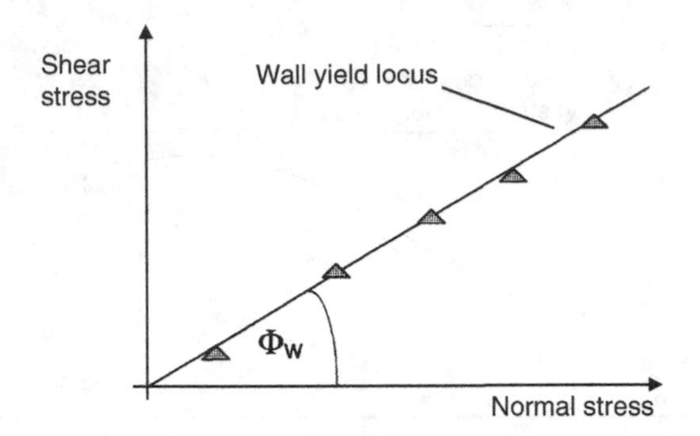

Figure 10.17 Kinematic angle of wall friction, Φ_w

Figure 10.18. It will be noticed that values of flow factor exist only in a triangular region; this defines the conditions under which mass flow is possible.

The following is an example of the use of these flow factor charts. Suppose that shear cell tests have given us δ and Φ_w equal to 30° and 19°, respectively, then entering the chart for conical hoppers with effective angle of friction $\delta = 30°$, we find that the limiting value of wall slope, θ, to ensure mass flow is 30.5° (point X in Figure 10.19). In practice it is usual to allow a safety margin of 3°, and so, in this case the semi-included angle of the conical hopper θ would be chosen as 27.5°, giving a hopper flow factor, $ff = 1.8$ (point Y, Figure 10.19).

10.6 SUMMARY OF DESIGN PROCEDURE

The following is a summary of the procedure for the design of conical hoppers for mass flows:

(i) Shear cell tests on powder give a family of yield loci.

(ii) Mohr's circle stress analysis gives pairs of values of unconfined yield stress, σ_y, and compacting stress, σ_C, and the value of the effective angle of internal friction, δ.

(iii) Pairs of values of σ_y and σ_C give the powder flow function.

(iv) Shear cell tests on the powder and the material of the hopper wall give the kinematic angle of wall friction, Φ_w.

(v) Φ_w and δ are used to obtain hopper flow factor, ff, and semi-included angle of conical hopper wall slope, θ.

(vi) Powder flow function and hopper flow factor are combined to give the stress corresponding to the critical flow – no flow condition, σ_{crit}.

(vii) σ_{crit}, $H(\theta)$ and bulk density, ρ_B, are used to calculate the minimum diameter of the conical hopper outlet B.

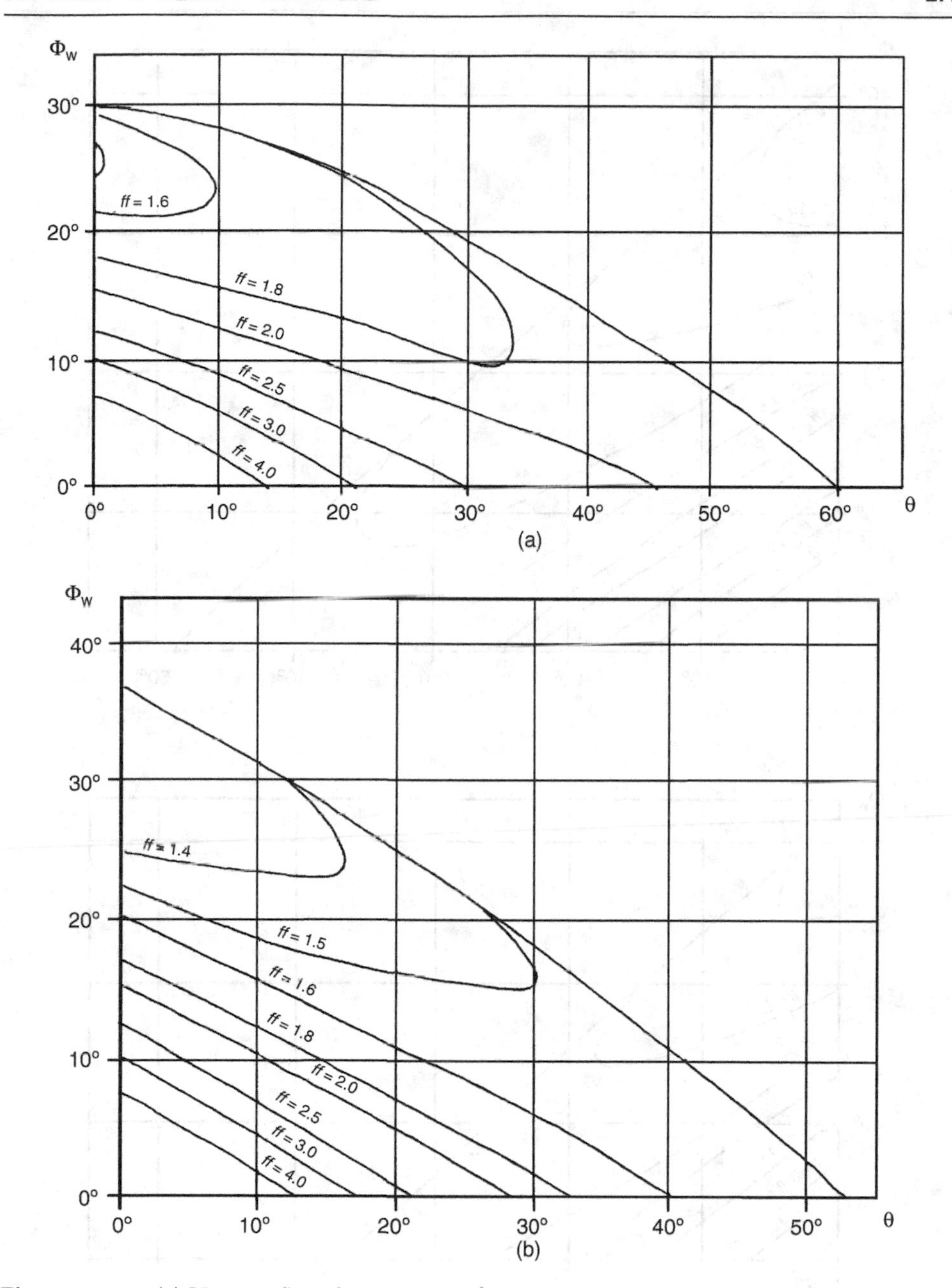

Figure 10.18 (a) Hopper flow factor values for conical channels, $\delta = 30°$. (b) Hopper flow factor values for conical channels, $\delta = 40°$. (c) Hopper flow factor values for conical channels, $\delta = 50°$. (d) Hopper flow factor values for conical channels, $\delta = 60°$

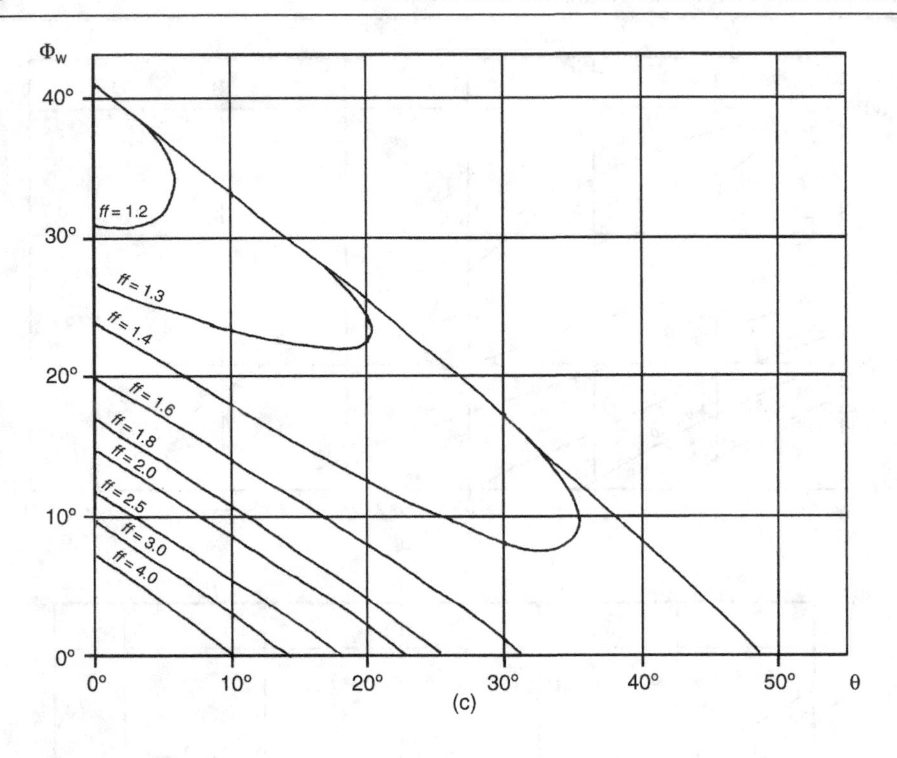

(c)

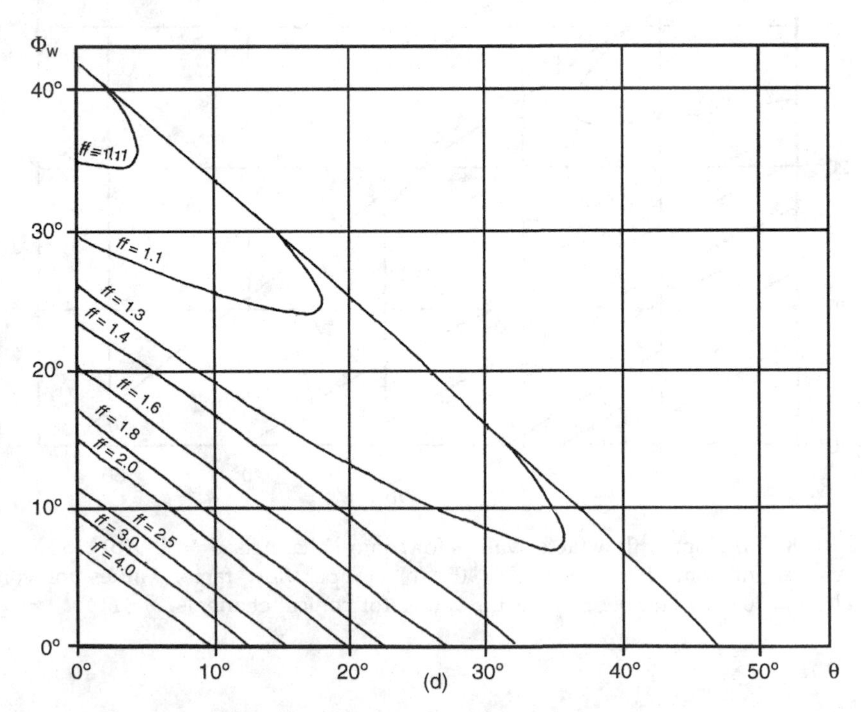

(d)

Figure 10.18 (*Continued*)

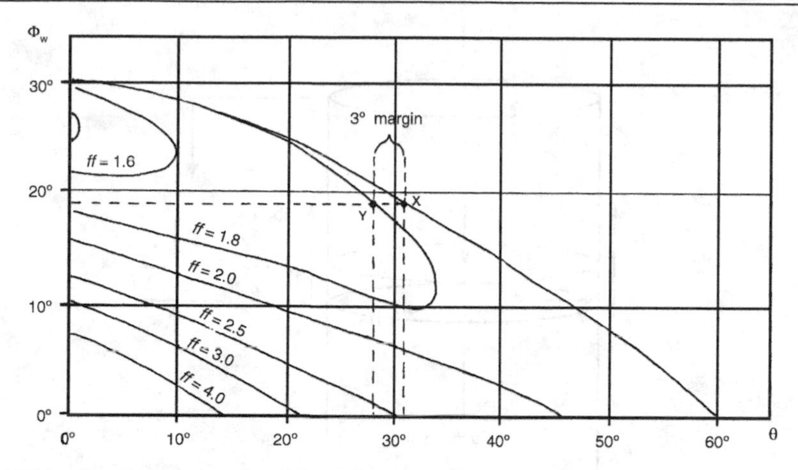

Figure 10.19 Worked example of the use of hopper flow factor charts. Hopper flow factor values for conical channels, $\delta = 30°$

10.7 DISCHARGE AIDS

A range of devices designed to facilitate flow of powders from silos and hoppers are commercially available. These are known as discharge aids or silo activators. These should not, however, be employed as an alternative to good hopper design.

Discharge aids may be used where proper design recommends an unacceptably large hopper outlet incompatible with the device immediately downstream. In this case the hopper should be designed to deliver uninterrupted mass flow to the inlet of the discharge aid, i.e. the slope of the hopper wall and inlet dimensions of the discharge aid are those calculated according to the procedure outlined in this chapter.

10.8 PRESSURE ON THE BASE OF A TALL CYLINDRICAL BIN

It is interesting to examine the variation of stress exerted on the base of a bin with increasing depth of powder. For simplicity we will assume that the powder is non-cohesive (i.e. does not gain strength on compaction). Referring to Figure 10.20, consider a slice of thickness ΔH at a depth H below the surface of the powder. The downward force is

$$\frac{\pi D^2}{4}\sigma_V \tag{10.7}$$

where D is the bin diameter and σ_V is the stress acting on the top surface of the slice. Assuming stress increases with depth, the reaction of the powder below the slice acts upwards and is

$$\frac{\pi D^2}{4}(\sigma_V + \Delta\sigma_V) \tag{10.8}$$

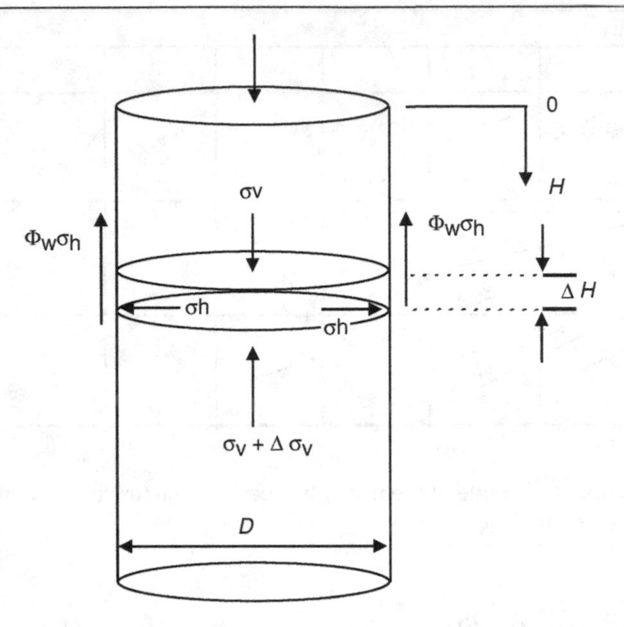

Figure 10.20 Forces acting on a horizontal slice of powder in a tall cylinder

The net upward force on the slice is then

$$\frac{\pi D^2}{4}\Delta\sigma_v \tag{10.9}$$

If the stress exerted on the wall by the powder in the slice is σ_h and the wall friction is $\tan\Phi_w$, then the friction force (upwards) on the slice is

$$\pi D\Delta H \tan\Phi_w\sigma_h \tag{10.10}$$

The gravitational force on the slice is

$$\frac{\pi D^2}{4}\rho_B g\Delta H \text{ acting downwards} \tag{10.11}$$

where ρ_B is the bulk density of the powder, assumed to be constant throughout the powder (independent of depth).

If the slice is in equilibrium the upward and downward forces are equated, giving

$$D\Delta\sigma_v + 4\tan\Phi_w\sigma_h\Delta H = D\rho_B g\Delta H \tag{10.12}$$

If we assume that the horizontal stress is proportional to the vertical stress and that the relationship does not vary with depth,

$$\sigma_h = k\sigma_v \tag{10.13}$$

and so as ΔH tends to zero,

$$\frac{d\sigma_v}{dH} + \left(\frac{4 \tan \Phi_w k}{D}\right)\sigma_v = \rho_B g \tag{10.14}$$

Noting that this is the same as

$$\frac{d\sigma_v}{dH}(e^{(4 \tan \Phi_w k/D)H}\sigma_v) = \rho_B g e^{(4 \tan \Phi_w k/D)H} \tag{10.15}$$

and integrating, we have

$$\sigma_v e^{(4 \tan \Phi_w k/D)H} = \frac{D\rho_B g}{4 \tan \Phi_w k} e^{(4 \tan \Phi_w k/D)H} + \text{constant} \tag{10.16}$$

If, in general the stress acting on the surface of the powder is σ_{v0}(at $H = 0$) the result is

$$\sigma_v = \frac{D\rho_B g}{4 \tan \Phi_w k}[1 - e^{-(4 \tan \Phi_w k/D)H}] + \sigma_{v0} e^{-(4 \tan \Phi_w k/D)H} \tag{10.17}$$

This result was first demonstrated by Janssen (1895).

If there is no force acting on the free surface of the powder, $\sigma_{v0} = 0$ and so

$$\sigma_v = \frac{D\rho_B g}{4 \tan \Phi_w k}(1 - e^{-(4 \tan \Phi_w k/D)H}) \tag{10.18}$$

When H is very small

$$\sigma_v \cong \rho_B H g$$
$$(\text{since for very small } z, e^{-z} \cong 1 - z) \tag{10.19}$$

equivalent to the static pressure at a depth H in fluid of density ρ_B.

When H is large, inspection of Equation (10.18) gives

$$\sigma_v \cong \frac{D\rho_B g}{4 \tan \Phi_w k} \tag{10.20}$$

and so the vertical stress developed becomes independent of depth of powder above. The variation in stress with depth of powder for the case of no force acting on the free surface of the powder ($\sigma_{v0} = 0$) is shown in Figure 10.21. Thus, contrary to intuition (which is usually based on our experience with fluids), the force exerted by a bed of powder becomes independent of depth if the bed is deep

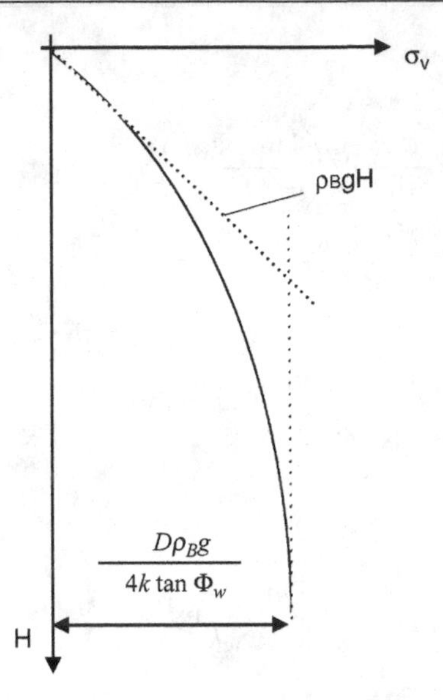

Figure 10.21 Variation in vertical pressure with depth of powder (for $\sigma_{v0} = 0$)

enough. Hence most of the weight of the powder is supported by the walls of the bin. In practice, the stress becomes independent of depth (and also independent of any load applied to the powder surface) beyond a depth of about 4*D*.

10.9 MASS FLOW RATES

The rate of discharge of powder from an orifice at the base of a bin is found to be independent of the depth of powder unless the bin is nearly empty. This means that the observation for a static powder that the pressure exerted by the powder is independent of depth for large depths is also true for a dynamic system. It confirms that fluid flow theory cannot be applied to the flow of a powder. For flow through an orifice in the flat-based cylinder, experiment shows that:

$$\text{mass flow rate}, M_p \propto (B - a)^{2.5} \text{for a circular orifice of diameter } B$$

where a is a correction factor dependent on particle size. [For example, for solids discharge from conical apertures in flat-based cylinders, Beverloo *et al.* (1961) give $M_p = 0.58\rho_B g^{0.5}(B - kx)^{2.5}$.]

For cohesionless coarse particles free falling over a distance h their velocity, neglecting drag and interaction, will be $u = \sqrt{2gh}$.

If these particles are flowing at a bulk density ρ_B through a circular orifice of diameter B, then the theortical mass flow rate will be:

$$M_p = \frac{\pi}{4}\sqrt{2}\rho_B g^{0.5} h^{0.5} B^2$$

The practical observation that flow rate is proportional to $B^{2.5}$ suggests that, in practice, particles only approach the free fall model when h is the same order as the orifice diameter.

10.10 CONCLUSIONS

Within the confines of a single chapter it has been possible only to outline the principles involved in the analysis of the flow of unaerated powders. This has been done by reference to the specific example of the design of conical hoppers for mass flow. Other important considerations in the design of hoppers such as time consolidation effects and determination of the stress acting in the hopper and bin wall have been omitted. These aspects together with the details of shear cell testing procedure are covered in texts specific to the subject. Readers wishing to pursue the analysis of failure (flow) in particulate solids in greater detail may refer to texts on soil mechanics.

10.11 WORKED EXAMPLES

WORKED EXAMPLE 10.1

The results of shear cell tests on a powder are shown in Figure 10W1.1. In addition, it is known that the angle of friction on stainless steel is 19° for this powder, and under flow conditions the bulk density of the powder is 1300 kg/m^3. A conical stainless steel hopper is to be designed to hold this powder.

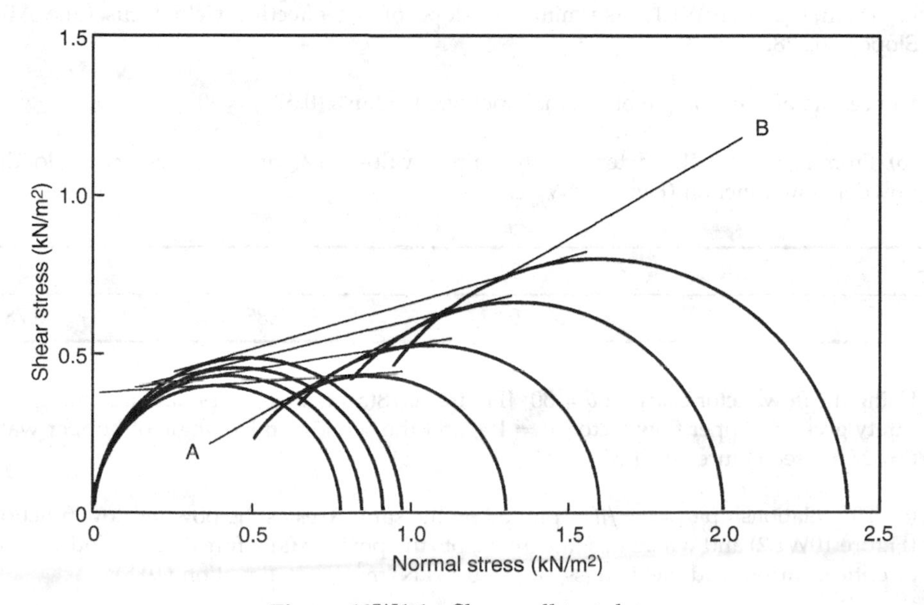

Figure 10W1.1 Shear cell test data

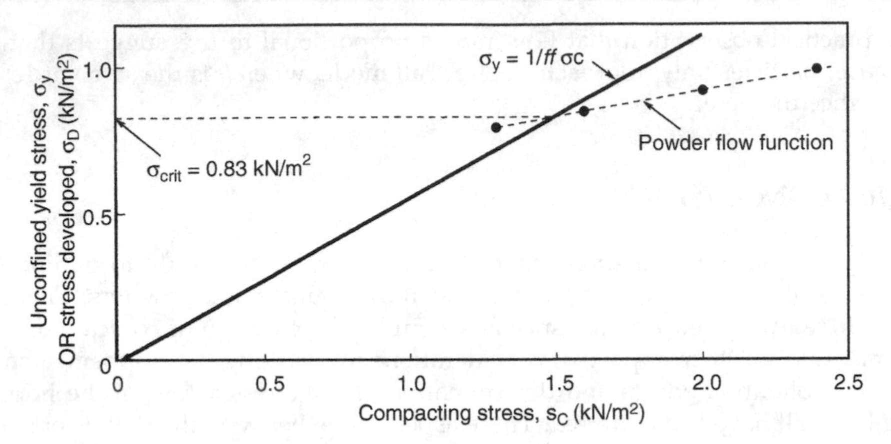

Figure 10W1.2 Determination of critical stress

Determine:

(a) the effective angle of internal friction;

(b) the maximum semi-included angle of the conical hopper which will confidently give mass flow;

(c) the minimum diameter of the circular hopper outlet necessary to ensure flow when the outlet slide valve is opened.

Solution

(a) From Figure 10W1.1, determine the slope of the effective yield locus (line AB). Slope = 0.578.

Hence, the effective angle of internal friction, $\delta = \tan^{-1}(0.578) = 30°$

(b) From Figure 10W1.1, determine the pairs of values of σ_C and σ_y necessary to plot the powder flow function (Figure 10W1.2).

σ_C	2.4	2.0	1.6	1.3
σ_y	0.97	0.91	0.85	0.78

Using the flow factor chart for $\delta = 30°$ [Figure 10.18(a)] with $\Phi_v = 19°$ and a 3° margin of safety gives a hopper flow factor, $ff = 1.8$, and the semi-included angle of hopper wall, $\theta = 27.5°$ (see Figure 10W1.3).

(c) The relationship $\sigma_y = \sigma_C/ff$ is plotted on the same axes as the powder flow function (Figure 10W1.2) and where this line intersepts the powder flow function we find a value of critical unconfined yield stress, $\sigma_{crit} = 0.83\,\text{kN/m}^2$. From Equation (10.5),

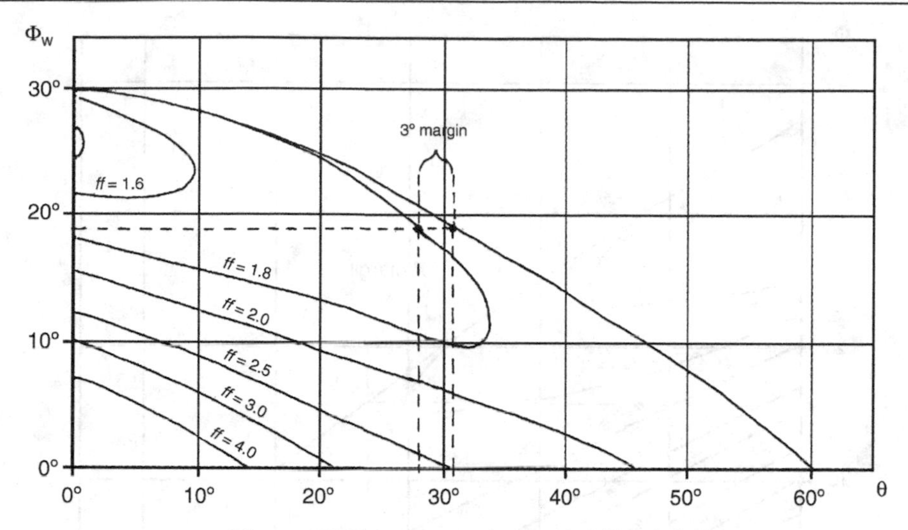

Figure 10W1.3 Determination of θ and *ff*

$$H(\theta) = 2.46 \text{ when } \theta = 27.5^\circ$$

and from Equation (10.4), the minimum outlet diameter for mass flow, B, is

$$B = \frac{2.46 \times 0.83 \times 10^3}{1300 \times 9.81} = 0.160\,\text{m}$$

Summarizing, then, to achieve mass flow without risk of blockage using the powder in question we require a stainless steel conical hopper with a maximum semi-included angle of cone, 27.5° and a circular outlet with a diameter of at least 16.0 cm.

WORKED EXAMPLE 10.2

Shear cell tests on a powder give the following information:
Effective angle of internal friction, $\delta = 40^\circ$
Kinematic angle of wall friction on mild steel, $\Phi_W = 16^\circ$
Bulk density under flow condition, $\rho_B = 2000\,\text{kg/m}^3$
The powder flow function which can be represented by the relationship, $\sigma_y = \sigma_C^{0.6}$, where σ_y is unconfined yield stress (kN/m^2) and σ_C is consolidating stress (kN/m^2)
Determine (a) the maximum semi-included angle of a conical mild steel hopper that will confidently ensure mass flow, and (b) the minimum diameter of circular outlet to ensure flow when the outlet is opened.

Solution

(a) With an effective angle of internal friction $\delta = 40^\circ$ we refer to the flow factor chart in Figure 10.18(b), from which at $\Phi_W = 16^\circ$ and with a safety margin of 3° we obtain the hopper flow factor, $ff = 1.5$ and hopper semi-included angle for mass flow, $\theta = 30^\circ$ (Figure 10W2.1).

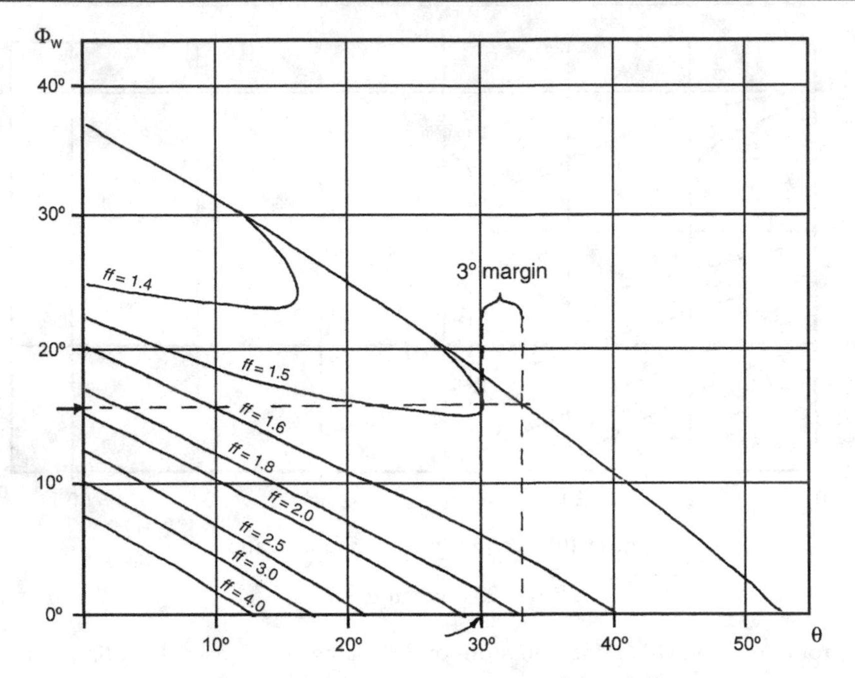

Figure 10W2.1 Determination of θ and ff

(b) For flow: $\frac{\sigma_C}{ff} > \sigma_y$ [Equation (10.3)]

but for the powder in question σ_y and σ_C are related by the material flow function: $\sigma_y = \sigma_C^{0.6}$.

Thus, the criterion for flow becomes

$$\left(\frac{\sigma_y^{1/0.6}}{ff}\right) > \sigma_y$$

and so the critical value of unconfined yield stress σ_{crit} is found when $\left(\frac{\sigma_y^{1/0.6}}{ff}\right) = \sigma_y$ hence, $\sigma_{crit} = 1.837\,\text{kN/m}^2$.

From Equation (10.5), $H(\theta) = 2.5$ when $\theta = 30°$ and hence, from Equation (10.4), minimum diameter of circular outlet,

$$B = \frac{2.5 \times 1.837 \times 10^3}{2000 \times 9.81} = 0.234\,\text{m}$$

Summarizing, mass flow without blockages is ensured by using a mild steel hopper with maximum semi-included cone angle 30° and a circular outlet diameter of at least 23.4 cm.

TEST YOURSELF

10.1 Explain with the aid of sketches what is meant by the terms *mass flow* and *core flow* with respect to solids flow in storage hoppers.

10.2 The starting point for the design philosophy presented in this chapter is the *flow-no flow criterion*. What is the flow-no flow criterion?

10.3 Which quantity describes the strength developed by a powder in an arch preventing flow from the base of a hopper? How is this quantity related to the hopper flow factor?

10.4 What is the *powder flow function?* Is the powder flow function dependent on (a) the powder properties, (b) the hopper geometry, (c) both the powder properties and the hopper geometry?

10.5 Show how the critical value of stress is determined from a knowledge of the *hopper flow factor* and the *powder flow function*.

10.6 What is meant by critical failure (yield) of a powder? What is its significance?

10.7 With the aid of a sketch plot of shear stress versus normal stress, show how the effective angle of internal of a powder is determined from a family of yield loci.

10.8 What is the kinematic angle of wall friction and how is it determined?

10.9 A powder is poured gradually into a measuring cylinder of diameter 3 cm. At the base of the cylinder is a load cell which measures the normal force exerted by the powder on the base. Produce a sketch plot showing how the normal force on the cylinder base would be expected to vary with powder depth, up to a depth of 18 cm.

10.10 How would you expect the mass flow rate of particulate solids from a hole in the base of a flat-bottomed container to vary with (a) the hole diameter and (b) the depth of solids?

EXERCISES

10.1 Shear cell tests on a powder show that its effective angle of internal friction is 40° and its powder flow function can be represented by the equation: $\sigma_y = \sigma_C^{0.45}$, where σ_y is the unconfined yield stress and σ_C is the compacting stress, both in kN/m^2. The bulk density of the powder is 1000 kg/m^3 and angle of friction on a mild steel plate is 16°. It is proposed to store the powder in a mild steel conical hopper of semi-included angle 30° and having a circular discharge opening of 0.30 m diameter. What is the critical outlet diameter to give mass flow? Will mass flow occur?

(Answer: 0.355 m; no flow.)

10.2 Describe how you would use shear cell tests to determine the effective angle of internal friction of a powder.

A powder has an effective angle of internal friction of 60° and has a powder flow function represented in the graph shown in Figure 10E2.1. If the bulk density of the powder is

$1500\,kg/m^3$ and its angle of friction on mild steel plate is 24.5°, determine, for a mild steel hopper, the maximum semi-included angle of cone required to safely ensure mass flow, and the minimum size of circular outlet to ensure flow when the outlet is opened.

(Answer: 17.5°; 18.92 cm.)

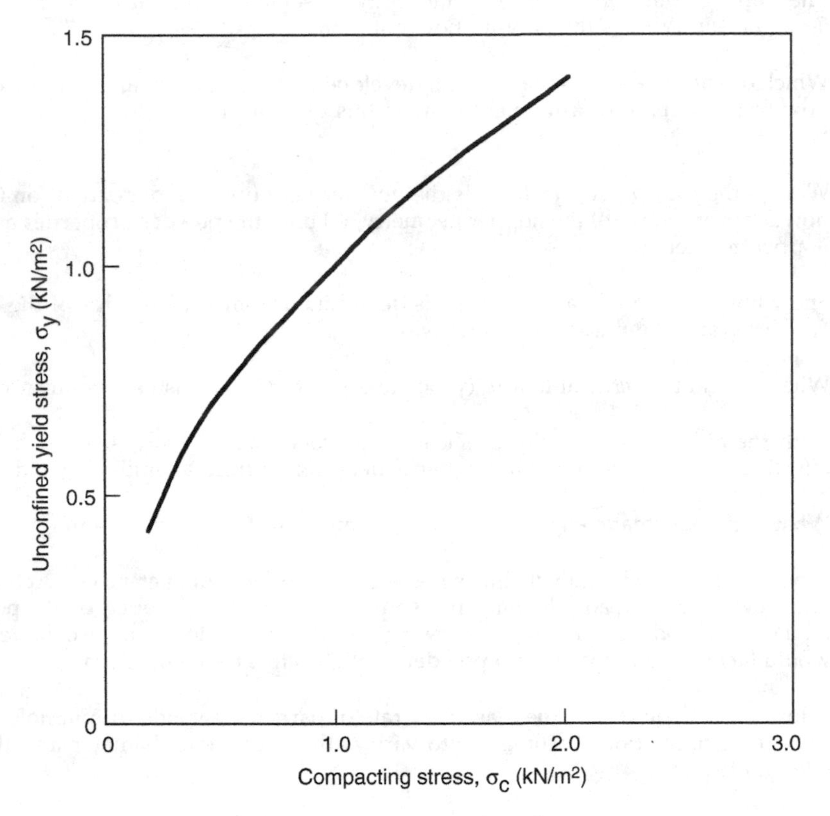

Figure 10E2.1 Powder flow function

10.3

(a) Summarize the philosophy used in the design of conical hoppers to ensure flow from the outlet when the outlet valve is opened.

(b) Explain how the powder flow function and the effective angle of internal friction are extracted from the results of shear cell tests on a powder.

(c) A firm having serious hopper problems takes on a chemical engineering graduate.

The hopper in question feeds a conveyor belt and periodically blocks at the outlet and needs to be 'encouraged' to restart. The graduate makes an investigation on the hopper, commissions shear cell tests on the powder and recommends a minor modification to the hopper. After the modification the hopper gives no further trouble and the graduate's reputation is established. Given the information below, what was the graduate's recommendation?

Existing design: Material of wall – mild steel

Semi-included angle of conical hopper – 33°

Outlet – circular, fitted with 25 cm diameter slide valve

Shear cell test data: Effective angle of internal friction, $\delta = 60°$

Angle of wall friction on mild steel, $\Phi_w = 8°$

Bulk density, $\rho_B = 1250\,kg/m^3$

Powder flow function: $\sigma_y = \sigma_C^{0.55}$ (σ_y and σ_C in kN/m^2)

10.4 Shear cell tests are carried out on a powder for which a stainless steel conical hopper is to be designed. The results of the tests are shown graphically in Figure 10E4.1. In addition it is found that the friction between the powder on stainless steel can be described by an angle of wall friction of 11°, and that the relevant bulk density of the powder is $900\,kg/m^3$.

(a) From the shear cell results of Figure 10E4.1, deduce the effective angle of internal friction δ of the powder.

(b) Determine:

(i) the semi-included hopper angle safely ensuring mass flow;

(ii) the hopper flow factor, *ff*.

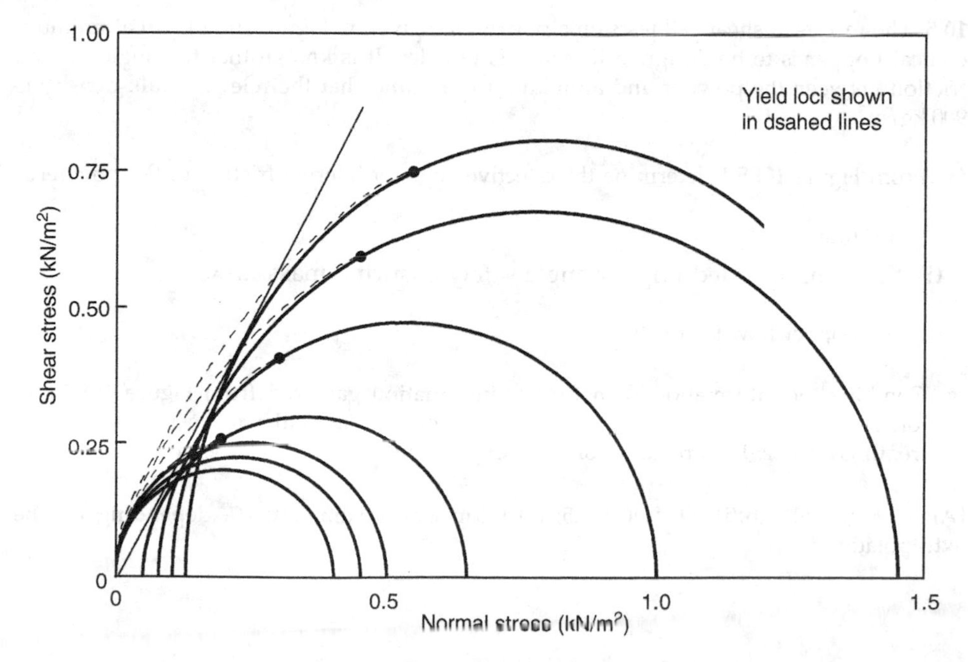

Figure 10E4.1 Shear cell test data

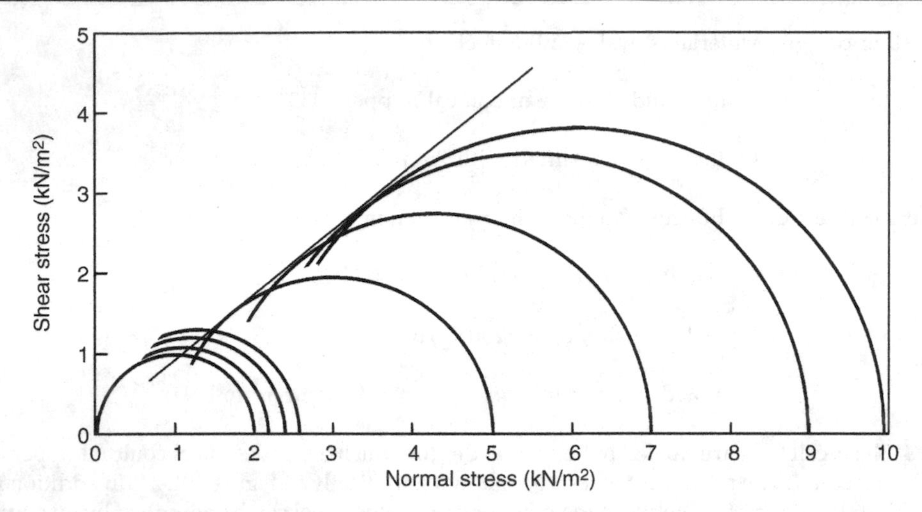

Figure 10E5.1 Shear cell test data

(c) Combine this information with further information gathered from Figure 10E4.1 in order to determine the minimum diameter of outlet to ensure flow when required. (*Note*: Extrapolation is necessary here.)

(d) What do you understand by 'angle of wall friction' and 'effective angle of internal friction'?

[Answer: (a) 60°; (b) (i) 32.5°, (ii) 1.29; (c) 0.110 m.]

10.5 The results of shear cell tests on a powder are given in Figure 10E5.1. An aluminium conical hopper is to be designed to suit this powder. It is known that the angle of wall friction between the powder and aluminium is 16° and that the relevant bulk density is 900 kg/m^3.

(a) From Figure 10E5.1 determine the effective angle of internal friction of the powder.

(b) Determine:

(i) the semi-included hopper angle safely ensuring mass flow;

(ii) the hopper flow factor, *ff*.

(c) Combine the information with further information gathered from Figure 10E5.1 in order to determine the minimum diameter of circular outlet to ensure flow when required. (*Note*: Extrapolation of these experimental results may be necessary.)

[Answer: (a) 40°; (b)(i) 29.5°; (ii) 1.5; (c) 0.5 m $\pm$ approximately 7% depending on the extrapolation.]

11 Mixing and Segregation

11.1 INTRODUCTION

Achieving good mixing of particulate solids of different size and density is important in many of the process industries, and yet it is not a trivial exercise. For free-flowing powders, the preferred state for particles of different size and density is to remain segregated. This is why in a packet of muesli the large particles come to the top as a result of the vibration caused by handling of the packet. An extreme example of this segregation is that a large steel ball can be made to rise to the top of a beaker of sand by simply shaking the beaker up and down – this has to be seen to be believed! Since the preferred state for free flowing powders is to segregate by size and density, it is not surprising that many processing steps give rise to segregation. Processing steps which promote segregation should not follow steps in which mixing is promoted. In this chapter we will examine mechanisms of segregation and mixing in particulate solids, briefly look at how mixing is carried out in practice and how the quality of a mixture is assessed.

11.2 TYPES OF MIXTURE

A perfect mixture of two types of particles is one in which a group of particles taken from any position in the mixture will contain the same proportions of each particle as the proportions present in the whole mixture. In practice, a perfect mixture cannot be obtained. Generally, the aim is to produce a random mixture, i.e. a mixture in which the probability of finding a particle of any component is the same at all locations and equal to the proportion of that component in the mixture as a whole. When attempting to mix particles which are not subject to segregation, this is generally the best quality of mixture that can be achieved. If the particles to be mixed differ in physical properties then segregation may occur. In this case particles of one component have a greater probability of being found

Introduction to Particle Technology - 2nd Edition Martin Rhodes

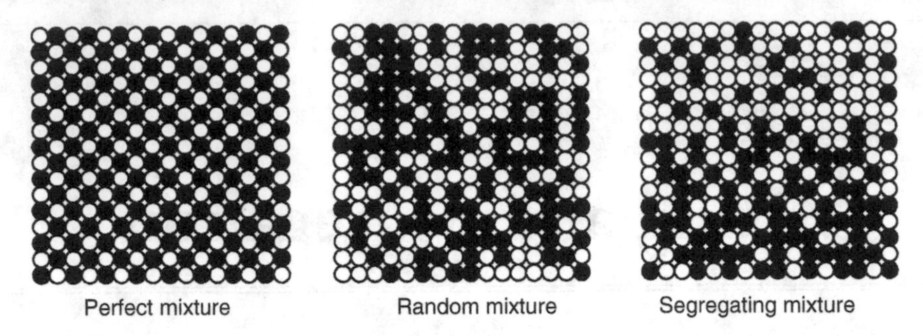

Figure 11.1 Types of mixture

in one part of the mixture and so a random mixture cannot be achieved. In Figure 11.1 examples are given of what is meant by perfect, random and segregating mixtures of two components. The random mixture was obtained by tossing a coin – heads gives a black particle at a given location and tails gives a white particle. For the segregating mixture the coin is replaced by a die. In this case the black particles differ in some property which causes them to have a greater probability of appearing in the lower half of the box. In this case, in the lower half of the mixture there is a chance of two in three that a particle will be black (i.e. a throw of 1, 2, 3 or 4) whereas in the upper half the probability is one in three (a throw of 5 or 6). It is possible to produce mixtures with better than random quality by taking advantage of the natural attractive forces between particles; such mixtures are achieved through 'ordered' or 'interactive' mixing (see below).

11.3 SEGREGATION

11.3.1 Causes and Consequences of Segregation

When particles to be mixed have the same important physical properties (size distribution, shape, density) then, provided the mixing process goes on for long enough, a random mixture will be obtained. However, in many common systems, the particles to be mixed have different properties and tend to exhibit segregation. Particles with the same physical property then collect together in one part of the mixture and the random mixture is not a natural state for such a system of particles. Even if particles are originally mixed by some means, they will tend to unmix on handling (moving, pouring, conveying, processing).

Although differences in size, density and shape of the constituent particles of a mixture may give rise to segregation, difference in particle size is by far the most important of these. Density difference is comparatively unimportant (see steel ball in sand example below) except in gas fluidization where density difference is more important than size difference. Many industrial problems arise from segregation. Even if satisfactory mixing of constituents is achieved in a powder mixing device, unless great care is taken, subsequent processing and handling of the mixture will result in demixing or segregation.

This can give rise to variations in bulk density of the powder going to packaging (e.g. not possible to fit 25 kg into a 25 kg bag) or, more seriously, the chemical composition of the product may be off specification (e.g. in blending of constituents for detergents or drugs).

11.3.2 Mechanisms of Segregation

Four mechanisms (Williams, 1990) of segregation according to size may be identified:

(1) *Trajectory segregation.* If a small particle of diameter x and density ρ_p, whose drag is governed by Stokes' law, is projected horizontally with a velocity U into a fluid of viscosity μ and density ρ_f, the limiting distance that it can travel horizontally is $U\rho_p x^2/18\mu$.

From Chapter 2, the retarding force on the particle $= C_D \frac{1}{2}\rho_f U^2\left(\frac{\pi x^2}{4}\right)$

Deceleration of the particle $= \dfrac{\text{retarding force}}{\text{mass of particle}}$

In Stokes' law region, $C_D = 24/Re_p$

Hence, deceleration $= \dfrac{18U\mu}{\rho_p x^2}$

From the equation of motion a particle with an initial velocity U and constant deceleration $18U\mu/\rho_p x^2$ will travel a distance $U\rho_p x^2/18\mu$ before coming to rest.

A particle of diameter $2x$ would therefore travel four times as far before coming to rest. This mechanism can cause segregation where particles are caused to move through the air (Figure 11.2). This also happens when powders fall from the end of a conveyor belt.

(2) *Percolation of fine particles.* If a mass of particles is disturbed in such a way that individual particles move, a rearrangement in the packing of the particles occurs. The gaps created allow particles from above to fall, and particles in some other place to move upwards. If the powder is composed of particles of different size, it will be easier for small particles to fall down and so there will be a tendency for small particles to move downwards leading to segregation. Even a very small difference in particle size can give rise to significant segregation.

Segregation by percolation of fine particles can occur whenever the mixture is disturbed, causing rearrangement of particles. This can happen during stirring, shaking, vibration or when pouring particles into a heap. Note that stirring, shaking and vibration would all be expected to promote mixing in

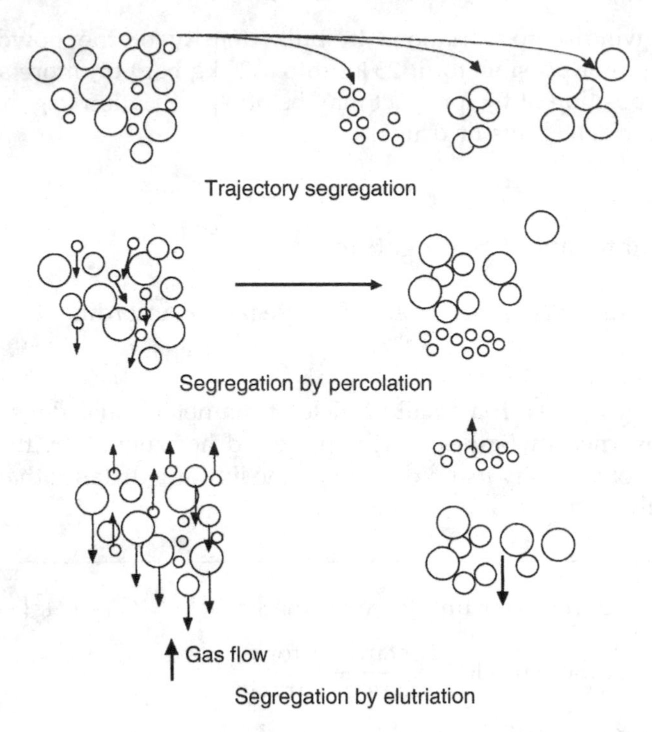

Figure 11.2 Mechanisms of segregation

liquids or gases, but cause segregation in free-flowing particle mixtures. Figure 11.3 shows segregation in the heap formed by pouring a mixture of two sizes of particles. The shearing caused when a particle mixture is rotated in a drum can also give rise to segregation by percolation.

Segregation by percolation occurs in charging and discharging storage hoppers. As particles are fed into a hopper they generally pour into a heap resulting in segregation if there is a size distribution and the powder is free-flowing. There are some devices and procedures available to minimize this effect if segregation is a particular concern. However, during discharge of a core flow hopper (see Chapter 10) sloping surfaces form, along which particles roll, and this gives rise to segregation in free-flowing powders. If segregation is a cause for concern, therefore, core flow hoppers should be avoided.

(3) *Rise of coarse particles on vibration.* If a mixture of particles of different size is vibrated the larger particles move upwards. This can be demonstrated by placing a single large ball at the bottom of a bed of sand (for example, a 20 mm steel ball or similarly sized pebble in a beaker of sand from the beach). On shaking the beaker up and down, the steel ball rises to the surface. Figure 11.4 show a series of photographs taken from a 'two-dimensional' version of the steel ball experiment. This so-called 'Brazil-nut effect' has received much attention in the literature over recent years,

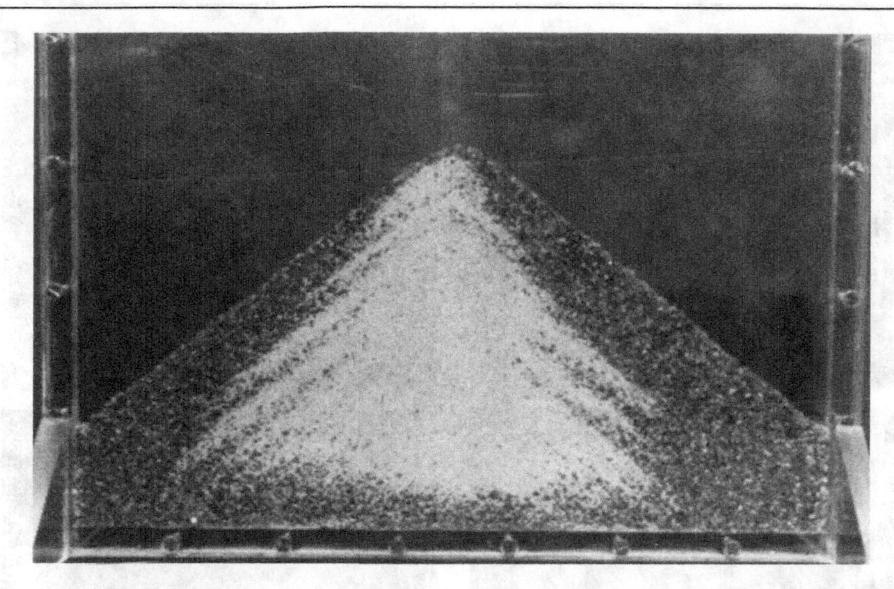

Figure 11.3 Segregation pattern formed by pouring a free-flowing mixture of two sizes of particles into a heap

but research and comment date back much further. The rise of the larger or denser 'intruder' within the bed of smaller particles has been explained in terms of creation and filling of voids beneath the intruder (Duran *et al.*, 1993; Jullien and Meakin, 1992; Rosato *et al.*, 1987, 2002; Williams, 1976;) and establishment of convection cells within the bed of small particles (Cooke *et al.*, 1996; Knight *et al.*, 1993, 1996; Rosato *et al.*, 2002). Observations that intruder rise times decrease with increasing intruder density (Liffman *et al.*, 2001; Mobius *et al.*, 2001; Shinbrot and Muzzie, 1998;) suggest that intruder inertia must play a part. However, Mobius *et al.* (2001) showed that this trend reverses for very low intruder densities suggesting that some other effect is present. Rhodes *et al.* (2003) suggested that this may be explained by a buoyancy effect created by the carrying pressure generated across the vibrated bed. Mobius *et al.* (2001) also demonstrated that interstitial gas plays a significant role in determination of the intruder rise time.

(4) *Elutriation segregation.* When a powder containing an appreciable proportion of particles under 50 μm is charged into a storage vessel or hopper, air is displaced upwards. The upward velocity of this air may exceed the terminal freefall velocity (see Chapter 2) of some of the finer particles, which may then remain in suspension after the larger particles have settled to the surface of the hopper contents (Figure 11.2). For particles in this size range in air the terminal freefall velocity will be typically of the order of a few centimetres per second and will increase as the square of particle diameter (e.g. for 30 μm sand particles the terminal velocity is 7 cm/s). Thus a pocket of fine particles is generated in the hopper each time solids are charged.

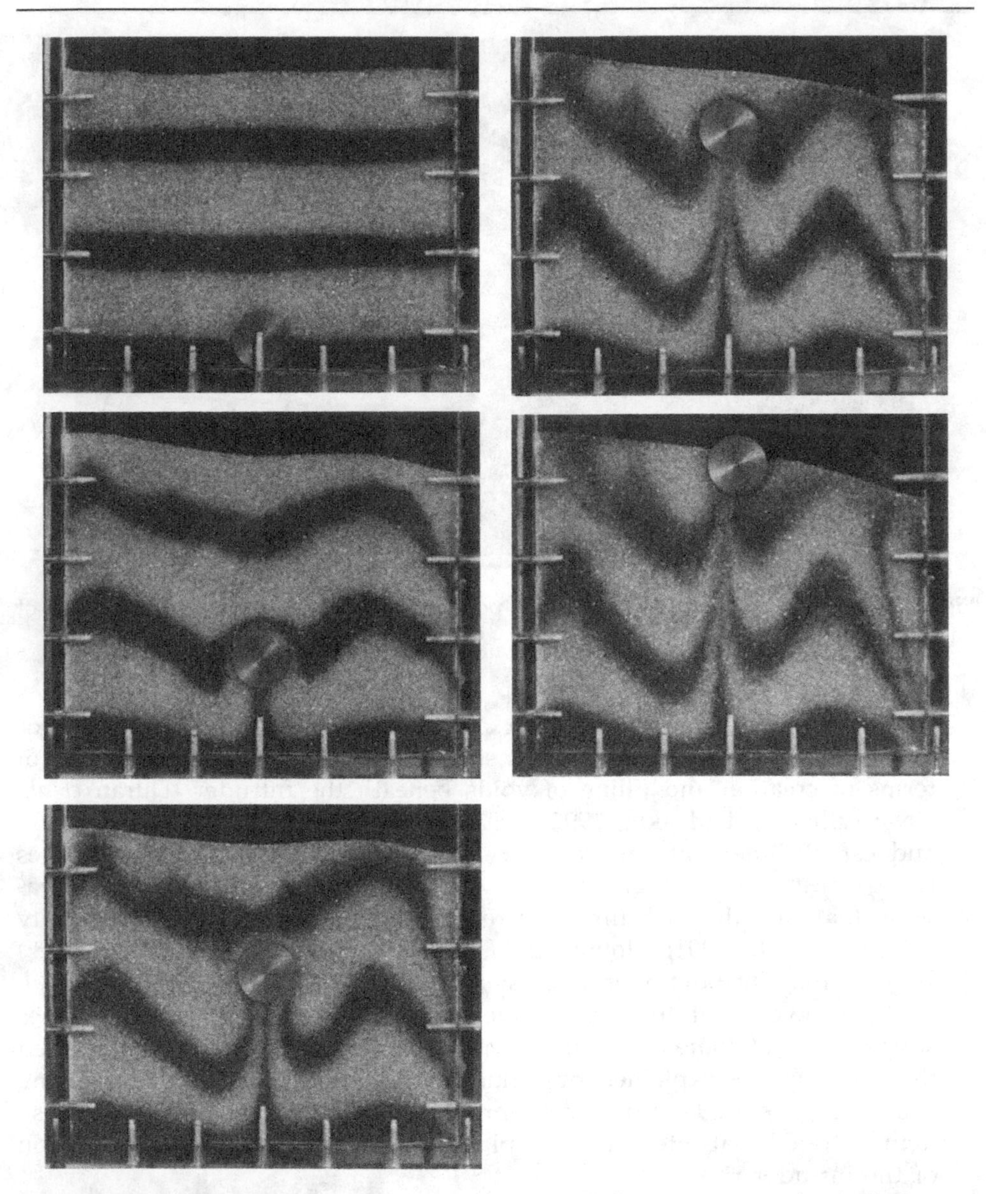

Figure 11.4 Series of photographs showing the rise of a steel disc through a bed of 2 mm glass spheres due to vibration (a 'two-dimensional' version of the rising steel ball experiment)

11.4 REDUCTION OF SEGREGATION

Segregation occurs primarily as a result of size difference. The difficulty of mixing two components can therefore be reduced by making the size of the components as similar as possible and by reducing the absolute size of both components. Segregation is generally not a serious problem when all particles are less than 30 μm (for particle densities in the range 2000–3000 kg/m^3). In such fine powders

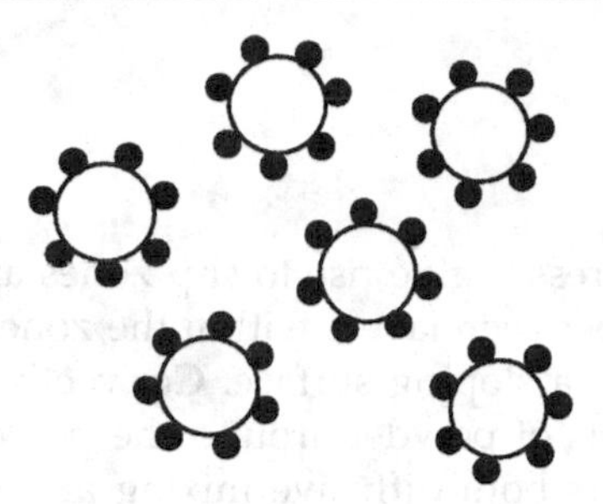

Figure 11.5 An ordered mixture of small particles on carrier particles

the interparticle forces (see Chapter 13) generated by electrostatic charging, van der Waals forces and forces due to moisture are large compared with the gravitational and inertial forces on the particles. This causes the particles to stick together preventing segregation as the particles are not free to move relative to one another. These powders are referred to as cohesive powders (Geldart's classification of powders for fluidization is relevant here – see Chapter 7). The lack of mobility of individual particles in cohesive powders is one reason why they give better quality of mixing. The other reason is that if a random mixture is approached, the standard deviation of the composition of samples taken from the mixture will decrease in inverse proportion to the number of particles in the sample. Therefore, for a given mass of sample the standard deviation decreases and mixture quality increases with decreasing particle size. The mobility of particles in free-flowing powders can be reduced by the addition of small quantities of liquid. The reduction in mobility reduces segregation and permits better mixing.

It is possible to take advantage of this natural tendency for particles to adhere to produce mixtures of quality better that random mixtures. Such mixtures are known as ordered or interactive mixtures; they are made up of small particles (e.g. $< 5\,\mu m$) adhered to the surface of a carrier particle in a controlled manner (Figure 11.5). By careful selection of particle size and engineering of interparticle forces, high quality mixtures with very small variance can be achieved. This technique is use in the pharmaceutical industry where quality control standards are exacting. For further details on ordered mixing and on the mixing of cohesive powders the reader is referred to Harnby *et al.* (1992).

If it is not possible to alter the size of the components of the mixture or to add liquid, then in order to avoid serious segregation, care should be taken to avoid situations which are likely to promote segregation. In particular pouring operations and the formation of a moving sloping powder surface should be avoided.

11.5 EQUIPMENT FOR PARTICULATE MIXING

11.5.1 Mechanisms of Mixing

Lacey (1954) identified three mechanisms of powder mixing:

(1) shear mixing;

(2) diffusive mixing;

(3) convective mixing.

In shear mixing, shear stresses give rise to slip zones and mixing takes place by interchange of particles between layers within the zone. Diffusive mixing occurs when particles roll down a sloping surface. Convective mixing is by deliberate bulk movement of packets of powder around the powder mass.

In free-flowing powders both diffusive mixing and shear mixing give rise to size segregation and so for such powders convective mixing is the major mechanism promoting mixing.

11.5.2 Types of Mixer

(1) *Tumbling mixers.* A tumbling mixer comprises a closed vessel rotating about its axis. Common shapes for the vessel are cube, double cone and V (Figure 11.6). The dominant mechanism is diffusive mixing. Since this can give rise to segregation in free-flowing powders the quality of mixture achievable with such powder in tumbling mixers is limited. Baffles may be installed in an attempt to reduce segregation, but have little effect.

(2) *Convective mixers.* In convective mixers circulation patterns are set up within a static shell by rotating blades or paddles. The main mechanism is convective mixing as the name suggests, although this is accompanied by some diffusive and shear mixing. One of the most common convective mixers is the ribbon blender in which helical blades or ribbons rotate on a horizontal axis in a static cylinder or trough (Figure 11.7). Rotational speeds are typically less than one revolution per second. A somewhat different type of convective mixer is the Nautamix (Figure 11.8) in which an Archimedean screw lifts material from the base of a conical hopper and progresses around the hopper wall.

(3) *Fluidized bed mixers.* These rely on the natural mobility afforded particles in the fluidized bed. The mixing is largely convective with the circulation

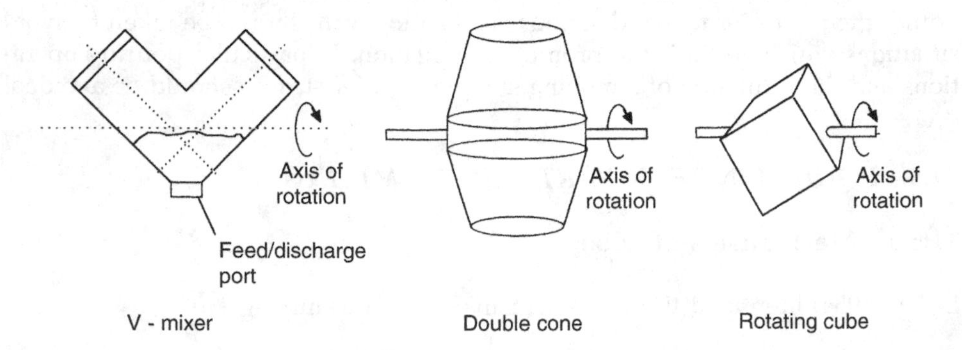

Figure 11.6 Tumbling mixers: V-mixer; double cone mixer; and rotating cube mixer

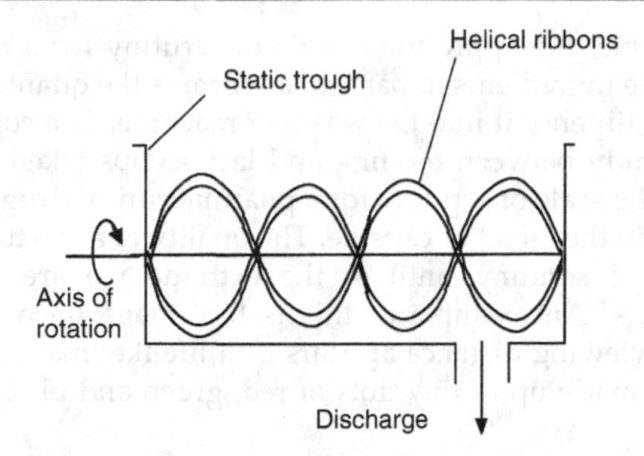

Figure 11.7 Ribbon blender

patterns set up by the bubble motion within the bed. An important feature of the fluidized bed mixer is that several processing steps (e.g. mixing, reaction, coating drying, etc.) may be carried out in the same vessel.

(4) *High shear mixers.* Local high shear stresses are created by devices similar to those used in comminution; for example, high velocity rotating blades, low velocity–high compression rollers (see Chapter 12). In the high shear mixers the emphasis is on breaking down agglomerates of cohesive powders rather than breaking individual particles. The dominant mechanism is shear mixing.

11.6 *ASSESSING THE MIXTURE*

11.6.1 Quality of a Mixture

The end use of a particle mixture will determine the quality of mixture required. The end use imposes a scale of scrutiny on the mixture. 'Scale of scrutiny' was a term used by Danckwerts (1953) meaning 'the maximum size of the regions of segregation in the mixture which would cause it to be regarded as imperfectly

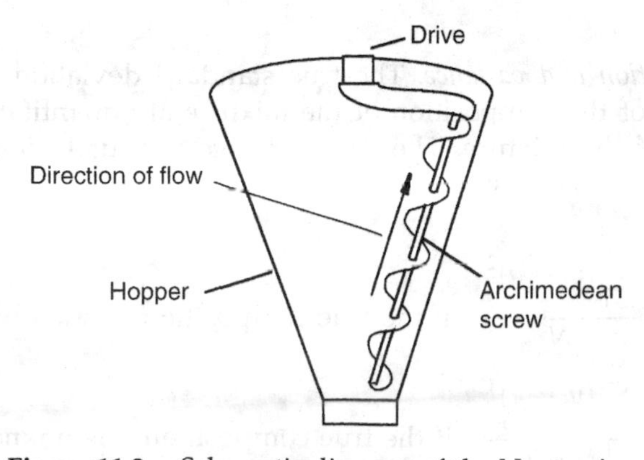

Figure 11.8 Schematic diagram of the Nautamixer

mixed'. For example, the appropriate scale of scrutiny for a detergent powder composed of active ingredients in particulate form is the quantity of detergent in the scoop used to dipense it into the washing machine. The composition should not vary significantly between the first and last scoops taken from the box. At another extreme the scale of scrutiny for a pharmaceutical drug is the quantity of material making up the tablet or capsule. The quality of a mixture decreases with decreasing scale of scrutiny until in the extreme we are scrutinizing only individual particles. An example of this is the image on a television screen, which at normal viewing distance appears as a lifelike image, but which under close 'scrutiny' is made up of tiny dots of red, green and blue colour.

11.6.2 Sampling

To determine the quality of a mixture it is generally necessary to take samples. In order to avoid bias in taking samples from a particulate mixture, the guidelines of sampling powders set out in Chapter 2 must be followed. The size of the sample required in order to determine the quality of the mixture is governed by the scale of scrutiny imposed by the intended use of the mixture.

11.6.3 Statistics Relevant to Mixing

It is evident that the sampling of mixtures and the analysis of mixture quality require the application of statistical methods. The statistics relevant to random binary mixtures are summarized below:

- *Mean composition.* The true composition of a mixture μ is often not known but an estimate $\bar{y}$ may be found by sampling. If we have N samples of composition y_1 to y_N in one component, the estimate of the mixture composition $\bar{y}$ is given by:

$$\bar{y} = \frac{1}{N}\sum_{i=1}^{N} y_i \tag{11.1}$$

- *Standard deviation and variance.* The true standard deviation, σ, and the true variance, σ^2, of the composition of the mixture are quantitative measures of the quality of the mixture. The true variance is usually not known but an estimate S^2 is defined as:

$$S^2 = \frac{\sum_{i=1}^{N}(y_i - \mu)^2}{N} \quad \text{if the true composition } \mu \text{ is known} \tag{11.2}$$

$$S^2 = \frac{\sum_{i=1}^{N}(y_i - \bar{y})^2}{N-1} \quad \text{if the true composition } \mu \text{ is unknown} \tag{11.3}$$

The standard deviation is equal to the square root of variance.

- *Theoretical limits of variance.* For a two-component system the theoretical upper and lower limits of mixture variance are:

  (a) upper limit (completely segregated) $\sigma_0^2 = p(1-p)$ (11.4)

  (b) lower limit (randomly mixed) $\sigma_R^2 = \dfrac{p(1-p)}{n}$ (11.5)

  where p and $(1-p)$ are the proportions of the two components determined from samples and n is the number of particles in each sample.

  Actual values of mixture variance lie between these two extreme values.

- *Mixing indices.* A measure of the degree of mixing is the Lacey mixing index (Lacey, 1954):

  $$\text{Lacey mixing index} = \frac{\sigma_0^2 - \sigma^2}{\sigma_0^2 - \sigma_R^2} \tag{11.6}$$

  In practical terms the Lacey mixing index is the ratio of 'mixing achieved' to 'mixing possible'. A Lacey mixing index of zero would represent complete segregation and a value of unity would represent a completely random mixture. Practical values of this mixing index, however, are found to lie in the range 0.75 to 1.0 and so the Lacey mixing index does not provide sufficient discrimination between mixtures.

  A further mixing index suggested by Poole *et al.* (1964) is defined as:

  $$\text{Poole } et\,al. \text{ mixing index} = \frac{\sigma}{\sigma_R} \tag{11.7}$$

  This index gives better discrimination for practical mixtures and approaches unity for completely random mixtures.

- *Standard error.* When the sample compositions have a normal distribution the sampled variance values will also have a normal distribution. The standard deviation of the variance of the sample compositions is known as the 'standard error' of the variance $E(S^2)$.

- *Tests for precision of mixture composition and variance.* The mean mixture composition and variance which we measure from sampling are only samples from the normal distribution of mixture compositions and variance values for that mixture. We need to be able to assign a certain confidence to this estimate and to determine its precision.

(*Note:* Standard statistical tables, which are widely and readily available, are a source of Student's *t*-values and χ^2 (chi squared) distribution values.)

Assuming that the sample compositions are normally distributed:

(1) *Sample composition*

Based on N samples of mixture composition with mean $\bar{y}$ and estimated standard deviation S, the true mixture composition μ may be stated with precision:

$$\mu = \bar{y} \pm \frac{tS}{\sqrt{N}} \tag{11.8}$$

where t is from Student's t-test for statistical significance. The value of t depends on the confidence level required. For example, at 95% confidence level, $t = 2.0$ for $N = 60$, and so there is a 95% probability that the true mean mixture composition lies in the range: $\bar{y} \pm 0.258\,S$. In other words, 1 in 20 estimates of mixture variance estimates would lie outside this range.

(2) *Variance*

(a) When more than 50 samples are taken (i.e. $n > 50$), the distribution of variance values can also be assumed to be normal and the Student's t-test may be used. The best estimate of the true variance σ^2 is then given by:

$$\sigma^2 = S^2 \pm [t \times E(S^2)] \tag{11.9}$$

The standard error of the mixture variance required in this test is usually not known but is estimated from:

$$E(S^2) = S^2\sqrt{\frac{2}{N}} \tag{11.10}$$

The standard error decreases as $1/\sqrt{N}$ and so the precision increases as $\sqrt{N}$.

(b) When less than 50 samples are taken (i.e. $n < 50$), the variance distribution curve may not be normal and is likely to be a χ^2 (chi squared) distribution. In this case the limits of precision are not symmetrical. The range of values of mixture variance is defined by lower and upper limits:

$$\text{lower limit} : \sigma_L^2 = \frac{S^2(N-1)}{\chi_\alpha^2} \tag{11.11}$$

$$\text{upper limit} : \sigma_U^2 = \frac{S^2(N-1)}{\chi_{1-\alpha}^2} \tag{11.12}$$

where α is the significance level [for a 90% confidence range, $\alpha = 0.5(1 - 90/100) = 0.05$; for a 95% confidence range, $\alpha = 0.5(1 - 95/100) = 0.025$]. The lower and upper χ^2 values, χ_α^2 and $\chi_{1-\alpha}^2$, for a given confidence level are found in χ^2 distribution tables.

11.7 WORKED EXAMPLES

WORKED EXAMPLE 11.1 (AFTER WILLIAMS, 1990)

A random mixture consists of two components A and B in proportions 60 and 40% by mass, respectively. The particles are spherical and A and B have particle densities 500 and 700 kg/m^3, respectively. The cumulative undersize mass distributions of the two components are shown in Table 11W.1.

If samples of 1 g are withdrawn from the mixture, what is the expected value for the standard deviation of the composition of the samples?

Table 11W1.1 Size distributions of particles A and B

Size x (μm)	2057	1676	1405	1204	1003	853	699	599	500	422
$F_A(x)$	1.00	0.80	0.50	0.32	0.19	0.12	0.07	0.04	0.02	0
$F_B(x)$			1.00	0.88	0.68	0.44	0.21	0.08	0	

Solution

The first step is to estimate the number of particles per unit mass of A and B. This is done by converting the size distributions into differential frequency number distributions and using:

$$\begin{pmatrix}\text{mass of particles}\\ \text{in each size range}\end{pmatrix} = \begin{pmatrix}\text{number of particles}\\ \text{in size range}\end{pmatrix} \times \begin{pmatrix}\text{mass of one}\\ \text{particle}\end{pmatrix}$$

$$\mathrm{d}m = \mathrm{d}n \quad \frac{\rho_p \pi x^3}{6}$$

where ρ_p is the particle density and x is the arithmetic mean of adjacent sieve sizes.

These calculations are summarized in Tables 11W1.2 and 11W1.3.

Table 11W.1.2 A particles

Mean size of range x (μm)	dm	dn
1866.5	0.20	117 468
1540.5	0.30	331 081
1304.5	0.18	309 681
1103.5	0.13	369 489
928	0.07	334 525
776	0.05	408 658
649	0.03	419 143
54.5	0.02	460 365
461	0.02	779 655
Totals	1.00	3.51×10^6

Table 11W.1.3 B particles

Mean size of range x (μm)	dm	dn
1866.5	0	0
1540.5	0	0
1304.5	0.12	0.147×10^6
1103.5	0.20	0.406×10^6
928	0.24	0.819×10^6
776	0.23	1.343×10^6
649	0.13	1.297×10^6
54.5	0.08	1.315×10^6
461	0	0
Totals	1.00	5.33×10^6

Thus $n_A = 3.51 \times 10^6$ particles per kg

and

$n_B = 5.33 \times 10^6$ particles per kg

And in samples of 1 g (0.001 kg) we would expect a total number of particles:

$$\begin{aligned} n &= 0.001 \times (3.51 \times 10^6 \times 0.6 + 5.33 \times 10^6 \times 0.4) \\ &= 4238 \text{ particles} \end{aligned}$$

And so, from Equation (11.5) for a random mixture,

$$\text{standard deviation, } \sigma = \sqrt{\frac{0.6 \times 0.4}{4238}} = 0.0075$$

WORKED EXAMPLE 11.2 (AFTER WILLIAMS, 1990)

Sixteen samples are removed from a binary mixture and the percentage proportions of one component by mass are:

$$41, 37, 41, 39, 45, 37, 39, 40$$
$$41, 43, 40, 38, 39, 37, 43, 40$$

Determine the upper and lower 95% and 90% confidence limits for the standard deviation of the mixture.

Solution

From Equation (11.1), the mean value of the sample composition is:

$$\bar{y} = \frac{1}{16}\sum_{i=1}^{16} y_i = 40\%$$

Since the true mixture composition is not known an estimate of the standard deviation is found from Equation (11.3):

$$S = \sqrt{\left[\frac{1}{16-1}\sum_{i=1}^{16}(y_i - 40)^2\right]} = 2.31$$

Since there are less than 50 samples, the variance distribution curve is more likely to be a χ^2 distribution. Therefore, from Equations (11.11) and (11.12):

$$\text{lower limit}: \quad \sigma_L^2 = \frac{2.31^2(16-1)}{\chi_\alpha^2}$$

$$\text{upper limit}: \quad \sigma_U^2 = \frac{2.31^2(16-1)}{\chi_{1-\alpha}^2}$$

At the 90% confidence level $\alpha = 0.05$ and so referring to the χ^2 distribution tables with 15 degrees of freedom $\chi_\alpha^2 = 24.996$ and $\chi_{1-\alpha}^2 = 7.261$.

Hence, $\sigma_L^2 = 3.2$ and $\sigma_U^2 = 11.02$.

At the 95% confidence level $\alpha = 0.025$ and so referring to the χ^2 distribution tables with 15 degrees of freedom $\chi_\alpha^2 = 27.49$ and $\chi_{1-\alpha}^2 = 6.26$.

Hence, $\sigma_L^2 = 2.91$ and $\sigma_U^2 = 12.78$.

WORKED EXAMPLE 11.3

During the mixing of a drug with an excipient the standard deviation of the compositions of 100 mg samples tends to a constant value of ± 0.005. The size distributions of drug (D) and excipient (E) are given in Table 11W3.1.

Table 11W3.1 Size distributions of drug and excipient

Size x (μm)	420	355	250	190	150	75	53	0
$F_D(x)$	1.00	0.991	0.982	0.973	0.964	0.746	0.047	0
$F_E(x)$	1.00	1.00	0.977	0.967	0.946	0.654	0.284	0

The mean proportion by mass of drug is know to be 0.2. The densities of drug and excipient are 1100 and 900 kg/m^3, respectively.

Determine whether the mixing is satisfactory (a) if the criterion is a random mixture and (b) if the criterion is an in-house specification that the composition of 95% of the samples should lie within +15% of the mean.

Solution

The number of particles of drug (Table 11W3.2) and excipient (Table 11W3.3) in each sample is first calculated as shown in Worked Example 11.1.

$$\text{Thus } n_D = 8.96 \times 10^9 \text{ particles per kg}$$

Table 11W3.2 Number of drug particles in each kg of sample

Mean size of range x (μm)	dm	dn
388	0.009	2.67×10^5
303	0.009	5.62×10^5
220	0.009	1.47×10^6
170	0.009	3.18×10^6
113	0.218	2.62×10^8
64	0.700	4.64×10^9
27	0.046	4.06×10^9
20	0.00	0
0	0.00	0
Totals	1.00	8.96×10^9

Table 11W3.3 Number of excipient particles in each kg of sample

Mean size of range x (μm)	dm	dn
388	0	0
303	0.023	1.75×10^6
220	0.010	1.99×10^6
170	0.021	9.07×10^6
113	0.292	4.29×10^8
64	0.374	3.03×10^9
27	0.28	3.02×10^{10}
20	0.00	0
0	0	0
Totals	1.00	3.37×10^{10}

and

$$n_E = 3.37 \times 10^{10} \text{ particles per kg}$$

And in samples of 1 g (0.001 kg) we would expect a total number of particles:

$$\begin{aligned} n &= 100 \times 10^{-6} \times (8.96 \times 10^9 \times 0.2 + 3.37 \times 10^{10} \times 0.8) \\ &= 2.88 \times 10^6 \text{ particles} \end{aligned}$$

And so, from Equation (11.5) for a random mixture,

$$\text{standard deviation, } \sigma_R = \sqrt{\frac{0.2 \times 0.8}{2.88 \times 10^6}} = 0.000235$$

Conclusion. The actual standard deviation of the mixture is greater than that for a random mixture and so the criterion of random mixing is not achieved.

For a normal distribution the in-house criterion that 95% of samples should lie within ±15% of the mean suggests that:

$$1.96\sigma = 0.15 \times 0.2$$

(since for a normal distribution 95% of the values lie within ± 1.96 standard deviations of the mean).

Hence, $\sigma = 0.0153$. So the in-house criterion is achieved.

TEST YOURSELF

1.1 Explain the difference between a *random mixture* and *perfect mixture*. Which of these two types of mixture is more likely to occur in an industrial process?

1.2 Explain how *trajectory segregation* occurs. Give examples of two practical situations that might give rise to trajectory segregation of powders in the process industries.

1.3 What type of segregation is produced when a free-flowing mixture of particles is poured into a heap? Describe the typical segregation pattern produced.

1.4 Explain how core flow of free-flowing particulate mixture from a hopper gives rise to a size-segregated discharge.

1.5 What is the *Brazil Nut Effect*? Under what conditions, relevant to the process industries, might it occur?

1.6 Explain why size segregation is generally not a problem if all components of the particulate mixture are smaller than around 30 μm.

1.7 Describe two types of industrially relevant mixer. Which mixing mechanism dominates in each type of mixer?

1.8 Explain what is meant by *scale of scrutiny* of a particulate mixture. What scale of scrutiny would be appropriate for (a) the active drug in powder mixture fed to the tabletting machine, (b) muesli breakfast cereal, (c) a health supplement fed to chickens?

1.9 For a two-component mixture, write down expressions for (a) *mean* composition, (b) *estimated variance* when the true mean is unknown, (c) upper and lower theoretical limits of mixture variance. Define all symbols used.

1.10 Explain how one would go about determining whether the mixture produced by an industrial process is satisfactory.

EXERCISES

11.1 Thirty-one samples are removed from a binary mixture and the percentage proportions of one component by mass are:

19, 22, 20, 24, 23, 25, 22, 18, 24, 21, 27, 22, 18, 20, 23, 19,

20, 22, 25, 21, 17, 26, 21, 24, 25, 22, 19, 20, 24, 21, 23

Determine the upper and lower 95% confidence limits for the standard deviation of the mixture.

(Answer: 0.355 to 0.595.)

11.2 A random mixture consists of two components A and B in proportions 30% and 70% by mass, respectively. The particles are spherical and components A and B have particle densities 500 and 700 kg/m^3, respectively. The cumulative undersize mass distributions of the two components are shown in Table 11E2.1.

Table 11E.2.1 Size distributions of particles A and B

Size x (μm)	2057	1676	1405	1204	1003	853	699	599	500	422	357
$F_A(x)$	1.00	1.00	0.85	0.55	0.38	0.25	0.15	0.10	0.07	0.02	0.00
$F_B(x)$			1.00	0.80	0.68	0.45	0.25	0.12	0.06	0.00	0.00

If samples of 5 g are withdrawn from the mixture, what is the expected value for the standard deviation of the composition of the samples?

(Answer: 0.0025.)

11.3 During the mixing of a drug with an excipient the standard deviation of the compositions of 10 mg samples tends to a constant value of ±0.005. The size distributions by mass of drug (D) and excipient (E) are given in Table 11E3.1.

Table 11E3.1 Size distributions of drug and excipient

Size x (μm)	499	420	355	250	190	150	75	53	0
$F_D(x)$	1.00	0.98	0.96	0.94	0.90	0.75	0.05	0.00	0.00
$F_E(x)$	1.00	1.00	0.97	0.96	0.93	0.65	0.25	0.05	0.00

The mean proportion by mass of drug is known to be 0.1. The densities of the drug and the excipient are 800 and 1000 kg/m^3, respectively.

Determine whether the mixing is satisfactory if:

(a) the criterion is a random mixture;

(b) the criterion is an in-house specification that the composition of 99% of the samples should lie within ±20% of the mean.

[Answer: (a) 0.00118, criterion not achieved; (b) 0.00775, criterion achieved.]

12
Particle Size Reduction

12.1 INTRODUCTION

Size reduction, or comminution, is an important step in the processing of many solid materials. It may be used to create particles of a certain size and shape, to increase the surface area available for chemical reaction or to liberate valuable minerals held within particles.

The size reduction of solids is an energy intensive and highly inefficient process: 5% of all electricity generated is used in size reduction; based on the energy required for the creation of new surfaces, the industrial scale process is generally less than 1% efficient. The two statements would indicate that there is great incentive to improve the efficiency of size reduction processes. However, in spite of a considerable research effort over the years, size reduction processes have remained stubbornly inefficient. Also, in spite of the existence of a well-developed theory for a strength and breakage mechanism of solids, the design and scale-up of comminution processes is usually based on past experience and testing, and is very much in the hands of the manufacturer of comminution equipment.

This chapter is intended as a introduction to the topic of size reduction covering the concepts and models involved and including a broad survey of practical equipment and systems. The chapter is divided into the following sections:

- particle fracture mechanisms;
- models for prediction of energy requirements and product size distribution;
- equipment: matching machine to material and duty.

Introduction to Particle Technology - 2nd Edition Martin Rhodes

12.2 PARTICLE FRACTURE MECHANISMS

Consider a crystal of sodium chloride (common salt) as a simple and convenient model of a brittle material. Such a crystal is composed of a lattice of positively charged sodium ions and negatively charged chloride ions arranged such that each ion is surrounded by six ions of the opposite sign. Between the oppositely charged ions there is an attractive force whose magnitude is inversely proportional to the square of the separation of the ions. There is also a repulsive force between the negatively charged electron clouds of these ions which becomes important at very small interatomic distances. Therefore two oppositely charged ions have an equilibrium separation such that the attractive and repulsive forces between them are equal and opposite. Figure 12.1 shows how the sum of the attractive and repulsive forces varies with changing separation of the ions. It can be appreciated that if the separation of the ions is increased or decreased by a small amount from the equilibrium separation there will be a resultant net force restoring the ions to the equilibrium position. The ions in the sodium chloride crystal lattice are held in equilibrium positions governed by the balance between attractive and repulsive forces. Over a small range of interatomic distances the relationship between applied tensile or compressive force and resulting change in ion separation is linear. That is, in this region (AB in Figure 12.1) Hooke's law applies: strain is directly proportional to applied stress. The Young's modulus of the material (stress/strain) describes this proportionality. In this Hooke's law range the deformation of the crystal is elastic, i.e. the original shape of the crystal is recovered upon removal of the stress.

In order to break the crystal it is necessary to separate adjacent layers of ions in the crystal and this involves increasing the separation of the adjacent ions beyond the region where Hooke's law applies, i.e. beyond point B in Figure 12.1 into the plastic deformation range. The applied stress required to induce this plastic

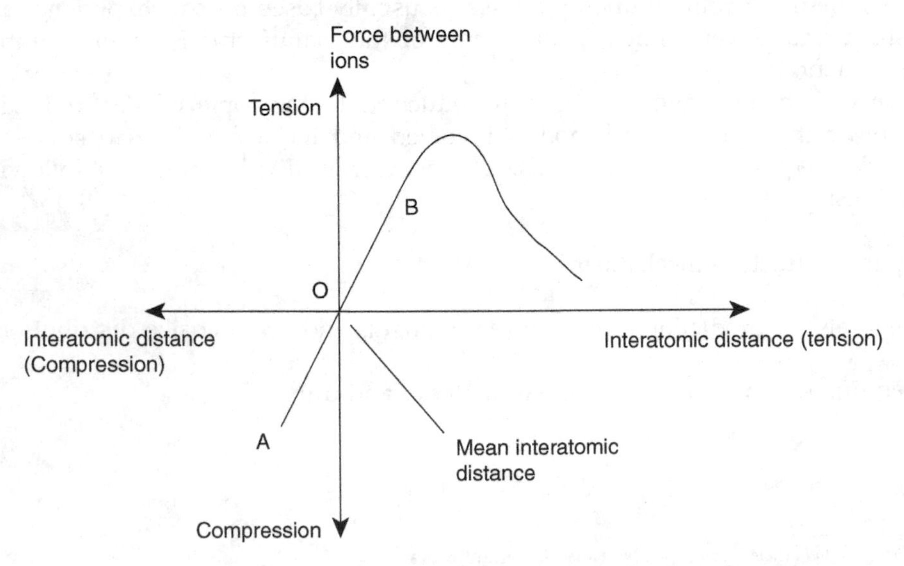

Figure 12.1 Force versus distance on an atomic scale

behaviour is known as the elastic limit or yield stress, and is sometimes defined as the material's strength. With a knowledge of the magnitude of attractive and repulsive forces between ions in such a crystal, it is therefore possible to estimate the strength of a salt crystal. One could assume first that under tensile stress all bonds in the crystal planes perpendicular to the applied stress are stretched until they simultaneously break and the material splits into many planes one atom thick–this gives a theoretical strength much greater than in reality. Alternatively, one could assume that only those bonds which are to be broken are stretched – this gives a theoretical strength much less than reality. In practice the true fracture mechanism for these materials turns out to be more involved and more interesting.

A body under tension stores energy–strain energy. The amount of strain energy stored by a brittle material under tension is given by the area under the appropriate stress–strain graph. This strain energy is not uniformly distributed throughout the body but is concentrated around holes, corners and cracks. Inglis (1913) proposed that the stress concentration factor, K, around a hole, crack or corner could be calculated according to the formula:

$$K = \left(1 + 2\sqrt{\frac{L}{R}}\right) \tag{12.1}$$

where L is the half the length of the crack, R is the radius of crack tip or hole and K is the stress concentration factor (local stress/mean stress in body).

Thus, for a round hole, $K = 3$.

For a 2μm long crack with tip radius equal to half the inter atomic distance ($R = 10^{-10}$ m), $K = 201$.

Griffith (1921) proposed that for a crack in the surface of a body to propagate the following criteria must be satisfied:

(1) The strain energy that would be released must be greater than the surface energy created.

(2) There must be a crack propagation mechanism available.

Griffith also pointed out that for a given mean stress applied to a body there should be a critical minimum crack length for which the stress concentration at the tip will just be sufficient to cause the crack to propagate. Under the action of this mean stress a crack initially longer than the critical crack length for that stress will grow longer and, since K increases as L increases, crack growth increases until the body is broken. As the crack grows, provided the mean stress remains constant, there is strain energy excess to that required to propagate that crack (since K is increasing). This excess strain energy is dissipated at the velocity of sound in the material to concentrate at the tips of other cracks, causing them to propagate. The rate of crack propagation is lower than the velocity of sound in the material and so other cracks begin to propagate before the first crack brings about failure. Thus, in brittle materials multiple fracture is common. If cracks in the surface of brittle materials can be avoided then the material strength would be near to the theoretical value. This can be demonstrated by heating a glass rod until it softens and then drawing it out to create a new surface. As soon as the rod is cooled it can withstand surprisingly high tensile stress, as demonstrated by bending the rod. Once the new surface is handled or even exposed to the normal environment for a short period, its tensile strength diminishes due to the formation of microscopic cracks in the surface. It has been shown that the surfaces of all materials have cracks in them.

Gilvary (1961) proposed the concept of volume, facial and edge flaws (cracks) in order to calculate the size distribution of breakage products. Assuming that all flaws were randomly distributed and independent of each other and that the initial stress system is removed once the first flaws begin to propagate, Gilvary showed that the product size distributions common to comminuted materials could be predicted. For example, if edge flaws dominate, then the common Rosin–Rammler distribution results.

Evans *et al.* (1961) showed that for a disc acted upon by opposing diametrical loads, there is a uniform tensile stress acting at 90° to the diameter. Under sufficiently high compressive loads, therefore, the resulting tensile stress could exceed the cohesive strength of the material and the disc would split across the diameter. Evans extended the analysis to three-dimensional particles to show that even when particles are stressed compressively, the stress pattern set up by virtue of the shape of the particle may cause it to fail in tension, whether cracks exist or not.

Cracks are less important for 'tough' materials (e.g. rubber, plastics and metals) since excess strain energy is used in deformation of the material rather than crack propagation. Thus in ductile metals, for example the stress concentration at the top of a crack will cause deformation of the material around the crack tip, resulting in a larger tip radius and lower stress concentration.

The observation that small particles are more difficult to break than large particles can be explained using the concept of failure by crack propagation. First, the length of a crack is limited by the size of the particle and so one would expect lower maximum stress concentration factors to be achieved in small particles. Lower stress concentrations mean that higher mean stresses have to be applied to the particles to cause failure. Secondly, the Inglis equation [Equation (12.1)] overpredicts K in the case of small particles since in these particles there is less room for the stress distribution patterns to develop. This effectively limits the maximum stress concentration possible and means that a higher mean stress is necessary to cause crack propagation. Kendal (1978) showed that as particle size decreases, the fracture strength increases until a critical size is reached when crack propagation becomes impossible. Kendal offered a way of predicting this critical particle size.

12.3 MODEL PREDICTING ENERGY REQUIREMENT AND PRODUCT SIZE DISTRIBUTION

12.3.1 Energy Requirement

There are three well-known postulates predicting energy requirements for particle size reduction. We will cover them in the chronological order in which they were proposed. Rittinger (1867) proposed that the energy required for particle size reduction was directly proportional to the area of new surface created. Thus, if the initial and final particle sizes are x_1 and x_2, respectively, then assuming a volume shape factor k_v independent of size,

$$\text{volume of initial particle} = k_v x_1^3$$
$$\text{volume of final particle} = k_v x_2^3$$

and each particle of size x_1 will give rise to x_1^3/x_2^3 particles of size x_2.

If the surface shape factor k_s is also independent of size, then for each original particle, the new surface created upon reduction is given by the expression:

$$\left(\frac{x_1^3}{x_2^3}\right) k_s x_2^2 - k_s x_1^2 \tag{12.2}$$

which simplifies to:

$$k_s x_1^3 \left(\frac{1}{x_2} - \frac{1}{x_1}\right) \tag{12.3}$$

Therefore,

new surface created per unit mass of original particles

$$= k_s x_1^3 \left(\frac{1}{x_2} - \frac{1}{x_1}\right) \times \text{(number of original particles per unit mass)}$$
$$= k_s x_1^3 \left(\frac{1}{x_2} - \frac{1}{x_1}\right) \times \left(\frac{1}{k_v x_1^3 \rho_p}\right)$$
$$= \frac{k_s}{k_v}\frac{1}{\rho_p}\left(\frac{1}{x_2} - \frac{1}{x_1}\right)$$

where ρ_p is the particle density. Hence assuming shape factors and density are constant, Rittinger's postulate may be expressed as:

$$\text{breakage energy per unit mass of feed, } E = C_R\left(\frac{1}{x_2} - \frac{1}{x_1}\right) \tag{12.4}$$

where C_R is a constant. If this is the integral form, then in differential form, Rittinger's postulate becomes:

$$\frac{dE}{dx} = -C_R \frac{1}{x^2} \tag{12.5}$$

However, since in practice the energy requirement is usually 200–300 times that required for creation of new surface, it is unlikely that energy requirement and surface created are related.

On the basis of stress analysis theory for plastic deformation, Kick (1885) proposed that the energy required in any comminution process was directly proportional to the ratio of the volume of the feed particle to the product particle. Taking this assumption as our starting point, we see that:

volume ratio, x_1^3/x_2^3, determines the energy requirement
(assuming shape factor is constant)

Therefore, size ratio, x_1/x_2 fixes the volume ratio, x_1^3/x_2^3, which determines the energy requirement. And so, if Δx_1 is the change in particle size,

$$\frac{x_2}{x_1} = \frac{x_1 - \Delta x_1}{x_1} = 1 - \frac{\Delta x_1}{x_1}$$

which fixes volume ratio, x_1^3/x_2^3, and determines the energy requirement.

And so, $\Delta x_1/x_1$ determines the energy requirement for particle size reduction from x_1 to $x_1 - \Delta x_1$. Or

$$\Delta E = C_K\left(\frac{\Delta x}{x}\right)$$

As $\Delta x_1 \to 0$, we have

$$\frac{dE}{dx} = C_K \frac{1}{x} \tag{12.6}$$

This is Kick's law in differential form (C_K is the Kick's law constant). Integrating, we have

$$E = C_K \ln\left(\frac{x_1}{x_2}\right) \tag{12.7}$$

This proposal is unrealistic in most cases since it predicts that the same energy is required to reduce 10 μm particles to 1 μm particles as is required to reduce 1 m boulders to 10 cm blocks. This is clearly not true and Kick's law gives ridiculously low values if data gathered for large product sizes are extrapolated to predict energy requirements for small product sizes.

Bond (1952) suggested a more useful formula, presented in its basic form in Equation (12.8a):

$$E = C_B\left(\frac{1}{\sqrt{x_2}} - \frac{1}{\sqrt{x_1}}\right) \tag{12.8a}$$

However, Bond's law is usually presented in the form shown in Equation (12.8b). The law is based on data which Bond obtained from industrial and laboratory scale processes involving many materials.

$$E_B = W_1\left(\frac{10}{\sqrt{X_2}} - \frac{10}{\sqrt{X_1}}\right) \tag{12.8b}$$

where E_B is the energy required to reduce the top particle size of the material from x_1 to x_2 and W_I is the Bond work index.

Since top size is difficult to define, in practice X_1 to X_2 are taken to be the sieve size in micrometres through which 80% of the material, in the feed and product, respectively, will pass. Bond attached particular significance to the 80% passing size.

Inspection of Equation (12.8b) reveals that W_I is defined as the energy required to reduce the size of unit mass of material from infinity to 100 μm in size. Although the work index is defined in this way, it is actually determined through laboratory scale experiment and assumed to be independent of final product size.

Both E_B and W_I have the dimensions of energy per unit mass and commonly expressed in the units kilowatt-hour per short ton (2000 lb) (1 kWh/short ton ≈ 4000 J/kg). The Bond work index, W_I, must be determined empirically. Some common examples are: bauxite, 9.45 kWh/short ton; coke from coal, 20.7 kWh/short ton; gypsum rock, 8.16 kWh/short ton.

Bond's formula gives a fairly reliable first approximation to the energy requirement provided the product top size is not less than 100 μm. In differential form Bond's formula becomes:

$$\frac{dE}{dx} = C_B \frac{1}{x^{3/2}} \tag{12.9}$$

Attempts have been made (e.g. Holmes, 1957; Hukki, 1961) to find the general formula for which the proposals of Rittinger, Kick and Bond are special cases. It can be seen from the results of the above analysis that these three proposals can be considered as being the integrals of the same differential equation:

$$\frac{dE}{dx} = -C \frac{1}{x^N} \tag{12.10}$$

with

$N = 2$ $\quad C = C_R$ for Rittinger

$N = 1$ $\quad C = C_K$ for Kick

$N = 1.5$ $\quad C = C_B$ for Bond

It has been suggested that the three approaches to prediction of energy requirements mentioned above are each more applicable in certain areas of product size. It is common practice to assume that Kick's proposal is applicable for large particle size (coarse crushing and crushing), Rittinger's for very small particle size (ultra-fine grinding) and the Bond formula being suitable for intermediate particle size–the most common range for many industrial grinding processes. This is shown in Figure 12.2, in which specific energy requirement is plotted against particle size on logarithmic scales. For Rittinger's postulate, $E \propto 1/x$ and so $\ln E \propto -1 \ln(x)$ and hence the slope is −1. For Bond's formula, $E \propto 1/x^{0.5}$ and so $\ln E \propto -0.5 \ln(x)$ and hence the slope is −0.5. For Kick's law, the specific energy requirement is

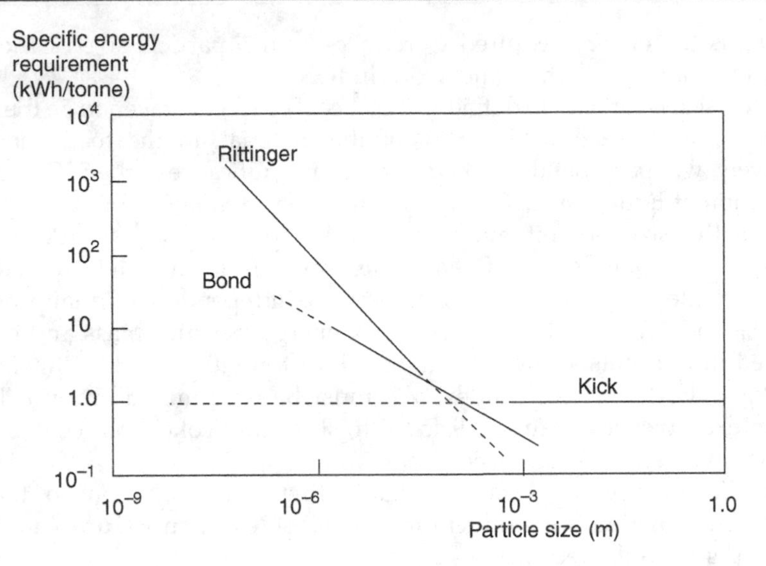

Figure 12.2 Specific energy requirement for breakage: relationship to laws of Rittinger, Bond and Kick

dependent on the reduction ratio x_1/x_2 irrespective of the actual particle size; hence the slope is zero.

In practice, however, it is generally advisable to rely on the past experience of equipment manufacturers and on tests in order to predict energy requirements for the milling of a particular material.

12.3.2 Prediction of the Product Size Distribution

It is common practice to model the breakage process in comminution equipment on the basis of two functions, the specific rate of breakage and the breakage distribution function. The specific rate of breakage S_j is the probability of a particle of size j being broken in unit time (in practice, 'unit time' may mean a certain number of mill revolutions, for example). The breakage distribution function $b(i, j)$ describes the size distribution of the product from the breakage of a given size of particle. For example, $b(i, j)$ is the fraction of breakage product from size interval j which falls into size interval i. Figure 12.3 helps demonstrate the meaning of S_j and $b(i, j)$ when dealing with 10 kg of monosized particles in size interval 1. If $S_1 = 0.6$ we would expect 4 kg of material to remain in size interval 1 after unit time. The size distribution of the breakage product would be described by the set of $b(i, j)$ values. Thus, for example, if $b(4, 1) = 0.25$ we would expect to find 25% by mass from size interval 1 to fall into size interval 4. The breakage distribution function may also be expressed in cumulative form as $B(i, j)$, the fraction of the breakage product from size interval j which falls into size intervals j to n, where n is the total number of size intervals. [$B(i, j)$ is thus a cumulative undersize distribution.]

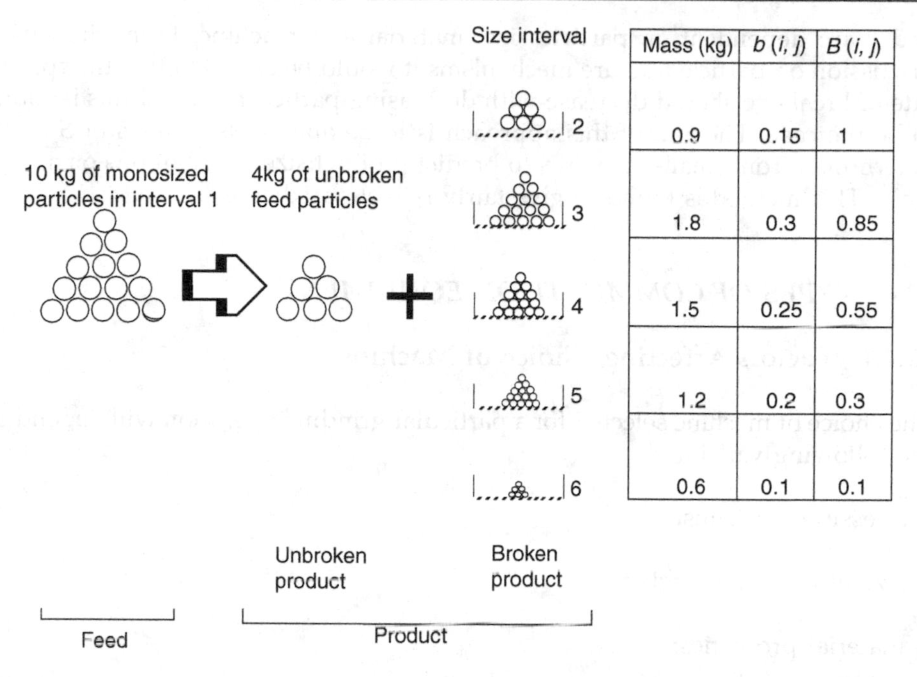

Figure 12.3 Meanings of specific rate of breakage and breakage distribution function

Thus, remembering that S is a *rate* of breakage, we have Equation (12.11), which expresses the rate of change of the mass of particles in size interval i with time:

$$\frac{dm_i}{dt} = \sum_{j=1}^{j=i-1} [b(i,j)S_j m_j] - S_i m_i \tag{12.11}$$

where

$$\sum_{j=1}^{j=i-1} [b(i,j)S_j m_j] = \text{mass broken into interval } i \text{ from all intervals of } j > i$$

$$S_i m_i = \text{mass broken out of interval } i$$

Since $m_i = y_i M$ and $m_j = y_j M$, where M is the total mass of feed material and y_i is the mass fraction in size interval i, then we can write a similar expression for the rate of change of mass fraction of material in size interval, i with time:

$$\frac{dy_i}{dt} = \sum_{j=1}^{j=i-1} [b(i,j)S_j y_j] - S_i y_i \tag{12.12}$$

Thus, with a set of S and b values for a given feed material, the product size distribution after a given time in a mill may be determined. In practice, both S

and b are dependent on particle size, material and machine. From the earlier discussion on particle fracture mechanisms it would be expected that the specific rate of breakage should decrease with decreasing particle size, and this is found to be the case. The aim of this approach is to be able to use values of S and b determined from small-scale tests to predict product size distributions on a large scale. This method is found to give fairly reliable predictions.

12.4 TYPES OF COMMINUTION EQUIPMENT

12.4.1 Factors Affecting Choice of Machine

The choice of machine selected for a particular grinding operation will depend on the following variables:

- stressing mechanism;
- size of feed and product;
- material properties;
- carrier medium;
- mode of operation;
- capacity;
- combination with other unit operations.

12.4.2 Stressing Mechanisms

It is possible to identify three stresses mechanisms responsible for particle size reduction in mills:

(1) Stress applied between two surfaces (either surface–particle or particle–particle) at low velocity, 0.01–10 m/s. Crushing plus attrition, Figure 12.4.

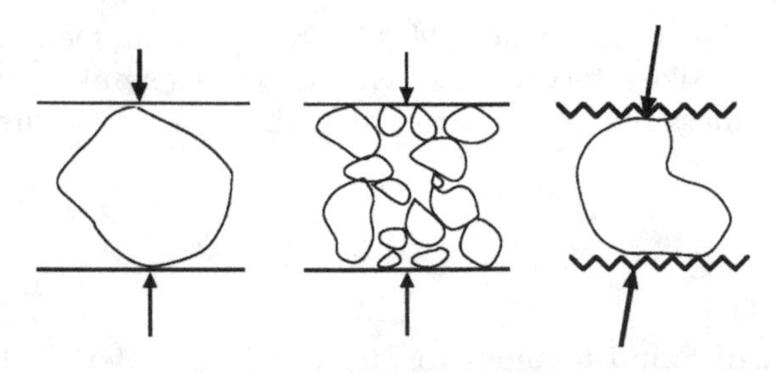

Figure 12.4 Stresses applied between two surfaces

Figure 12.5 Stresses applied at a single solid surface

(2) Stress applied at a single solid surface (surface–particle or particle–particle) at high velocity, 10–200 m/s. Impact fracture plus attrition (Figure 12.5).

(3) Stress applied by carrier medium–usually in wet grinding to bring about disagglomeration.

An initial classification of comminution equipment can be made according to the stressing mechanisms employed, as follows.

Machines using mainly mechanism 1, crushing

The *jaw crusher* behaves like a pair of giant nutcrackers (Figure 12.6). One jaw is fixed and the other, which is hinged at its upper end, is moved towards and away from the fixed jaw by means of toggles driven by an eccentric. The lumps of material are crushed between the jaws and leave the crusher when they are able to pass through a grid at the bottom.

The *gyratory crusher*, shown in Figure 12.7, has a fixed jaw in the form of a truncated cone. The other jaw is a cone which rotates inside the fixed jaw on an eccentric mounting. Material is discharged when it is small enough to pass through the gap between the jaws.

In the *crushing roll* machine two cylindrical rolls rotate in opposite directions, horizontally and side by side with an adjustable gap between them (Figure 12.8). As the rolls rotate, they drag in material which is choke-fed by gravity so that

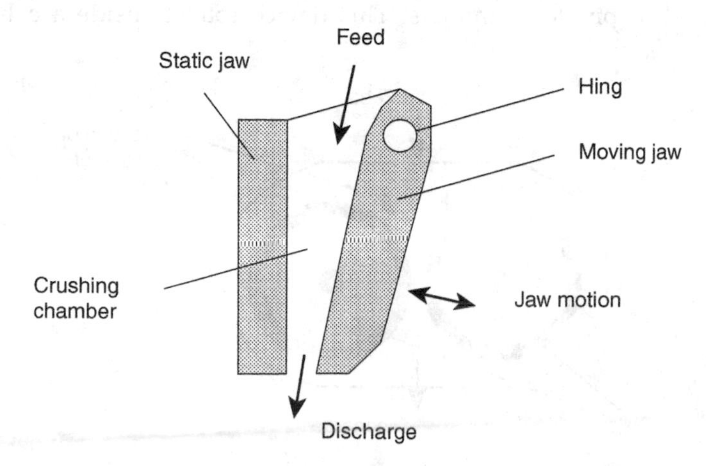

Figure 12.6 Schematic diagram of a jaw crusher

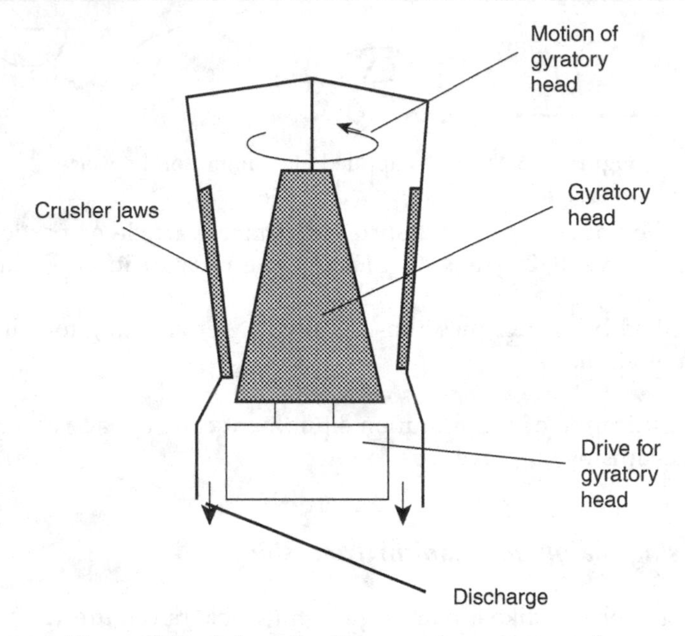

Figure 12.7 Schematic diagram of gyratory crusher

particle fracture occurs as the material passes through the gap between the rolls. The rolls may be ribbed to give improved purchase between material and rolls.

In the *horizontal table mill,* shown in Figure 12.9, the feed material falls on to the centre of a circular rotating table and is thrown out by centrifugal force. In moving outwards the material passes under a roller and is crushed.

Machines using mainly mechanism 2, high velocity impact

The *hammer mill,* shown in Figure 12.10, consists of a rotating shaft to which are attached fixed or pivoted hammers. This device rotates inside a cylinder. The

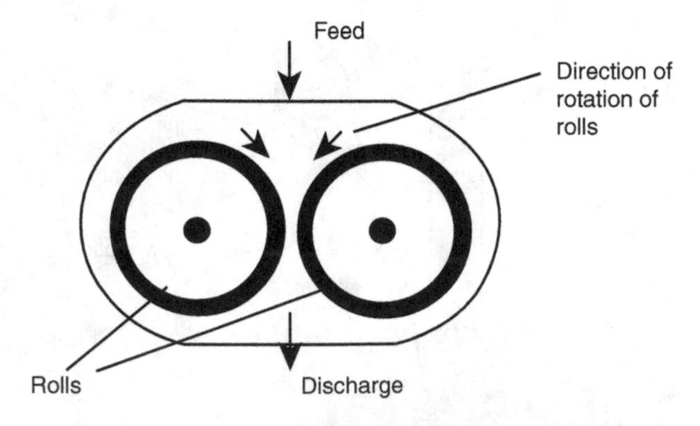

Figure 12.8 Schematic diagram of crushing rolls

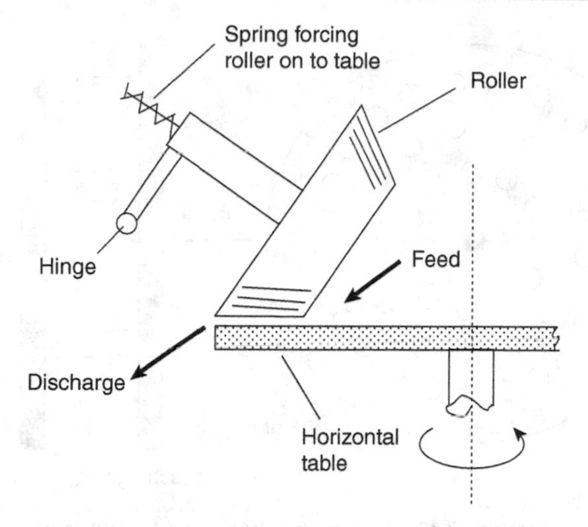

Figure 12.9 Schematic diagram of horizontal table mill

particles are fed into the cylinder either by gravity or by gas stream. In the gravity-fed version the particles leave the chamber when they are small enough to pass through a grid at the bottom.

A *pin mill* consists of two parallel circular discs each carrying a set of projecting pins (Figure 12.11). One disc is fixed and the other rotates at high speed so that its

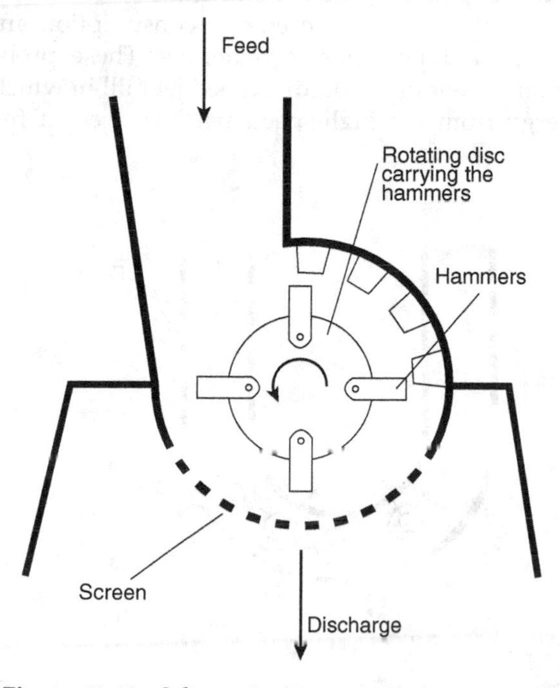

Figure 12.10 Schematic diagram of a hammer mill

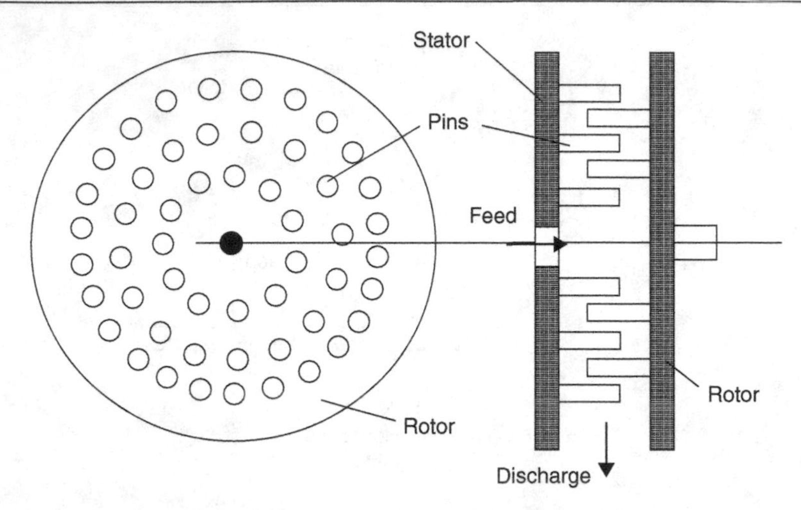

Figure 12.11 Schematic diagram of a pin mill

pins pass close to those on the fixed disc. Particles are carried in air into the centre and as they move radially outwards are fractured by impact or by attrition.

The *fluid energy mill* relies on the turbulence created in high velocity jets of air or steam in order to produce conditions for interparticle collisions which bring about particle fracture. A common form of fluid energy mill is the loop or oval jet mill shown in Figure 12.12. Material is conveyed from the grinding area near the jets at the base of the loop to the classifier and exit situated at the top of the loop. These mills have a very high specific energy consumption and are subject to extreme wear when handling abrasive materials. These problems have been overcome to a certain extent in the fluidized bed jet mill in which the bed is used to absorb the energy from the high-speed particles ejected from the grinding zone.

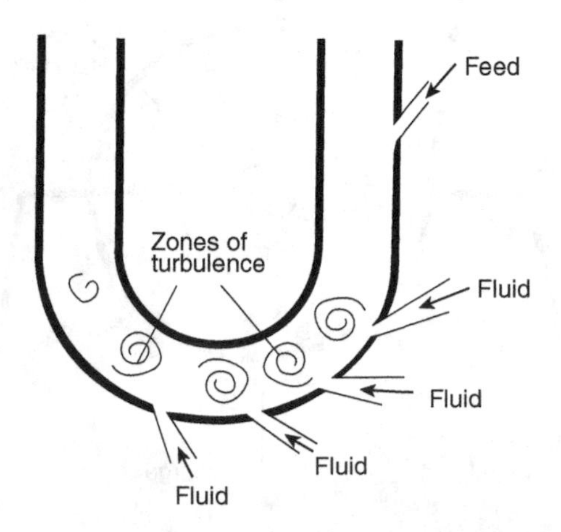

Figure 12.12 Schematic diagram of a fluid energy mill

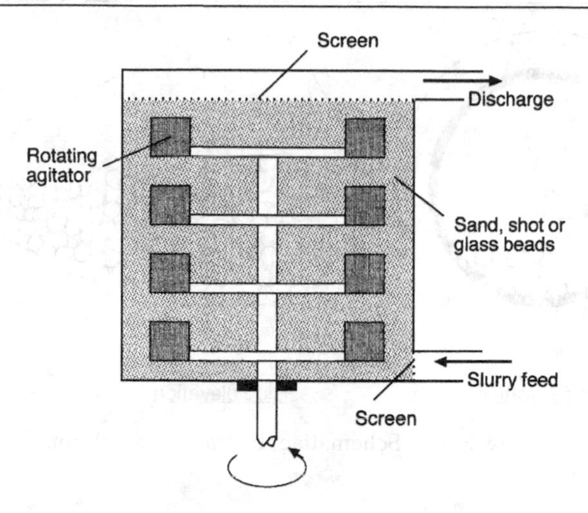

Figure 12.13 Schematic diagram of a sand mill

Machines using a combination of mechanisms 1 and 2, crushing and impact with attrition

The *sand mill*, shown in Figure 12.13, is a vertical cylinder containing a stirred bed of sand, glass beads or shot. The feed, in the form of a slurry, is pumped into the bottom of the bed and the product passes out at the top through a screen which retains the bed material.

In the *colloid mill*, the feed in the form of a slurry passes through the gap between a male, ribbed cone rotating at high speed and a female static cone (Figure 12.14).

The *ball mill*, shown in Figure 12.15, is a rotating cylindrical or cylindrical-conical shell about half filled with balls of steel or ceramic. The speed of rotation of the cylinder is such that the balls are caused to tumble over one another

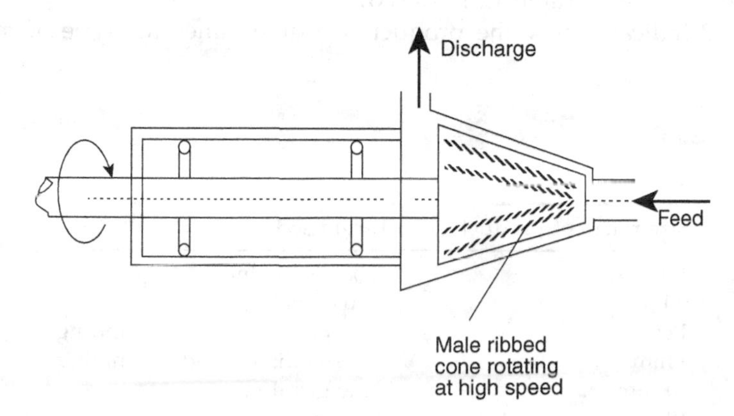

Figure 12.14 Schematic diagram of a colloid mill

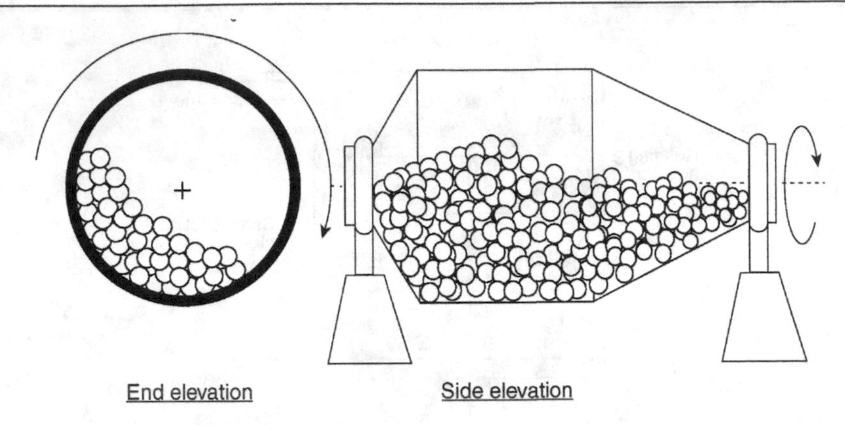

Figure 12.15 Schematiac diagram of a ball mill

without causing cascading. This speed is usually less than 80% of the critical speed which would just cause the charge of balls and feed material to be centrifuged. In continuous milling the carrier medium is air, which may be heated to avoid moisture which tends to cause clogging. Ball mills may also be used for wet grinding with water being used as the carrier medium. The size of balls is chosen to suit the desired product size. The conical section of the mill shown in Figure 12.15 causes the smaller balls to move towards the discharge end and accomplish the fine grinding. Tube mills are very long ball mills which are often compartmented by diaphragms, with balls graded along the length from large at the feed end to small at the discharge end.

12.4.3 Particle Size

Although it is technically interesting to classify mills according to the stressing mechanisms it is the size of the feed and the product size distribution required which in most cases determine the choice of a suitable mill. Generally the terminology shown in Table 12.1 is used.

Table 12.2 indicates how the product size determines the type of mill to be used.

Table 12.1 Terminology used in comminution

Size range of product	Term used
1–0.1 m	Coarse crushing
0.1 m	Crushing
1 cm	Fine crushing, coarse grinding
1 mm	Intermediate grinding, milling
100 μm	Fine grinding
10 μm	Ultrafine grinding

Table 12.2 Categorizing comminution equipment according to product size

Down to 3 mm	3 mm–50 μm	<50 μm
Crushers	Ball mills	Ball mills
Table mills	Rod mills	Vibration mills
Edge runner mills	Pin mills	Sand mills
	Tube mills	Perl mills
	Vibration mills	Colloid mills
		Fluid energy mills

12.4.4 Material Properties

Material properties affect the selection of mill type, but to a lesser extent than feed and product particle size. The following material properties may need to be considered when selecting a mill:

- *Hardness.* Hardness is usually measured on the Mohs' scale of hardness where graphite is ranked 1 and diamond is ranked 10. The property hardness is a measure of the resistance to abrasion.

- *Abrasiveness.* This is linked closely to hardness and is considered by some to be the most important factor in selection of commercial mills. Very abrasive materials must generally be ground in mills operating at low speeds to reduce wear of machine parts in contact with the material (e.g. ball mills).

- *Toughness.* This is the property whereby the material resists the propagation of cracks. In tough materials excess strain energy brings about plastic deformation rather than propagation of new cracks. Brittleness is the opposite of toughness. Tough materials present problems in grinding, although in some cases it is possible to reduce the temperature of the material, thereby reducing the propensity to plastic flow and rendering the material more brittle.

- *Cohesivity/adhesivity.* The properties whereby particles of material stick together and to other surfaces. Cohesivity and adhesivity are related to moisture content and particle size. Decrease of particle size or increasing moisture content increases the cohesivity and adhesivity of the material. Problems caused by cohesivity/adhesivity due to particle size may be overcome by wet grinding.

- *Fibrous nature.* Materials of a fibrous nature are a special case and must be comminuted in shredders or cutters which are based on the hammer mill design.

- *Low melting point.* The heat generated in a mill may be sufficient to cause melting of such materials causing problems of increased toughness and increased cohesivity and adhesivity. In some cases, the problem may be overcome by using cold air as the carrier medium.

- *Other special properties.* Materials which are thermally sensitive and have a tendency to spontaneous combustion or high inflammability must be ground using an inert carrier medium (e.g. nitrogen). Toxic or radioactive materials must be ground using a carrier medium operating on a closed circuit.

12.4.5 Carrier Medium

The carrier medium may be a gas or a liquid. Although the most common gas used is air, inert gases may be used in some cases as indicated above. The most common liquid used in wet grinding is water although oils are sometimes used. The carrier medium not only serves to transport the material through the mill but, in general, transmits forces to the particles, influences friction and hence abrasion, affects crack formation and cohesivity/adhesivity. The carrier medium can also influence the electrostatic charging and the flammability of the material.

12.4.6 Mode of Operation

Mills operate in either batch or continuous mode. Choice between modes will be based on throughput, the process and economics. The capacity of batch mills varies from a few grammes on the laboratory scale to a few tonnes on a commercial scale. The throughput of continuous milling systems may vary from several hundred grammes per hour at laboratory scale to several thousand tonnes per hour at industrial scale.

12.4.7 Combination with other Operations

Some mills have a dual purpose and thus may bring about drying, mixing or classification of the material in addition to its size reduction.

12.4.8 Types of Milling Circuit

Milling circuits are either 'open circuit' or 'closed circuit'. In open circuit milling (Figure 12.16) the material passes only once through the mill, and so the only controllable variable is the residence time of the material in the mill. Thus the

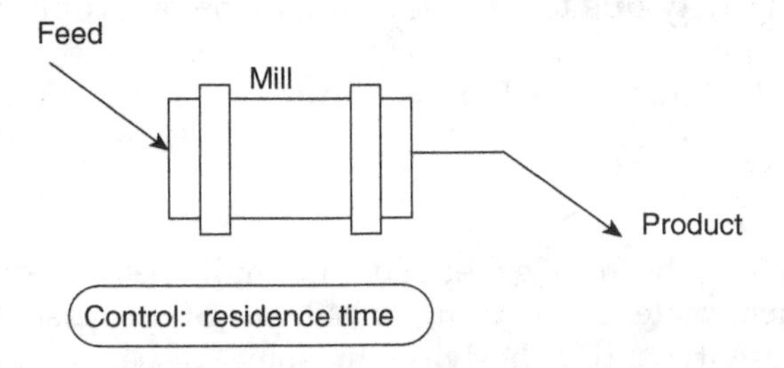

Figure 12.16 Open circuit milling

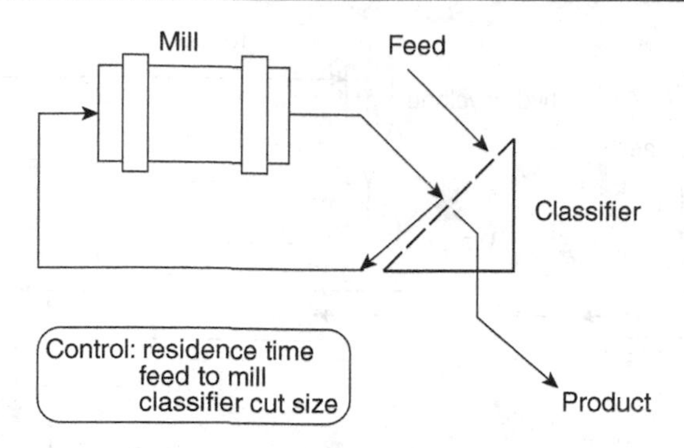

Figure 12.17 Closed circuit milling

product size and distribution may be controlled over a certain range by varying the material residence time (thoughput), i.e. feed rate governs product size and so the system is inflexible.

In closed circuit milling (Figure 12.17) the material leaving the mill is subjected to some form of classification (separation according to particle size) with the oversize being returned to the mill with the feed material. Such a system is far more flexible since both product mean size and size distribution may be controlled.

Figures 12.18 and 12.19 show the equipment necessary for feeding material into the mill, removing material from the mill, classifying, recycling oversize material and removing product in the case of dry and wet closed milling circuits, respectively.

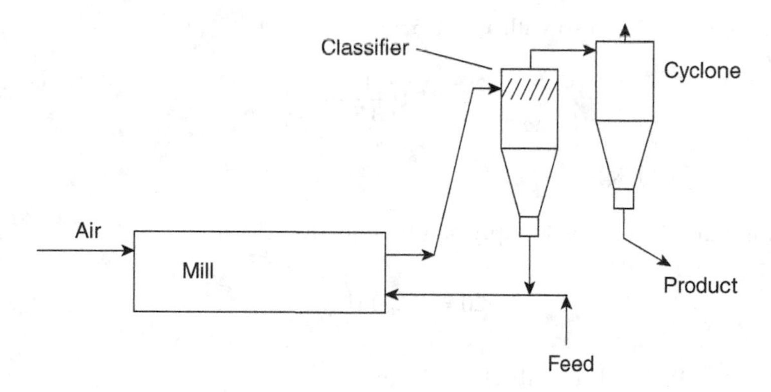

Figure 12.18 Dry milling: closed circuit operation

12.5 *WORKED EXAMPLES*

WORKED EXAMPLE 12.1

A material consisting originally of 25 mm particles is crushed to an average size of 7 mm and requires 20 kJ/kg for this size reduction. Determine the energy required to crush the

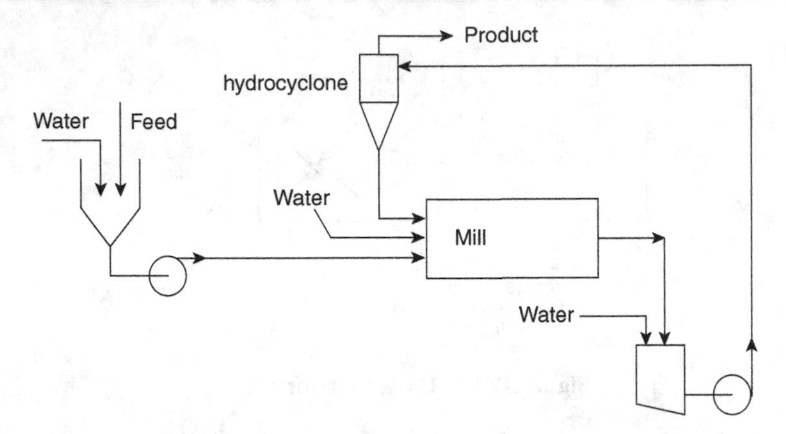

Figure 12.19 Wet milling: closed circuit operation

material from 25 mm to 3.5 mm assuming (a) Rittinger's law, (b) Kick's law and (c) Bond's law.

Solution

(a) Applying Rittinger's law as expressed by Equation (12.4):

$$20 = C_R\left(\frac{1}{7} - \frac{1}{25}\right)$$

hence $C_R = 194.4$ and so with $x_2 = 3.5\,\text{mm}$,

$$E = 194.4\left(\frac{1}{3.5} - \frac{1}{25}\right)$$

hence $E = 47.8\,\text{kJ/kg}$

(b) Applying Kick's law as expressed by Equation (12.7):

$$20 = -C_K \ln\left(\frac{7}{25}\right)$$

hence $C_K = 15.7$ and so with $x_2 = 3.5\,\text{mm}$,

$$E = -15.7 \ln\left(\frac{3.5}{25}\right)$$

hence $E = 30.9\,\text{kJ/kg}$

(c) Applying Bond's law as expressed by Equation (12.8a):

$$20 = C_B\left(\frac{1}{\sqrt{7}} - \frac{1}{\sqrt{25}}\right)$$

Table 12W2.1 Specific rates of breakage and breakage distribution function for the ball mill

Size interval (μm)	212–150	150–106	106–75	75–53	53–37	37–0
Interval no.	1	2	3	4	5	6
S_j	0.7	0.6	0.5	0.35	0.3	0
$b(1, j)$	0	0	0	0	0	0
$b(2, j)$	0.32	0	0	0	0	0
$b(3, j)$	0.3	0.4	0	0	0	0
$b(4, j)$	0.14	0.2	0.5	0	0	0
$b(5, j)$	0.12	0.2	0.25	0.6	0	0
$b(6, j)$	0.12	0.2	0.25	0.4	1.0	0

hence $C_B = 112.4$ and so with $x_2 = 3.5\,\text{mm}$,

$$E = 112.4\left(\frac{1}{\sqrt{3.5}} - \frac{1}{25}\right)$$

hence $E - 37.6\,\text{kJ/kg}$

WORKED EXAMPLE 12.2

Values of breakage distribution function *b(i, j)* and specific rates of breakage S_j for a particular material in a ball mill are shown in Table 12W2.1. To test the validity of these values, a sample of the material with the size distribution indicated in Table 12W2.2 is to be ground in a ball mill. Use the information in these tables to predict the size distribution of the product after one minute in the mill. (*Note*: S_j values in Table 12W2.1 are based on 1 minute grinding time.)

Solution

Applying Equation (12.12):

Change of fraction in interval 1

$$\frac{dy_1}{dt} = 0 - S_1 y_1 = 0 - 0.7 \times 0.2$$
$$= -0.14$$

Hence, new $y_1 = 0.2 - 0.14 = 0.06$

Table 12W2.2 Feed size distribution

Interval no. (j)	1	2	3	4	5	6
Fraction	0.2	0.4	0.3	0.06	0.04	0

Change of fraction in interval 2

$$\begin{aligned}\frac{dy_2}{dt} &= b(2,1)S_1y_1 - S_2y_2 \\ &= (0.32 \times 0.7 \times 0.2) - (0.6 \times 0.4) \\ &= -0.1952\end{aligned}$$

Hence new $y_2 = 0.4 - 0.1952 = 0.2048$

Change of fraction in interval 3

$$\begin{aligned}\frac{dy_3}{dt} &= [b(3,1)S_1y_1 + b(3,2)S_2y_2] - S_3y_3 \\ &= [(0.3 \times 0.7 \times 0.2) + (0.4 \times 0.6 \times 0.4)] - (0.5 \times 0.3) \\ &= -0.012\end{aligned}$$

Hence, new $y_3 = 0.3 - 0.012 = 0.288$

Similarly for intervals 4, 5 and 6:

$$\begin{aligned}\text{new } y_4 &= 0.1816 \\ \text{new } y_5 &= 0.1429 \\ \text{new } y_6 &= 0.1227\end{aligned}$$

Checking:

Sum of predicted product interval mass fractions $= y_1 + y_2 + y_3 + y_4 + y_5 + y_6 = 1.000$

Hence product size distribution:

Interval no. (j)	1	2	3	4	5	6
Fraction	0.06	0.2048	0.288	0.1816	0.1429	0.1227

TEST YOURSELF

12.1 Explain why in brittle materials multiple fracture is common.

12.2 Using the concept of failure by crack propagation, explain why small particles are more difficult to break than large particles.

12.3 Summarize three different models for predicting the energy requirement associated with particle size reduction. Over what size ranges might each model be most appropriately applied?

12.4 Define *selection function* and *breakage function*, functions used in the prediction of product size distribution in a size reduction process.

12.5 Explain in words the meaning for the following equation.

$$\frac{dy_i}{dt} = \sum_{j=1}^{j=i-1} [b(i,j)S_j y_j] - S_i y_i$$

12.6 Describe the operation of the *hammer mill*, the *fluid energy mill* and the *ball mill*. In each case identify the dominant stressing mechanism responsible for particle breakage.

12.7 Under what conditions might wet grinding be used?

12.8 List five material properties that would influence selection of a mill type.

EXERCISES

12.1 (a) Rittinger's energy law postulated that the energy expended in crushing is proportional to the area of new surface created. Derive an expression relating the specific energy consumption in reducing the size of particles from x_1 to x_2 according to this law.

(b) Table 12E1.1 gives values of specific rates of breakage and breakage distribution functions for the grinding of limestone in a hammer mill. Given that values of specific rates of breakage are based on 30 s in the mill at a particular speed, determine the size distribution of the product resulting from the feed described in Table 12E1.2 after 30 s in the mill at this speed.

(Answer: 0.12, 0.322, 0.314, 0.244.)

Table 12E1.1 Specific rates of breakage and breakage distribution function for the hammer mill

Interval (μm) Interval no. j	106–75 1	75–53 2	53–37 3	37–0 4
S_j	0.6	0.5	0.45	0
$b(1, j)$	0	0	0	0
$b(2, j)$	0.4	0	0	0
$b(3, j)$	0.3	0.6	0	0
$b(4, j)$	0.3	0.4	1.0	0

Table 12E1.2 Feed size distribution

Interval	1	2	3	4
Fraction	0.3	0.5	0.2	0

12.2 Table 12E2.1 gives information gathered from tests on the size reduction of coal in a ball mill. Assuming that the values of specific rates of breakage, S_j, are based on 25 revolutions of the mill at a particular speed, predict the product size distribution resulting from the feed material, details of which are given in Table 12E2.2 after 25 revolutions in the mill at that speed.

(Answer: 0.125, 0.2787, 0.2047, 0.1661, 0.0987, 0.0779, 0.04878.)

Table 12E2.1 Results of ball mill tests on coal

Interval (μm) Interval no. j	300–212 1	212–150 2	150–106 3	106–75 4	75–53 5	53–37 6	37–0 7
S_j	0.5	0.45	0.42	0.4	0.38	0.25	0
$b(1, j)$	0	0	0	0	0	0	0
$b(2, j)$	0.25	0	0	0	0	0	0
$b(3, j)$	0.24	0.29	0	0	0	0	0
$b(4, j)$	0.19	0.27	0.33	0	0	0	0
$b(5, j)$	0.12	0.2	0.3	0.45	0	0	0
$b(6, j)$	0.1	0.16	0.25	0.3	0.6	0	0
$b(7, j)$	0.1	0.08	0.12	0.25	0.4	1.0	0

Table 12E2.2 Feed size distribution

Interval	1	2	3	4	5	6	7
Fraction	0.25	0.45	0.2	0.1	0	0	0

12.3 Table 12E3.1 gives information on the size reduction of a sand-like material in a ball mill. Given that the values of specific rates of breakage S_j are based on five revolutions of the mill, determine the size distribution of the feed materials shown in Table 12E3.2 after five revolutions of the mill.

(Answer: 0.0875, 0.2369, 0.2596, 0.2115, 0.2045.)

Table 12E3.1 Results of ball mill tests

Interval (μm) Interval no. (j)	150–106 1	106–75 2	75–53 3	53–37 4	37–0 5
S_j	0.65	0.55	0.4	0.35	0
$b(1, j)$	0	0	0	0	0
$b(2, j)$	0.35	0	0	0	0
$b(3, j)$	0.25	0.45	0	0	0
$b(4, j)$	0.2	0.3	0.6	0	0
$b(5, j)$	0.2	0.25	0.4	1.0	0

Table 12E3.2 Feed size distribution

Interval	1	2	3	4	5
Fraction	0.25	0.4	0.2	0.1	0.05

12.4 Comminution processes are generally less than 1% efficient. Where does all the wasted energy go?

13

Size Enlargement

Karen Hapgood and Martin Rhodes

13.1 INTRODUCTION

Size enlargement is the process by which smaller particles are put together to form larger masses in which the original particles can still be identified. Size enlargement is one of the single most important process steps involving particulate solids in the process industries. Size enlargement is mainly associated with the pharmaceutical, agricultural and food industries, but also plays an important role in other industries including minerals, metallurgical and ceramics.

There are many reasons why we may wish to increase the mean size of a product or intermediate. These include reduction of dust hazard (explosion hazard or health hazard), to reduce caking and lump formation, to improve flow properties, increase bulk density for storage, creation of non-segregating mixtures of ingredients of differing original size, to provide a defined metered quantity of active ingredient (e.g. pharmaceutical drug formulations), and control of surface to volume ratio (e.g. in catalyst supports).

Methods by which size enlargement is brought about include granulation, compaction (e.g. tabletting), extrusion, sintering, spray drying and prilling. Agglomeration is the formation of agglomerates or aggregates by sticking together of smaller particles and granulation is agglomeration by agitation methods. Since it is not possible in the confines of a single chapter to cover all size enlargement methods adequately, the focus of this chapter will be on granulation, which will serve as an example.

In this chapter the different types of interparticle forces and their relative importance as a function of particle size are summarized. Liquid bridge forces, which are specifically important to granulation processes, are covered in more detail. The rate processes important to granulation (wetting, growth, consolidation

Introduction to Particle Technology - 2nd Edition Martin Rhodes

and attrition) are reviewed and population balance included in order to develop a simple model for simulation for the granulation process. To conclude, a brief overview of industrial granulation equipment is also given.

13.2 INTERPARTICLE FORCES

13.2.1 van der Waals Forces

There exist between all solids molecularly based attractive forces collectively known as van der Waals forces. The energy of these forces is of the order of 0.1 eV and decreases with the sixth power of the distance between molecules. The range of van der Waals forces is large compared with that of chemical bonds. The attractive force, F_{vw}, between a sphere and a plane surface as a result of van der Waals forces was derived by Hamaker (1937) and is usually presented in the form:

$$F_{vw} = \frac{K_H R}{6y^2} \tag{13.1}$$

where K_H is the Hamaker constant, R is the radius of the sphere and y is the gap between the sphere and the plane.

13.2.2 Forces due to Adsorbed Liquid Layers

Particles in the presence of a condensable vapour will have a layer of adsorbed vapour on their surface. If these particles are in contact a bonding forces results from the overlapping of the adsorbed layers. The strength of the bond is dependent on the area of contact and the tensile strength of the adsorbed layers. The thickness and strength of the layers increase with increasing partial pressure of the vapour in the surrounding atmosphere. According to Coelho and Harnby (1978) there is a critical partial pressure at which the adsorbed layer bonding gives way to liquid bridge bonding.

13.2.3 Forces due to Liquid Bridges

In addition to the interparticle forces resulting from adsorbed liquid layers, as described above, even in very small proportions the presence of liquid on the surface of particles affects the interparticle forces by the smoothing effect it has on surface imperfections (increasing particle–particle contact) and its effect of reducing the interparticle distance. However, these forces are usually negligible in magnitude compared with forces resulting when the proportion of liquid present is sufficient to form interparticle liquid bridges. Newitt and Conway-Jones (1958) identified four types of liquid states depending on the proportion of liquid present between groups of particles. These states are known as pendular, funicular,

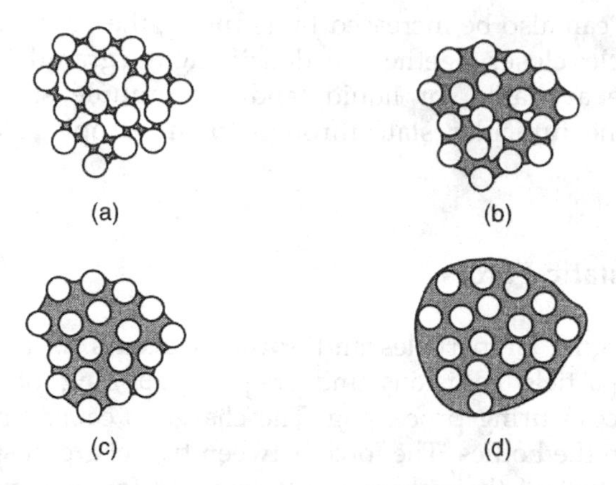

Figure 13.1 Liquid bonding between particles: (a) pendular; (b) funicular; (c) capillary; (d) droplet

capillary and droplet. These are shown in Figure 13.1. In the pendular state liquid is held as a point contact in a bridge neck between particles. The liquid bridges are all separate and independent of each other. The interstitial space between the particles is the porosity or voidage. The strong boundary forces resulting from the surface tension of the liquid draw the particles together. In addition there is capillary pressure resulting from the curved liquid surfaces of the bridge. The capillary pressure is given by:

$$p_c = \gamma\left(\frac{1}{r_1} - \frac{1}{r_2}\right) \tag{13.2}$$

where γ is the liquid surface tension and r_1 and r_2 are the radii of curvature of the liquid surfaces. If the pressure in the liquid bridge is less than the ambient pressure, the capillary pressure and surface tension boundary attractive force are additive.

As the proportion of liquid to particles is increased the liquid is free to move and the attractive force between particles decreases (funicular). When there is sufficient liquid to completely fill the interstitial pores between the particles (capillary) the granule strength falls further as there are fewer curved liquid surfaces and fewer boundaries for surface tension forces to act on. Clearly when the particles are completely dispersed in the liquid (droplet) the strength of the structure is very low.

In the pendular state increasing the amount of liquid present has little effect on the strength of the bond between the particles until the funicular state is achieved. However, increasing the proportion of liquid increases the resistance of the bond to rupture since the particles can be pulled further apart without rupture of bridges. This has practical implication for granulation processes; pendular bridges give rise to strong granules in which the quantity of liquid is not critical but should be less than that required to move into the funicular and capillary regimes.

The saturation can also be increased by reducing the voidage or porosity and moving the particles closer together (i.e. densifying the granule). This reduces the open pore space available for liquid, and the granule saturation gradually increases from the funicular state through to the droplet state as shown in Figure 13.1.

13.2.4 Electrostatic Forces

Electrostatic charging of particles and surfaces occurs as a result of friction caused by interparticle collisions and frequent rubbing of particles against equipment surfaces during processing. The charge is caused by the transfer of electrons between the bodies. The force between two charged spheres is proportional to the product of their charges. Electrostatic forces may be attractive or repulsive, do not require contact between particles and can act over relatively long distances compared with adhesional forces which require contact.

13.2.5 Solid Bridges

Granules formed by liquid bridges are usually not the end product in a granulation process. More permanent bonding within the granule is created by solid bridges formed as liquid is removed from the original granule. Solid bridges between particles may take three forms: crystalline bridges, liquid binder bridges and solid binder bridges. If the material of the particles is soluble in the liquid added to create granules, crystalline bridges may be formed when the liquid evaporates. The process of evaporation reduces the proportion of liquid in the granules producing high strength pendular bridges before crystals form. Alternatively the liquid used initially to form the granules may contain a binder or glue which takes effect upon evaporation of solvent.

In some cases a solid binder may be used. This is a finely ground solid which reacts with the liquid present to produce a solid cement to hold the particles together.

13.2.6 Comparison and Interaction between Forces

In practice all interparticle forces act simultaneously. The relative importance of the forces varies with changes in particle properties and with changes in the humidity of the surrounding atmosphere. There is considerable interaction between the bonding forces. For example, in aqueous systems adsorbed moisture can considerably increase van der Waals forces. Adsorbed moisture can also reduce interparticle friction and potential for interlocking, making the powder more free-flowing. Electrostatic forces decay rapidly if the humidity of the surrounding air is increased.

A powder which in a dry atmosphere exhibits cohesivity due to electrostatic charging may become more free-flowing as humidity of the atmosphere is

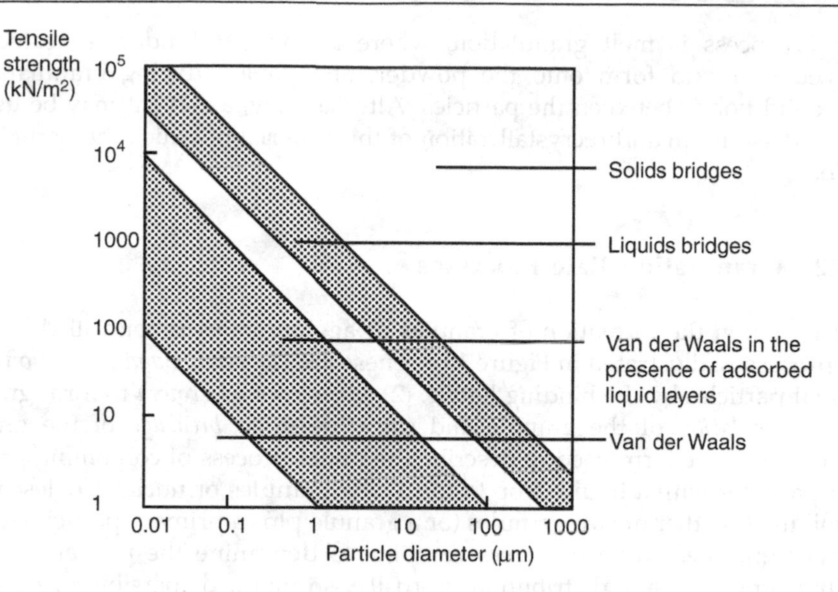

Figure 13.2 Theoretical tensile strength of agglomerates with different bonding mechanisms (after Rumpf, 1962)

increased. If humidity is further increased liquid bridge formation can result in a return to cohesive behaviour. This effect has been reported in powder mixing studies by Coelho and Harnby (1978) and Karra and Fuerstenau (1977).

Figure 13.2, after Rumpf (1962), illustrates the relative magnitude of the different bonds discussed above as a function of particle size. We see that van der Waals forces become important only for particles below 1 μm in size, adsorbed vapour forces are relevant below 80 μm and liquid bridge forces are active below about 500 μm.

13.3 *GRANULATION*

13.3.1 Introduction

Granulation is particle size enlargement by sticking together smaller particles using agitation to impart energy to particles and granules. The most common type of granulation method is wet granulation where a liquid binder is distributed over the bed to initiate granule formation. The resulting assembly of particles is called a 'granule' and consists of the primary particles arranged as a three-dimensional porous structure. Important properties of granules include their size, porosity, strength, flow properties, dissolution, and composition uniformity.

The motion of the particles and granules in the granulator results in collisions which produce growth by coalescence and coating. In general the individual component particles are held together by a liquid (glue-like) or solid (cement-like) binder which is sprayed into the agitated particles in the granulator. A hybrid process is melt granulation, where a polymer binder is heated and sprayed in liquid form onto the powder, and cooled during granulation to form solid bonds between the particles. Alternatively, a solvent may be used to induce dissolution and recrystallization of the material of which the particles are made.

13.3.2 Granulation Rate Processes

In granulation the formation of granules or agglomerates is controlled by three rate processes, illustrated in Figure 13.3. These are (1) *wetting and nucleation* of the original particles by the binding liquid, (2) *coalescence* or *growth* to form granules plus *consolidation* of the granule and (3) attrition or *breakage* of the granule. Nucleation is the term used to describe the initial process of combining primary solid particles with a liquid drop to form new granules or nuclei. Coalescence is the joining together of two granules (or a granule plus a primary particle) to form a larger granule. These processes combine to determine the properties of the product granule (size distribution, porosity, strength, dispersibility, etc.). The

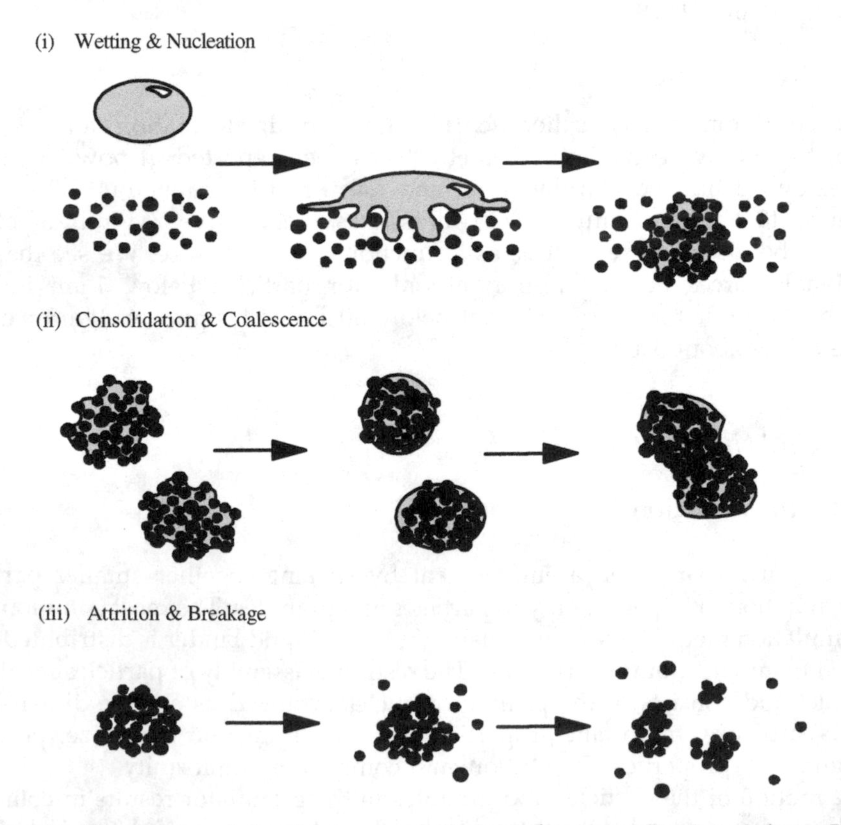

Figure 13.3 Summary of the three granulation mechanisms. Reprinted from *Powder Technology*, **117** (1–2), Iveson *et al.*, 'Nucleation, growth and breakage phenomena in agitated wet granulation processes: a review', pp. 3–39, Copyright (2001), with permission from Elsevier

final granule size in a granulation process is controlled by the competing mechanisms of growth, breakage and consolidation.

Wetting and Nucleation

Wetting is the process by which air within the voids between particles is replaced by liquid. Ennis and Litster (1997) stress the important influence which the extent and rate of *wetting* have on product quality in a granulation process. For example, poor wetting can result in much material being left ungranulated and requiring recycling. When granulation is used to combine ingredients, account must be taken of the different wetting properties which components of the final granule may have.

Wetting is governed by the surface tension of the liquid and the contact angle it forms with the material of the particles. The rate at which wetting occurs is important in granulation. An impression of this rate is given by the Washburn equation [Equation (13.3)] for the rate of penetration of liquid of viscosity μ and surface tension γ into a bed of powder when gravity is not significant:

$$\frac{dz}{dt} = \frac{R_p \gamma \cos\theta}{4\mu z} \tag{13.3}$$

where t is the time, z is the penetration distance of the liquid into the powder and θ is the dynamic contact angle of the liquid with the solid of the powder. R_p is the average pore radius which is related to the packing density and the size distribution of the powder. Thus in granulation, the factors controlling the rate of wetting are adhesive tension ($\gamma \cos\theta$), liquid viscosity, packing density and the size distribution.

The Washburn test requires some specialized testing equipment to perform, and an alternate test is the drop penetration time test which is more directly related to the wetting of spray drops into a granulating powder (Hapgood *et al.*, 2003). A drop of known volume V_d is gently placed onto a small powder bed with porosity ε_b and the time taken for the drop to completely sink into the powder bed is measured. The drop penetration time, t_p, is given by:

$$t_p = 1.35 \frac{V_d^{2/3} \mu}{\varepsilon_b^2 R \gamma \cos\theta} \tag{13.4}$$

The Washburn test and drop penetration time test are closely related, but the latter is simpler to perform as a screening and investigation test when developing or troubleshooting a granulation process.

In general, improved wetting is desirable. It gives a narrower granule size distribution and improved product quality through better control over the granulation process. In practice, the rate of wetting will significantly influence the extent of wetting of the powder mass, especially where evaporation of binder solvent takes place simultaneously with wetting. The brief analysis above shows us that the rate of wetting is increased by reducing viscosity, increasing surface tension, minimizing contact angle and increasing the size of pores within the powder. Viscosity is determined by the binder concentration and the operating temperature. As concentration changes with solvent evaporation, the binder viscosity will increase. Small particles give small pores and large particles give large pores. Also, a wider particle size distribution will give rise to smaller pores. Large pores ensure a high rate of liquid penetration but give rise to a lower extent of wetting.

In addition to the wetting characteristics, the drop size and overall distribution of liquid are crucial parameters in granulation. If one drop is added to a granulator, only one nucleus granule will be formed, and the size of the nucleus granule will be proportional to the size of the drop (Waldie, 1991). During atomization of the fluid onto the powder, it is important the conditions in the spray zone balance the rate of incoming fluid drops with the rate of penetration into the powder and/or removal of wet powder from the spray zone. Ideally, each spray droplet should land on the powder without touching other droplets and sink quickly into the powder to form a new nucleus granule. These ideal conditions are called the 'drop controlled nucleation regime' and occur at low penetration time (described above) and low dimensionless spray flux Ψ_a, which is given by (Litster *et al.*, 2001):

$$\psi_a = \frac{3Q}{2v_s w_s d_d} \tag{13.5}$$

where Q is the solution flow rate, v_s is the powder velocity in the spray zone, d_d is the average drop diameter and w_s is the width of the spray. The dimensionless spray flux is a measure of the density of drops falling on the powder surface. At low spray flux ($\psi_a \ll 1$) drop footprints will not overlap and each drop will form a separate nucleus granule. At high spray flux ($\psi_a \approx 1$) there will significant overlap of drops hitting the powder bed. Nuclei granules formed will be much larger and their size will no longer be a simple function of the original drop size. At a given spray flux value, the fraction of the powder surface that is wetted by spray drops as it passes beneath the spray zone (f_{wet}) is given by (Hapgood *et al.*, 2004):

$$f_{\text{wet}} = 1 - exp(-\Psi_a) \tag{13.6}$$

and the fraction of nuclei f_n formed by n drops can be calculated using (Hapgood *et al.*, 2004):

$$f_n = \exp(-4\Psi_a)\left[\frac{(4\Psi_a)^{n-1}}{(n-1)!}\right] \tag{13.7}$$

The dimensionless spray flux parameter can be used both as a scale-up parameter and as a parameter to estimate nuclei starting sizes for population balance modelling (see Section 13.3.3). When combined with the drop penetration time, Ψ_a forms part of a nucleation regime map (see Figure 13.4) (Hapgood

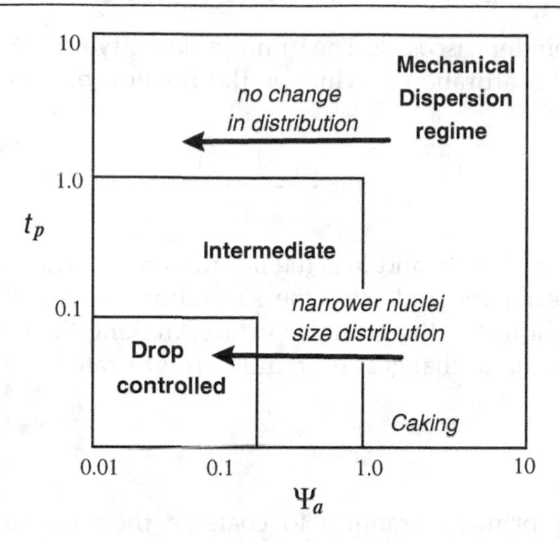

Figure 13.4 Nucleation regime map. For ideal nucleation in the drop controlled regime, low Ψ_a and low t_p are required. Reprinted from *AIChE Journal*, **49** (2), Hapgood *et al.*, 'Nucleation regime map for liquid bound granules', pp. 350–361. Reproduced with permission. Copyright 2003 American Institute of chemical Engineers (AIChE)

et al., 2003). Three nucleation regimes have been defined: *drop controlled, shear controlled,* and an *intermediate* zone. Drop controlled nucleation occurs when one drop forms one nucleus and should occur when there is both:

(1) Low spray flux ψ_a – the spray density is low and relatively few drops overlap;

(2) Fast penetration time t_p – the drop must penetrate completely into the powder bed before it touches either other drops on the powder surface or new drops arriving from the spray.

If *either* criterion is not met, powder mixing and shear characteristics will dominate: this is the mechanical *dispersion regime.* Viscous or poorly wetting binders are slow to flow through the powder pores and form nuclei. Drop coalescence on the powder surface (also known as 'pooling') may occur and create a very broad nuclei size distribution. In the mechanical dispersion regime, the liquid binder can only be dispersed by powder shear and agitation.

Granule Consolidation

Consolidation is the term used to describe the increase in granule density caused by closer packing of primary particles as liquid is squeezed out as a result of collisions. Consolidation can only occur whilst the binder is still liquid. Consolidation determines the porosity and density of the final granules. Factors influencing the rate and degree of consolidation include particle size, size distribution and binder viscosity. The granule porosity ε, and the liquid level w, control the granule saturation s, which is the fraction of pore space filled with liquid:

$$s = \frac{w\rho_s(1-\varepsilon)}{\rho_l\varepsilon} \tag{13.8}$$

where ρ_s is the solid density and ρ_l is the liquid density. The saturation increases as the porosity decreases, and once the saturation exceeds 100%, further consolidation pushes liquid to the granule surface, making the surface wet. Surface wetness causes dramatic changes in granule growth rates (see below).

Growth

For two colliding primary granules to coalesce their kinetic energy must be dissipated and the strength of the resulting bond must be able to resist the external forces exerted by the agitation of the powder mass in the granulator. Granules which are able to deform readily will absorb the collisional energy and create increased surface area for bonding. As granules grow so do the internal forces trying to pull the granule apart. It is possible to predict a critical maximum size of granule beyond which coalescence is not possible during collision (see below).

Ennis and Litster (1997) suggest a rationale for interpreting observed granule growth regimes in terms of collision physics. Consider collision between two rigid granules (assume low deformability) of density ρ_g, each coated with a layer of thickness h of liquid of viscosity μ, having a diameter x and approach velocity V_{app}. The parameter which determines whether coalescence will occur is a Stokes number Stk:

$$Stk = \frac{\rho_g V_{app} x}{16\mu} \tag{13.9}$$

This Stokes number is different from that used in cyclone scale-up in Chapter 9. The cyclone scale-up Stokes number Stk_{50} incorporates the dimensionless ratio of particle size to cyclone diameter, i.e.:

$$Stk_{50} = \frac{\rho v x_{50}^2}{18\mu D} = \left(\frac{x_{50}}{D}\right)\frac{\rho v x_{50}}{18\mu}$$

where v is the characteristic gas velocity.

The Stokes number is a measure of the ratio of collisional kinetic energy to energy dissipated through viscous dissipation. For coalescence to occur the Stokes number must be less than a critical value Stk^*, given by:

$$Stk^* = \left(1+\frac{1}{e}\right)\ln\left(\frac{h}{h_a}\right) \tag{13.10}$$

where e is the coefficient of restitution for the collision and h_a is a measure of surface roughness of the granule.

Based on this criterion, three regimes of granule growth are identified for batch systems with relatively low agitation intensity. These are the non-inertial, inertial and coating regimes. Within the granulator at any time there will be a distribution of granule sizes and velocities which gives rise to a distribution of Stokes numbers for collisions. In the non-inertial regime Stokes number is less than Stk^* for all granules and primary particles and practically all collisions result in coalescence. In this regime therefore the growth rate is largely independent of liquid viscosity, granule or primary particle size and kinetic energy of collision. The rate of wetting of the particles controls the rate of growth in this regime.

As granules grow some collisions will occur for which the Stokes number exceeds the critical value. We now enter the inertial regime in which the rate of growth is dependent on liquid viscosity, granule size and collision energy. The proportion of collisions for which the Stokes number exceeds the critical value increases throughout this regime and the proportion of successful collisions decreases. Once the average Stokes number for the powder mass in the granulator is comparable with the critical value, granule growth is balanced by breakage and growth continues by coating of primary particles onto existing granules, since these are the only possible successful collisions according to our criterion. This simple analysis breaks down when granule deformation cannot be ignored, as in high agitation intensity systems (Ennis and Litster, 1997).

The two types of growth behaviour that occur depending on granule deformation are shown in Figure 13.5 (Iveson *et al.*, 2001). Steady growth occurs when the granule size increase is roughly proportional to granulation time i.e. a plot of granule size versus time is linear. Induction growth occurs when there is a long period during which no increase in size occurs. The granules form and consolidate, but do not grow further until the granule porosity is reduced enough to squeeze liquid to the surface. This excess free liquid on the granules causes sudden coalescence of many granules and a rapid increase in granule size (Figure 13.5).

Deformation during granule collisions can be characterized by a Stokes deformation number, St_{def} relating the kinetic energy of collision to the energy dissipated during granule deformation (Tardos *et al.*, 1997):

$$St_{\text{def}} = \frac{\rho_g U_c^2}{2Y_d} \tag{13.11}$$

where U_c is the representative collision velocity in the granulator and ρ_g is the average granule density and Y_d is the dynamic yield stress of the granule, respectively. Both Y_d and ρ_g are strong functions of granule porosity and the granulation formulation properties, and are often evaluated at the maximum granule density, when the granules are strongest. This occurs when the granule porosity reaches a minimum value ε_{min} after which the granule density remains constant.

All types of granule growth can be described using the saturation and deformation St_{def} and the granule growth regime map (Hapgood *et al.*, 2007)

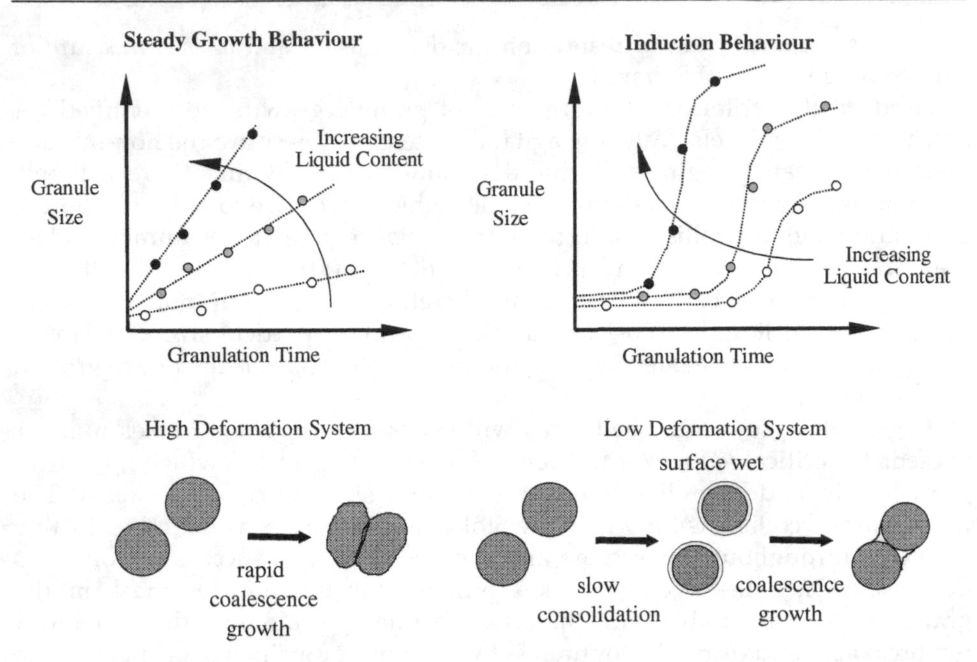

Figure 13.5 Schematic of the two main different types of granule growth and the way that they depend on the deformability of the granules. Reprinted from *Handbook of Powder Technology*, Vol. 11, eds Salman *et al.*, 'Granule rate processes', Hapgood *et al.*, p. 933, Copyright (2007) Elsevier

shown in Figure 13.6. At very low liquid contents, the product is similar to a dry powder. At slightly higher granule saturation, granule nuclei will form, but there is insufficient moisture for these nuclei to grow any further.

For systems with high liquid content, the behaviour depends on the granule strength and St_{def}. A weak system will form a slurry, an intermediate strength

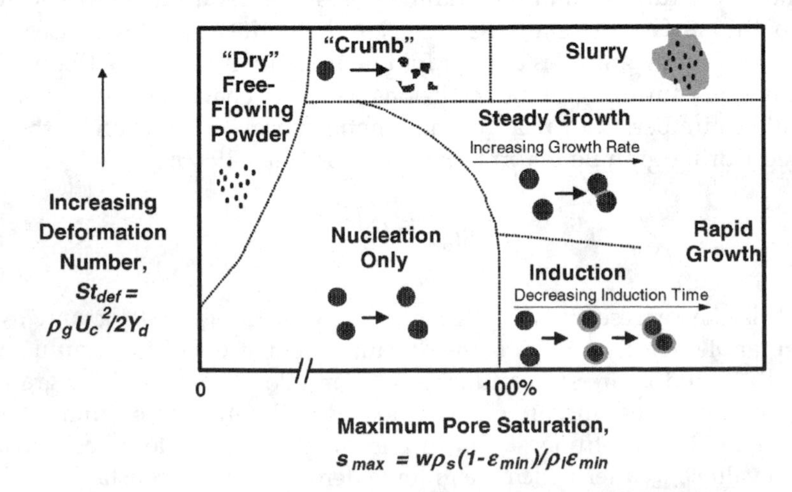

Figure 13.6 Granule growth regime map. Reprinted from *Handbook of Powder Technology*, Vol. 11, eds Salman *et al.*, 'Granule rate processes', Hapgood *et al.*, p. 933, Copyright (2007) Elsevier

system will display steady growth, and a strong system (low St_{def}) will exhibit induction time behaviour. At extremely high liquid saturations, rapid growth occurs where the granules grow in size extremely quickly and any induction time is reduced to zero. The granule growth regime map has been successfully validated for several formulations in mixers, fluid beds and tumbling granulators (Iveson *et al.*, 2001).

Granule breakage

Granule breakage also takes place during the granulation process. Breakage (also called fragmentation) is the fracture of a granule to form two or more pieces. Attrition (also called erosion) is the reduction in size of a granule by loss of primary particles from its surface. Empirical and theoretical approaches exist for modelling the different breakage mechanisms (Ennis and Litster, 1997). In practice, breakage may be controlled by altering the granule properties (e.g. increase fracture toughness and increase resistance to attrition) and by making changes to the process (e.g. reduce agitation intensity).

13.3.3 Simulation of the Granulation Process

As in the processes of comminution and crystallization, the simulation of granulation hinges on the population balance. The population balance tracks the size distribution (by number, volume or mass) with time as the process progresses. It is a statement of the material balance for the process at a given instant. In the case of granulation the instantaneous population balance equation is often written in terms of the number distribution of the volume of granules, $n(v, t)$ (rather than granule diameter since volume is assumed to be conserved in any coalescence). $n(v, t)$ is the number frequency distribution of granule volume at time t. Its units are number of granules per unit volume size increment, v, per unit volume of granulator, V. In words it is written as:

$$\begin{pmatrix}\text{rate of}\\ \text{increase}\\ \text{of number}\\ \text{of granules}\\ \text{in size}\\ \text{interval}\\ v \text{ to } v+dv\\ \text{(1)}\end{pmatrix} = \begin{pmatrix}\text{rate of}\\ \text{in flow}\\ \text{of granules}\\ \text{in size}\\ \text{interval}\\ v \text{ to } v+dv\\ \text{(2)}\end{pmatrix} - \begin{pmatrix}\text{rate of}\\ \text{outflow}\\ \text{of granules}\\ \text{in size}\\ \text{interval}\\ v \text{ to } v+dv\\ \text{(3)}\end{pmatrix} + \begin{pmatrix}\text{rate at which}\\ \text{granules}\\ \text{enter size}\\ \text{range}\\ v \text{ to } v+dv\\ \text{by growth}\\ \text{(4)}\end{pmatrix} - \begin{pmatrix}\text{rate at}\\ \text{which}\\ \text{granules}\\ \text{leave size}\\ \text{range}\\ v \text{ to } v+dv\\ \text{by breakage}\\ \text{(5)}\end{pmatrix} \tag{13.12}$$

Term (4) may be expanded to account for the different growth mechanisms and term (5) may be expanded to include the different mechanisms by which breakage occurs.

For a constant volume granulator the terms in the population balance equation become:
Term (1)

$$\frac{\partial n(v,t)}{\partial t} \tag{13.13}$$

Term (2)–term (3)

$$\frac{Q_{in}}{V} n_{in}(v) - \frac{Q_{out}}{V} n_{out}(v) \tag{13.14}$$

Term (4)

$$\begin{pmatrix}\text{net rate of}\\ \text{growth}\\ \text{by coating}\end{pmatrix} + \begin{pmatrix}\text{rate growth}\\ \text{by nucleation}\end{pmatrix} + \begin{pmatrix}\text{rate of growth}\\ \text{by coalescence}\end{pmatrix}$$

Growth by coating causes granules to grow into and out of the size range v to $v + dv$.

$$\text{Hence, the net rate of growth by coating} = \frac{\partial G(v)n(v,\,t)}{\partial v} \tag{13.15}$$

$$\text{The rate of growth by nucleation} = B_{nuc}(v) \tag{13.16}$$

$G(v)$ is the volumetric growth rate constant for coating, $B_{nuc}(v)$ is the rate constant for nucleation. It is often acceptable to assume that $G(v)$ is proportional to the available granule surface area; this is equivalent to assuming a constant linear growth rate, $G(x)$. In reference to granulation processes, the use of the term 'nucleation' is sometimes confusing. Sastry and Loftus (1989) suggest that in a granulation process there is in general a continuous phase and a particulate phase. The continuous phase may, for example, be a solution, slurry or very small particles; the particulate phase is made up of original particles and granules. Nuclei are formed from the continuous phase and then become part of the particulate phase. What constitutes the continuous phase depends on the nature of the granulation, and the cut-off between continuous and particulate phases may be arbitrary. It will be appreciated therefore that the form of the granulation rate constant may vary considerably depending on the definitions of continuous and particulate phases.

The rate of growth of granules by coalescence may be written as (Randolph and Larson, 1971):

$$\left[\frac{1}{2}\int_0^v \underset{\text{(i)}}{\beta(u, v-u, t)n(u,\,t)n(v-u,t)\,du}\right] + \left[\int_0^\infty \underset{\text{(ii)}}{\beta(u,v,t)\,n(u,t)\,n(v,t)\,du}\right] \tag{13.17a}$$

Term (i) is the rate of formation of granules of size v by coalescence of smaller granules. Term (ii) is the rate at which granules of size v are lost by coalescence to form larger granules. β is called the coalescence kernel. The rate of coalescence of two granules of volume u and $(v - u)$ to form a new granule of volume v is assumed to be directly proportional to the product of the number densities of the starting granules:

$$\begin{bmatrix} \text{collision rate of granules} \\ \text{of volume } u \text{ and}(v-u \end{bmatrix} \propto \begin{pmatrix} \text{number density of} \\ \text{granules of volume} u \end{pmatrix} \times \begin{bmatrix} \text{number density of} \\ \text{granules of volume} \\ v-u \end{bmatrix}$$

The number densities are time dependent in general and so $n(u, t)$, the number density of granules of volume u at time t and $n(v - u, t)$ is the number density of granules of volume $(v - u)$ at time t. The constant in this proportionality is β, the coalescence kernel or coalescence rate constant, which is in general assumed to be dependent on the volumes of the colliding granules. Hence, $\beta(u, v - u, t)$ is the coalescence rate constant for collision between granules of volume u and $(u - v)$ at time t.

The above assumes a pseudo-second order process of coalescence in which all granules have an equal opportunity to collide with all other particles. In real granulation systems this assumption does not hold and collision opportunities are limited to local granules. Sastry and Fuerstenau (1970) suggested that for a batch granulation system, which was effectively restricted in space, the appropriate form for terms (i) and (ii) was:

$$\left[\frac{1}{2N(t)}\int_0^{\mathrm{V}} \underset{\text{(i)}}{\beta(u, v-u, t)n(u, t)n(v-u, t)\mathrm{d}u}\right] + \left[\frac{1}{N(t)}\int_0^{\infty} \underset{\text{(ii)}}{\beta(u, v, t)\, n\,(u, t)\, n\,(v, t)\mathrm{d}u}\right] \tag{13.17b}$$

where $N(t)$ is the total number of granules in the system at time t.

The integrals in Equation (13.17b) account for all the possible collisions and the $\frac{1}{2}$ in term (i) of this equation ensures that collisions are only counted once.

In practice, many coalescence kernels are determined empirically and based on laboratory or plant data specific to the granulation process and the product. More recent, physically derived kernels (Litster and Ennis, 2004) have been tested in laboratories and over time are expected to slowly replace empirically derived kernels in industrial models.

According to Sastry (1975) the coalescence kernel is best expressed in two parts:

$$\beta(u, v) = \beta_0\beta_1(u, v) \tag{13.18}$$

β_0 is the coalescence rate constant which determines the rate at which successful collisions occur and hence governs average granule size. It depends on solid and liquid properties and agitation intensity. $\beta_1(u, v)$ governs the functional dependency of the kernel on the sizes of the coalescing granules, u and v. $\beta_1(u, v)$ determines the shape of the size distribution of granules. Various forms of β have been published; Ennis and Litster (1997) suggest the form shown in

Equation (13.19), which is consistent with the granulation regime analysis described above:

$$\beta(u,v) = \begin{cases} \beta_0, w < w^* \\ 0, w > w^* \end{cases} \text{ where } w = \frac{(uv)^a}{(u+v)^b} \tag{13.19}$$

w^* is the critical average granule volume in a collision and corresponds to the critical Stokes number value Stk^*. From the definition the Stokes number given in Equation (13.10), the critical diameter x^* would be:

$$x^* = \frac{16\mu\, Stk^*}{\rho_{gr} V_{app}} \tag{13.20}$$

and so, assuming spherical granules,

$$w^* = \frac{\pi}{6}\left(\frac{16\mu\, Stk^*}{\rho\, V_{app}}\right)^3 \tag{13.21}$$

The exponents α and β in Equation (13.19) are dependent on granule deformability and on the granule volumes u and v. In the case of small feed particles in the non-inertial regime, β reduces to the size-independent rate constant β_0 and the coalescence rate is independent of granule size. Under these conditions the mean granule size increases exponentially with time. Coalescence stops ($\beta = 0$) when the critical Stokes number is reached.

Using this approach, Adetayo and Ennis (1997) were able to demonstrate the three regimes of granulation (nucleation, transition and coating) traditionally observed in drum granulation and to model a variety of apparently contradictory observations.

Term (5)

$$\begin{pmatrix}\text{rate of breakage} \\ \text{by attrition}\end{pmatrix} + \begin{pmatrix}\text{rate of breakage} \\ \text{by fragmentation}\end{pmatrix}$$
$$\text{rate of breakage by attrition} = \frac{\partial A(v)n(v,t)}{\partial v} \tag{13.22}$$

$A(v)$ is the rate constant for attrition. The rate of attrition depends on the material to be granulated, the binder, the type of equipment used for granulation and the intensity of agitation. The rate of breakage may also be accounted for by the use of selection and breakage function as used in the simulation of population balances in comminution (see Chapter 12).

13.3.4 Granulation Equipment

Three categories of granulator are in common use: tumbling, mixer and fluidized granulators. Typical product properties, scale and applications of each of these granulators are summarized in Table 13.1 (after Ennis and Litster, 1997).

Table 13.1 Types of granulator, their features and applications (after Ennis and Litster, 1997)

Method	Product granule size(mm)	Granule density	Scale/ throughput	Comments	Typical applications
Tumbling (disc, drum)	0.5–20	Moderate	0.5–800 t/h	Very spherical granules	Fertilizers, iron ore, agricultural chemicals
Mixer (continuous and batch high shear)	0.1–2	Low	<50 t/h or <200 kg/ batch	Handles cohesive materials well	Chemicals, detergents, pharmaceuticals, ceramics
Fluidized (bubbling beds, spouted beds)	0.1–2	High	<500 kg/batch	Good for coating, easy to scale up	Continuous (fertilizers, detergents) batch (pharmaceuticals, agricultural chemicals)

Tumbling granulators

In tumbling granulators a tumbling motion is imparted to the particles in an inclined cylinder (drum granulator) or pan (disc granulator, Figure 13.7). Tumbling granulators operate in continuous mode and are able to deal with large throughputs (see Table 13.1). Solids and liquid feeds are delivered continuously to the granulator. In the case of the disc granulator the tumbling action gives rise to a

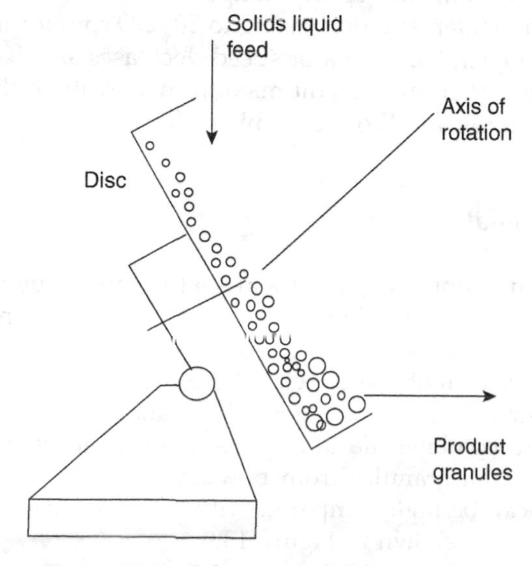

Figure 13.7 Schematic of disc granulator

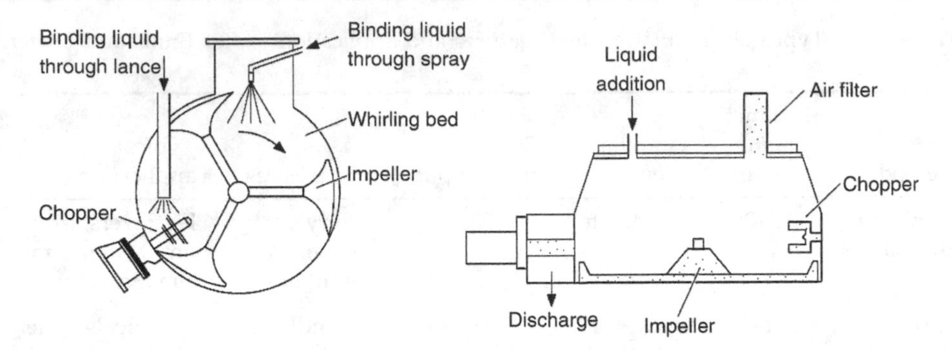

Figure 13.8 Schematic of a horizontal axis (a) and vertical axis (b) mixer granulator. Reprinted from *Perry Chemical Engineers' Handbook*, 7th Edition, 'Section 20: size enlargement', Ennis and Litster, pp. 20–77. Copyright (1977) with permission of The McGraw-Hill Companies

natural classification of the contents according to size. Advantage is taken of this effect and the result is a product with a narrow size distribution.

Mixer Granulators

In mixer granulators the motion of the particles is brought about by some form of agitator rotating at low or high speed on a vertical or horizontal axis. Rotation speeds vary from 50 revolutions per minute (rpm) in the case of the horizontal pug mixers used for fertilizer granulation to over 3000 pm in the case of the vertical Schugi high shear continuous granulator used for detergents and agricultural chemicals. For vertical axis mixers used by the pharmaceutical industry (Figure 13.8), impeller speeds range from 500 to 1500 rpm for mixers less than 30 cm in diameter, and decreasing to 50–200 rpm for mixers larger than 1 m in diameter. In general, the agitator speed decreases as mixer scale increases, in order to maintain either (a) constant maximum velocity at the blade tip or (b) constant mixing patterns and Froude number.

Fluidized bed granulators

In fluidized bed granulators the particles are set in motion by fluidizing air. The fluidized bed may be either a bubbling or spouted bed (see Chapter 7) and may operate in batch or continuous mode. Liquid binder and wetting agents are sprayed in atomized form above or within the bed. The advantages which this granulator has over others include good heat and mass transfer, mechanical simplicity, ability to combine the drying stage with the granulation stage and ability to produce small granules from powder feeds. However, running costs and attrition rates can be high compared with other devices. A typical spouted bed granulator circuit is shown in Figure 13.9.

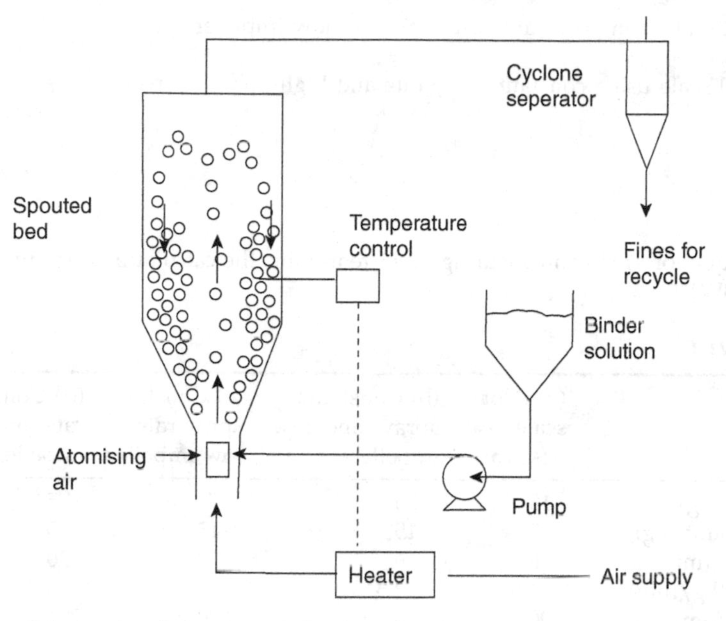

Figure 13.9 Schematic of a spouted fluidized bed granulator circuit (Liu and Litster, 1993)

For further details of industrial equipment for granulation and other means of size enlargement the reader is referred to Capes (1980), Ennis and Litster (1997) or Litster and Ennis (2004).

13.4 WORKED EXAMPLES

WORKED EXAMPLE 13.1

A pharmaceutical product is being scaled up from a pilot scale mixer granulator with a batch size of 15 kg to a full scale mixer granulator with a 75 kg batch size. In the pilot scale mixer, 3 kg of water is added to the mixer over 6 min. through a nozzle producing 200 μm diameter spray drops across a 0.2 m wide spray. During scale-up, the ratio of liquid to dry powder is kept constant but the solution flowrate can be scaled to maintain constant spray time or constant spray rate through the nozzle. If the flow rate is increased to maintain constant spray time, the new nozzle produces 400 μm drops over a 0.3 m wide spray zone. Powder velocity in the spray zone is currently 0.7 m/s at pilot scale. At full scale, the powder velocities are 0.55 m/s and 1 m/s at the 'low' and 'high' impeller speeds, respectively. Calculate the change in dimensionless spray flux for the following cases:

(a) base case at pilot scale;

(b) scale-up to full scale using spray time of 6 min and low impeller speed;

(c) full scale using constant spray rate and low impeller speed;

(d) full scale using constant spray rate and high impeller speed.

Solution

Using Equation (13.5) and ensuring consistent units, the calculations are summarized in Table 13W1.1.

Table 13W1.1

Scale-up approach	(a) Pilot scale base case	(b) Constants spray time, low impeller	(c) Constant spray rate, low impeller	(d) Constant spray rate, high impeller
Batch size (kg)	15	75	75	75
Spray amount (kg)	3	15	15	15
Spray time (min)	6	6	30	30
Flow rate (kg/min)	0.5	2.5	0.5	0.5
Drop size (μm)	200	400	200	200
Spray width (m)	0.2	0.3	0.2	0.2
Impeller speed (rpm)	216	108	108	220
Powder velocity (m/s)	0.7	0.55	0.55	1
Spray flux Ψ_a	0.45	0.95	0.57	0.31

Worked Example 13.2

When the spray flux is 0.1, 0.2, 0.5 and 1.0:

(a) Calculate the fraction of the spray zone wetted.

(b) Calculate the number of nucleus formed from only one drop (f_1).

(c) If the drop controlled regime was defined as the spray flux at which 50% or more of the nuclei are formed from a single drop, what would be the critical value of spray flux and what fraction of the spray zone would be wetted?

Solution

Using Equation (13.6) for (a) and (c), solutions are shown in Table 13W2.1. To calculate the fraction of nuclei formed from a single drop for (b), use Equation (13.7) with $n=1$, which simplifies to $f_1 = \exp(-4\Psi_a)$.

TEST YOURSELF

13.1 Redraw Figure 13.1 showing how saturation increases as the porosity of the granule decreases.

13.2 How do the rate of wetting and drop penetration time change with (a) increase in viscosity, (b) decrease in surface tension, (c) increase in contact angle?

13.3 If the flow rate of solution to be added to the granulator doubles, how much wider does the spray zone need to be to maintain constant spray flux? If the spray cannot be adjusted, what does the bed velocity need to be to maintain constant spray flux?

13.4 At a spray flux of 0.1, calculate the number of nuclei formed from on drop, two drops and three drops, etc. What fraction of the powder surface will be wetted by the spray?

13.5 Explain the non-inertial, inertial, and coating regimes of granule granule growth. What happens to the maximum granule size as (a) the approach velocity increases, (b) the viscosity increases?

13.6 Explain the difference between the steady growth and the induction growth regimes on the granule growth regime map.

13.7 Explain how induction growth is linked to granule porosity and staturation.

13.8 Explain the five terms in the granulation population balance.

EXERCISES

13.1 In the pharmaceutical industry, any batch that deviates from the set parameters is designated as an 'atypical' batch and must be investigated before the product can be released. You are a pharmaceutical process engineer, responsible for granulating a pharmaceutical product in a 600 litre mixer containing 150 kg of dry powders. It was noticed while manufacturing a new batch that the liquid delivery stage ended earlier than usual and the batch contained larger granules than normal. During normal production, the impeller speed is set to 90 rpm and water is added at a flowrate of 2 litre/m n through a nozzle producing an average drop size of 400 μm. Due to an incorrect setting, the actual flow rate used in the atypical batch was 3.5 litre/min and the actual drop size was

Table 13W2.1

Ψ_a	$f_{wet}(\%)$	$f_1(\%)$
0.1	10	67
0.17	16	51
0.2	18	45
0.5	39	14
1	63	2

estimated at 250 μm. The spray width and powder surface velocity were unaffected and remained constant at 40 cm and 60 cm/s, respectively. Calculate the dimensionless spray flux for the normal case and the atypical batch, and explain why this would have created larger granules.

(Answer: normal 0.52; atypical, 1.46.)

13.2 Calculate the fraction or percentage of nuclei formed from one drop, two drops and three drops, etc. at spray flux values of (a) 0.05, (b) 0.3, (c) 0.8. What fraction of the powder surface will be wetted by the spray?

[Answer: (a) 0.82, 0.16, 0.02, 0.0, 0.0; (b) 0.30, 0.36, 0.22, 0.09, 0.03; (c) 0.04, 0.13, 0.21, 0.22, 0.18.]

14

Health Effects of Fine Powders

14.1 INTRODUCTION

When we think of the health effects of fine powders we might think first of the negative effects related to inhalation of particles which have acute or chronic (asbestos fibres, coal dust) effects on the lungs and the body. However, with the invention of the metred dose inhaler in 1955, the widespread use of fine particle drugs delivered directly to the lungs for the treatment of asthma began. Pulmonary delivery, as this method is called, is now a widespread method of drug delivery. In this chapter, therefore, we will look at both the negative and positive health effects of fine powders, beginning with a description of the respiratory system and an analysis of the interaction of the system with fine particles.

14.2 THE HUMAN RESPIRATORY SYSTEM

14.2.1 Operation

The requirement for the human body to exchange oxygen and carbon dioxide with the environment is fulfilled by the respiratory system. Air is delivered to the lungs via the nose and mouth, the pharynx and larynx and the trachea. Beyond the trachea, the single airway branches to produce the two main bronchi which deliver air to each of two lungs. Within each lung the bronchi branch repeatedly to produce many smaller airways called bronchioles, giving an inverted tree-like structure (Figure 14.1). The upper bronchioles and bronchi are lined with specialized cells some of which secrete mucus whilst others have hairs or cilia which beat causing an upward flow of mucus along the walls. The bronchioles

Introduction to Particle Technology - 2nd Edition Martin Rhodes

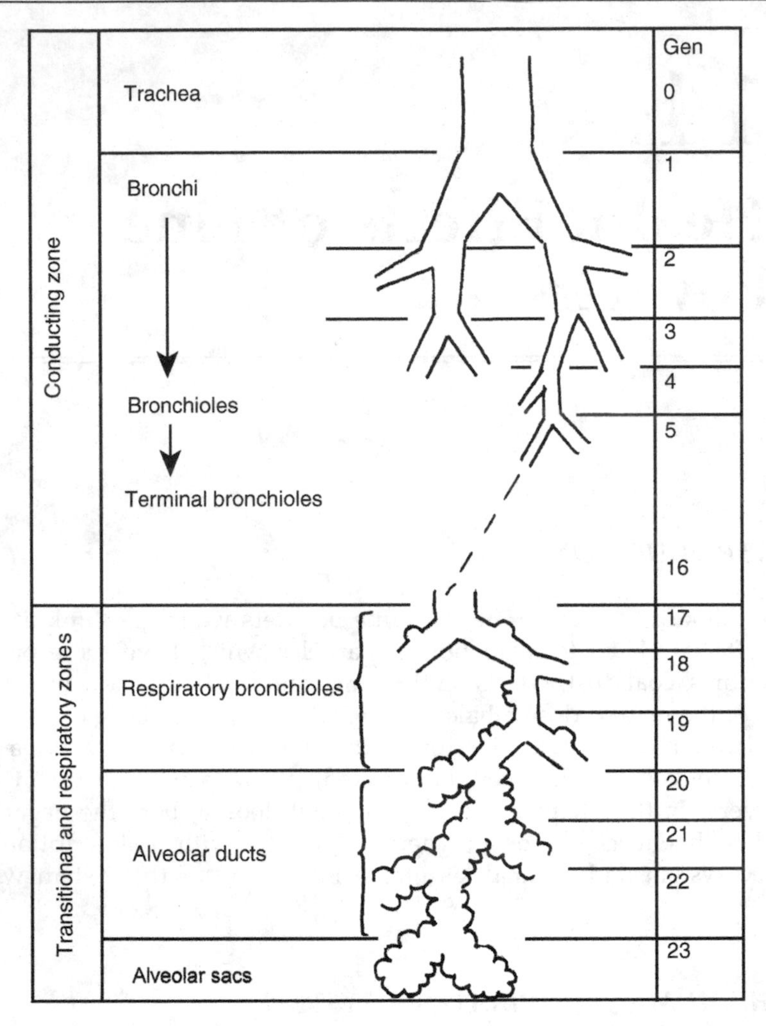

Figure 14.1 Human tracheobronchial tree. Reprinted from *Particulate Interactions in Dry Powder Formulations of Inhalation*, Xian Ming Zeng, Gary Martin and Christopher Marriott, Copyright 2001, Taylor & Francis with permission

lead to alveolar ducts (or alveoli) which terminate in alveolar sacs. In the adult male there are about 300 million alveoli, each approximately 0.2 mm in diameter. The walls of the alveoli have a rich supply of blood vessels. Oxygen from the air diffuses across a thin membrane around the alveolar sacs into the blood and carbon dioxide from the blood diffuses in the reverse direction.

Air is forced in and out of the lungs by the movement of the diaphragm muscle beneath the thoracic cage or chest cavity. The lungs are sealed within the chest cavity, so that as the diaphragm muscle moves down, the pressure within the cavity falls below atmospheric, causing air to be drawn into the lungs. The lungs expand, filling the chest cavity. As the diaphragm muscle moves up, the lungs are squeezed and the air within them is expelled to the environment.

Table 14.1 Characteristics of the respiratory tract, based on steady flow of 60 litre/min. Reprinted from *Particulate Interactions in Dry Powder Formulations of Inhalation*, Xian Ming Zeng, Gary Martin and Christopher Marriott, Copyright 2001, Taylor & Frances with permission

Part	Number	Diameter (mm)	Length (mm)	Typical air velocity (m/s)	Typical residence time (s)
Nasal airways		5–9		9	
Mouth	1	20	70	3.2	0.022
Pharynx	1	30	30	1.4	0.021
Trachea	1	18	120	4.4	0.027
Two main bronchi	2	13	37	3.7	0.01
Lobar bronchi	5	8	28	4.0	0.007
Segmental bronchi	18	5	60	2.9	0.021
Bronchioles	504	2	20	0.6	0.032
Secondary bronchioles	3024	1	15	0.4	0.036
Terminal bronchioles	12 100	0.7	5	0.2	0.023
Alveolar ducts	8.5×10^5	0.8	1	0.0023	0.44
Alveolar sacs	2.1×10^7	0.3	0.5	0.0007	0.75
Alveoli	5.3×10^8	0.15	0.15	0.00004	4

14.2.2 Dimensions and Flows

A systematic description commonly used is that of Weibel (1963). In this treatment the respiratory system is considered as a sequence of regular branches or dichotomies. The tree starts with the trachea, as 'generation' 0, leading to the two main bronchi, as 'generation' 1. Within the lungs the branching continues until generation 23 representing the alveolar sacs. According to Weibel's model, at generation n there are 2^n tubes. So, for example, at generation 16 (the terminal bronchioles) there are 65 536 tubes of diameter 0.6 mm.

Table 14.1 is based more on direct measurement and gives the typical dimensions of the component of the airways together with typical air velocities and residence times.

The nasal airways are quite tortuous and change diameters several times in their path. The lower section is lined with hairs. The narrowest section is the nasal valve, which has a cross-sectional area of 20–60 mm^2 (equivalent to 5 – 9 mm diameter) and typically accounts for 50% of the resistance to flow in the nasal airways. Typical air flow rates in the adult nasal airways range from 180 ml/s during normal breathing to 1000 ml/s during a strong sniff. This gives velocities in the nasal airways as high as 9 m/s during normal breathing and 50 m/s during a strong sniff.

The mouth leading to the pharynx and larynx presents a far smoother path for the flow of air and offers lower resistance. However, whilst moving through the pharynx and larynx the airstream is subject to some sharp changes in direction. A typical air velocity in the mouth during normal breathing is 3 m/s.

We see from Table 14.1 that due mainly to the continuous branching, the air velocity in the airways decreases at the start of the bronchioles. The result is that the residence time in the different sections of the airways is of the same order until the alveolar region is reached, where the residence time increases significantly.

14.3 INTERACTION OF FINE POWDERS WITH THE RESPIRATORY SYSTEM

Airborne particles entering the respiratory tract will be deposited on the surface of any part of the airways with which they come into contact. Because the surfaces of the airways are always moist, there is a negligible chance of a particle becoming re-entrained in the air after once having made contact with a surface. Particles which do not make contact with the airway surfaces will be exhaled. The deposition of particles in the airways is a complicated process. Zeng *et al.* (2001) identify five possible mechanisms contributing to the deposition of particles carried into the respiratory tract. These are sedimentation, inertial impaction, diffusion, interception and electrostatic precipitation. After introducing each mechanism, their relative importance in each part of the respiratory tract will be discussed.

14.3.1 Sedimentation

Particles sediment in air under the influence of gravity. As discussed in Chapter 2, a particle in a static fluid of infinite extent will accelerate from rest under the influence of the gravity and buoyancy forces (which are constant) and the fluid drag force, which increases with relative velocity between the particle and the fluid. When the upward-acting drag and buoyancy forces balance the gravity force, a constant, terminal velocity is reached. For particles of the size of relevance here (less than 40 μm), falling in air, the drag force will be given by Stokes' law [Equation (2.3)] and the terminal velocity will be given by [Equation (2.13)]:

$$U_T = \left[\frac{x^2(\rho_p - \rho_f)g}{18\mu}\right]$$

(For particles of diameter less that the mean free path of air, 0.0665 μm, a correction factor must be applied. Particles as small as this will not be considered here.) Particles in the size range under consideration accelerate rapidly to their terminal velocity in air. For example, a 40 μm particle of density 1000 kg/m^3 accelerates to 99% of its terminal velocity (47 mm/s) in 20 ms and travels a distance less than 1 mm in doing so. So, for these particle in air, we can take the sedimentation velocity as equivalent to the terminal velocity. The above analysis is relevant to particles falling in stagnant air or air in laminar flow. If the air flow becomes turbulent, as it may do in certain airways at higher breathing rates, the characteristic velocity fluctuations increase the propensity of particles to deposit.

14.3.2 Inertial Impaction

Inhaled air follows a tortuous path as it makes its way through the airways. Whether the airborne particles follow the path taken by the air at each turn depends on the balance between the force required to cause the particle to change direction and the fluid drag available to provide that force. If the drag force is sufficient to cause the required change in direction, the particle will follow the gas and will not be deposited.

Consider a particle of diameter x travelling at a velocity U_p in air of viscosity μ within an airway of diameter D. Let us consider the extreme case where this particle is required to make a 90° change in direction. The necessary force F_R is that required to cause the particle to stop and then be re-accelerated to velocity U_p. The distance within which the particle must stop is of the order of the airway of diameter D and so:

$$\text{work done} = \text{force} \times \text{distance} = \text{particle kinetic energy}$$

$$F_R D = 2\left(\frac{1}{2} m U_p^2\right) \tag{14.1}$$

(The 2 on the right-hand side is there because we need to re-accelerate the particle.)

$$\text{Therefore, required force, } F_R = \frac{\pi}{6} x^3 \rho_p U_p^2 \frac{1}{D} \tag{14.2}$$

The force available is the fluid drag. Stokes' law [Equation (2.3)] will apply to the particles under consideration and so:

$$\text{Available drag force, } F_D = 3\pi\mu x U_p$$

The rationale for using U_p as the relevant relative velocity is that this represents the maximum relative velocity that would be experienced by the particle as it attempts to continue in a straight line.

The ratio of the force required F_R to the force available F_D is then:

$$\frac{x^2 \rho_p U_p}{18\mu D} \tag{14.3}$$

The greater the value of this ratio is above unity, the greater will be the tendency for particles to impact with the airway walls and so deposit. The further the value is below unity, the greater will be the tendency for the particles to follow the gas. This dimensionless ratio is known as the Stokes number:

$$Stk = \frac{x^2 \rho_p U_p}{18\mu D} \tag{14.4}$$

(We also met the Stokes number in Chapter 9, where it was one of the dimensionless numbers used in the scale up of gas cyclones for the separation of particles from gases. There are obvious similarities between the collection of particles in a gas cyclone and 'collection' of particles in the airways of the respiratory system. The Stokes number we met in Chapter 13, describing collision between granules, is not readily comparable with the one used here.)

14.3.3 Diffusion

The motion of smaller airborne particles is influenced by the bombardment of air molecules. This is known as Brownian motion and results in a random motion of the particles. Since the motion is random, the particle displacement is expressed as a root mean square:

$$\text{Root mean square displacement in time } t, L = \sqrt{6\alpha t} \tag{14.5}$$

where α is the diffusion coefficient given by:

$$\alpha = \frac{kT}{3\pi\mu x} \tag{14.6}$$

for a particle of diameter x in a fluid of viscosity μ at a temperature T. k is the Boltzmann constant, which has the value 1.3805×10^{-23} J/K.

14.3.4 Interception

Interception is deposition of the particle by reason of its size and shape compared with the size of the airway.

14.3.5 Electrostatic Precipitation

Particles and droplets may become electrostatically charged, particularly during the dispersion stage, by interaction with each other or with nearby surfaces. It has been speculated that charged particle could induce opposite charges on the walls of some airways, resulting in the particles being attracted to the walls and deposited.

14.3.6 Relative Importance of These Mechanisms Within the Respiratory Tract

Under the humid conditions found with the respiratory tract, it is most likely that any charges on particles will be quickly dissipated and so electrostatic precipitation is unlikely to play a significant role in particle deposition anywhere in the airways.

Table 14.2 A comparison of displacements due to sedimentation and diffusion for particles of density 1000 kg/m^3 in air at 30°C, with a density 1.21 kg/m^3 and viscosity 1.81×10^{-5} Pa s

Particle diameter (μm)	Particle terminal velocity (m/s)	Displacement in 1 s due to sedimentation (μm)	Displacement in 1 s due to Brownian motion (μm)
50	7.5×10^{-2}	75000	1.7
30	2.7×10^{-2}	27000	2.2
20	1.2×10^{-2}	12000	2.7
10	3.0×10^{-3}	3000	3.8
5	7.5×10^{-4}	750	5.4
2	1.2×10^{-4}	120	8.5
1	3.0×10^{-5}	30	12.0
0.5	7.5×10^{-6}	7.5	17.0
0.3	2.7×10^{-6}	2.7	21.9
0.2	1.2×10^{-6}	1.2	26.8
0.1	3.0×10^{-7}	0.3	37.9

Deposition by interception is also not significant, since the particles of interest are far smaller than the airways.

We will now consider the relative importance of the other three mechanisms, *sedimentation, inertial impaction* and *diffusion*, in the various parts of the respiratory tract. First, it is interesting to compare, for different particle sizes, the displacements due to sedimentation and diffusion under the conditions typically found within the respiratory tract. Table 14.2 makes this comparison for particles of density 1000 kg/m^3 in air at a temperature of 30°C.

From Table 14.2 we can see that diffusion does not become significant compared with sedimentation until particle size falls below 1 μm. In the case where Brownian motion was to act downwards with the sedimentation, the minimum displacement would occur for particles of around 0.5 μm in diameter. This would suggest that particles around this diameter would have the least chance of being deposited by these mechanisms. We see that particles smaller than about 10 μm would require significant residence times to travel far enough to be deposited. For example, in Table 14.3, we look at the time required for particles to be displaced a distance equivalent to the airway diameter, due to the combined effect of sedimentation and diffusion. In practice, sedimentation becomes a significant mechanism for deposition only in the smaller airways and in the alveolar region, where air velocities are low, airway dimensions are small and air residence times are relatively high.

Now we will consider the importance of inertial impaction as a mechanism for causing deposition. In Table 14.4 we have calculated [using Equation (14.4)] the Stokes numbers for particles of different sizes in the various areas of the respiratory duct, based on the information provided in Table 14.1. This calculation assumes that the particle velocity is the same as the air velocity in the relevant part of the respiratory tract.

Table 14.3 Time required for particles of different diameters to be displaced a distance equivalent to the airway diameter, due to the combined effect of sedimentation and diffusion

Airway component	Diameter (mm)	Typical air residence time (s)	Required particle component residence time (s)		
			5 μm	1 μm	0.5 μm
Two main bronchi	13	0.01	17.2	309	580
Bronchioles	2	0.032	2.6	48	90
Secondary bronchioles	1	0.036	1.3	24	45
Terminal bronchioles	0.7	0.023	1	17	30
Alveolar ducts	0.8	0.44	1	19	35
Alveolar sacs	0.3	0.75	0.4	7	13
Alveoli	0.15	4	0.2	3.5	6.6

Recalling that the further the value of the Stokes number is above unity, the greater will be the tendency for particles to impact with the airway walls and so deposit, we see that particles as small as 50 μm are likely to be deposited in the mouth or nose, pharynx, larynx and trachea. These figures are for steady breathing; higher rates will give higher Stokes numbers and so would cause smaller particles to be deposited at a given part of the tract.

Whilst moving through the mouth or nose, pharynx and larynx, the airstream is subject to some sharp changes in direction. This induces turbulence and instabilities which increase the chances of particle deposition by inertial impaction.

Table 14.4 Stokes number for a range of different size particles in each section of the respiratory tract. Also typical Reynolds number for flow in each section of the respiratory tract

Region	Stokes number for particles of different sizes						
	1 μm	5 μm	10 μm	20 μm	50 μm	100 μm	*Re* flow
Nose	3.1×10^{-3}	7.7×10^{-4}	0.31	1.2	7.7	31	5415
Mouth	4.9×10^{-4}	1.2×10^{-4}	0.05	0.2	1.2	5	4254
Pharynx	1.5×10^{-4}	3.7×10^{-5}	0.01	0.1	0.4	1	2865
Trachea	7.6×10^{-4}	1.9×10^{-4}	0.08	0.3	1.9	8	5348
2 main bronchi	8.7×10^{-4}	2.2×10^{-4}	0.09	0.3	2.2	9	3216
Lobar bronchi	1.5×10^{-3}	3.8×10^{-4}	0.15	0.6	3.8	15	2139
Segmental bronchi	1.8×10^{-3}	4.4×10^{-4}	0.18	0.7	4.4	18	955
Bronchioles	9.6×10^{-4}	2.4×10^{-4}	0.10	0.4	2.4	10	84
Secondary bronchioles	1.3×10^{-3}	3.2×10^{-4}	0.13	0.5	3.2	13	28
Terminal bronchioles	9.5×10^{-4}	2.4×10^{-4}	0.10	0.4	2.4	10	10
Alveolar ducts	8.7×10^{-6}	2.2×10^{-6}	0.0009	0.003	0.022	0	0
Alveolar sacs	6.8×10^{-6}	1.7×10^{-6}	0.0007	0.003	0.017	0	0
Alveoli	7.7×10^{-7}	1.9×10^{-7}	0.0001	0.003	0.002	0	0

In summary, inertial impaction is responsible for the deposition of the larger airborne particles and this occurs mainly in the upper airways. In practice, therefore, we find that only those particles smaller than about 10 μm will travel beyond the main bronchi. Such particles have a decreasing propensity to be deposited by inertial impaction the further they travel into the lungs, but are more likely to be deposited by the action of sedimentation and diffusion as they reach the smaller airways and the alveolar region, where air velocities are low, airway dimensions are small and air residence times are relatively high.

14.4 PULMONARY DELIVERY OF DRUGS

Target areas: The delivery of drugs for treatment of lung diseases (asthma, bronchitis, etc.) via aerosols direct to the lungs is attractive for number of reasons. Compared with other methods of drug delivery (oral, injection) pulmonary delivery gives a rapid, predictable onset of drug action with minimum dose and minimum side effects. The adult human lung has a very large surface area for drug absorption (typically 120 m^2). The alveoli wall membrane is permeable to many drug molecules and has rich blood supply. This makes so-called aerosol delivery attractive for the delivery of drugs for treatment of other illnesses. (Aerosols are suspensions of liquid drops or solid particles in air or other gas.) In pulmonary delivery using aerosols, the prime targets for the aerosol particles are the alveoli, where the conditions for absorption are best. In practice, particles larger than about 2 μm rarely reach the alveoli and particles smaller than about 0.5 μm may reach the alveoli but are exhaled without deposition (Smith and Bernstein, 1996; see also analysis above).

Surprisingly, the use of aerosols for medicinal use on humans dates back thousands of years. These include volatile aromatic substances such as thymol, menthol and eucalyptus and smoke generated from the burning of leaves. Nebulizers – devices which generate a fine mist of aqueous drug solution – have been used in western hospitals for over 100 years, although the modern nebulizer bears little resemblance to the earlier devices. The modern nebulizer (Figure 14.2) is used where patients cannot use other devices or when large doses of drugs are required. Portable nebulizers have been developed, although these devices still require a source of power for air compression.

In the 1950s the metered-dose inhaler (MDI) was invented. This was the precursor to the modern MDI which has been used widely by asthma sufferers for many years – despite its disadvantages (Zeng *et al.*, 2001). In the MDI the drug, which is dispersed or dissolved in a liquid propellant, is held in a small pressurized container (Figure 14.3). Each activation of the device releases a metered quantity of propellant carrying a predetermined dose of drug. The liquid in the droplets rapidly vaporizes leaving solid particles. The high velocity of discharge from the container means that many particles impact on the back of the throat and are therefore ineffective. Another disadvantage is that the patient must, usually in a stressed condition, coordinate the activation of the MDI with inspiration.

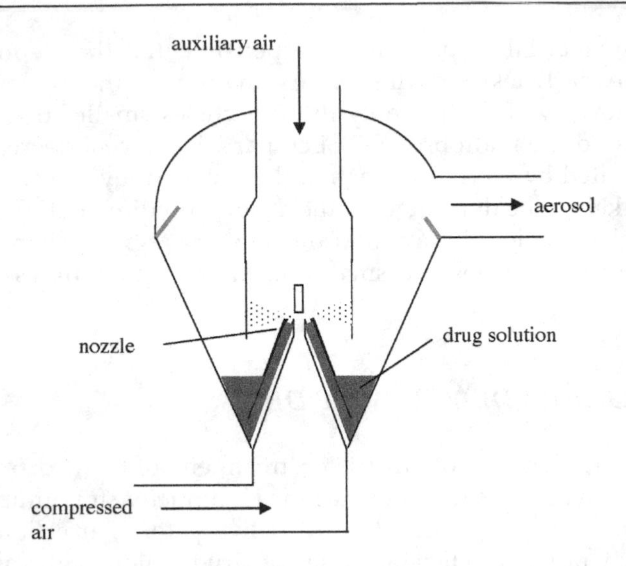

Figure 14.2 Schematic of a nebulizer The compressed air expands as it leaves the nozzle. This causes reduced pressure which induces the drug solution to flow up and out of the nozzle where it atomized by contact with the air stream. Copyright (1996) from *Lung Biology in Health and Disease*, Vol. 94, Dalby *et al.*, *Inhalation Aerosols*, p. 452. Reproduced by permission of Routledge/Taylor & Francis Group, LLC

The third common type of device for pulmonary drug delivery is the dry powder inhaler (DPI). This device now has many forms, some of which appear remarkably simple in their design (Figure 14.4). Particular effort has gone into the formulation of the powder. In most cases the drug is of the order a few micrometre in size and is adhered – usually by natural forces – to inert 'carrier

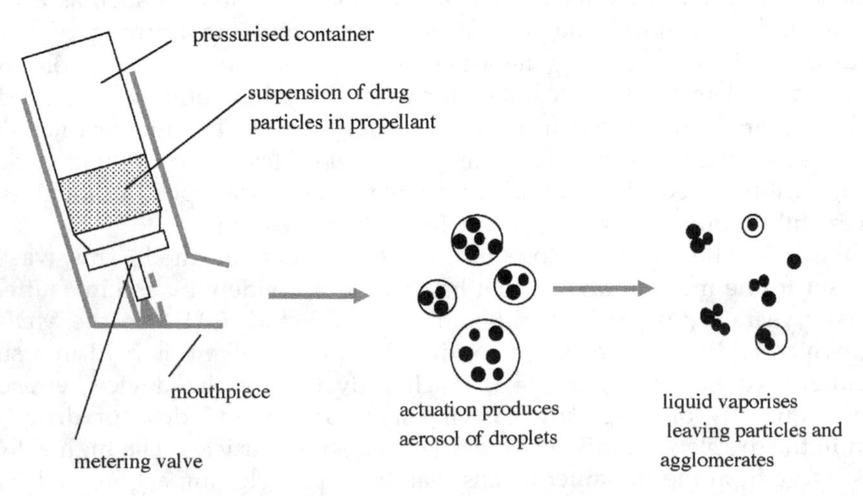

Figure 14.3 Metered-dose inhaler. Copyright (1996) from *Lung Biology in Health and Disease*, Vol. 94, Dalby *et al.*, *Inhalation Aerosols*, p. 452. Reproduced by permission of Routledge/Taylor & Francis Group, LLC

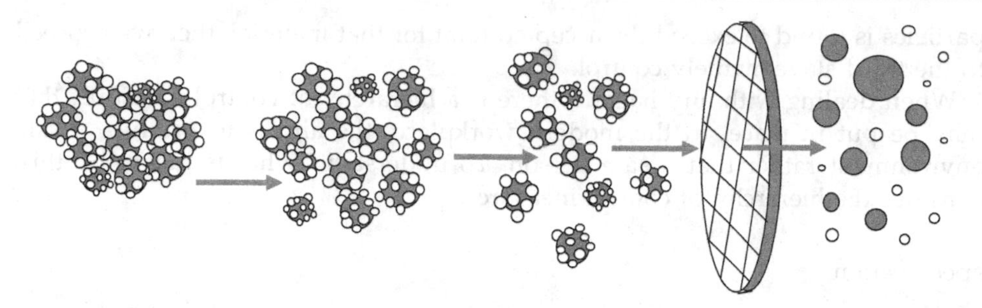

Figure 14.4 Carrier particles in a dry powder inhaler. The powder may be initially loosely compacted, but by the shearing action of the air stream and impingement on a screen the agglomerates are dispersed and the drug particles dislodged from the carrier particles. Copyright (1996) from *Lung Biology in Health and Disease*, Vol. 94, Dalby *et al.*, *Inhalation Aerosols*, p. 452. Reproduced by permission of Routledge/Taylor & Francis Group, LLC

particles', which are much larger in size (100 μm or more). The carrier particles are required for two reasons: because the required quantity of drug is so small that it would be difficult to package; and because such a fine powder would be very cohesive – therefore difficult to handle and unlikely to easily disperse in air. The intention is that the carrier particles should be left in the device, in the back of the mouth or the upper airways and that the drug particles should detach during inspiration and travel to the lower airways before deposition.

14.5 HARMFUL EFFECTS OF FINE POWDERS

Particles which find their way into the lungs can also have a negative effect on health. In the workplace and in our everyday lives, exposure to fine particle aerosols should be avoided. History has shown that exposure to coal dust, silica dust and asbestos dust, for example, can have disastrous effects on the health of workers many years after exposure. Many other workplace dusts, less well known, have been found to have negative health effects.

As discussed above, particles smaller than 10 μm present the greatest risk, since they can penetrate deep into the lungs – offering the greatest chance of chemical absorption into the blood as well as physical interaction with the lungs.

If a dust hazard is suspected in the workplace, the first step is to monitor the working environment to determine the exposure of the worker. One of the better methods of achieving this is for the worker to wear a portable sampling device, which gives a measure of the type of particles and their size distribution in the air immediately around the worker. Such devices usually sample at a typical respiration rate and velocity and some devices are designed to capture directly only the respirable particles (particles capable of reaching the alveoli) or the inhalable particles (particles capable of being inhaled).

The second step in dealing with the potential dust hazard is for the results of the monitoring process to be compared with the accepted workplace standards for the particulate materials in question. If the concentration of respirable particles is found to exceed the accepted limit for that material, then we proceed to the third stage, namely control.

When dealing with any hazard, there is a hierarchy of control measures that may be put in place. In the modern workplace, the aim is to produce a safe environment rather that a safe person. To understand what is meant by this, consider the hierarchy of control measures:

specification;

substitution;

containment;

ventilation;

reduced time of exposure;

protective equipment.

In designing a process, an engineer or scientist should aim for control measures at the top of the hierarchy. Only as a last resort should the measures lower down in the hierarchy be used.

Specification: Devising an alternative process which does not include this hazard. In the case of dust hazard, this may mean using a completely different wet process. For example, using wet milling rather than dry milling. Granulation may be an option – with loose granules, the large surface area provided by fine particles can still be made available, but without the associated dust hazard.

Substitution: Replacing the hazardous material with a non-hazardous material, for example; by using wood fibres instead of asbestos in the manufacture of building products.

Containment: Designing the process using equipment which ensures that hazardous materials are contained and do not, under normal operation, escape into the environment.

An example is using pneumatic transport rather than conveyors belts or other mechanical conveyors in transporting powders within the workplace. Also, use fully enclosed conveyor belts.

Ventilation: Accepting that the hazardous material is present in the workplace environment and creating airflows to draw the material away from workers or reduce its concentration in the environment.

Reduced time of exposure: Accepting that the hazardous material is present in the workplace environment and reducing the time spent by each worker in that environment.

Protective equipment: Accepting that the worker must work in an environment where the hazardous material is present and providing suitable protective equipment for the worker to wear. Examples for controlling dust hazard (in order of decreasing efficacy) are: air-line helmets – clean air is provided under pressure via flexible tubing to a full headset worn my the worker; positive pressure sets – a pump and filter worn by the worker provides air to a headset which may partially or fully enclose the worker's head; airstream helmets – a pump and filter fitted to a hard hat with a visor such that the filtered air stream is blown across the worker's face; ori-nasal respirators – a well-fitting rubber or plastic mask covering the nose and mouth and fitted with efficient filters suitable for the material in question; disposable facemasks – mask made of filter material covering the nose and mouth and usually not well-fitting.

TEST YOURSELF

14.1 Make a diagrammatic sketch of the human lung, labelling the following regions: alveoli sacs, bronchi, trachea, respiratory brochioles.

14.2 In the human respiratory systems what are the typical diameters and air velocities in the trachea, terminal brochioles, main bronchi?

14.3 List the mechanisms by which inhaled particles may become deposited on the walls of the airways in the respiratory systems. Which mechanism dominates for particles smaller than around 10 μm in the upper airways, and why?

14.4 What is the Stokes number and what does it tell us about the likelihood of particles being deposited on the walls on the respiratory system?

14.5 In which part of the respiratory system is sedimentation important as a mechanism for depositing particles?

14.6 For pulmonary delivery of drugs, what size range of particles is desirable and why?

14.7 Describe the construction and operation of three types of device for the delivery of respirable drugs.

14.8 What is a carrier particle and why is it needed?

14.9 What steps should be taken to determine whether a dust in the workplace has the potential to cause a health hazard through inhalation?

14.10 Explain what is meant by the hierarchy of control measures when applied to the control of a dust hazard.

EXERCISES

14.1 Determine, by calculation, the likely fate of a 20 μm particle of density 2000 kg/m^3 suspended in the air inhaled by a human at a rate giving rise to the following velocities in parts of the respiratory system:

Part	Number	Diameter (mm)	Length (mm)	Typical air velocity (m/s)	Typical residence time (s)
Mouth	1	20	70	3.2	0.022
Pharynx	1	30	30	1.4	0.021
Trachea	1	18	120	4.4	0.027
Two main bronchi	2	13	37	3.7	0.01

14.2 Given the following information, in which region of the respiratory tract is a 3 μm particle of density $1500\,kg/m^3$ most likely to be deposited and by which mechanism? Support your conclusion by calculation.

Part	Diameter (mm)	Length (mm)	Typical air velocity (m/s)	Typical residence time (s)
Trachea	18	120	4.4	0.027
Bronchioles	2	20	0.6	0.032
Terminal bronchioles	0.7	5	0.2	0.023
Alveolar ducts	0.8	1	0.0023	0.44
Alveoli	0.15	0.15	0.00004	4

14.3 Compare the Stokes numbers for 2, 5, 10 and 40 μm particles of density $1200\,kg/m^3$ in air passing through the nose. What conclusions do you draw regarding the likelihood of deposition of these particles in the nose?

Data:

Characteristic velocity in the nose: 9 m/s

Characteristic diameter of the airway in the nose: 6 mm

Viscosity of air: 1.81×10^{-5} Pa s.

Density of air: $1.21\,kg/m^3$

14.4 Carrier particles are used in dry powder inhalers. What is a carrier particle? What is the role of a carrier particle? Why are carrier particles needed in these inhalers?

14.5 With reference to the control of dusts as a health hazard, explain what is meant by the *hierarchy of controls.*

14.6 The required dose of a particulate drug of particle size 3 μm and particle density $1000\,kg/m^3$ is 10 μg. Estimate the number of particles in this dose and the volume occupied by the dose, assuming a voidage of 0.6.

(Answer: 7×10^5; $0.25\,mm^3$.)

15

Fire and Explosion Hazards of Fine Powders

15.1 INTRODUCTION

Finely divided combustible solids, or dusts, dispersed in air can give rise to explosions in much the same way as flammable gases. In the case of flammable gases, fuel concentration, local heat transfer conditions, oxygen concentration and initial temperature all affect ignition and resulting explosion characteristics. In the case of dusts, however, more variables are involved (e.g. particle size distribution, moisture content) and so the analysis and prediction of dust explosion characteristics is more complex than for the flammable gases. Dust explosions have been known to give rise to serious property damage and loss of life. Most people are probably aware that dust explosions have occurred in grain silos, flour mills and in the processing of coal. However, explosions of dispersions of fine particles of metals (e.g. aluminium), plastics, sugar and pharmaceutical products can be particularly potent. Process industries where fine combustible powders are used and where particular attention must be directed towards control of dust explosion hazard include: plastics, food processing, metal processing, pharmaceuticals, agricultural, chemicals and coal. Process steps where fine powders are heated have a strong association with dust explosion; examples include dilute pneumatic conveying and spray drying, which involves heat and a dilute suspension.

In this chapter, the basics of combustion are outlined followed by the fundamentals specific to dust explosions. The measurement and application of dust explosion characteristics, such as ignition temperature, range of flammable concentrations minimum ignition energy are covered. Finally the available methods for control of dust explosion hazard are discussed.

15.2 COMBUSTION FUNDAMENTALS

15.2.1 Flames

A flame is a gas rendered luminous by emission of energy produced by chemical reaction. In a stationary flame (for example a candle flame or gas stove flame) unburned fuel and air flow into the flame front as combustion products flow away from the flame front. A stationary flame may be from either premixed fuel and air, as observed in a Bunsen burner with the air hole open, or by diffusion of air into the combustion zone, as for a Bunsen burner with the air hole closed.

When the flame front is not stationary it is called an explosion flame. In this case the flame front passes through a homogeneous premixed fuel–air mixture. The heat released and gases generated result in either an uncontrolled expansion effect or, if the expansion is restricted, a rapid build-up of pressure.

15.2.2 Explosions and Detonations

Explosion flames travel through the fuel–air mixture at velocities ranging from a few metres per second to several hundreds of metres per second and this type of explosion is called a deflagration. Flame speeds are governed by many factors including the heat of combustion of the fuel, the degree of turbulence in the mixture and the amount of energy supplied to cause ignition. It is possible for flames to reach supersonic velocities under some circumstances. Such explosions are accompanied by pressure shock waves, are far more destructive and called detonations. The increased velocities result from increased gas densities generated by pressure waves. It is not yet understood what conditions give rise to detonations. However, in practice it is likely that all detonations begin as deflagrations.

15.2.3 Ignition, Ignition Energy, Ignition Temperature – a Simple Analysis

Ignition is the self-propagation of a combustion reaction through a fuel–air mixture after the initial supply of energy. Ignition of a fuel–air mixture can be analysed in a manner similar to that used for runaway reactions (thermal explosions). Consider an element of fuel air mixture of volume V and surface area A, in which the volumetric concentration of fuel is C. If the temperature of the fuel–air mixture in the element is T_i and if the rate at which heat is lost from the element to the surroundings (at temperature T_s) is governed by a heat transfer coefficient h, then the rate of heat loss to surroundings, Q_s is:

$$Q_s = hA(T_i - T_s) \tag{15.1}$$

The variation of the combustion reaction rate with temperature will be governed by the Arrhenius equation. For a reaction which is first order in fuel concentration:

$$-V\rho_{\mathrm{m_{fuel}}}\frac{\mathrm{d}C}{\mathrm{d}t} = VC\rho_{\mathrm{m_{fuel}}}Z\exp\left(-\frac{E}{RT}\right) \tag{15.2}$$

where Z is the pre-exponential coefficient, E is the reaction activation energy, R is the ideal gas constant and $\rho_{\mathrm{m_{fuel}}}$ is the molar density of the fuel.

The rate Q_{a} at which heat is absorbed by the fuel–air mixture in the element is:

$$Q_{\mathrm{a}} = V\frac{\mathrm{d}T}{\mathrm{d}t}[C\rho_{\mathrm{m_{fuel}}}C_{\mathrm{p_{fuel}}} + (1-C)\rho_{\mathrm{m_{air}}}C_{\mathrm{p_{air}}}] \tag{15.3}$$

where $C_{\mathrm{p_{fuel}}}$ and $C_{\mathrm{p_{air}}}$ are the molar specific heat capacities of the fuel and air and $\rho_{\mathrm{m_{fuel}}}$ and $\rho_{\mathrm{m_{air}}}$ are the molar densities of the fuel and air, respectively.

If Q_{input} is the rate at which heat energy is fed into the element from outside, then the heat balance for the element becomes:

$$\underset{(1)}{Q_{\mathrm{input}}} + \underset{(2)}{(-\Delta H)VC\rho_{\mathrm{m_{fuel}}}Z\exp\left(-\frac{E}{RT_{\mathrm{R}}}\right)} = \underset{(3)}{V\frac{\mathrm{d}T_{\mathrm{R}}}{\mathrm{d}t}[C\rho_{\mathrm{m_{fuel}}}C_{\mathrm{p_{fuel}}} + (1-C)\rho_{\mathrm{m_{air}}}C_{\mathrm{p_{air}}}]} + \underset{(4)}{hA(T_{\mathrm{Ri}} - T_{\mathrm{s}})} \tag{15.4}$$

It is instructive to analyse this heat balance graphically for the steady-state condition [term (3) is zero]. We do this by plotting the rates of heat loss to the surroundings and the rate of heat generation by the combustion reaction as a function of temperature. The former of course results in a straight line of slope hA and intercept of the temperature axis T_{s}. The rate of heat generation by reaction within the element is given by term (2) and results in an exponential curve. A typical plot is shown in Figure 15.1. Analysing this type of plot gives us insight into the meaning of ignition, ignition energy, ignition temperature, etc.

Consider initially the case where Q_{input} is zero. Referring to the case shown in Figure 15.1, we see that at an initial element temperature T_{i} the rate of heat loss from the element is greater than the rate of heat generation and so the temperature of the element will decrease until point A is reached. Any initial temperature between T_{B} and T_{A} will result in the element cooling to T_{A} . This is a stable condition.

If, however, the initial temperature is greater than T_{B} , the rate of heat generation within the element will be always greater than the rate of heat loss to the surroundings and so the element temperature will rise, exponentially. Thus initial temperatures beyond T_{B} give rise to an unstable condition. T_{B} is the *ignition temperature*, T_{ig}, for the fuel–air mixture in the element. *Ignition energy* is the energy that we must supply from the outside in order to raise the mixture from its initial temperature T_{i} to the ignition temperature T_{ig}. Since the element is continuously losing energy to the surroundings, the ignition energy will actually be a rate of energy input, Q_{input} . This raises the heat generation curve by an amount Q_{input} , reducing the value of T_{ig} (Figure 15.2). The conditions for heat transfer from the element to the surroundings are obviously important in determining the ignition

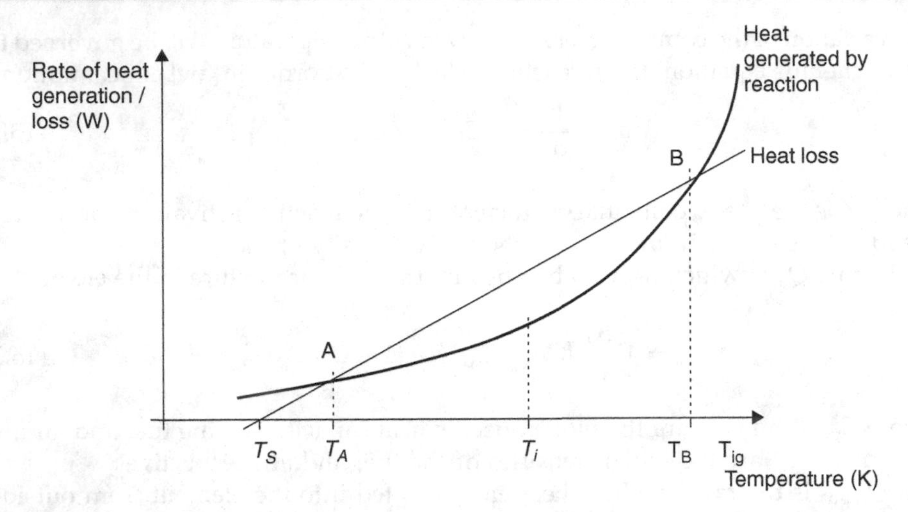

Figure 15.1 Variation of rates of heat generation and loss with temperature of element.

temperature and energy. There are cases where the heat loss curve will be always lower than the heat generation curve (Figure 15.3). Under such circumstances the mixture may self-ignite; this is referred to as auto-ignition or spontaneous ignition.

In many combustion systems there is an appreciable interval (from milliseconds to several minutes) between arrival at the ignition temperature and the apparent onset of ignition. This is known as the *ignition delay*. It is not well understood, but based on the analysis above it is related to the time required for the reaction in the element to go to completion once the ignition temperature has been reached.

If there is adequate oxygen for combustion and the fuel concentration in the element is increased the rate of heat generation by combustion will also increase.

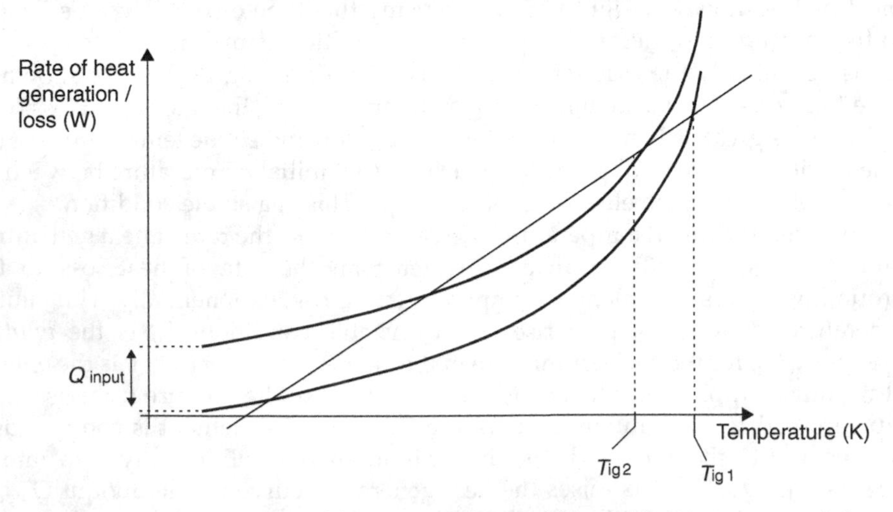

Figure 15.2 Variation in rates of heat generation and loss with temperature of element; the effect on ignition temperature of adding energy.

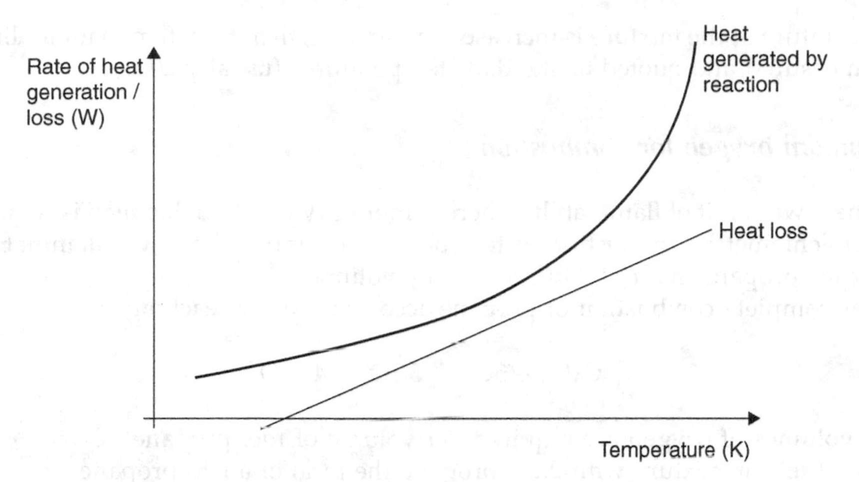

Figure 15.3 Variation in rates of heat generation and loss with temperature of element; rate of heat generation always greater that rate of heat loss.

Whether the ignition temperature and energy will be affected depends on the relative physical properties (specific heat capacity and conductivity) of the fuel and air.

15.2.4 Flammability Limits

From the above analysis it can be seen that below a certain fuel concentration ignition will not occur, since the rate of heat generation within the element is insufficient to match the rate of heat loss to the surroundings (T_{ig} is never reached). This concentration is known as the lower flammability limit C_{fL} of the fuel–air mixture. It is generally measured under standard conditions in order to give reproducible heat transfer conditions. At C_{fL} the oxygen is in excess. As the fuel concentration is increased beyond C_{fL} the amount of fuel reacting per unit volume of mixture and the quantity of heat released per unit volume will increase until the stoichiometric ratio for the reaction is reached. For fuel concentration increase beyond the stoichiometric ratio, the oxygen is limiting and so the amount of fuel reacting per unit volume of mixture and the quantity of heat released per unit volume decrease with fuel concentration. A point is reached when the heat release per unit volume of mixture is too low to sustain a flame. This is the upper flammability limit, C_{fU}. This is the concentration of fuel in the fuel–air mixture above which a flame cannot be propagated. For many fuels the amount of fuel reacting per unit volume of fuel–air mixture (and hence the heat release per unit volume mixture) at C_{fL} is similar to that at C_{fU} showing that heat release per unit volume of fuel–air mixture is important in determining whether a flame is propagated. The differences between the values at C_{fL} and C_{fU} are due to the different physical properties of the fuel–air mixtures under these conditions.

Thus, in general, there is a range of fuel concentration in air within which a flame can be propagated. From the analysis above it will be apparent that this range will widen (C_{fL} will decrease and C_{fU} will increase) as the initial temperature of the mixture is increased. In practice, therefore, flammability limits are measured and quoted at standard temperatures (usually 20°C).

Minimum oxygen for combustion

At the lower limit of flammability there is more oxygen available than is required for stoichiometric combustion of the fuel. For example, the lower flammability limit for propane in air at 20°C is 2.2% by volume.

For complete combustion of propane according to the reaction:

$$C_3H_8 + 5O_2 \rightarrow 3CO_2 + 4H_2O$$

five volumes of oxygen are required per volume of fuel propane.

In a fuel–air mixture with 2.2% propane the ratio of air to propane is:

$$\frac{100 - 2.2}{2.2} = 44.45$$

and since air is approximately 21% oxygen, the ratio of oxygen to propane is 9.33. Thus in the case of propane at the lower flammability limit oxygen is in excess by approximately 87%.

It is therefore possible to reduce the concentration of oxygen in the fuel–air mixture whilst still maintaining the ability to propagate a flame. If the oxygen is replaced by a gas which has similar physical properties (nitrogen for example) the effect on the ability of the mixture to maintain a flame is minimal until the stoichiometric ratio of oxygen to fuel is reached. The oxygen concentration in the mixture under these conditions is known as the minimum oxygen for combustion (MOC). Minimum oxygen for combustion is therefore the stoichiometric oxygen equivalent to the lower flammability limit. Thus

$$\text{MOC} = C_{fL} \times \left(\frac{\text{mol } O_2}{\text{mol fuel}}\right)_{\text{stoich}}$$

For example, for propane, since under stoichiometric conditions five volumes of oxygen are required per volume of fuel propane,

$$\text{MOC} = 2.2 \times 5 = 11\% \text{ oxygen by volume}$$

15.3 COMBUSTION IN DUST CLOUDS

15.3.1 Fundamentals Specific to Dust Cloud Explosions

The combustion rate of a solid in air will in most cases be limited by the surface area of solid presented to the air. Even if the combustible solid is in particulate form a few millimetres in size, this will still be the case. However, if the particles of solid are small enough to be dispersed in air without too much propensity to settle, the reaction rate will be great enough to permit an explosion flame to propagate. For a dust explosion to occur the solid material of which the particles are composed must be combustible, i.e. it must react exothermically with the oxygen in air. However, not all combustible solids give rise to dust explosions.

To make our combustion fundamentals considered above applicable to dust explosions we need only to add in the influence of particle size on reaction rate. Assuming the combustion reaction rate is now determined by the surface area of solid fuel particles (assumed spherical) exposed to the air, the heat release term (2) in Equation (15.4) becomes:

$$(-\Delta H)VC\left(\frac{6}{x}\right)Z'\exp\left[-\frac{E}{RT_{\mathrm{R}}}\right] \qquad (15.5)$$

where x is the particle size and $\rho_{\mathrm{m_{fuel}}}$ is the molar density of the solids fuel.

The rate of heat generation by the combustion reaction is therefore inversely proportional to the dust particle size. Thus, the likelihood of flame propagation and explosion will increase with decreasing particle size. Qualitatively, this is because finer fuel particles:

- more readily form a dispersion in air;
- have a larger surface area per unit mass of fuel;
- offer a greater surface area for reaction (higher reaction rate, as limiting);
- consequently generate more heat per unit mass of fuel;
- have a greater heat-up rate.

15.3.2 Characteristics of Dust Explosions

Consider design engineers wishing to know the potential fire and explosion hazards associated with a particulate solid made or used in a plant which they are designing. They are faced with the same problem which they face in gathering any 'property' data of particulate solids; unlike liquids and gases there are few published data, and what is available is unlikely to be relevant. The particle size distribution, surface properties and moisture content all influence the potential fire hazard of the powder, so unless the engineers can be sure that their powder is identical in every way to the powder used for the published data, they must have the explosion characteristics of the powder tested. Having made the decision, the engineers must ensure that the sample given to the test laboratory is truly representative of the material to be produced or used in the final plant.

Although there has been recent progress towards uniform international testing standards for combustible powders, there remain some differences. However, most tests include an assessment of the following explosion characteristics:

- minimum dust concentration for explosion;
- minimum energy for ignition;
- minimum ignition temperature;
- maximum explosion pressure;
- maximum rate of pressure rise during explosion;
- minimum oxygen for combustion.

An additional classification test is sometimes used. This is simply a test for explosibility in the test apparatus, classifying the dust as able or unable to ignite and propagate a flame in air at room temperature under test conditions. This test in itself is not very useful particularly if the test conditions differ significantly from the plant conditions.

15.3.3 Apparatus for Determination of Dust Explosion Characteristics

There are several different devices for determination of dust explosion characteristics. All devices include a vessel which may be open or closed, an ignition source which may be an electrical spark or electrically heated wire coil and a supply of air for dispersion of the dust. The simplest apparatus is known as the vertical tube apparatus and is shown schematically in Figure 15.4. The sample dust is placed in the dispersion cup. Delivery of dispersion air to the cup is via a solenoid valve. Ignition may be either by electrical spark across electrodes or by heated coil. The vertical tube apparatus is used for the classification test and for determination of minimum dust concentration for explosion, minimum energy for ignition and in a modified form for minimum oxygen for combustion.

A second apparatus known as the 20 litre sphere is used for determination of maximum explosion pressure and maximum rate of pressure rise during explosion. These give an indication of the severity of explosion and enable the design of explosion protection equipment. This apparatus, which is shown schematically in Figure 15.5, is based on a spherical 20 litre pressure vessel fitted with a pressure transducer. The dust to be tested is first charged to a reservoir and then blown by air into the sphere via a perforated dispersion ring. The vessel pressure is reduced to about 0.4 bar before the test so that upon injection of the dust, the pressure rises to atmospheric. Ignition is by a pyrotechnical device with a standard total energy (typically 10 kJ) positioned at the centre of the sphere. The delay between dispersion of the dust and initiation of the ignition source has been found to affect the results. Turbulence caused by the air injection influences the rate of the combustion

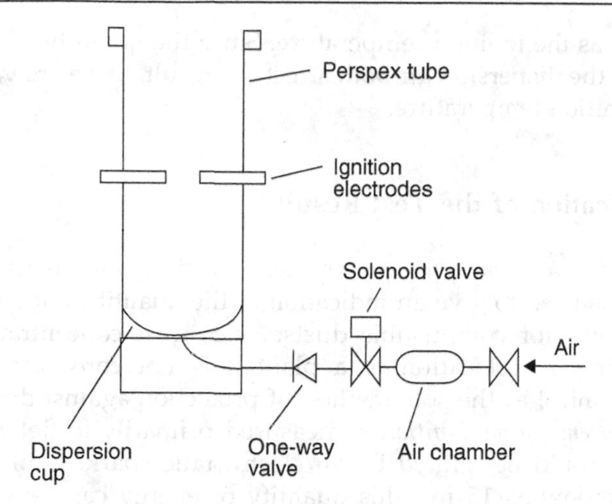

Figure 15.4 Vertical tube apparatus for determination of dust explosion characteristics.

reaction. A standard delay of typically 60 ms is therefore employed in order to ensure the reproducibility of the test. There is also a 1 m^3 version of this apparatus.

The third basic test device is the Godbert–Greenwald furnace apparatus, which is used to determine the minimum ignition temperature and the explosion characteristics at elevated temperatures. The apparatus includes a vertical electrically heated furnace tube which can be raised to controlled temperatures up to 1000°C. The dust under test is charged to a reservoir and then dispersed through the tube. If ignition occurs the furnace temperature is lowered in 10°C steps until ignition does not occur. The lowest temperature at which ignition

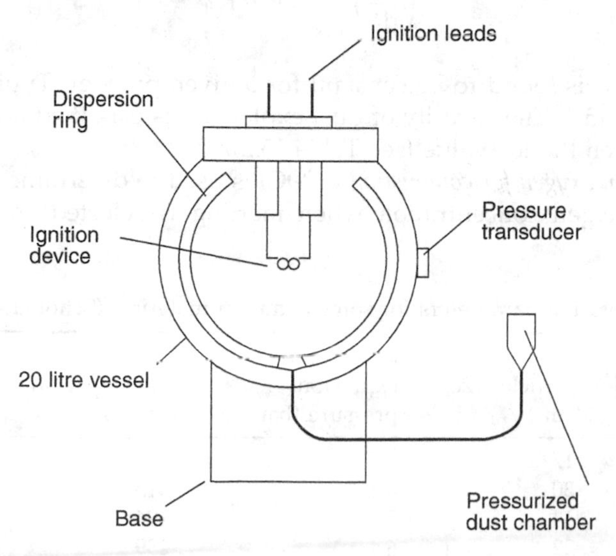

Figure 15.5 The 20 litre sphere apparatus for determination of dust explosion characteristics (after Lunn, 1992).

occurs is taken as the ignition temperature. Since the quantity of dust used and the pressure of the dispersion air both affect the result, these are varied to obtain a minimum ignition temperature.

15.3.4 Application of the Test Results

The *minimum dust concentration for explosion* is measured in the vertical tube apparatus and is used to give an indication of the quantities of air to be used in extraction systems for combustible dusts. Since dust concentrations can vary widely with time and location in a plant it is not considered wise to use concentration control as the sole method of protection against dust explosion.

The *minimum energy for ignition* is measured primarily to determine whether the dust cloud could be ignited by an electrostatic spark. Ignition energies of dusts can be as low as 15 mJ; this quantity of energy can be supplied by an electrostatic discharge.

The *minimum ignition temperature* indicates the maximum temperature for equipment surfaces in contact with the powder. For new materials it also permits comparison with well-known dusts for design purposes. Table 15.1 gives some values of explosion parameters for common materials.

The *maximum explosion pressure* is usually in the range 8–13 bar and is used mainly to determine the design pressure for equipment when explosion containment or protection is opted for as the method of dust explosion control.

The *maximum rate of pressure rise* during explosion is used in the design of explosion relief. It has been demonstrated that the maximum rate of pressure rise in a dust explosion is inversely proportional to the cube root of the vessel volume, i.e.

$$\left(\frac{dP}{dt}\right)_{max} = V^{1/3} K_{St} \tag{15.6}$$

The value of K_{St} is found to be constant for a given powder. Typical values are given in Table 15.1. The severity of dust explosions is classified according to the St class based on the K_{St} value (see Table 15.2).

The *minimum oxygen for combustion* (MOC) is used to determine the maximum permissible oxygen concentration when inerting is selected as the means of

Table 15.1 Explosion parameters for some common materials (Schofield, 1985).

Dust	Mean particle size (μm)	Maximum explosion pressure (bar)	Maximum rate of pressure rise (bar/s)	K_{St}
Aluminium	17	7.0	572	155
Polyester	30	6.1	313	85
Polyethylene	14	5.9	494	134
Wheat	22	6.1	239	65
Zinc	17	4.7	131	35

Table 15.2 Dust explosion classes based on 1 m^3 test apparatus.

Dust explosion class	K_{St} (bar m/s)	Comments
St 0	0	Non-explosible
St 1	0–200	Weak to moderately explosible
St 2	200–300	Strongly explosible
St 3	>300	Very strongly explosible

controlling the dust explosion. Organic dusts have an MOC of about 11% if nitrogen is the diluent and 13% in the case of carbon dioxide. Inerting requirements for metal dusts are more stringent since MOC values for metals can be far lower.

15.4 CONTROL OF THE HAZARD

15.4.1 Introduction

As with the control of any process hazard, there is a hierarchy of approaches that can be taken to control dust explosion hazard. These range from the most desirable strategic approach of changing the process to eliminate the hazardous powder altogether to the merely tactical approach of avoiding ignition sources. In approximate order of decreasing strategic component, the main approaches are listed below:

- change the process to eliminate the dust;
- design the plant to withstand the pressure generated by any explosion;
- remove the oxygen by complete inerting;
- reduce oxygen to below MOC;
- add moisture to the dust;
- add diluent powder to the dust;
- detect start of explosion and inject suppressant;
- vent the vessel to relieve pressure generated by the explosion;
- control dust concentration to be outside flammability limits;
- minimize dust cloud formation;
- Exclude ignition sources.

15.4.2 Ignition Sources

Excluding ignition sources sounds a sensible policy. However, statistics of dust explosions indicate that in a large proportion of incidents the source of ignition was unknown. Thus, whilst it is good policy to avoid sources of ignition as far as possible, this should not be relied on as the sole protection mechanism. It is interesting to look briefly at the ignition sources which have been associated with dust explosions.

- *Flames*. Flames from the burning of gases, liquids or solids are effective sources of ignition for flammable dust clouds. Several sources of flames can be found in a process plant during normal operation (e.g. burners, pilot flames, etc.) and during maintenance (e.g. welding and cutting flames). These flames would usually be external to the vessels and equipment containing the dust. To avoid exposure of dust clouds to flames, therefore, good housekeeping is required to avoid a build-up of dust which may generate a cloud and a good permit-to-work system should be in place to ensure a safe environment before maintenance commences.

- *Hot surfaces*. Careful design is required to ensure that surfaces likely to be in contact with dust do not reach temperatures which can cause ignition. Attention to detail is important; for example ledges inside equipment should be avoided to prevent settling of dust and possible self ignition. Dust must not be able to build up on hot or heated surfaces, otherwise surface temperatures will rise as heat dissipation from the surface is reduced. Outside the vessel care must also be taken; for example if dust is allowed to settle on electric motor housings, overheating and ignition may occur.

- *Electric sparks*. Sparks produced in the normal operation of electrical power sources (by switches, contact breakers and electric motors) can ignite dust clouds. Special electrical equipment is available for application in areas where there is a potential for dust explosion hazard. Sparks from electrostatic discharges are also able to ignite dust clouds. Electrostatic charges are developed in many processing operations (particularly those involving dry powders) and so care must be taken to ensure that such charges are led to earth to prevent accumulation and eventual discharge. Even the energy in the charge developed on a process operator can be sufficient to ignite a dust cloud.

- *Mechanical sparks and friction*. Sparks and local heating caused by friction or impact between two metal surfaces or between a metal surface and foreign objects inadvertently introduced into the plant have been known to ignite dust clouds.

15.4.3 Venting

If a dust explosion occurs in a closed vessel at 1 atm, the pressure will rise rapidly (up to and sometimes beyond 600bar/s) to a maximum of around 10 bar.

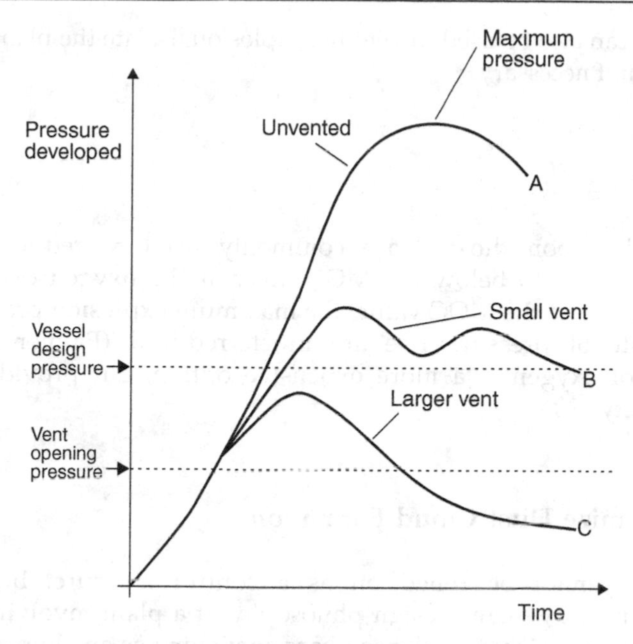

Figure 15.6 Pressure variation with time for dust explosions (A) unvented, (B) vented with inadequate vent area and (C) vented with adequate vent area (after Schofield, 1985).

If the vessel is not designed to withstand such a pressure, deformation and possible rupture will occur. The principle of explosion venting is to discharge the vessel contents through an opening or vent to prevent the pressure rising above the vessel design pressure. Venting is a relatively simple and inexpensive method of dust explosion control but cannot be used when the dust, gas or combustion products are toxic or in some other way hazardous, or when the rate of pressure rise is greater than 600bar/s (Lunn, 1992). The design of vents is best left to the expert although there are published guides (Lunn, 1992). The mass and type of the vent determine the pressure at which the vent opens and the delay before it is fully open. These factors together with the size of the vent determine the rate of pressure rise and the maximum pressure reached after the vent opens. Figure 15.6 shows typical pressure rise profiles for explosions in a vessel without venting and with vents of different size.

15.4.4 Suppression

The pressure rise accompanying a dust explosion is rapid but it can be detected in time to initiate some action to suppress the explosion. Suppression involves discharging a quantity of inert gas or powder into the vessel in which the explosion has commenced. Modern suppression systems triggered by the pressure rise accompanying the start of the explosion have response times of the order of a few milliseconds and are able to effectively extinguish the explosion. The fast acting trigger device can also be used to vent the explosion, isolate the plant item or shut down the plant if necessary.

15.4.5 Inerting

Nitrogen and carbon dioxide are commonly used to reduce the oxygen concentration of air to below the MOC. Even if the oxygen concentration is not reduced as far as the MOC value, the maximum explosion pressure and the maximum rate of pressure rise are much reduced (Palmer, 1990). Total replacement of oxygen is a more expensive option, but provides an added degree of safety.

15.4.6 Minimize Dust Cloud Formation

In itself this cannot be relied on as a control measure, but should be incorporated in the general design philosophy of a plant involving flammable dusts. Examples are (1) use of dense phase conveying as an alternative to dilute phase, (2) use of cyclone separators and filters instead of settling vessels for separation of conveyed powder from air, and (3) avoiding situations where a powder stream is allowed to fall freely through air (e.g. in charging a storage hopper). Outside the vessels of the process good housekeeping practice should ensure that deposits of powder are not allowed to build up on ledges and surfaces within a building. This avoids secondary dust explosions caused when these deposits are disturbed and dispersed by a primary explosion or shock wave.

15.4.7 Containment

Where plant vessels are of small dimensions it may be economic to design them to withstand the maximum pressure generated by the dust explosion (Schofield and Abbott, 1988). The vessel may be designed to contain the explosion and be replaced afterwards or to withstand the explosion and be reusable. In both cases design of the vessel and its accompanying connections and ductwork is a specialist task. For large vessels the cost of design and construction to contain dust explosions is usually prohibitive.

15.5 WORKED EXAMPLES

WORKED EXAMPLE 15.1

It is proposed to protect a section of duct used for pneumatically transporting a plastic powder in air by adding a stream of nitrogen. The air flow rate in the present system is

1.6 m^3/s and the air carries 3% powder by volume. If the minimum oxygen for combustion (by replacement of oxygen with nitrogen) of the powder is 11% by volume, what is the minimum flow rate of nitrogen which must be added to ensure safe operation?

Solution

The current total air flow of 1.6 m^3/s includes 3% by volume of plastic powder and 97% air (made up of 21% oxygen and 79% nitrogen by volume). In this stream the flow rates are therefore:

powder: 0.048m^3/s

oxygen: 0.3259m^3/s

nitrogen: 1.226m^3/s

At the limit, the final concentration of the flowing mixture should be 11% by volume. Hence, using a simple mass balance assuming constant densities,

$$\frac{\text{volume flow } O_2}{\text{total volume flow}} = \frac{0.3259}{1.6 + n} = 0.11$$

from which, the minimum required flow rate of added nitrogen, $n = 1.36\,\text{m}^3/\text{s}$.

WORKED EXAMPLE 15.2

A combustible dust has a lower flammability concentration limit in air at 20°C of 0.9% by volume. A dust extraction system operating at 2 m^3/s is found to have a dust concentration of 2% by volume. What minimum flow rate of additional air must be introduced to ensure safe operation?

Solution

Assume that the dust explosion hazard will be reduced by bringing the dust concentration in the extract to below the lower flammability limit. In 2 m^3/s of extract, the flow rates of air and dust are

air: 1.96 m^3/s

dust: 0.04 m^3/s

At the limit, the dust concentration after addition of dilution air will be 0.9%, hence:

$$\frac{\text{volume of dust}}{\text{total volume}} = \frac{0.04}{2 + n} = 0.09$$

from which the minimum required flow rate of added dilution air, $n = 2.44\,\text{m}^3/\text{s}$.

WORKED EXAMPLE 15.3

A flammable dust is suspended in air at a concentration within the flammable limits and with an oxygen concentration above the minimum oxygen for combustion. Sparks generated by a grinding wheel pass through the suspension at high speed, but no fire or explosion results. Explain why.

Solution

In this case it is likely that the temperature of the sparks will be above the measured minimum ignition temperature and the energy available is greater than the minimum ignition energy. However, an explanation might be that the heat transfer conditions are unfavourable. The high speed sparks have insufficient contact time with any element of the fuel–air mixture to provide the energy required for ignition.

WORKED EXAMPLE 15.4

A fine flammable dust is leaking from a pressurized container at a rate of 2 litre/min into a room of volume 6 m^3 and forming a suspension in the air. The minimum explosible concentration of the dust in air at room temperature is 2.22% by volume. Assuming that the dust is fine enough to settle only very slowly from suspension, (a) what will be the time from the start of the leak before explosion occurs in the room if the air ventilation rate in the room is 4 m^3/h, and (b) what would be the minimum safe ventilation rate under these circumstances?

Solution

(a) Mass balance on the dust in the room:

$$\begin{pmatrix}\text{rate of}\\ \text{accumulation}\end{pmatrix} = \begin{pmatrix}\text{rate of flow}\\ \text{into the room}\end{pmatrix} - \begin{pmatrix}\text{rate of flow out}\\ \text{of room with air}\end{pmatrix}$$

assuming constant gas density,

$$V\frac{dC}{dt} = 0.12 - 4C$$

where 0.12 is the leak rate in m^3/h, V is the volume of the room and C is the dust concentration in the room at time t.

Rearranging and integrating with the initial condition, $C = 0$ at $t = 0$,

$$t = -1.5\ln\left(\frac{0.12 - 4C}{0.12}\right)\text{ h}$$

Assuming the explosion occurs when the dust concentration reaches the lower flammability limit, 2.22%

$$\text{time required} = 2.02\text{ h}.$$

(b) To ensure safety, the limiting ventilation rate is that which gives a room dust concentration of 2.22% at steady state (i.e. when $dC/dt = 0$). Under this condition,

$$0 = 0.12 - FC_{fL}$$

hence, the minimum ventilation rate, $F = 5.4\,m^3/h$.

WORKED EXAMPLE 15.5

Table 15W5.1 Combustion data for various fuels.

Substance	Lower flammability limit (vol % of fuel in air) (20°C and 1 bar)	Upper flammability limit (vol % of fuel in air)	Standard enthalpy of reaction (MJ/kmol)
Benzene	1.4	8.0	−3302
Ethanol	3.3	19	−1366
Methanol	6	36.5	−764
Methane	5.2	33	−890

Using the information in Table 15W5.1, calculate the heat released per unit volume for each of the fuels at their lower and upper flammability limits. Comment on the results of your calculations.

Solution

As the concentration of fuel in the fuel–air mixture increases, the fuel burned per unit volume of mixture increases, and hence, the heat released per unit volume of mixture increases, until the point is reached when the fuel and air are in stoichiometric proportions C_{Fstoic}. Beyond this point, there is insufficient oxygen to combust all the fuel in the mixture. For fuel concentrations beyond C_{Fstoic}, therefore, it is the quantity of oxygen per unit volume of mixture which dictates the quantity of fuel burned and hence the heat released per unit volume of mixture.

Below C_{Fstoic}

Heat release does not occur below the lower flammability limit C_{FL} since flame propagation does not occur.

At C_F volume of fuel in $1m^3$ mixture $= C_F$

Assuming ideal gas behaviour, molar density $= \frac{n}{V} = \frac{P}{RT}$

At 20°C and 1 bar pressure, molar density $= 0.0416\,kmol/m^3$

$$\text{So, heat released per m}^3\text{mixture} = (-\Delta H_c) \times C_F \times 0.0416 \qquad (15W5.1)$$

Hence, at C_{FL} the heat released per m^3 mixture = $(-\Delta H_c) \times C_{FL} \times 0.0416$

The results for the listed fuels are shown in Table 15W5.2.

Table 15W5.2 Heat released per m^3 fuel–air mixture at C_{FL} for various fuels.

Fuel	C_{FL}	$-\Delta H_c$ (MJ/kmol)	Heat released per m^3 mixture (MJ/m^3)
Benzene	0.014	3302	1.92
Ethanol	0.033	1366	1.87
Methanol	0.06	764	1.91
Methane	0.052	890	1.93

For the fuels listed in Table 15W5.2 the values of heat released per unit volume of mixture at the lower flammability limit are quite similar (1.91 MJ/m^3 $\pm$2%) demonstrating that it is the heat released per unit volume which determines whether a flame will propagate in the fuel–air mixture.

Below C_{Fstoic}

When C_F increases beyond C_{Fstoic} oxygen is limiting. Therefore, the first step is to determine the oxygen concentration (kmol/m^3 mixture) for a given C_F.

For 1 m^3 of mixture: volume of fuel = C_F

and volume of air = $1 - C_F$

Taking air as 21 % oxygen and 79 % nitrogen, by volume,

volume of oxygen = $0.21(1 - C_F)$

Assuming ideal gas behaviour, 1 m^3 at 20°C and 1 bar holds 0.0416 kmol (see above),

Moles of oxygen per m^3 in a fuel–air mixture at C_F, $n_{O_2} = 0.0416 \times 0.21(1 - C_F)$ kmol/m^3

Hence, moles of fuel reacting per m^3 in a fuel–air mixture at C_F

$$= n_{O_2} \times \text{mol fuel reacting with every mol of oxygen}$$

$$= n_{O_2} \times \frac{\text{stoichiometric coefficient of fuel}}{\text{stoichiometric coefficient of oxygen}}$$

Hence, beyond C_{Fstoic} heat released per m^3 fuel–air mixture

$$= 0.0416 \times 0.21(1 - C_F) \times R_{ST} \times (-\Delta H_c) \quad (15W5.2)$$

where

$$R_{ST} = \frac{\text{stoichiometric coefficient of fuel}}{\text{stoichiometric coefficient of oxygen}}$$

Stoichiometric coefficients:

$$\text{Benzene}: C_6H_6 + 7.5O_2 \rightarrow 6CO_2 + 3H_2O, \ \text{so}\ R_{ST} = 0.1333$$
$$\text{Ethanol}: C_2H_5OH + 3O_2 \rightarrow 2CO_2 + 3H_2O, \ \text{so}\ R_{ST} = 0.3333$$
$$\text{Methanol}: CH_3OH + 1.5O_2 \rightarrow CO_2 + 2H_2O, \ \text{so}\ R_{ST} = 0.6666$$
$$\text{Methane}: CH_4 + 2O_2 \rightarrow CO_2 + 2H_2O, \ \text{so}\ R_{ST} = 0.5$$

Table 15W5.3 gives the calculated values of heat released per m^3 fuel–air mixture at the upper flammability limit ($C_F = C_{FU}$) for these fuels [based on Equation (15W5.2)].

Table 15W5.3 Heat released per m^3 fuel–air mixture at C_{FU} for various fuels.

Fuel	C_{FU}	$-\Delta H$ (MJ/kmol)	R_{ST}	Heat released per m^3 mixture (MJ/m^3)
Benzene	0.08	3302	0.1333	3.54
Ethanol	0.19	1366	0.3333	3.22
Methanol	0.365	764	0.6666	2.83
Methane	0.33	890	0.50	2.60

Amongst these fuels, the values for heat released per m^3 of fuel–air mixture at the upper flammability limit are of the same order ($3.05 \pm 16\%$). Also, we note that the values at the upper flammability limit are somewhat higher (but certainly the same order of magnitude) than the values at the lower flammability limit. These differences are likely to be due to the different physical properties (conductivity, specific heat capacity, for example) of the fuel–air mixture at *low fuel concentrations* compared with the physical properties at *higher fuel concentrations*.

WORKED EXAMPLE 15.6

Based on the information for methane in Table 15W5.1, produce a plot of heat released per m^3 of mixture as a function of fuel concentration for methane–air mixtures at atmospheric pressure and 20°C. In the light of this, explain why fuels have an upper flammability limit.

Solution

For $C_F < C_{FL}$ and $C_F > C_{FU}$ no heat will be released since no combustion takes place.

For $C_{FL} < C_F < C_{Fstoic}$ (the fuel-limiting range) the heat released per m^3 of fuel–air mixture will be described by (See example 15.5):

$$(-\Delta H_c) \times C_F \times 0.0416 \qquad (15W5.1)$$

For methane, the heat released per m^3 of fuel–air mixture becomes:
37.02 C_F MJ/m^3.

For $C_{Fstoic} < C_F < C_{FU}$ (the oxygen-limiting range) the heat released per m^3 of fuel–air mixture will be described by (See example 15.5):

$$0.0416 \times 0.21(1 - C_F) \times R_{ST} \times (-\Delta H_c) \qquad (15W5.2)$$

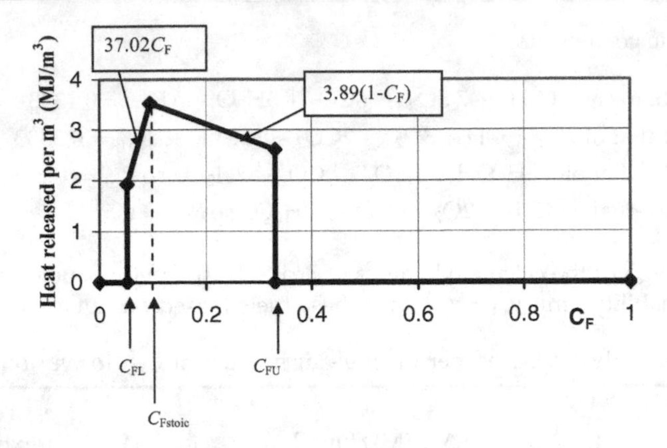

Figure 15W6.1 The heat released per m^3 of methane–air mixtures as a function of methane concentration C_F.

For methane, the heat release per m^3 of fuel–air mixture becomes:
$3.89(1 - C_F)\ \text{MJ/m}^3$

Determine C_{Fstoic}:

$$\text{For } 1\ \text{m}^3 : C_F = \frac{\text{volume of fuel}}{\text{total volume of mixture}} = \frac{\text{vol fuel}}{\text{vol fuel} + \text{vol } O_2 + \text{vol } N_2}$$

Taking air as 21 mol% oxygen and 79 mol% nitrogen,

$$C_F = \frac{\text{vol fuel}}{\text{vol fuel} + \text{vol } O_2 + \text{vol } N_2} = \frac{\text{vol fuel}}{\text{vol fuel} + 4.762\text{vol } O_2}$$

At stoichiometric conditions (assuming ideal gas behaviour),

$$\frac{\text{vol fuel}}{\text{vol } O_2} = \frac{\text{mol fuel}}{\text{mol } O_2} = R_{ST}$$

$$\text{So} : C_{Fstoic} = \frac{\text{vol fuel}}{\text{vol fuel} + 4.762\left(\dfrac{\text{vol fuel}}{R_{ST}}\right)} = \frac{R_{ST}}{R_{ST} + 4.762}$$

For methane this gives $C_{Fstoic} = 0.095$

Plotting these gives Figure 15W6.1.

TEST YOURSELF

15.1 List five process industries which process combustible particulate materials and in which there is therfore the potential for dust explosion.

15.2 What is meant by the terms *lower flammability limit* and *upper flammability limit*?

15.3 The lower flammability limits for benzene, methanol and methane are 1.4, 6.0 and 5.2 vol %, respectively. However, for each of these fuels, the heat generated per unit volume of fuel–air mixture at the lower flammability limit is approximately 1.92 MJ/m^3. Explain the significance of these statements.

15.4 Explain why, for a suspension of combustible dust in air, the likelihood of flame propagation and explosion increases with decreasing particle size.

15.5 List and define five explosion characteristics that are determined experimentally in the 20 litre sphere test apparatus.

15.6 In decreasing order of desirability, list five approaches to reducing the risk of dust cloud explosion.

15.7 Explain why a policy of eliminating ignition sources from a process plant handling combustible powders cannot be relied upon as the sole measure taken to prevent a dust explosion.

15.8 What factors must be taken into account when designing a vent to protect a vessel from the effects of a dust explosion?

EXERCISES

15.1 It is proposed to protect a section of duct used for pneumatically transporting a food product in powder form in air by adding a stream of carbon dioxide. The air flow rate in the present system is 3 m^3/s and the air carries 2% powder by volume. If the minimum oxygen for combustion (by replacement of oxygen with carbon dioxide) of the powder is 13% by volume, what is the minimum flow rate of carbon dioxide which must be added to ensure safe operation?

(Answer: 1.75 m^3/s.)

15.2 A combustible dust has a lower flammability limit in air at 20°C of 1.2% by volume. A dust extraction system operating at 3 m^3/s is found to have a dust concentration of 1.5% by volume. What minimum flow rate of additional air must be introduced to ensure safe operation?

(Answer: 0.75 m^3/s.)

15.3 A flammable pharmaceutical powder suspended in air at a concentration within the flammable limits and with an oxygen concentration above the minimum oxygen for combustion flows at 40 m/s through a tube whose wall temperature is greater than the measured ignition temperature of the dust. Give reasons why ignition does not necessarily occur.

15.4 A fine flammable plastic powder is leaking from a pressurized container at a rate of 0.5 litre/min into another vessel of volume 2 m^3 and forming a suspension in the air in the vessel. The minimum explosible concentration of the dust in air at room temperature is 1.8% by volume. Stating all assumptions, estimate:

(a) the delay from the start of the leak before explosion occurs if there is no ventilation;

(b) the delay from the start of the leak before explosion occurs if the air ventilation rate in the second vessel is $0.5\,m^3/h$;

(c) the minimum safe ventilation rate under these circumstances.

[(Answer: (a) 1.2 h; (b) 1.43 h; (c) $1.67\,m^3/h$).]

15.5 Using the information in Table 15E5.1, calculate the heat released per unit volume for each of the fuels at their lower and upper flammability limits. Comment on the results of your calculations.

Table 15E5.1 Combustion data for various fuels.

Substance	Lower flammability limit (vol % of fuel in air) (20°C and 1 bar)	Upper flammability limit (vol % of fuel in air)	Standard enthalpy of reaction (MJ/kmol)
Cyclohexane	1.3	8.4	−3953
Toluene	1.27	7	−3948
Ethane	3	12.4	−1560
Propane	2.2	14	−2220
Butane	1.8	8.4	−2879

15.6 Based on the information for propane in Table 15E5.1, produce a plot of heat released per m^3 of mixture as a function of fuel concentration for propane–air mixtures at atmospheric pressure and 20°C. In the light of this, explain why fuels have an upper flammability limit.

16
Case Studies

16.1 *CASE STUDY 1*

High Windbox Pressure in a Fluidized Bed Roaster

A fluidized bed for roasting mineral ores operates satisfactorily when fed continuously with ore A. When the feed is switched to ore B, however, the windbox pressure gradually rises and on occasions reaches unacceptable levels. When this happens action is taken to prevent the blower which provides the fluidizing air from overloading and perhaps tripping. The action usually taken is to switch the feed back to ore A. When this is done the windbox pressure gradually falls back to its original value.

Solids leave the fluidized bed by overflowing a weir. The internal configuration of the fluidized bed is shown diagrammatically in Figure 16.1.

Ores A and B change particle size during the roasting process and so a representative measure of the particle properties can only be obtained by sampling the roasted product overflowing the weir. The measured particle properties of the ores at this point are given in Table 16.1 The particles in ores A and B are found to have similar shape and the particle size distributions are also similar.

The bed operating conditions are independent of the ore being roasted (see Table 16.2). Further details of bed geometry are shown in Table 16.3.

Possible causes of the increase in windbox pressure

The pressure in the windbox is determined by the pressure drop for flow of gas downstream of the windbox. Downstream of the windbox is the distributor, the

Introduction to Particle Technology - 2nd Edition Martin Rhodes

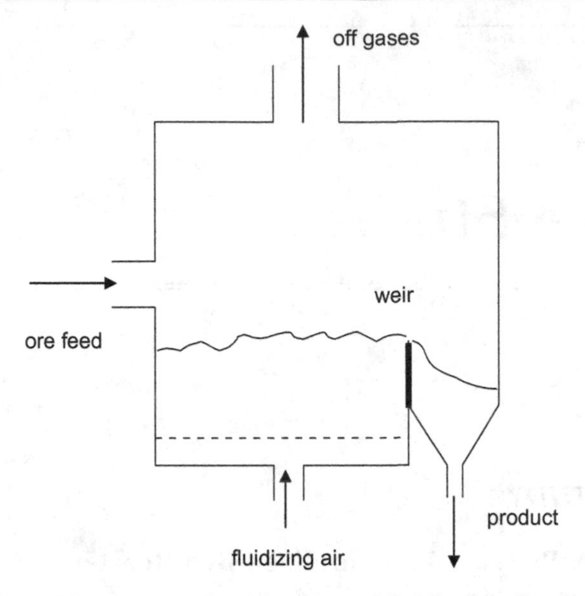

Figure 16.1 Configuration of fluidized bed

Table 16.1 Particle properties of ores A and B

Ore	Mean size (μm)	Particle density (kg/m^3)
A	130	3500
B	350	3500

Table 16.2 Fluidized bed operating conditions

Bed temperature (°C)	877
Superficial gas velocity (m/s)	0.81
Gas viscosity (kg/ms)	4.51×10^{-5}
Gas density (kg/m^3)	0.307

Table 16.3 Fluidized bed details

Bed diameter (m)	6.4
Nozzle diameter (mm)	6
Number of nozzles	3000
Air velocity in bed (m/s)	0.78
Bed height (weir setting) (mm)	1200

fluidized bed, gas cyclones and a baghouse (filter). Measurements of pressure drops across the gas cyclones and the baghouse are found to change little when the ore feed to changed from A to B, so the cause is not here.

The distributor is equipped with 3000 6 mm nozzles. If some of the nozzles were becoming fully or partially blocked then this would lead to a rise in distributor pressure since the gas volume flow delivered to the bed is effectively constant. However, the fact that the windbox pressure gradually returns to normal when the feed is switched back to ore A suggests that nozzle blockage is not the most obvious cause of the problem (unless the blockage is reversible, which seems unlikely).

This leaves us with an increase in bed pressure drop as the most likely cause of the observed increase in windbox pressure. Let us look more closely at bed pressure drop:

In a fluidized bed, the buoyant weight of the particles is supported by the upwardly flowing gas, yielding the expression:

$$\Delta p_{\text{bed}} = H(1 - \varepsilon)(\rho_p - \rho_f)g \tag{7.2}$$

A further level of detail is introduced by considering the make-up of the bed – a mixture of bubbles that contain essentially no particles and a dense emulsion phase consisting of solids and gas in intimate contact (from the two phase theory of fluidization – see Chapter 7). The overall bed density is thus a function of the density of the particulate solids, the proportion of solids and gas in the emulsion phase and the proportion of the bed occupied by bubbles.

$$\Delta p_{\text{bed}} = H(1 - \varepsilon_{\text{p}})(1 - \varepsilon_{\text{B}})(\rho_{\text{p}} - \rho_{\text{f}})g \tag{16.1}$$

where ε_p is the average particulate phase voidage and ε_B is the fraction of the bed occupied by the bubbles.

In the specific situation of the case under consideration, two simplifications to the above can be made: The bed height H is fixed by the overflow weir setting. Hence, the bed is at its minimum height and fixed there as windbox pressure continues to increase. This implies that for analysis, a fixed bed height can be assumed (1.2 m in this case). A further simplification is that the particle density of both ores A and B are the same (3500 kg/m^3).

The bed height and particle density can therefore be eliminated as contributing factors in the windbox pressure increase, since they are the same for both ores. We therefore turn our attention to the bubble fraction and particulate phase voidage in the fluidized bed. These properties are a combined function of the gas flow rate through the bed and the particle shape, particle size and particle size distribution.

Since both the volumetric flow of air to the fluidized bed and the bed temperature are maintained constant during normal operation and the times of excessive windbox pressure, both the superficial gas velocity and the gas transport properties would be the same for both ores A and B (see Table 16.2).

The remaining variables by which bed density (and hence bed pressure drop) is influenced therefore relate directly to particle size and shape.

As indicated above, whether operating with ore A or B the particle size distributions in the overflow have approximately the same shape. However, the mean particle size in the overflow typically increases from around 130 μm to 350 μm during the high windbox pressure periods.

Referring to Equation (16.2), we therefore need to assess the effect of an increasing mean particle size (i.e. from a mean size of 130 μm to 350 μm) on:

ε_p (average particulate phase voidage) and

ε_B (bubble fraction)

As a first approximation, we will the simple two phase theory of fluidization (see Chapter 7), which means that the average particulate phase voidage will be taken as the voidage at minimum fluidization ε_{mf} and the bubble fraction will be given by Equation (7.28):

$$\varepsilon_B = \frac{(U - U_{mf})}{\bar{U}_B} \tag{7.28}$$

where $\bar{U}_B$ is the mean bubble rise velocity in the bed.

$\bar{U}_B$ is taken as the arithmetic mean of the bubble rise velocity calculated at the distributor ($L = 0$) and at the bed surface ($L = H$) from Equations (7.31) and (7.32) for Group B powders:

$$d_{Bv} = \frac{0.54}{g^{0.2}}(U - U_{mf})^{0.4}(L + 4N^{-0.5})^{0.8} \text{ (Darton } et\ al.\text{, 1977)} \tag{7.31}$$

$$U_B = \Phi_B(gd_{Bv})^{0.5} \text{ (Werther, 1983)} \tag{7.32}$$

where

$$\left\{ \begin{array}{ll} \Phi_B = 0.64 & \text{for } D \leq 0.1\text{m} \\ \Phi_B = 1.6D^{0.4} & \text{for } 0.1 < D \leq 1\text{m} \\ \Phi_B = 1.6 & \text{for } D > 1\text{m} \end{array} \right\} \tag{7.33}$$

N is the orifice density of the distributor (in this case, 93 nozzles per square metre).

D is the bed diameter or smallest cross-sectional dimension (in this case 6.4 m). The depth of the bed in normal operation is 1.2 m (dictated by the weir setting). In order to do these calculations, we need values for U_{mf} and ε_{mf}. U_{mf} is estimated using the Wen and Yu equation (Wen and Yu, 1966), which is appropriate for both ores since they are Group B materials. ε_{mf} is estimated using information, given for a range of particles of different size and shape, in Kunii and Levenspiel (1990) (page 69, Table C1.3). Such measurements, performed at the normal operating temperature of the fluidized bed gave the results shown in Table 16.4.

Table 16.4 Estimated values of U_{mf} and ε_{mf} for ores A and B

Ore	U_{mf} (m/s)	ε_{mf}
A	0.008	0.56
B	0.056	0.495

Table 16.5 Results of calculations of bed pressure drops

Value	Ore A	Ore B
Bubble size at distributor ($L = 0$) (m)	0.155	0.151
Bubble size at distributor ($L = 1.0$) (m)	0.459	0.448
Bubble velocity at distributor ($L = 0$) (m/s)	1.97	1.95
Bubble velocity at distributor ($L = 1.0$) (m/s)	3.40	3.36
Mean bubble rise velocity, $\bar{U}_B$ (m/s)	2.68	2.65
Bubble fraction	0.299	0.284
Mean bed density (kg/m^3)	1234	1446
Particulate phase voidage	0.56	0.495
Bed pressure drop (Pa)	12 107	14 181

The resulting calculations are shown in Table 16.5.

We see that the predicted increase in bed pressure drop when changing the feed ore from A to B is 2074 Pa. In practice, the increase was a little more than predicted and it turns out that the blower was under rated for the job. As a result the blower came close to its upper limit on a number of occasions. The analysis shows that the most likely cause of the increase in bed pressure drop in switching from ore A to ore B is the increase in bed pressure drop caused by an increase in mean bed density.

So what is the solution? One option is to reduce the bed pressure drop by 2074 Pa when feeding ore B. This could be done by reducing the weir height so that the bed depth reduces. This means a 14.6% (2074/14181) reduction in bed level. This change would, of course, reduce the residence time of the ores in the bed and may not be acceptable. However, the change could be made and the product tested. If the tests show that reduction in residence time is not acceptable, then the blower must be uprated to enable it to deliver the required 14 181 Pa bed pressure drop with some to spare.

16.2 *CASE STUDY 2*

Inappropriate Use of an L-valve

There is a market for high purity calcium carbonate and one method of making this is to calcine limestone by heating it to between 800°C and 900°C to remove the CO_2 and produce a pure lime which is then put into a tank containing demineralised water. The combustion gas rich in CO_2 is stripped of its dust

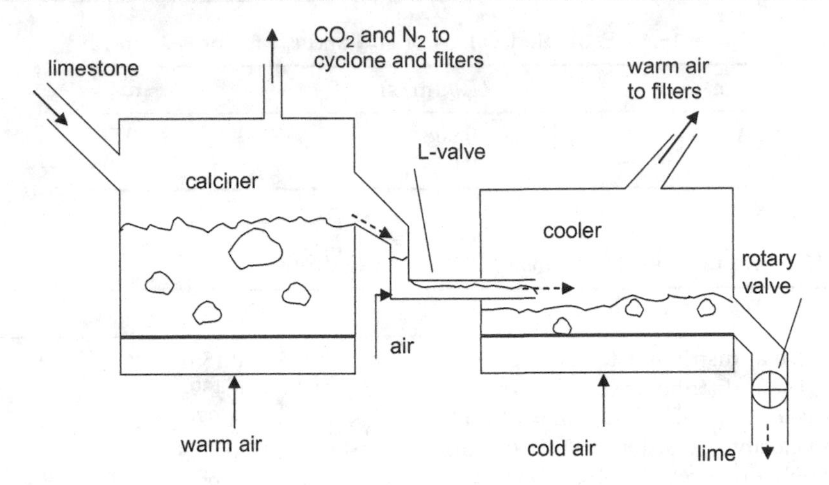

Figure 16.2 Original layout of calciner and cooler with L-valve

content by passing it through cyclones, cooled, given a final cleaning in a baghouse and then introduced through spargers to the bottom of the tank containing the slaked lime to give a very pure $CaCO_3$ suitable for use as an additive in pharmaceutical and food products.

A new plant using this process (Figure 16.2), was designed to calcine crushed limestone by feeding it to a fluidized bed heated by burning natural gas. The lime so produced is transferred to a long rectangular cooler fluidized by cold air. The hot combustion gases enriched by the CO_2 from the calciner are dedusted in a cyclone and cooled in a heat exchanger by the air that fluidizes the bed of lime and limestone. The cooled gases then pass into a baghouse. Some attrition of the lime particles occurs in the fluidized bed lime cooler and these entrained particles are removed in another baghouse before the warm air is discharged to a chimney. The lime product is removed continuously from the end of the cooler remote from the feed point and, because its temperature is below 200°C, a rotary valve can be used to control the removal rate. The hot lime particles overflow from the calciner onto an inclined chute leading to the cooler, but because of the high solids temperature (800–900°C) the use of a motorized rotary valve to control the flow and provide a gas seal was impossible. The designers therefore decided to use an L-valve to provide both a seal against hot gas passing to the fluidized cooler and to control the solids discharge rate.

Brief description of how the L-valve works

The principle of the L-valve has been around for a long time. The first patent for a means for controlling the flow of particulate solids from one vessel to another, or into a pipeline, using only small flow rates of gas, was taken out by ICI in the 1930s. Interest in controlling the flow of *fine* powders (<100 μm) in dilute phase conveying systems grew considerably with the advent of Fluid Bed Catalytic Cracking, but it took the large synthetic fuels R&D programme in the USA in the late 1970s and in the 1980s to stimulate interest in controlling the flow of relatively *coarse* solids at high temperatures, and this led to the re-invention of the L-valve.

Experimental work carried out by Knowlton and co-workers (Knowlton and Hirsan, 1978) at the Institute of Gas Technology in Chicago established some of the empirical principles of L-valves, sometimes called *non-mechanical valves*, but the first paper setting out a step-by-step design procedure (based on experiments with sands in 40, 70 and 100 μm diameter valves) was written by Geldart and Jones (1991). Arena *et al.* (1998) added more data to the literature and confirmed that the basic approach of the earlier workers is sound.

Although the L-valve appears to be quite a simple device, its hydrodynamics are quite complicated, and only a basic explanation is given below.

Knowlton and Hirsan (1978) established that the best position for the injection of the aeration gas is in the vertical leg (sometimes called the *downcomer* or *standpipe*) about 1.5 pipe diameters above the centreline of the horizontal leg. This is because gas is needed to reduce friction between the particles and the pipe at the inside corner of the elbow, and if the gas is injected on or below the centreline, less is available to perform this function. This gas then streams along the top of the bed of powder lying in the horizontal leg carrying particles with it in a mode of transport that resembles dense phase pneumatic conveying. As the aeration gas flow is increased the depth of the flowing stream of powder increases and the mass flow rate of solids increases. Eventually the whole depth of powder may become active with the solids velocity increasing from bottom to top of the horizontal leg.

Moving the solids requires a force to overcome friction and that is provided by a pressure drop along the horizontal pipe. This pressure drop is a function of the solids rate, the pipe diameter, and the mean size and bulk density of the powder. The pressure is a maximum at the point of gas injection. The *minimum* gas flow rate to initiate solids flow depends on the minimum fluidization velocity of the powder, U_{mf}, and when this minimum gas flow rate is exceeded, solids flow starts at the *minimum controllable solids flux*, *G*. The solids flux then increases linearly with the ratio U_{ext}/U_{mf}, where U_{ext} is the volumetric flow rate of aeration gas divided by the cross-sectional area of the horizontal leg. However, there is limit to the maximum solids flow rate that can be attained and this depends on the pressure balance. If the gas finds it is easier to flow upwards through the bed of solids in the standpipe it will do so and less will be available to transport the solids horizontally. Since the particles have the same characteristics in both vertical and horizontal legs, the resistance to gas flow in the standpipe depends primarily on the depth of powder in it. As a rule of thumb, the height of the standpipe must be *at least* equal to the length of the horizontal leg. In practice, the criterion is that the pressure drop in the horizontal section (that required to move the solids) divided by the depth of solids in the standpipe must be lower than the pressure gradient in a fluidized bed at minimum fluidization. If it is equal to, or higher than this, the solids in the vertical leg may start to fluidize (since the applied pressure gradient is greater than that required for fluidization) and flow instabilities will occur. These take the form of a cessation of solids flow followed by flushing of the solids through the L-valve.

One of the problems of operating an L-valve system is that small changes in the aeration gas flow rate can cause quite large variations in the solids flow rate so a method of control is needed to match the solids efflux rate to the rate at which the solids enter the standpipe. This might take the form of pressure measurements in the standpipe linked to a flow controller on the aeration air. Fortunately, provided the standpipe is long enough, it is possible to operate an L-valve in automatic mode. The aeration rate is set at a value high enough to fluidize the solids above the aeration point whilst the solids below are in packed bed mode. If the solids flow rate into the standpipe increases, the level of fluidized solids in the standpipe rises and more gas flows downwards causing the solids discharge rate to increase. If the solids rate decreases the level of solids in the standpipe falls, more gas flows upwards and less downwards, causing the solids discharge rate to decrease.

Unfortunately, in the application shown in Figure 16.2, because of limitations in the vertical height available between the overflow weir in the side of the calciner and the lime cooler the solids transfer system installed could not operate as a stable L-valve because (a) the vertical leg was shorter than the horizontal leg, and (b) the aeration point was not positioned above the centreline of the elbow. Moreover, the discharge end of the horizontal section was higher than the level of the fluidized bed in the cooler. It also turned out that the valve diameter (200 mm) was so large that the minimum controllable solids flow rate was almost seven times the design production rate. Reducing the aeration rate below the minimum value resulted in no solids flow at all, so in order to make the system work at all, the operators of the plant installed an additional aeration line on the centreline of the horizontal leg and then operated the aeration system intermittently. When the bed level in the calciner became too high and the red hot solids filled the overflow chute, the aeration air was turned on fully, causing the solids in the short vertical leg to fluidize. This altered the pressure balance in the system and caused the entire contents of the L-valve to suddenly flush into the cooler, leaving a clear passage for the hot combustion gases to blow through to the space above the bed in the cooler. This resulted in the destruction of the baghouse when it caught fire.

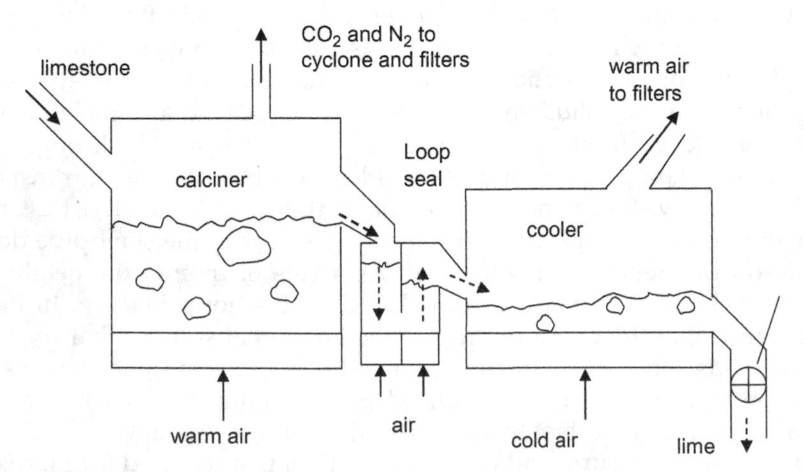

Figure 16.3 Modified calciner and cooler layout using a loop seal to transfer solids and prevent gas flow from cooler to calciner

The solution to the problem was to replace the L-valve by a loop seal (Figure 16.3), which is essentially a fluidized bed divided by a vertical baffle except for a gap at the bottom to allow solids to pass between the two side-by-side beds, fluidized by separate air supplies. Hot lime particles fall into the bed nearest to the calciner, flow under the vertical baffle into the second bed adjacent to the cooler driven by the higher pressure in the first leg of the loop. The solids in the second bed then overflow into the cooler. This system operates like two arms of a manometer and there are fluidized solids present at all times in the loop preventing hot gas from passing direct into the cooler.

16.3 CASE STUDY 3

Fluidized Bed Dryer

Solving one problem may cause another that could not have been anticipated.

One stage in the manufacture of a water treatment chemical involved feeding cube-shaped, slightly damp particles into a fluidized bed dryer. The distributor was rectangular in plan view, about 3 m long by 1 m wide and contained many drilled holes. The fluidizing air was supplied by a centrifugal fan and passed through a heat exchanger before entering the plenum chamber through a duct positioned about 1 m from the solids feed point. The fluidized solids formed a bed about 1 cm deep. The warm product overflowed from the far end of the bed into a chute which directed them to a bagging machine where the product was packaged into 50 kg batches. In the customers' warehouses the paper or plastic sacks were stored on top of each other before use, and in some sacks the product particles became stuck so firmly together that they formed a solid block that resembled a tombstone and so the particles could not be poured out.

Although the damp feed was fed to the dryer and removed continuously, samples taken at the solids exit showed that the moisture content was unacceptably variable. Visual inspection of the fluidized bed through several of the hatches in the wall of the steel expanded section above the bed surface revealed that the bed was unevenly fluidized with large areas of unfluidized solids remaining virtually motionless on the plate.

In poorly fluidized beds it is common for the defluidized and well fluidized regions to change randomly and it seemed likely that from time to time that static areas having higher moisture contents might suddenly become mobile and flush through to the bed exit. Measurements of the pressure drop across the air distributor plate were made in the absence of a bed and found to be only 5 mm water gauge (~49 Pa), much too low for a system having a very small aspect ratio (bed depth/bed length or for a cylindrical bed, bed depth/bed diameter). The consultant recommended that the distributor plate be replaced by one having a pressure drop at least equal to that across the fluidized solids.

This was done, and when the dryer was started up again, visual observation showed that the entire bed was uniformly well fluidized. However, it was

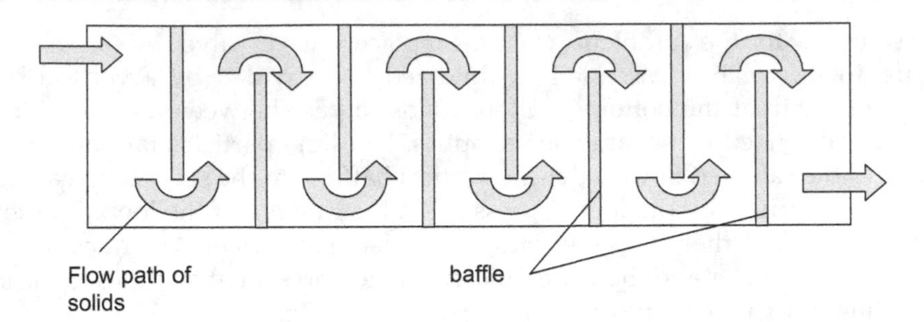

Figure 16.4 Plan view of dryer with baffles inserted

found that virtually none of the samples of the product had a moisture content within specification. With hindsight it was deduced that the defluidized regions had acted like baffles in causing the well-fluidized solids to take a tortuous path along the plate, so increasing their residence time in the system to allow some particles to dry thoroughly. When the plate was changed to give a really well fluidized bed, all the particles flowed too rapidly through the equipment to give a low moisture content. One solution was to insert a series of baffles (see Figure 16.4) at right angles to the direction of solids flow, which would cause the bed to operate nearer to plug flow, increasing the mean residence time and reducing the range of residence times. An alternative solution would be to insert a weir at the solids exit so as to increase the mean residence time in the dryer and rely on the length to width ratio of the bed to minimize backmixing of solids against the solids flow direction – and hence reduce the range of residence times.

16.4 CASE STUDY 4

Aeration of a Hopper Leads to Air Shortage at a Coal Plant

Peter Arnold, University of Wollongong, Australia

In a coal preparation plant three 'fine' coal feed silos were fed with a common rotary plough feeder – a device for delivering coal from the silo at a regular rate. Arching of coal at the hopper outlet resulted in problems achieving stable discharge of the silos. To overcome the arching problems at the hopper outlet a sophisticated aeration system had been installed to the outlet slot. Since the coal was size was 'minus 12 mm' (i.e. it has passed though a 12 mm sieve) a lot of air was required for the aeration system. The result was that the rest of the coal plant was starved of air and plans were being made to increase the compressed air capacity. The injected air actually had little effect on the arching problem. The problem was the result of a simple mismatch between the hopper outlet and the feeder caused by the presence of a number of structural columns protruding into the outlet area. At intervals along the

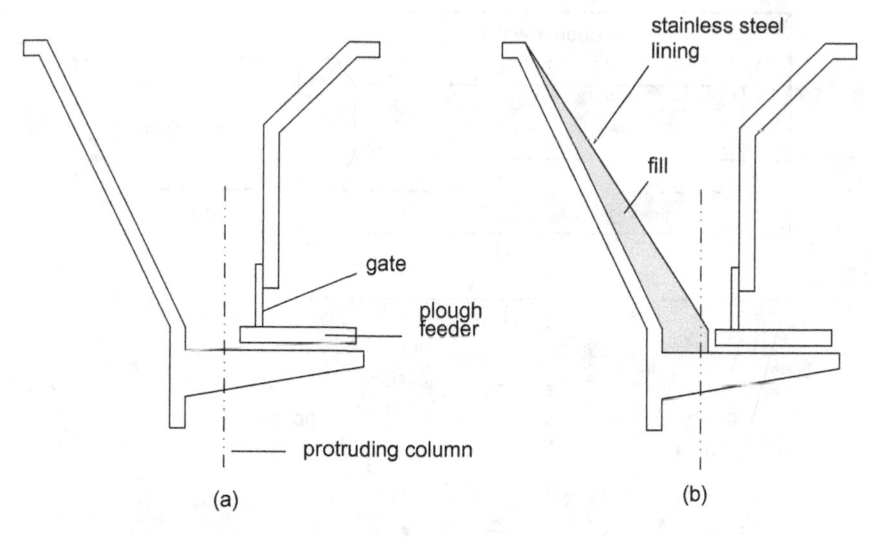

Figure 16.5 (a) Profile through the hopper slot outlet showing the offending columns and (b) modifications to the geometry

slot, the structural columns protruded out from the hopper back wall and prevented the plough feeder from fully activating the hopper outlet slot. The plough blades had to avoid the columns and hence could not plough coal off the full shelf width, effectively leaving coal to build up on the unploughed shelf between the protruding columns. The solution was to implement the philosophy 'if you can't get the plough to the back wall of the hopper then bring the back wall to the plough!' (Figure 16.5). Bringing the back wall towards the plough feeder has reduced the steepness of the hopper wall and so if that is all we did there would be a risk of compounding the arching problem. However, to overcome this potential risk the new hopper wall is clad with stainless steel, which would have far lower wall friction than the original concrete wall. As a result of this simple modification the flow was stabilized and the aeration air was no longer needed.

16.5 *CASE STUDY 5*

Limestone Hopper Extension Overloads Feeder

Peter Arnold, University of Wollongong, Australia

A limestone hopper at a cement plant was duplicated and uprated (the hopper slot was lengthened in an attempt to increase delivery rate). As Figure 16.6 shows, the hopper was wedge-shaped and fed by an apron feeder. The hopper fed from the rear of the feeder and required the feed material to be dragged

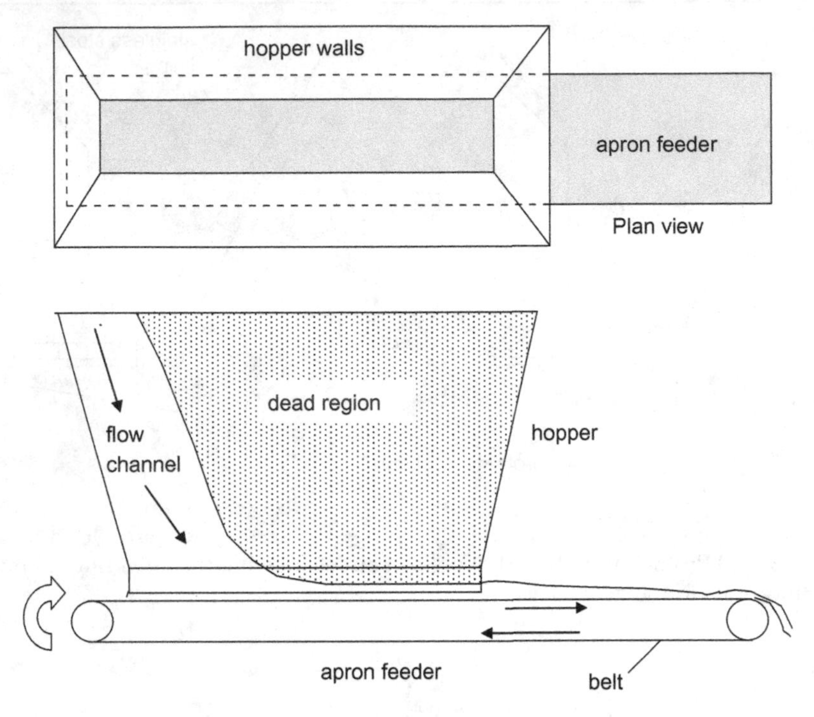

Figure 16.6 Initial design of limestone hopper with apron feeder. A high proportion of the length of the hopper slot is not active

under 'dead' or stagnant material, thereby overloading the feeder. In an initial attempt to overcome this problem, higher capacity motors and gearboxes were tried with the result that the apron feeder itself was overstressed. The solution finally adopted was a redesign of the hopper–feeder interface to achieve a fully active hopper outlet (Figure 16.7). The outlet slot was redesigned to increase along its length. This eliminated the region of stagnant solids, reducing the force applied to the feeder and so overcoming the problems of overloading the apron feeder. The feeder power required after the modification was less than that originally specified.

16.6 CASE STUDY 6

The Use of Inserts in Hoppers

Lyn Bates, Ajax Equipment, UK

It may seem strange that placing obstructions in the flow path of bulk solids can offer any form of advantage but, in practice, a wide range of operating benefits can be gained by the use of inserts in bulk storage installations. The fact is that the change of flow path created by an insert forms new flow channel shapes that

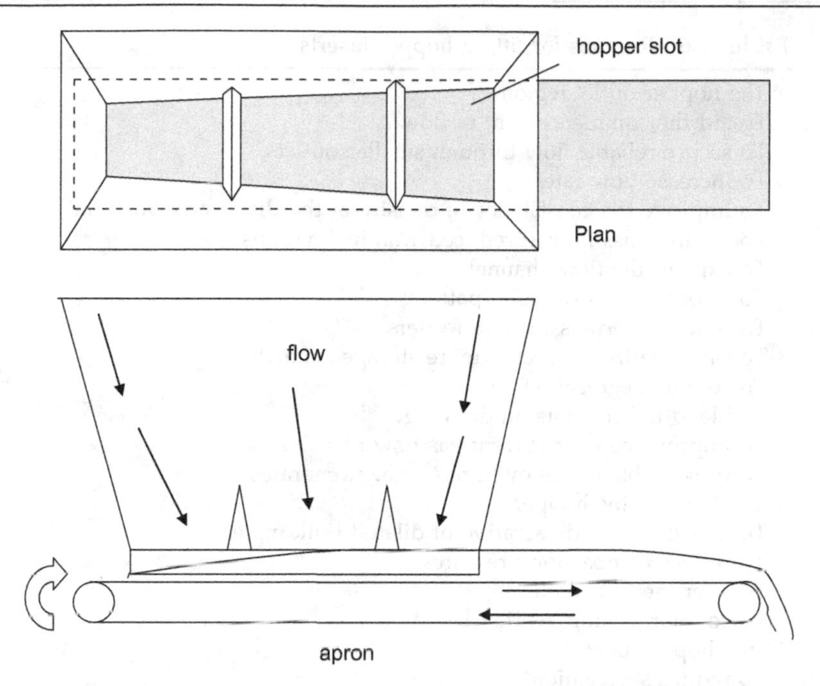

Figure 16.7 Plan view of the solution to the problem. Note that the width of the outlet slot is redesigned to increase along its length. This eliminates the dead region and so avoids the problems of overloading the apron feeder

have more favourable flow characteristics than the original. Some modifications are simple and self-evident, like inserting a device to catch lumps before they reach and block the silo outlet. Others are more sophisticated and call for a greater degree of technical competence and expertise, although the principles involved can be readily understood from a basic appreciation of the technology. There are many different reasons for fitting inserts into storage silos; a summary of the main ones is given in Table 16.6. It must be pointed out that this is not a field for the amateur. The obvious task of securing the required performance may itself require a degree of expertise. It is even more important to ensure that these fittings do not introduce adverse consequences to the operation or integrity of the equipment, to the condition or fitness for use of the product, to hygiene, maintenance, or other factors that may detract from overall efficiency and economics of the plant.

The economic benefit of incorporating an insert in an original design is rarely considered in the primary hopper selection process of determining the appropriate flow regime for a bulk storage facility even though, at this stage, they enlarge the scope for optimization of an integrated design. By contrast, retrofits are usually compromised by the constraints of the installed plant. Whilst these may limit the options available, it is commonly found that inserts offer one of the most promising routes available to address many of the problems encountered in bulk storage applications.

Table 16.6 Reasons for fitting hopper inserts

Reasons for fitting hopper inserts
In the hopper outlet region
To aid the commencement of flow
To secure reliable flow through smaller outlets
To increase flow rates
To improve the consistency of density of the discharged material
To secure mass flow at reduced wall inclinations
To expand the flow channel
To improve the extraction pattern
To reduce overpressures on feeders
To save headroom/secure more storage capacity
To counter segregation
To blend the contents on discharge
To improve counter-current gas flow distribution
To prevent blockages by lumps or agglomerates
In the body of the hopper
To accelerate the de-aeration of dilated bulk material
To reduce compaction pressures
To alter the flow pattern
To counter 'caking' tendencies
At the hopper inlet
To reduce segregation
To reduce particle attrition

An understanding of the flow regimes, mechanisms of stress systems and behaviour characteristics of loose solids, provide a background for the selection of an appropriate insert type for specific functions and offers a sound basis for their design. The performance of inserts interacts with the hopper geometry, the type of feeder or discharge control, ambient and operating conditions. Design considerations for inserts, hopper and feeder therefore require an overall systems approach. With rare exceptions, most insert techniques have been developed and introduced by industry, rather than arising from fundamental research. Some forms of insert are covered by patent, registered designs or similar restrictions. Availability of design data on applications is therefore somewhat restricted. There remains open a wide avenue of research to optimize the design of different techniques and develop an improved basis for their selection and integration with hopper forms and flow aid techniques.

It will be seen that many interests are served by the placement of inserts in the outlet regions of hoppers. This is because the flow pattern governing the behaviour of material within the hopper is initiated in this region. Whereas the outlet region is relatively small, divergence of the hopper walls quickly attains a size of cross-section where flow problems are not an issue. It is therefore commonly possible to fit flow-modifying inserts in this larger cross-section above the outlet that do not in themselves cause any flow impediment, but usefully change the flow pattern in the hopper and shelter the underlying regions to considerable advantage.

By way of example, two situations are described that show how the overall behaviour of material in storage can be dramatically changed by comparatively small inserts. More particularly, the properties of the discharged product are made more suitable for the specific requirements.

Loss-in-weight make-up

One means of controlling the rate at which solids are fed to a step in a process is known as loss-in-weight feeding. The storage silo holding the solids is mounted on load cells and the weight of the solids in the vessel is continually monitored as solids are discharged, so that the amount delivered or rate of delivery is known.

Loss-in-weight feeders require regular replenishment within a short period. In this particular case, the unit handling delivering a fine powder feed under weight control over a period of about 2 minutes was required to be re-filled from a conical, mass flow hopper with quick-acting slide valve in 10–15 s. The problem was that re-fill was very slow to start; it took over a minute for the powder to emerge (Figure 16.8). Also, when it did emerge, the powder was so highly aerated that it flushed through the feeder and was not in a suitable condition for the next step in the process. Headroom, holding capacity, supply route and general plant construction was, of course fixed in stone, so changes were only permissible within the existing equipment geometry.

Fine powders usually have highly variable flow conditions. When settled, they are cohesive and it is often difficult to initiate flow. One reason is that the close proximity of the particles enables various molecular scale forces to attain significance to resist particle separation. Fine cohesive powders therefore can exhibit significant strength as compaction is increased. Expansion is also inhibited by the rate at which air can percolate through the fine interstices between the particles of fine powders. However, once a fine powder becomes expanded and flow does occur, aeration tends to be excessive and the powder may become effectively self-fluidized. De-aeration of the powder in such a state is slow, again due to the minute size of the escape passages, so the bulk may take some time to settle to a stable condition. Fluidized products have no strength to resist

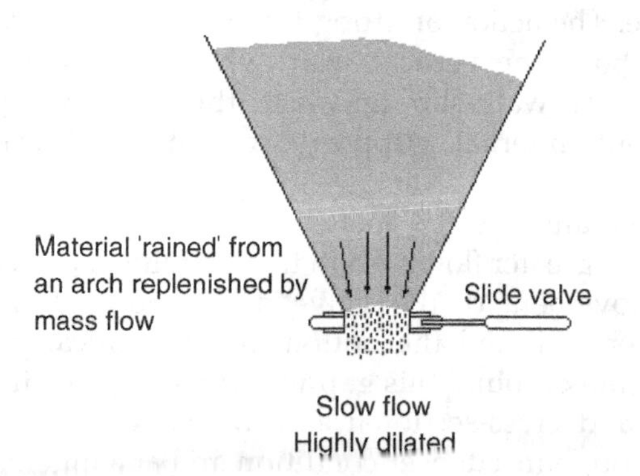

Figure 16.8 Original arrangement of loss-in-weight make-up hopper

deformation but are subjected to hydrostatic pressure that, combined with their loose nature, makes them very searching for flow routes.

Every converging flow channel develops a velocity gradient across its cross-section because, even in a mass flow channel, it is far easier for material immediately above the outlet to move down than it is for product alongside to move diagonally towards the outlet. This faster moving region in the feed hopper creates a depression in the local surface that was more readily filled by the fresh powder than by the more stable material previously loaded into the loss-in-weight hopper. As this loose material progresses down towards the outlet its hydrostatic pressure resists the entry of less free-flowing product to the more rapid region of flow, to further increase the flow velocity differential. The resulting small cross-section of rapidly moving fluid material quickly penetrates the depth of the bed of powder to flush, without control, from the machine outlet.

The retrofit task was twofold. It was necessary to accelerate the re-fill to deliver powder quicker, and at the same time deliver the material in a more stable condition. The first thing to recognize is that, provided flow takes place, the rate of flow from mass flow hoppers is invariably slower than from hoppers with the same size outlets that are not mass flow. The basic reason is that particles disengage from the mass to accelerate in 'free fall' from a higher dynamic arch in non-mass flow hoppers, so are travelling faster as they pass through the outlet. It should also be noted that a flow stream travelling in an unconfined condition can attain a high velocity, and it is therefore possible to achieve large rates of flow through relatively small orifices. To secure a much higher flow rate in a less dilate condition through the same size of opening therefore calls for an increase in the area of the initial flow outlet and/or disturbance of the settle bulk to encourage initial failure of the bulk state.

The solution adopted was to replace the quick acting slide valve with an internal 'cylinder and cone' valve that fitted into the conical hopper some distance above the hopper outlet (Figure 16.9). This comprises a short cylinder with a superimposed cone, the bottom edge of which seats on the inner conical wall of the hopper. This fitting was lifted by means of an air cylinder attached to the top of the hopper. The action of lifting the insert exposed an annular gap. The immediate flow channel above this circular slot had a vertical inner surface and a conical outer shape. The action of lifting the valve disrupted the bulk resting on the cone section above the vertical wall, whilst movement of the cylindrical surface reduced local wall slip and left the local wedge-shaped annulus of product without internal support free to fall away with minimum deformation.

As the area of the annular gap increases as the square of the diameter, it is practical to provide a greater flow area in this annulus than was given by the final outlet. Upward movement of the valve disrupted material that was resting between the hopper wall and the bottom of the conical portion of the valve. The collapse of the bulk around this gap was therefore precipitated and occurred at a greater span and cross-sectional area than the flow annulus exposed, so material flowed through in a denser condition and at a much greater rate than the original opening. The resulting flow was focused through the final hopper outlet

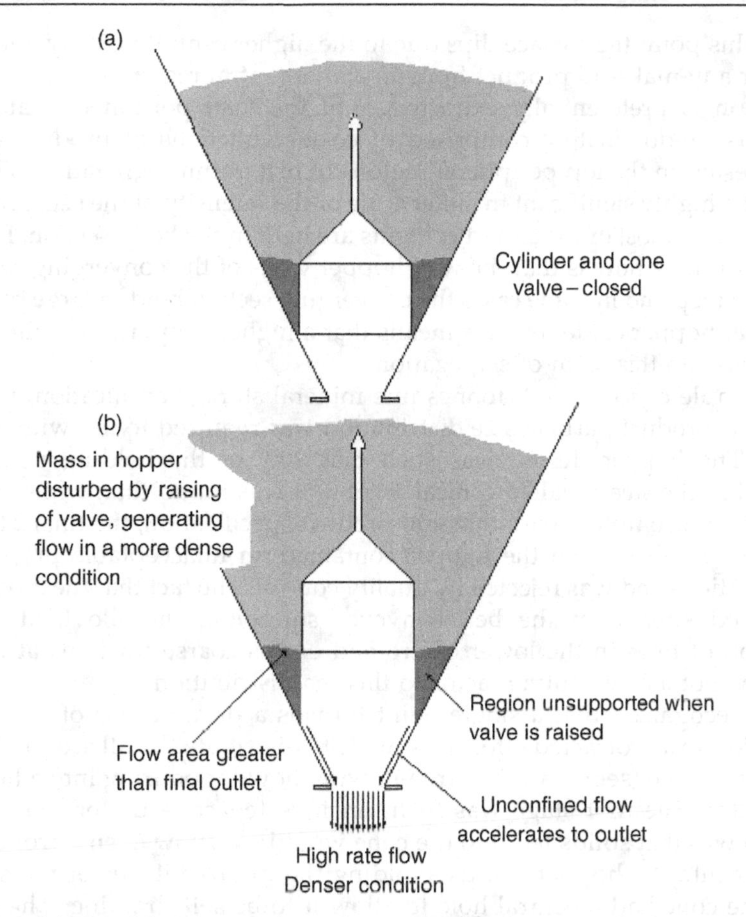

Figure 16.9 Modified loss-in-weight make-up hopper showing cylinder and cone insert closed (a) and open (b)

as a virtually unconfined stream, so accelerating to a velocity that allowed the higher rate to pass without either restraint or excessive dilation. As a result, the make up quantity to the loss-in-weight hopper was supplied in a few seconds by the bulk material in a stable, non-fluid, flow condition.

Segregation in mass flow hopper

It is commonly suggested that a mass flow hopper will overcome segregation problems. The thinking is that any segregation taking place during the filling of a mass flow hopper will be redressed by re-mixing of the solids from all regions of the cross sectional area as the hopper discharges. Mass flow will restore the blend of the silo cross section while extraction takes material from the parallel body section of the silo and the surface level moves down uniformly. The problem arises when the level of the contents fall to the hip of the conical transition.

Around this point the surface dips due to the higher central velocity and there is no longer any make-up product from an above central region. Progressively, the core region is preferentially extracted until the final portion of material discharged is predominately comprised of coarser fractions of product that previously rested in the top peripheral regions of maximum segregation. This effect may not be highly significant in either scale or the intensity of the segregation in a tall silo, where most of the stored contents are held in the body section. However, a feature of mass flow is that the wall hopper walls of the converging section are relatively steep and in some cases the converging section holds a large proportion of the total hopper contents. This means that a higher proportion of the contents is vulnerable to this form of segregation.

An example concerns a 20 tonnes fine mineral storage application, where the discharged product particle size distribution was required to fall within narrow bounds. The hopper design was such that 90% of the holding volume was contained in the steep-walled conical section. It was found that, whilst the input supply stream complied with the tight product specification, the final 2 tonnes of product discharged from the hopper contained an unacceptable proportion of coarse fractions and was rejected by quality control. The fact that the cross section of the feed stream on the belt conveyor supplying the silo held a higher proportion of fines in the lower centre and excess coarse fractions at the outer edges was not a contributing factor to this maldistribution.

It was recognized that a single point fill was a prime cause of repose slope segregation so a protracted effort was made to distribute the fill around the cross section of the silo (segregation of free-flowing powder pouring into a heap – see Chapter 11). The first stage was to fit an inverted cone under the inlet. The intention was that solids fed into the cone would overflow evenly from its large perimeter into the hopper – thus avoiding the disadvantages of the single fill point. The cone had a central hole to allow it to be self-draining. The problem encountered was that the repose pile of solids in the inverted cone, which formed a reasonably concentric conical heap, would not spill evenly over the circular ridge in the manner of a liquid weir as expected. It was first thought that the circumferential bias was due to an eccentric fill point. However, after working on this meticulously to direct the feed to a precise central location, it was seen that the repose surface of the cone did not grow as a smooth rate but increased as a succession of radial avalanche surges down the path of least resistance.

The deposit of each surge increased the local resistance in the formative pile stage, so the next surge found a different route around the periphery of the pile. Alas, when the pile filled the inverted cone, the surge that first overflowed the rim did not build up the local surface to offer increased resistance, so following surges took the same biased path.

Protracted efforts were made to obtain an even overflow around the circumference by fitting a castellated rim with adjustable weir gates and successive 'tweaking' of their settings. Many miserable and dusty hours later, defeat was eventually conceded as it slowly occurred to the engineers that the path that offered the least resistance would always attract the following surges down the same part of the repose slope and so the overflow from the cone would never be even.

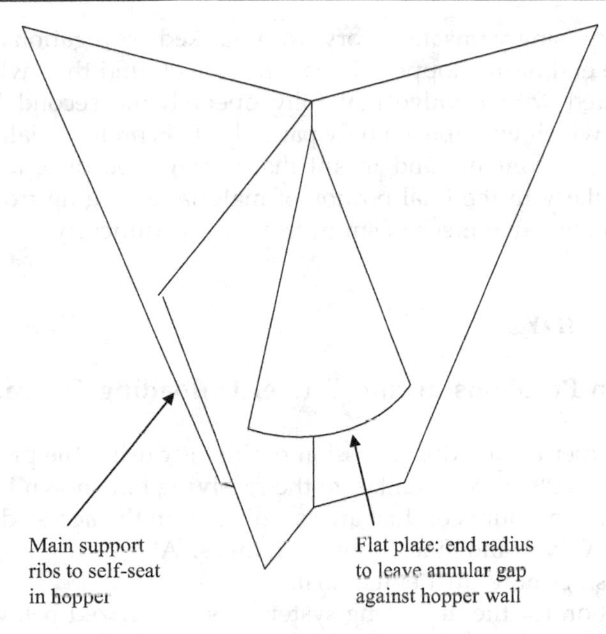

Figure 16.10 Central insert to expand flow channel

Time to call a specialist!
The equipment layout did not permit the adoption of an effective fill geometry that would counter segregation, so attention was directed to modifying the discharge pattern. An insert system (Figure 16.10) was developed that prevented central discharge and provided instead a 'tributary' collecting system. This allowed multiple flow channels to develop at various radial and circumferential locations in proportion to the relevant cross section of the silo that they served. An important feature of this construction was that baffles above the outlet port prevented the development of preferential flow from any of the sectors. The equipment included a triangular, pyramid shaped insert that had three ribs to self-seat the device in the bottom cone of the hopper above the outlet. The walls of this pyramid left spaces at the bottom for one-third of the outflow to pass through each opening.

Three sets of plates were placed on the hopper cone, the lower edges of which rested against the faces of the central pyramid insert to allow discharge from underneath the plates. These plates were spaced apart by ribs, to provide extra flow routes for the hopper contents to pass from different radii of the hopper cross-section to the openings provided by the insert.

The initial draw down pattern was completely even until the surface reached the top of the conical section. Thereafter, the surface profile reflected multiple, preferential extraction points from different areas of the silo cross-section, but as these represented the correct proportion of the different grades of fractions, the discharged product was of the required particle size distribution.

The importance of detail and comprehensive route review was emphasized when a second hopper of similar construction was similarly adapted. Initial results were surprisingly unsatisfactory, with marked segregation as the contents approached the end of the hopper discharge. It was found that, whereas the first hopper was fitted with a valve that fully opened, the second hopper had a different valve, which was opened only partially. This preferentially emptied one side of the hopper contents and meant that segregation once again became a problem particularly in the final portion of material emerging from the hopper. Modification to the valve mechanism overcame this difficulty.

16.7 *CASE STUDY 7*

Dust Emission Problems during Tanker Unloading Operations

A company is experiencing dust emission problems during the pneumatic lifting of soda ash from a 28 m^3 road tanker to the receiving bin shown in Figure 16.11. For example, large amounts of dust are escaping from the access door, silo-saver (pressure relief valve) and filter housing flanges. Also, the bin is occasionally over-filled (causing more dust emissions).

The specification for the unloading system is summarized below:

- solids conveying rate, $m_s \approx 22$ t/h;
- air mass flow rate, $m_f \approx 0.433$ kg/s (from blower performance curves, which show intake volume $\approx 21.7\,m^3/min$ at air density $\rho_f = 1.2\,kg/m^3$);
- inside diameter of pipe, $D = 150$ mm; total length of conveying pipeline, $L = 35$ m (including total length of vertical lift, $L_v = 20$ m);
- maximum air temperature (into receiving bin), $t \approx 70$ °C;

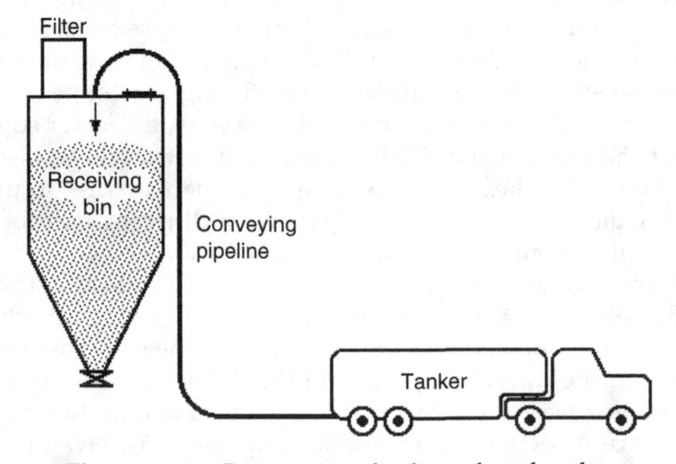

Figure 16.11 Pneumatic unloading of road tanker

- particle size range of soda ash, $50 \le d \le 1000\ \mu m$;
- median particle diameter, $d_{50} = 255\ \mu m$;
- solids particle density, $\rho_s = 2533\ kg/m^3$;
- loose-poured bulk density, $\rho_{bl} = 1040\ kg/m^3$;
- volumetric capacity of tanker = 28 m^3;
- initial pressure in tanker = 100 kPa *g* (i.e. before opening discharge valve/s);
- steady-state pressure drop = 70 kPa (as seen by the blower);
- blower is positioned on site – pressurizes the tanker vessel, which acts like a blowtank.

The existing dust filter has a filtration area of 15 m^2 (Figure 16.12) and is fitted with a fan and polyester needle-felt bags. From performance curves, the fan can extract ≈33.3 m^3/min of air at 150 mm H_2O = 1.5 kPa. At first glance, this appears to cover quite adequately the blower capacity of 21.7 m^3/min. However, transient effects and filtration efficiency also have to be evaluated.

Evaluation of Unloading Operation

Dust emissions occur during both the end-of-cycle purge sequence and especially the subsequent clean-blow cycle (required to clean out the tanker). Note:

- dust emissions do not occur during steady-state operation (namely, constant pressure, air flow, conveying rate);
- both 'air-only' operations are unavoidable (i.e. a new 'self-cleaning' tanker cannot be used).

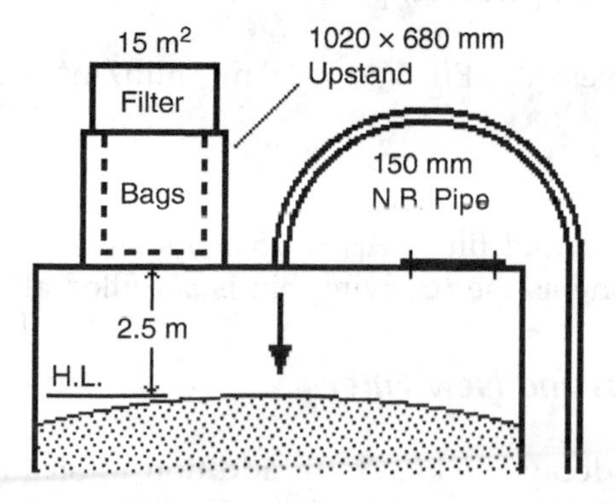

Figure 16.12 Receiving bin with existing 15 m^2 filter

The company was asked to take pressure measurements with respect to time. The pressure decay in the tanker (during either air-only operation) was approximately linear and the maximum rate was found equal to $\approx$ (90 kPa g – 35 kPa g)/(11 s) = 5 kPa/s. Note, this 'supply' of air may occur during an end-of-cycle purge sequence and hence, is additional to the steady-state air flow rate of 0.433 kg/s. The filter has to cope with this maximum possible flow rate of air (including dust loading).

Steady-State Operating Conditions

The following steady-state parameters are calculated from the specification given above:

- density of air in bin, $\rho_{fe} = 1.028\ kg/m^3$;
- air flow, $m_f = 0.433\ kg/s$;
- flow rate of air into bin, $Q_{fe} \approx 25.3\ m^3/min$;
- air velocity into bin, $V_{fe} = 23.8\ m/s$. This is not considered excessive for dilute-phase conveying of soda ash, as long as sufficient volumetric capacity is available in the bin for the air to expand;
- existing filter opening (aperture) $= 0.7\ m^2$.

Therefore, the velocity of air into the filter upstand, $V_{up} = 0.6\ m/s$. Based on experience with similar materials and on the results of free settling tests, this is considered to be quite acceptable and the dust loading should be only moderate. However, if the bin is filled above the high-level mark, which is 2.5 m from the top of the bin, then additional turbulence may cause increased loading on the filter – this is considered in more detail later (when selecting a new filter). For the present study, it is assumed that filling ceases when the high-level is reached (i.e. ideal situation). Using appropriate values of ideal filtration velocity and correction factors for temperature, particle size, dust loading, etc., then:

- effective filtration velocity, Eff. $V_{filt} = 2.0\ m^3/min/m^2$ = 'air-to-cloth' ratio;
- minimum filter area, $A_{filt} = 13\ m^2$.

Hence, the existing fan and filter ($A_{filt} = 15\ m^2$) should be adequate for steady-state operation (as long as the receiving bin is *not* filled above its high level).

Air-Only Operations and New Filter

Maximum pressure decay $= 5\ kPa/s$ (in addition to steady-state flow – worst case scenario). This equates to an additional air flow of $\approx 1.42\ kg/s$.

Hence, maximum $Q_{fe} = 108\ m^3/min >$ existing filter capacity (i.e. a larger filter is needed).

Assume new filter area $45\ m^2$ with fan ($108\ m^3/min$ at 150 mm $H_2O = 1.5$ kPa).

Standard aperture opening = $1.84\ m^2$.

This will generate $V_{up} \approx 1$ m/s, which should be sufficient to provide above moderate dust loading (especially if the bin is filled above the designated high level). Also, due to high-velocity purging (and particle attrition), the dust level may increase by up to 5% (especially in the 3 – 10 μm range).

Using appropriate values of ideal filtration velocity and correction factors, then:

- effective filtration velocity, Eff. $V_{filt} = 1.6\ m^3/min \cdot m^2$;
- minimum filter area, $A_{filt} \approx 68\ m^2$ (worst case scenario).

Hence, at least $50\ m^2$ or preferably $60\ m^2$ of filter area would be required (with a larger capacity fan). However, if the following improvements are implemented, then it may be possible to select a smaller filter.

Modifications to Filter and Bin

To reduce dust loading and improve filtration efficiency, the following modifications are recommended.

An upstand with a large as possible aperture (i.e. opening) should be custom-built for the new filter (e.g. for a $45\ m^2$ filter, the suggested 2005 mm × 1220 mm aperture provides a clearance of ≈200 mm between the bags and the side walls of the upstand). Also, the upstand should be tall enough to provide a minimum of 1.5 m clearance between the underside of the bags and the top of the bin. These modifications, together with a larger-diameter tee-bend at the end of the pipeline to reduce conveying velocity, will reduce the dust loading (by reducing upflow velocity, allowing more time for particle settling and allowing space for the new velocity profile to develop in the upstand). This also reduces any damage to the bags caused by turbulence (the splashing of particles upwards on to the filter surface – due to high velocity particle stream leaving the pipeline impinging on the material surface in the bin).

The maximum design pressure of the bin also should be determined, so that a suitable (and reliable) silo-saver can be installed. Assuming a maximum possible pressure of say, 5 kPa (i.e. before buckling), the silo-saver should be set at 4 kPa. Note, a large-flow unit should be selected (e.g. 150 or 200 mm nominal bore).

Based on the above improvements, the filter requirements were re-calculated: larger aperture opening $2.45\ m^2$. This will generate $V_{up} \approx 0.7$ m/s (maximum), which is quite acceptable. The dust loading should be moderate.

Using appropriate values of ideal filtration velocity and correction factors, then:

- effective filtration velocity, Eff. $V_{filt} = 2.22\ m^3/min/m^2$.
- minimum filter area, $A_{filt} = 49\ m^2$ (worst case scenario).

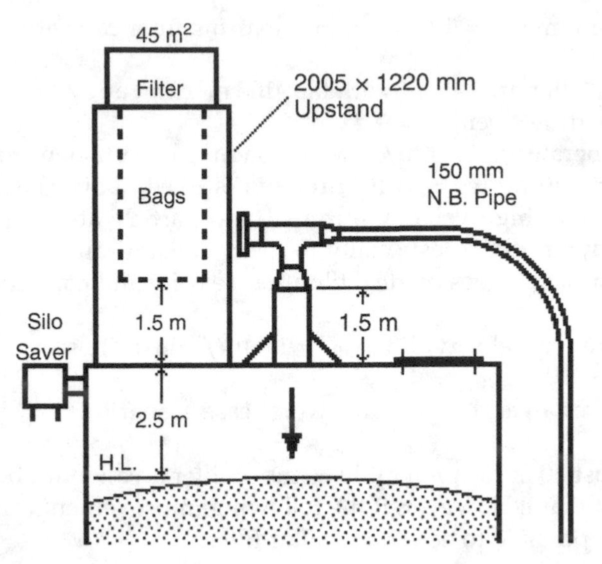

Figure 16.13 Modified Receiving bin with new 45 m^2 filter

Hence, with a fan rated at say, 108 m^3/min at 150 mm H_2O=1.5 kPa, a 45 m^2 filter should be adequate, as shown in Figures 16.13 and 16.14, which includes the modifications to the upstand and pipeline discussed previously. However, if the bin is over-filled, there is still every possibility that material could block

Figure 16.14 Large-diameter tee-bend at end of pipeline

or restrict the flow path of the air, hence causing over-pressurization of the bin (and subsequent dust emissions). To alleviate this situation, either this practice has to be stopped or avoided (e.g. by making sure there is sufficient capacity before the tanker arrives) or the volumetric capacity of the bin has to be increased.

One final matter of consideration is the selection of the filter material. With the larger upstand shown in Figure 16.13, it is recommended to use bags with a Teflon coating, which should be better suited to higher temperatures and also provide better cleaning and filtration efficiency. With the bottom of the bags at least 1.5 m from the top of the bin and assuming that the bin is not over-filled, the filter surfaces should be 'safe' from damage due to particle impact.

Outcome

The above modifications and recommendations were pursued by the company, resulting in the complete elimination of dust emissions. Work in the vicinity of the soda ash silo now proceeds without interruption (i.e. irrespective of tanker unloading operations). This project was so successful, the company incorporated the same principles and criteria into the design of a new production plant, as shown in Figure 16.15.

Figure 16.15 Pneumatic tanker unloading system at new production plant

16.8 CASE STUDY 8

Pneumatic Conveying and Injection of Mill Scale

Oxygen lances often are used in steel plants for hot metal pretreatment. With the increasing cost of supplying compressed oxygen for such applications, one company investigated the possibility of recycling and injecting mill scale (i.e. the waste material from rolling operations) into the hot metal launders or directly into the ladles. Chemical tests demonstrated that the mill scale contained sufficient oxygen for this purpose. Hence, it was required to design a pneumatic conveying system according to the following specification:

- mill scale (sub 10 mm; loose powred bulk density, $\rho_{bl} = 3500\,kg/m^3$; particle density, $\rho_S = 6000\,kg/m^3$);
- mass of mill scale required for each tapping operation 300 – 3000 kg (depending on hot metal properties);
- each batch size is to be delivered evenly during the tapping operation (usually 8 min);
- noise generated during transportation must be less than 80 dBA at a distance of 1 m (due to the pipe work having to travel alongside a control room).

The above material delivery requirements equated to a steady-state conveying solids mass flow rate, $m_s = 2.25 - 22.5$ t/h, depending on the level of oxidation required. To determine reliable operating conditions (e.g. minimum air flows and operating pressures) and generate sufficient steady-state data for modelling purposes, several tests was carried out on fresh samples of mill scale using the following test rig: Tandem $0.9\,m^3$ bottom-discharge blowtanks (Figure 16.16); mild steel pipeline (inside diameter of pipe, $D = 105$ mm; total length of conveying pipeline, $L = 108$ m; total length of vertical lift, $L_v = 7$ m; number of 90° bends, $N_b = 5$).

Note, the bottom-discharge blow tanks used for this test program were not considered appropriate for the final application (e.g. higher air flows needed to cope with the heavy and flooding nature of this material; insufficient feed rate control accuracy); however, they were found adequate for the purpose of generating steady-state operating conditions.

Using a sound meter, sound levels also were recorded at various stages during the test program. The readings were taken out in the open on top of the laboratory roof approximately 1 m from the pipeline and found to be 66 dBA (background noise prior to conveying) and 72 dBA (during conveying). The latter level was considered sufficiently low for the final application.

By employing the test-design procedure described by Pan and Wypych (1992), models were developed for: horizontal straight sections of pipe; vertical straight

Figure 16.16 Tandem 0.9 m^3 blowtanks used for the test program

sections of pipe; 1 m radius 90° bends. The resulting 'PCC Model' and the Weber A4 Model (Wypych *et al.*, 1990) were employed to predict lines of constant m_s for the test rig pipeline and the various pipeline configurations being considered for the two submerged arc furnaces.

To establish optimal operating conditions, it was realised that air flows could be minimized by metering or controlling the flow rate of material into the pipeline. This also would meet the requirement of achieving a turndown ratio of 10:1 on conveying rate (namely, $m_s = 22.5 - 2.25$ t/h). The screw-feeding blow tank (Wypych, 1995) shown in Figure 16.17 was found suitable for this material and application.

Using the above models and method of feeding, the following optimal operating conditions were predicted for the extreme case of maximum pipeline length and maximum throughput. Note, a stepped-diameter pipeline was selected to minimize pressure drop, air mass flow rate and hence, conveying air velocities.

$D = 102/128$ mm; $L = 120$ m, $L_v = 22$ m; $N_b = 5$; $t = 20$°C

$m_s = 22.5$ t/h; air mass flow rate, $m_f = 0.46$ kg/s

Total pipeline air pressure drop, $\Delta p_t = 180$ kPa(PCC Model)–205 kPa (WeberA4Model)

Superfacial air velocity, $V_f = 16$–38 m/s($D = 102$ mm) and 24 -30 m/s($D = 128$ mm)

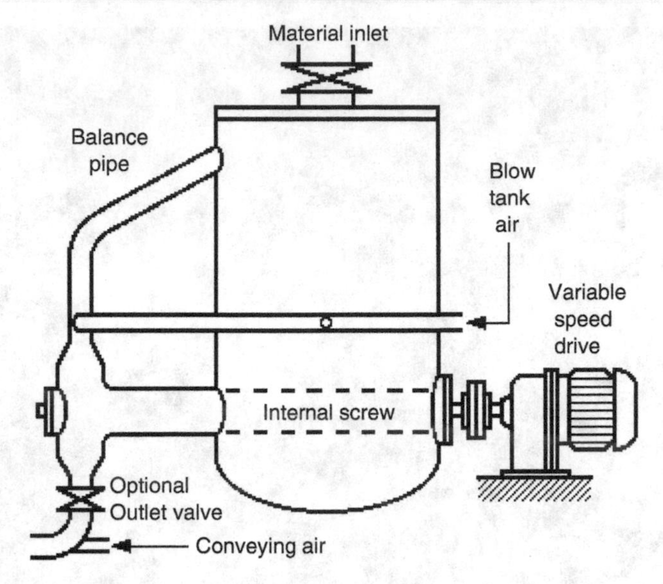

Figure 16.17 General arrangement of screw-feeding blow tank

Based on these concept design parameters, the plant was designed, installed and commissioned and has been operating successfully ever since. The operators confirmed that the blow tank pressure reached approximately 200 kPa *g* when $m_s = 22.5\,\text{t/h}$ was selected for the furthest tapping point (namely, $L = 120\text{m}$).

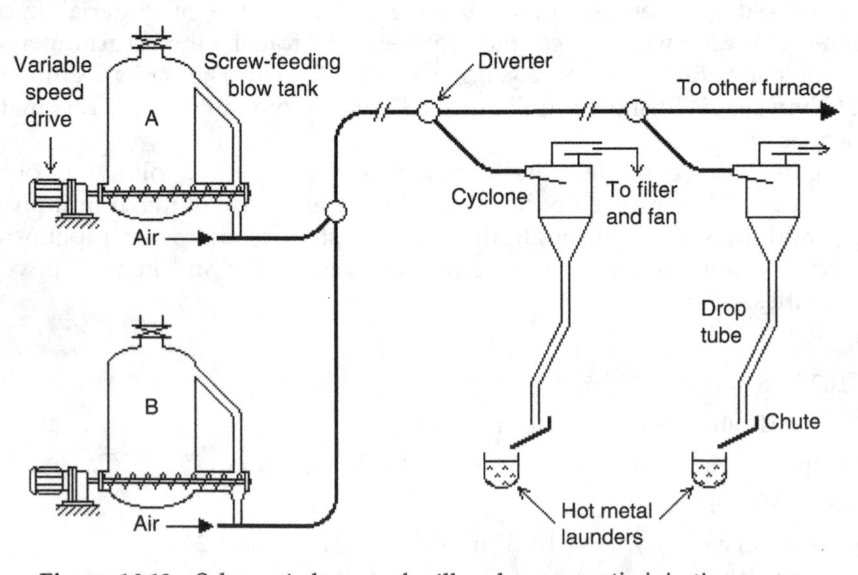

Figure 16.18 Schematic layout of mill scale pneumatic injection system

Figure 16.19 Screw-feeding blow tank

Figure 16.20 Fused alumina-zircon bends

A schematic layout of the plant is shown in Figure 16.18 and one of the screw-feeding blow tanks is shown in Figure 16.19 Special fused alumina-zircon bends (Figure 16.20) were selected to handle this extremely dense and abrasive material.

Notation

a	exponent in Equation (13.13)	—
a	surface area of particles per unit bed volume	m^2
A	cross-sectional area	—
A	cyclone body length (Figure 9.5)	m
A	surface area of element of fuel–air mixture	m^2
A_f	flow area occupied by fluid	m^2
A_p	flow area occupied by particles	m^2
$A(v)$	rate constant for granule attrition	m^3/s
Ar	Archimedes number $\left[Ar = \frac{\rho_f(\rho_p - \rho_f)gx^3}{\mu^2}\right]$	—
b	exponent [Equation (13.13)]	—
B	length of the cyclone cone (Figure 9.5)	m
B	minimum diameter of circular hopper outlet	m
$B_{nuc}(v)$	rate of granule growth by nucleation	$m^{-6}\,s^{-1}$
c	particle suspension concentration	kg/m^3
C	fuel concentration in fuel–air mixture	m^3 fuel/m^3
C	length of cylindrical part of cyclone (Figure 9.5)	m
C	particle volume fraction	—
C	suspension concentration	—
C_B	concentration of the reactant in the bubble phase at height h	mol/m^3
C_B	suspension concentration in downflow section of thickener	—
C_{BH}	concentration of reactant leaving the bubble phase	mol/m^3
C_D	drag coefficient [Equation (2.5)]	—
C_F	concentration of feed suspension	—
C_{fL}	lower flammability limit	m^3 fuel/m^3
C_{fU}	upper flammability limit	m^3 fuel/m^3
C_g	specific heat capacity of gas	$J/kg \cdot K$
C_H	concentration of reactant leaving the reactor	mol/m^3

Introduction to Particle Technology - 2nd Edition Martin Rhodes

C_L	suspension concentration in underflow of thickener	—
C_P	concentration of reactant in the particulate phase	mol/m^3
$C_{P_{air}}$	molar specific heat capacity of air	J/kmol·K
$C_{P_{fuel}}$	molar specific heat capacity of fuel	J/kmol·K
C_S	sediment concentration	—
C_T	suspension concentration in upflow section of thickener	—
C_V	suspension concentration in overflow of thickener	—
C_0	concentration of reactant at distributor	mol/m^3
C_0	initial concentration of suspension	—
C_1	constant [Equation (6.22)]	s/m^6
D	diameter of bed, bin, cyclone (Figure 9.5), pipe, tube [Equation (6.2)], vessel	—
d_{Bv}	equivalent volume diameter of a bubble	m
$d_{B_{vmax}}$	equivalent volume diameter of a bubble (maximum)	m
d_{Bvs}	equivalent volume diameter of a bubble at the surface	m
d_d	average drop diameter	m
D_e	equivalent tube diameter [Equation (6.3)]	m
(dp/dt)max	maximum rate of pressure rise in dust explosion	bar/s
e	coefficient of restitution for granule collision	—
E	diameter of solids outlet of cyclone (Figure 9.5)	m
E	reaction activation energy in Arrhenius equation [Equation (15.2)]	s^{-1}
$E(S^2)$	standard error (standard deviation of variance of a sample)	—
E_T	total separation efficiency	—
Eu	Euler number	—
F	cumulative frequency undersize	μm^{-1}
F	volumetric flow rate of feed suspension to thickener	—
f^*	friction factor for packed bed flow	—
F_D	total drag force	N
ff	hopper flow factor	—
f_g	Fanning friction factor	—
F_{gw}	gas-to-wall friction force per unit volume of pipe	Pa/m
F_N	number of particles per unit volume of pipe	m^{-3}
f_p	solids-to-wall friction factor [Equation (8.19)]	—
f_n	fraction of nuclei formed by n drops	—
F_p	form drag force	N
F_{pw}	solids-to-wall friction force per unit volume of pipe	Pa/m
F_s	drag force due to shear	N
F_v	force exerted by particles on the pipe wall per unit volume of pipe	N/m^3
F_{vw}	van der Waals force between a sphere and a plane	N
f_w	wall factor, U_D/U_∞	—
f_{wet}	fraction of the powder surface welted by spray drops	m^2/m^2

$f(x)$	differential size distribution $\mathrm{d}F/\mathrm{d}x$	m^{-1}or $\mu\mathrm{m}^{-1}$
g	acceleration due to gravity	$\mathrm{m/s^2}$
G	solids mass flux $= M_p/A$	$\mathrm{kg/m^2 \cdot s}$
$G(v)$	volumetric growth rate constant for coating	$\mathrm{m^3/s}$
$g(x)$	weighting function [Equation (1.5)]	—
$G(x)$	grade efficiency	—
$G(x)$	linear growth rate constant for coating	m/s
h	depth of sampling tube below surface [Equation (1.15)]	m
h	heat transfer coefficient	$\mathrm{W/m^2 \cdot K}$
h	height of interface from base of vessel	m
h	thickness of liquid coating on a granule	m
H	height of bed	m
H	height of powder in bin	m
H	height of solids in standpipe	m
h_a	measure of roughness of granule surface	m
H_c	depth of filter cake	m
H_e	equivalent length of flow paths through packed bed [Equation (6.3)]	m
H_{eq}	depth of cake equivalent to medium resistance	m
h_{gc}	gas convective heat transfer coefficient	$\mathrm{W/m^2 \cdot K}$
h_{gp}	gas-to-particle heat transfer coefficient	$\mathrm{W/m^2 \cdot K}$
H_m	thickness of filter medium	m
h_{max}	maximum bed-to-surface heat transfer coefficient	$\mathrm{W/m^2 \cdot K}$
H_{mf}	height of bed at incipient fluidization	m
h_{pc}	particle convective heat transfer coefficient	$\mathrm{W/m^2 \cdot K}$
h_r	radiative heat transfer coefficient	$\mathrm{W/m^2 \cdot K}$
h_0	initial height of interface from base of vessel	m
h_1	height defined in Figure 3.9	m
$H(\theta)$	function given by $H(\theta) = 2.0 + \theta/60$ [Equation (10.5)]	—
j	reaction order	—
J	length of gas outlet pipe of cyclone (Figure 9.5)	m
k	reaction rate constant per unit volume of solids	$\mathrm{mol/m^3 \cdot s}$
K	height of inlet of cyclone (Figure 9.5)	m
K_C	interphase mass transfer coefficient per unit bubble volume	$\mathrm{s^{-1}}$
k_g	gas conductivity	$\mathrm{W/m \cdot K}$
K_H	Hamaker constant [Equation (13.1)]	Nm
K_{ih}^*	elutriation constant for size range x_i at height h above distributor	$\mathrm{kg/m^2 \cdot s}$
$K_{i\infty}^*$	elutriation constant for size range x_i above TDH	$\mathrm{kg/m^2 \cdot s}$
K_{St}	proportionality constant [Equation (15.6)]	bar m·s
K_1	constant [Equation (6.3)]	—
K_2	constant [Equation (6.5)]	—
K_3	constant [Equation (6.8)]	—
L	height above the distributor	m

L	length of pipe	m
L	volumetric flow rate of underflow suspension from thickener	m^3/s
L	width of inlet of cyclone (Figure 9.5)	m
L_H	length of horizontal pipe	m
L_v	length of vertical pipe	m
M	mass flow rate of solids feed to separation device	kg/s
M	mass of solids in the bed	kg
M_B	mass of solids in the bed	kg
m_{B_i}	mass fraction of size range x_i in the bed	—
M_C	mass flow rate of coarse product at solids discharge	kg/s
M_f	mass flow rate of fines product at gas discharge	kg/s
M_f	gas mass flow rate	kg/s
MOC	minimum oxygen for combustion	$m^3 O_2/m^3$
M_P	solids mass flow rate	kg/s
n	exponent in Richardson–Zaki equation [Equation (3.24)]	—
n	number of cyclones in parallel	—
n	number of particles in a sample	—
N	diameter of gas outlet of cyclone (Figure 9.5)	m
N	number of granules per unit volume in the system	—
N	number of holes per unit area in the distributor	m^{-2}
N	number of samples	—
$N(t)$	total number of granules in the system at time t	—
Nu	Nusselt number ($h_{gp}x/k_g$)	—
Nu_{max}	Nusselt number corresponding to h_{max}	—
$n(v,t)$	number density of granule volume v at time t	m^{-6}
p	pressure	Pa
p	proportion of component in a binary mixture	—
p_c	capillary pressure [Equation (13.2)]	Pa
Pr	Prandtl number ($C_g\mu/k_g$)	—
p_s	pressure difference [Equation (6.29)]	Pa
q	gas flow rate	m^3/s
Q	volume flow rate	m^3/s
Q	volume flow rate of gas into bed ($= UA$)	m^3/s
Q	volume flow rate of suspension to thickener	m^3/s
Q_a	rate of heat absorption by element of fuel-air mixture	W
Q_f	volume flow rate of gas/fluid	m^3/s
Q_{in}	volume flow into granulator	m^3/s
Q_{input}	rate of heat input to element of fuel–air mixture	W
Q_{mf}	volume flow rate of gas into bed at $U_{mf}(= U_{mf}A)$	m^3/s
Q_{out}	volume flow out of granulator	m^3/s
Q_p	volume flow rate of particles/solids	m^3/s
Q_s	rate of heat loss to surroundings	W
r	radius of curved liquid surface	m
R	radius of cyclone body	m

R	radius of sphere	m
R	universal gas constant	J/kmol·K
R'	drag force per unit projected area of particle	N/m^2
r_c	filter cake resistance [Equation (6.17)]	m^{-2}
r_1, r_2	radii of curvature of two liquid surfaces	m
Re^*	Reynolds number for packed bed flow [Equation (6.12)]	—
Re_{mf}	Reynolds number at incipient fluidization ($U_{mf}x_{sv}\rho_f/\mu$)	—
Re_p	single particle Reynolds number [Equation (1.4)]	—
R_i	rate of entrainment of solids in size range x_i	kg/s
r_m	filter medium resistance [Equation (6.24)]	m^{-2}
R_p	average pore radius	m
s	granule saturation	—
S	total surface area of population of particles	m^2
S	estimate of standard deviation of a mixture composition	—
S_v	surface area of particles per unit volume of particles	m^2/m^3
S^2	estimate of variance of mixture composition	—
S_B	surface area of particles per unit volume of particles	m^2/m^3
St_{def}	Stokes deformation number	—
Stk	Stokes number [Equation (13.4)]	—
Stk^*	critical Stokes number for coalescence	—
Stk_{50}	Stokes number for x_{50}	—
t	time	s
tp	drop penetration time	s
T	reaction temperature	K
TDH	transport disengagement height	m
T_g	gas temperature	K
T_{ig}	ignition temperature	K
T_s	solids temperature	K
u	granule volume	m^3
u and v	size of two coalescing granules	m
U	superficial gas velocity ($= Q_f/A$)	m/s
U_B	mean bubble rise velocity	m/s
U_c	representative collision velocity in the granulator	m/s
U_{ch}	choking velocity (superficial)	m/s
U_D	velocity in a pipe of diameter D	m/s
U_f	actual or interstitial gas velocity	m/s
U_{fH}	interstitial gas velocity in horizontal pipe	m/s
U_{fs}	superficial fluid velocity	m/s
U_{fv}	interstitial gas velocity in vertical pipe	m/s
U_i	actual interstitial velocity of fluid	m/s
U_{int}	interface velocity	m/s
U_m	superficial velocity at which h_{max} occurs	m/s

U_{mb}	superficial gas velocity at minimum bubbling	m/s
U_{mf}	superficial gas velocity at minimum fluidization	m/s
U_{ms}	minimum velocity for slugging	m/s
U_p	actual particle or solids velocity	m/s
U_{pH}	actual solids velocity in horizontal pipe	m/s
U_{ps}	superficial particle velocity	m/s
U_{pv}	actual solids velocity in vertical pipe	m/s
U_r	radial gas velocity	m/s
U_R	radial gas velocity at cyclone wall	m/s
U_{rel}	relative velocity ($= U_{slip} = U_f = U_p$)	m/s
$\|U_{rel}\|$	magnitude of U_{rel}	m/s
U_{relT}	relative velocity at terminal freefall	m/s
U_{salt}	saltation velocity (superficial)	m/s
U_{slip}	slip velocity ($U_f - U_p$)	m/s
U_T	single particle terminal velocity	m/s
$U_{T2.7}$	single particle terminal velocity for a particle 2.7 times the mean size	m/s
U_{Ti}	single particle terminal velocity for particle size x_i	m/s
U_θ	particle tangential velocity at radius r	m/s
$U_{\theta R}$	particle tangential velocity at cyclone wall	m/s
U_∞	velocity in an infinite fluid	m/s
v	characteristic gas velocity based on D	m/s
v	granule volume	m^3
v_s	powder velocity in the spray zone	m/s
V	granulator volume	m^3
V	volume of element of fuel–air mixture	m^3
V	volume of filtrate passed	m^3
V	volumetric flow rate of overflow suspension from thickener	m^3/s
V_{app}	approach velocity of granules	m/s
Vd	volume of a drop	m^3
V_{eq}	volume of filtrate that must pass in order to create a cake	m^3
w	average granule volume defined in Equation (13.13)	m^3
w	liquid level on dry mass basis	g liquid/g dry powder
w_s	width of the spray	m
$w*$	critical average granule volume for coalescence	m^3
x	granule or particle diameter	m
$\bar{x}$	mean particle diameter	m
$x*$	critical average granule volume for coalescence	m
$\bar{x}_a$	arithmetic mean diameter (Table 1.4)	m
$\bar{x}_{aN}$	arithmetic mean of number distribution	m
$\bar{x}_{aS}$	arithmetic mean of surface distribution	m
$\bar{x}_C$	cubic mean diameter (Table 1.4)	m
x_{crit}	critical particle size for separation [Equation (9.20)]	m
$\bar{x}_g$	geometric mean diameter (Table 1.4)	m

$\bar{x}_h$	harmonic mean diameter (Table 1.4)	m
$\bar{x}_{hV}$	harmonic mean of volume distribution	m
$\bar{x}_{NL}$	number-length mean	m
$\bar{x}_{NS}$	number-surface mean	m
$\bar{x}_p$	mean sieve size of a powder	m
x_p	sieve size	m
$\bar{x}_q$	quadratic mean diameter (Table 1.4)	m
$\bar{x}_{qN}$	quadratic mean of number distribution	m
x_s	equivalent surface sphere diameter	m
x_{SV}	surface-volume diameter (diameter of a sphere having the same surface/volume ratio as the particle)	m
x_v	volume diameter (diameter of a sphere having the same volume as the particle)	m
x_{50}	cut size (equiprobable size)	m
y	gap between sphere and plane [Equation (13.1)]	m
Y_d	dynamic yield stress of the granule	—
Υ	factor [Equation (7.30)]	—
y_i	composition of sample number i	—
z	log x	—
z	depth of penetration of liquid into powder mass	m
Z	Arrhenius equation pre-exponential constant [Equation (15.2)]	s^{-1}
Z'	Arrhenius equation pre-exponential constant [Equation (15.5)]	mol fuel/ (m^2 fuel s)
α_s	factor relating linear dimension of particle to its surface area	—
α_V	factor relating linear dimension of particle to its volume	—
α	significance level	—
β	coalescence kernel or rate constant	s^{-1}
β	$(U - U_{mf})/U$	—
β_0	coalescence rate constant [Equation (13.12)]	—
$\beta_1(u, v)$	coalescence rate constant [Equation (13.12)]	—
$\beta(u, v-u, t)$	coalescence rate constant for granules of volume u and $(u-v)$ at time t	s^{-1}
γ	surface tension	N/m
$\gamma \cos\theta$	adhesive tension	mN/m
δ	effective angle of internal friction	deg
Δp	static pressure drop	Pa
$(-\Delta p)$	pressure drop across bed/cake	Pa
$(-\Delta p_c)$	pressure drop across cake	Pa
ε	granule porosity or voidage	—
ε_b	porosity of voidage of powder bed	—
ε_{min}	minimum granule porosity or voidage reached during granulation	—

θ	dynamic contact angle of the liquid with the powder	—
μ	liquid viscosity	Pa s
ρ_g	granule density	kg/m^3
ρ_l	liquid density	kg/m^3
ρ_s	solid density	kg/m^3
Ψ_a	dimensionless spray flux	—

References

Abrahamsen, A. R. and Geldart, D. (1980) Behaviour of gas fluidized beds of fine powder. Part 1. Homogenous Expansion, *Powder Technol.*, **26**, 35.

Adetayo, A. A. and Ennis, B. J. (1997) 'Unifying approach to modelling granule coalescence mechanisms', *AIChE J*, **43** (4), 927–934.

Allen, T. (1990) *Particle Size Measurement*, 4th Edition, Chapman & Hall, London.

Arena U., Cammmarota, A. and Mastellone, M.L. (1998) 'The influence of operating parameters on the behaviour of a small diameter L-valve' in *Fluidization IX, Proceedings of the Ninth Foundation Conference on Fluidization*, eds L.S. Fan and T.M. Knowlton, Engineering Foundation, New York, pp. 365–372.

Baeyens, J. and Geldart, D. (1974) 'An investigation into slugging fluidized beds', *Chem. Eng. Sci*, **29**, 255.

Barnes, H. A., Hutton, J. F. and Walters, K. (1989) An Introduction to Rheology, Elsevier, Amsterdam.

Baskakov, A. P. and Suprun, V. M. (1972) 'The determination of the convective component of the coefficient of heat transfer to a gas in a fluidized bed', *Int. Chem. Eng.*, **12**, 53.

Batchelor, G. K. (1977) 'The effect of Brownian motion on the bulk stress in a suspension of spherical particles, *J. Fluid Mech.*, **83**, 97–117.

Beverloo, W. A., Leniger, H. A. and Van de Velde, J. (1961) The Flow of Granual Solids Through Onfices', *Chem. Eng. Sci.*, **15**, issues 3-4, September 1961, 260–269.

Bodner, S. (1982) *Proceedings of International Conference on Pneumatic Transport Technology*, Powder Advisory Centre, London.

Bond, F. C. (1952) 'The third theory of comminution', *Mining Eng. Trans. AIME*, **193**, 484–494.

Botterill, J. S. M. (1975) *Fluid Bed Heat Transfer*, Academic Press, London.

Botterill, J. S. M. (1986) 'Fluid bed heat transfer' in *Gas Fluidization Technology*, ed. D. Geldart, John Wiley & Sons, Ltd, Chichester, Chapter 9.

Brown, G.G, Katz, D., Foust, A.S. and Schneidewind, R. (1950) *Unit Operations*, John Wiley & Sons, Ltd, New York, Chapman & Hall, London.

Capes, C. E. (1980) *Particle Size Enlargement*, Vol. 1, *Handbook of Powder Technology*, Elsevier, Amsterdam.

Introduction to Particle Technology - 2nd Edition Martin Rhodes

Carman, P. C. (1937) Fluid flow through granular beds, *Trans. Inst. Chem. Eng.*, **15**, 150–166.

Chhabra, R. P. (1993) *Bubbles, Drops and Particles in Non-Newtonian Fluids*, CRC Press, Boca Raton, FL.

Clift, R., Grace, J. R. and Weber, M. E. (1978) *Bubbles, Drops and Particles*, Academic Press, London.

Coelho, M. C. and Harnby, N. (1978) 'The effect of humidity on the form of water retention in a powder', *Powder Technol.*, **20**, 197.

Cooke, W., Warr, S., Huntley, J. M. and Ball, R. C. (1996) 'Particle size segregation in a two-dimensional bed undergoing vertical vibration', *Phys. Rev. E*, **53**(3), 2812–2822.

Coulson, J. M. and Richardson, J. F. (1991) *Chemical Engineering, Volume 2: Particle Technology and Separation Processes*, 4th Edition, Pergamon, Oxford.

Dalby, R.N., Tiano, S.L. and Hickey, A.J. (1996) 'Medical devices for the delivery of therapeutic aerosol to the lungs' in *Inhalation Aerosols*, ed. A. J. Hickey, Vol. 94, *Lung Biology in Health and Disease*, ed. C. Lenfant, Marcel Dekker, New York, pp. 441–473.

Danckwerts, P. V. "Definition and Measurement of some characteristics of Mixtures", Applied Scientific Research, VA3, N5, 1952, p385–390.

Darcy, H. P. G. (1856) Les fontaines publiques de la ville de Dijon. *Exposition et applications à suivre et des formules à employer dans les questions de distribution d'eau.* Victor Dalamont.

Darton, R. C., La Nauze, R. D., Davidson, J. F. and Harrison, D. (1977) 'Bubble growth due to coalescence in fluidised beds', *Trans. Inst. Chem. Eng.*, **55**, 274.

Davidson, J. F. and Harrison, D. (1971) *Fluidization*, Academic Press, London.

Davies, R., Boxman, A. and Ennis, B.J. (1995), Conference Summary Paper: Control of Particulate Processes III, *Powder Technol.* **82**, 3–12.

Dixon, G. (1979) 'The impact of powder properties on dense phase flow, in *Proceedings of International Conference on Pneumatic Transport*, London.

Dodge, D.W. and Metzner, A.B. (1959) 'Turbulent flow of non-Newtonian systems', *AIChE J.*, **5**, 89.

Duran, J., Rajchenbach, J. and Clement, E. (1993) 'Arching effect model for particle size segregation', *Phys. Rev. Lett.*, **70**(16), 2431–2434.

Durand, R. and Condolios, E. (1954) 'The hydraulic transport of coal' in *Proceedings of Colloquim on Hydraulic Transport of Coal*, National Coal Board, London.

Einstein, A. (1906) 'Eine neue Bestimmung der Molekuldimension', *Ann. Physik*, **19**, 289–306.

Einstein, A. (1956) *Investigations in the Theory of the Brownian Movement*, Dover, New York.

Ennis, B. J. and Litster, J. D. (1997) 'Section 20: size enlargement' in *Perry's Chemical Engineers' Handbook*, 7th Edition, McGraw-Hill, New York, pp. 20–77.

Ergun, S. (1952) 'Fluid flow through packed columns', *Chem. Eng. Prog.*, **48**, 89–94.

Evans, I., Pomeroy C. D. and Berenbaum, R. (1961) The compressive strength of coal, *Colliery Eng.*, 75–81, 123–127, 173–178.

Flain, R. J. (1972) 'Pneumatic conveying: how the system is matched to the materials', *Process Eng.*, Nov, p 88–90.

Francis, A. W. (1933) 'Wall effects in falling ball method for viscosity', *Physics*, **4**, 403.

Franks, G. V. and Lange, F. F. (1996) 'Plastic-to-brittle transition of saturated, alumina powder compacts', *J. Am. Ceram. Soc.*, **79** (12), 3161–3168.

Franks, G. V., Zhou, Z., Duin, N. J. and Boger, D. V. (2000) 'Effect of interparticle forces on shear thickening of oxide suspensions', *J. Rheol.*, **44** (4), 759–779.

Fryer, C. and Uhlherr, P. H. T. (1980) *Proceedings of CHEMECA '80, the 8th Australian Chemical Engineering Conference*, Melbourne, Australia.

Geldart, D. (1973) 'Types of gas fluidisation', *Powder Technol.*, **7**, 285–292.

Geldart, D. (ed.) (1986) *Gas Fluidization Technology*, John Wiley & Sons, Ltd, Chichester.

Geldart, D. (1990) 'Estimation of basic particle properties for use in fluid–particle process calculations', *Powder Technol.*, **60**, 1.

Geldart, D. (1992) *Gas Fluidization Short Course*, University of Bradford, Bradford.

Geldart, D. and Abrahamsen, A. R. (1981) 'Fluidization of fine porous powders', *Chem. Eng. Prog. Symp. Ser.*, **77**(205), 160.

Geldart, D. and Jones, P. (1991) 'Behaviour of L-valves with granular powders', *Powder Technol.*, **67**, 163–174.

Geldart, D., Cullinan, J., Gilvray, D., Georghiades, S. and Pope, D. J. (1979) 'The effects of fines on entrainment from gas fluidised beds', *Trans. Inst. Chem. Eng.*, **57**, 269.

Gillies, R.G., Schaan, J., Sumner, R.J., McKibben, M.J. and Shook, C.A. (2000) 'Deposition velocities for Newtonian slurries in turbulent flow', *Can. J. Chem. Eng.*, **78**, 704.

Gilvary, J. J. (1961) 'Fracture of brittle solids. I. Distribution function for fragment size in single fracture', *J. Appl. Phys.*, **32**, 391–399.

Grace, J. R., Avidan A. A. and Knowlton T. M. (eds) (1997) *Circulating Fluidized Beds*, Blackie Academic and Professional, London.

Gregory, J. (2006) *Particles in Water, Properties and Processes*, CRC Press, Taylor & Francis, Boca Raton, FL.

Griffith, A. A. (1921) "The phenomena of Rupture and Flow in Solids", Philosophical Transactions of the Royal Society of London, V221, 21 Oct. 1920, p. 163–198. *Phil. Trans. R. Soc.*, **221**,163.

Haider, A. and Levenspiel, O. (1989) 'Drag coefficient and terminal velocity of spherical and no-spherical particles', *Powder Technol.*, **58**, 63 – 70.

Hamaker, H. C. (1937) 'The London–van der Waals attraction between spherical particles', *Physica*, **4**, 1058.

Hanks, RW. and Ricks, B.L. (1974) "Laminar-turbulent transition in flow of pseudoplastic fluids with yield stress", *J. Hydronautics*, **8**, 163.

Hapgood, K.P., Litster, J.D. and Smith, R. (2003) 'Nucleation regime map for liquid bound granules', *AIChE J.*, **49**(2), 350–361.

Hapgood, K.P., Litster, J.D. White, E.T. *et al.* (2004) 'Dimensionless spray flux in wet granulation: Monte-Carlo simulations and experimental validation', *Powder Technol.*, **141** (1–2), 20–30.

Hapgood, K.P., Iveson, S.M., Litster, J.D. and Liu, L. (2007) 'Granule rate processes' in *Granulation, Handbook of Powder Technology, Vol. 11*, eds A.D. Salman, M.J. Hounslow and J.P.K. Seville, Elsevier, London, p. 933.

Harnby, N., Edwards, M. F. and Nienow, A. W. (1992) *Mixing in the Process Industries*, 2nd Edition, Butterworth-Heinemann, London.

Hawkins, A. E. (1993) *The Shape of Powder Particle Outlines*, Research Studies Press, John Wiley & Sons, Ltd, Chichester.

Hiemenz, P. C. and Rajagopalan, R. (1997) *Principles of Colloid and Surface Chemistry*, 3rd Edition, Marcel Dekker, New York.

Hinkle, B. L. (1953) *PhD Thesis*, Georgia Institute of Technology.

Holmes, J. A. (1957) 'Contribution to the study of comminution – modified form of Kick's law', *Trans. Inst. Chem. Eng.*, **35**, (2), 125–141.

Horio, M., Taki, A., Hsieh, Y. S. and Muchi, I. (1980) 'Elutriation and particle transport through the freeboard of a gas–solid fluidized bed', in *Fluidization*, eds J. R. Grace and J. M. Matsen, Engineering Foundation, New York, p. 509.

Hukki, R. T. (1961) 'Proposal for Solomonic settlement between the theories of von Rittinger, Kick and Bond', *Trans. AIME*, **220**, 403–408.

Hunter, R. J. (2001) *Foundations of Colloid Science*, 2nd Edition, Oxford University Press, Oxford.

Inglis, C. E. (1913) 'Stress in a plate due to the presence of cracks and sharp corners', *Proc. Inst. Nab. Arch.*

Israelachvili, J. N. (1992) *Intermolecular and Surface Forces*, 2nd Edition, Academic Press, London.

Iveson, S.M., Litster, J.D., Hapgood, K.P. and Ennis, B.J. (2001) 'Nucleation, growth and breakage phenomena in agitated wet granulation processes: a review', *Powder Technol.*, **117**(1–2), 3–39.

Janssen, H. A. (1895) 'Tests on grain pressure silos', *Z. Ver. Deutsch Ing.*, **39**(35), 1045–1049.

Jenike, A. W. (1964) 'Storage and flow of solids', *Bull. Utah Eng. Exp. Station*, **53**(123), 26.

Johnson, S.B., Franks, G. V., Scales, P. J. and Healy, T. W. (1999) 'The binding of monovalent electrolyte ions on alpha-alumina. II. The shear yield stress of concentrated suspensions', *Langmuir*, **15**, 2844–2853.

Johnson, S. B., Franks, G. V., Scales, P. J., Boger, D. V. and Healy, T. W. (2000) 'Surface chemistry–rheology relationships in concentrated mineral suspensions', *Int. J. Mineral Process.*, **58**, 267-304.

Jones, D. A. R., Leary, B. and Boger, D. V. (1991) 'The rheology of a concentrated colloidal suspension of hard spheres', *J. Colloid Interface Sci.*, **147**, 479–495.

Jullien, R. and Meakin, P. (1992) 'Three-dimensional model for particle size segregation by shaking', *Phys. Rev. Lett.*, **69**, 640–643.

Karra, V. K. and Fuerstenau, D. W. (1977) 'The effect of humidity on the trace mixing kinetic in fine powders', *Powder Technol.*, **16**, 97.

Kendal, K. (1978) 'The impossibility of comminuting small particles by compression', *Nature*, **272**, 710.

Khan, A. R and Richardson, J. F. (1989) 'Fluid–particle interactions and flow characteristics of fluidised beds and settling suspensions of spherical particles', *Chem. Eng. Commun.*, **78**, 111.

Khan, A. R., Richardson, J. F. and Shakiri, K. J. (1978) in *Fluidization, Proceedings of the Second Engineering Foundation Conference*, eds J. F. Davidson and D. L. Keairns, Cambridge University Press, p. 375.

Kick, F. (1885) *Das Gasetz der proportionalen Widerstände und seine Anwendung*, Leipzig.

Klintworth, J. and Marcus, R. D. (1985) 'A review of low-velocity pneumatic conveying systems', *Bulk Solids Handling*, **5**,(4), 747–753.

Knight, J. B., Jaeger, H. M. and Nagel, S. R. (1993) 'Vibration-induced separation in granular media: the convection connection', *Phys. Rev. Lett.*, **70**, 3728–3731.

Knight, J. B., Ehrichs, E. E., Kuperman, V. Y., Flint, J. K., Jaeger, H. M. and Nagel, S. R. (1996) 'Experimental study of granular convection', *Phys. Rev. E*, **54**, 5726–5738.

Knowlton, T. M. (1986) 'Solids transfer in fluidized systems', in *Gas Fluidization Technology*, ed. D. Geldart, John Wiley & Sons, Ltd, Chichester, Chapter 12.

Knowlton, T. M. (1997) 'Standpipes and non-mechanical valves', notes for the continuing education course *Gas Fluidized Beds: Design and Operation*, Department of Chemical Engineering, Monash University.

Knowlton, T.M. and Hirsan, I. (1978) 'L-valves characterized for solids flow', *Hydrocarbon Process.*, **57**, 149–156.

Konno, H. and Saito, S. J. (1969) 'Pneumatic conveying of solids through straight pipes', *Chem. Eng. Jpn*, **2**, 211–217.

Konrad, K. (1986) 'Dense phase conveying: a review', *Powder Technol.*, **49**, 1–35.

Kozeny, J. (1927) ''Capillary Motion of Water in Soils'', Sitzungsberichte der Akactemie der Wissenschaften in Wien, Mathematisch-Naturwissenschaftliche Klasse, V136, N 5-6, p 271–306. *Sitzb. Akad. Wiss.*, **136**, 271–306.

Kozeny, J. (1933) *Z. Pfl.-Ernahr. Dung. Bodenk*, **28A**, 54–56.

Krieger, I. M. and Dougherty, T. J. (1959) 'A mechanism for non-Newtonian flow in suspensions of rigid spheres', *Trans. Soc. Rheol.*, **3**, 137–152.

Kunii, D. and Levenspiel, O. (1969) *Fluidization Engineering*, John Wiley & Sons, Ltd, Chichester.

Kunii, D. and Levenspiel, O. (1990) *Fluidization Engineering*, 2nd Edition, John Wiley & Sons, Ltd, Chichester.

Lacey, P. M. C. (1954) 'Developments in the theory of particulate mixing', *J. Appl. Chem.*, **4**, 257.

Leung, L. S. and Jones, P. J. (1978) in *Proceedings of Pneumotransport 4 Conference*, BHRA Fluid Engineering, Paper Dl.

Liffman, K., Muniandy, K., Rhodes, M. J., Gutteridge, D. and Metcalfe, G. A. (2001) 'A general segregation mechanism in a vertically shaken bed', *Granular Matter*, **3**(4), 205–214.

Litster, J.D., Hapgood, K.P., Michaels, J.N. *et al.* (2001) 'Liquid distribution in wet granulation: dimensionless spray flux', *Powder Technol.*, **114**(1–3), 32–39.

Litster, J.D. and Ennis, B.J. (2004) *The Science and Engineering and Granulation Processes*, Kluwer Academic Publishers, Dordrecht.

Liu, L. X. and Litster, J. (1993) 'Coating mass distribution from a spouted bed seed coater: experimental and modelling studies', *Powder Technol.*, **74**, 259.

Lunn, G. (1992) *Guide to Dust Explosion, Prevention and Protection, Part I, Venting*, Institution of Chemical Engineers, Rugby.

Mainwaring, N. J. and Reed, A. R. (1987) 'An appraisal of Dixon's slugging diagram for assessing the dense phase transport potential of bulk solid materials', in *Proceedings of Pneumatech 3*, pp. 221–234.

Mills, D. (1990) *Pneumatic Transport Design Guide*, Butterworth, London.

Mobius, E., Lauderdale, B. E., Nagel, S. R. and Jaeger, H. M. (2001) 'Size separation of granular particles', *Nature*, **414**, 270.

Newitt, D. M. and Conway-Jones J. M. (1958) 'A contribution to the theory and practice of granulation', *Trans. Inst. Chem. Eng.*, **36**, 422–442.

Niven, R. W., 'Atomization and Nebulizers' in Inhalation Aerosols, Anthony J. Hickey, Marcel Dekker, New York, 1996–Vol. 94 of Lung Biology in Health and Disease, Ed. Claude Lenfant, pp. 273–312.

Orcutt, J. C., Davidson J. F. and Pigford, R. L. (1962) ''Reaction Time Distributions in Fluidized Catalytic Reactors''. *CEP Symp. Ser.*, **58**(38), 1.

Palmer, K. N. (1990) 'Explosion and fire hazards of powders', in *Principles of Powder Technology*, ed. M. J. Rhodes, John Wiley & Sons, Ltd, Chichester, pp. 299–334.

Pan, R. and Wypych, P.W. (1992) 'Scale-up procedures for pneumatic conveying design', *Powder Handling Process.*, **4**(2), 167–172.

Perrin, J. (1990) *Les Atomes 1913* (in French), *Atoms* (English translation), Ox Bow Press, Woodbridge, CT.

Perry, R. H. and Green, D. (eds) (1984) *Perry's Chemical Engineers' Handbook*, 6th Edition, McGraw-Hill, New York.

Poole, K. R., Taylor, R. F. and Wall, G. P. (1964) 'Mixing powders to fine-scale homogeneity: studies of batch mixing', *Trans. Inst. Chem. Eng.*, **42**, T305.

Punwani, D. V., Modi, M. V. and Tarman, P. B. (1976) Paper presented at the *International Powder and Bulk Solids Handling and Processing Conference*, Chicago.

Quemada, D. (1982) *Lecture Notes in Physics: Stability of Thermodynamic Systems*, eds J. Cases-Vasquez and J. Lebon, Springer, Berlin, p. 210.

Randolph, A. D. and Larson M. A. (1971) *Theory of Particulate Processes*, Academic Press, London.

Rhodes, M., Takeuchi, S., Liffman, K. and Muniandy, K. (2003) 'Role of interstitial gas in the Brazil Nut Effect', *Granular Matter*, **5**, 107–114.

Richardson, J. F. and Zaki, W. N. (1954) 'Sedimentation and fluidization', *Trans. Inst. Chem. Eng.*, **32**, 35.

Rittinger, R. P. von (1867) *Textbook of Mineral Dressing*, Ernst and Korn, Berlin.

Rizk, F. (1973) *Dr-Ing. Dissertation*, Technische Hochschule Karlsruhe.

Rosato, A. D., Strandburg, K. J., Prinz, F. and Swendsen, R. H. (1987) 'Why the Brazil nuts are on top: size segregation of particulate matter by shaking', *Phys. Rev. Lett.*, **58**, 1038–1040.

Rosato, A. D., Blackmore, D. L., Ninghua Zhang and Yidan Lan. (2002) 'A perspective on vibration-induced size segregation of granular materials', *Chem. Eng. Sci.*, **57**, 265–275.

Rumpf, H. (1962) in *Agglomeration*, ed. W. A. Krepper, John Wiley & Sons, Ltd, New York, p. 379.

Sastry, K. V. S. (1975) 'Similarity of size distribution of agglomerates during their growth by coalescence in granulation of green pelletization', *Int. J. Min. Process.*, **2**, 187.

Sastry, K. V. S. and Fuerstenau D. W. (1970) 'Size distribution of agglomerates in coalescing disperse systems', *Ind. Eng. Chem. Fundam.*, **9**, (1), 145.

Sastry, K. V. S. and Fuerstenau, D. W. (1977) in *Agglomeration 77*, ed. K. V. S. Sastry, AIME, New York, p. 381.

Sastry, K. V. S. and Loftus K. D. (1989) 'A unified approach to the modeling of agglomeration processes', *Proceedings of the 5th International Symposium on Agglomeration*, I. Chem. E., Rugby, p. 623.

Schiller, L. and Naumann, A. (1993) 'Über die grundlegenden Berechnungen der Schwerkraftaufbereitung', *Z. Ver. Deutsch Ing.*, **77**, 318.

Schofield, C. (1985) *Guide to Dust Explosion, Prevention and Protection, Part I, Venting*, I. Chem. E., Rugby.

Schofield, C. and Abbott, J. A. (1988) *Guide to Dust Explosion, Prevention and Protection, Part II, Ignition Prevention, Containment, Suppression and Isolation*, I. Chem. E., Rugby.

Shinbrot, T. and Muzzio, F. (1998) 'Reverse buoyancy in shaken granular beds', *Phys. Rev. Lett.*, **81**, 4365–4368.

Smith, S.J. and Bernstein, J.A. (1996) 'Therapeutic use of lung aerosols' in *Inhalation Aerosols* ed. A. J. Hickey, Vol. 94, *Lung Biology in Health and Disease*, ed. C. Lenfant, Dekker, New York, pp. 233–269.

Stokes, G. G. (1851) 'On the effect of the internal friction of fluids on the motion of Pendulums', *Trans. Cam. Phil. Soc.*, **9**, 8.

Svarovsky, L. (1981) *Solid–Gas Separation*, Elsevier, Amsterdam.

Svarovsky, L. (1986) 'Solid–gas separation', in *Gas Fluidization Technology*, ed. D. Geldart, John Wiley & Sons, Ltd, Chichester, pp. 197–217.

Svarovsky, L. (1990) 'Solid–gas separation', in *Principles of Powder Technology*, ed. M. J. Rhodes, John Wiley & Sons, Ltd, Chichester, pp. 171–192.

Tardos, G.I., Khan, M.I. and Mort, P.R. (1997) 'Critical parameters and limiting conditions in binder granulation of fine powders', *Powder Technol.*, **94**, 245–258.

Toomey, R. D. and Johnstone, H. F. (1952) 'Gas fluidization of solid particles', *Chem. Eng. Prog.*, **48**, 220–226.

Tsuji, Y. (1983) Recent Studies of Pneumatic Conveying in Japan *Bulk Solids Handling*, **3**, 589–595.

Waldie, B. (1991) 'Growth mechanism and the dependence of granule size on drop size in fluidised bed granulation', *Chem. Eng. Sci.* **46**(11), 2781–2785.

Wasp, E.J., Kenny, J.P. and Gandhi, R.L. (1977) 'Solid–liquid flow – slurry pipeline transportation', in *Series on Bulk Materials Handling*, Trans Tech Publications, Clausthal, Germany Vol. 1 (1975/77), No. 4.

Weibel, E.R. (1963) *Morphometry of the Human Lung*, Springer-Verlag, Berlin.

Wen, C. Y. and Yu, Y. H. (1966) 'A generalised method for predicting minimum fluidization velocity', *AIChE J.*, **12**, 610.

Werther, J. (1983) 'Hydrodynamics and mass transfer between the bubble and emulsion phases in fluidized beds of sand and cracking catalyst', in *Fluidization*, eds D. Kunii and R. Toei, Engineering Foundation, New York, p. 93.

Williams, J. C. (1976) 'The segregation of particulate materials: a review', *Powder Technol.*, **15**, 245–251.

Williams, J. C. (1990) 'Mixing and segregation in powders' in *Principles of Powder Technology*, ed. M. J. Rhodes, John Wiley & Sons, Ltd, Chichester, Chapter 4.

Wilson, K. C. (1981) 'Analysis of slip of particulate mass in a horizontal pipe', *Bulk Solids Handling*, **1**, 295–299.

Wilson, K.C. and Judge, D.G. (1976) 'New techniques for the scale-up of pilot plant results to coal slurry pipelines', in *Proceedings of the International Symposium on Freight Pipelines*, University of Pennsylvania, pp. 1–29.

Woodcock, C. R. and Mason, J. S. (1987) *Bulk Solids Handling*, Chapman and Hall, London.

Wypych, P.W. (1995) 'Latest developments in the pneumatic pipeline transport of bulk solids, in Proceedings of the *5th International Conference on Bulk Materials Storage, Handling and Transportation*, IEAust, Newcastle, Vol. 1, pp. 47–56.

Wypych, P.W., Kennedy, O.C. and Arnold, P.C. (1990) 'The future potential of pneumatically conveying coal through pipelines', *Bulk Solids Handling* **10**(4), 421–427.

Yagi, S. and Muchi, I. (1952) *Chem. Eng., (Jpn)*, **16**, 307.

Zabrodsky, S.S. (1966) *Hydrodynamics and Heat Transfer in Fluidized Beds*, MIT Press, Cambridge, MA.

Zeng, X. M., Martin, G. and Marriott, C. (2001) *Particulate Interactions in Dry Powder Formulations of Inhalation', Taylor & Francis, London.*

Zenz, F. A. (1964) 'Conveyability of materials of mixed particle size', *Ind. Eng. Fund.*, **3**(1), 65–75.

Zenz, F. A. (1983) 'Particulate solids – the third fluid phase in chemical engineering', *Chem. Eng*, **Nov**, 61–67.

Zenz, F. A. and Weil, N. A. (1958) 'A theoretical–empirical approach to the mechanism of particle entrainment from fluidized beds', *AIChE J*, **4**, 472.

Zhou, Y., Gan, Y., Wanless, E. J., Jameson, G. J. and Franks, G. V. (2008) 'Influence of bridging forces on aggregation and consolidation of silica suspensions', submitted.

Zhou, Z., Scales, P J. and Boger, D. V. (2001) 'Chemical and physical control of the rheology of concentrated metal oxide suspensions', *Chem. Eng. Sci.*, **56**, 2901–2920.

Index

References to figures are given in italic type. References to tables are given in bold type.

9 780470 014288